TRANSPARENCY MASTERS

TO ACCOMPANY

Human Anatomy and Physiology

THIRD EDITION

ELAINE N. MARIEB

The Benjamin/Cummings Publishing Company, Inc.

Redwood City, California ■ Menlo Park, California
Reading, Massachusetts ■ New York ■ Don Mills, Ontario
Wokingham, U.K. ■ Amsterdam ■ Bonn ■ Sydney
Singapore ■ Tokyo ■ Madrid ■ San Juan

ISBN 0-8053-4287-7

2 3 4 5 6 7 9 10-CRS-98-97-96-95

The Benjamin/Cummings Publishing Company, Inc.
390 Bridge parkway
Redwood City, CA 94065

Examples of selected interrelationships among body organ systems

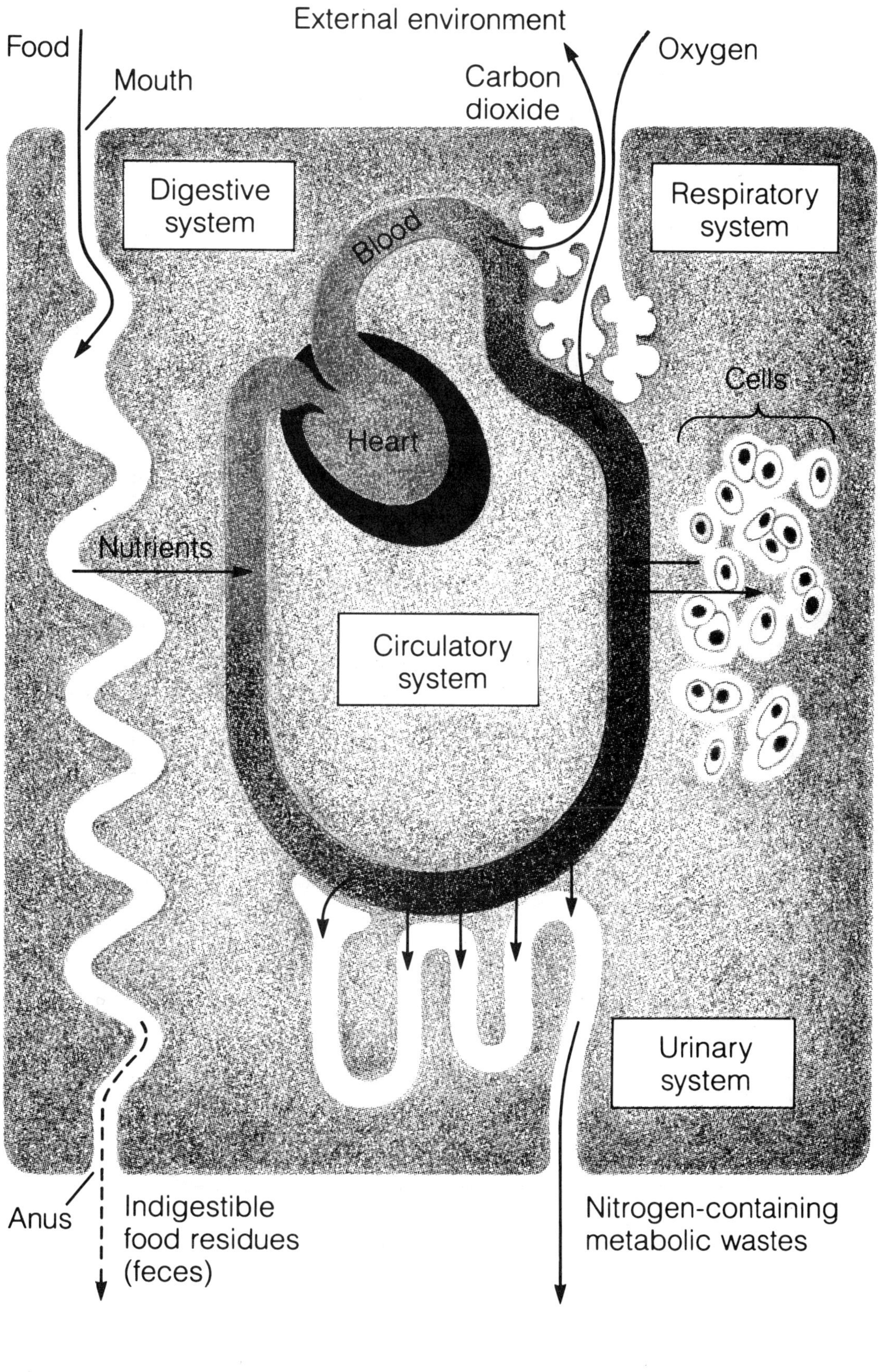

Schematic representation of the elements of a control system

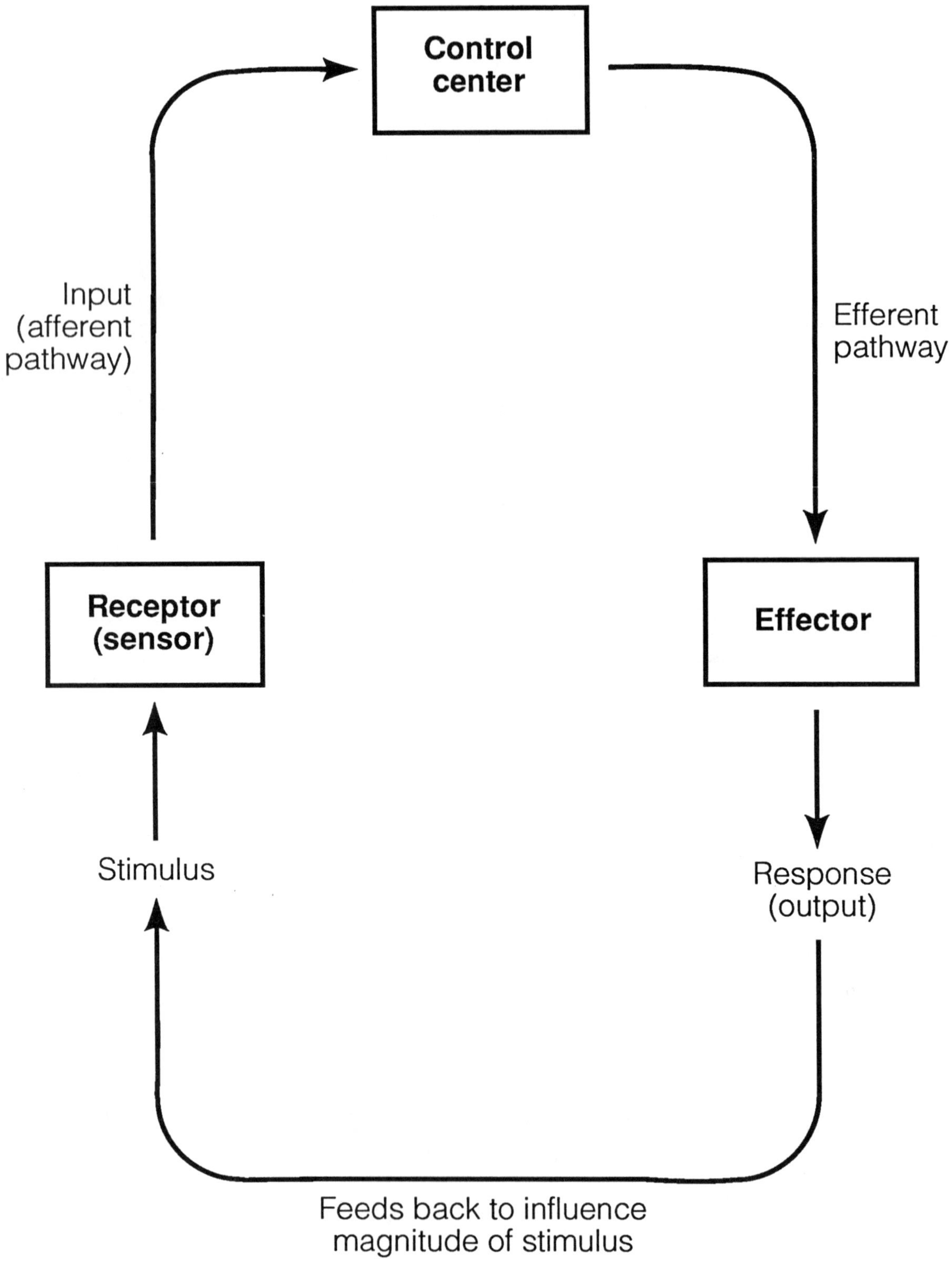

Regulation of blood glucose levels by a negative feedback mechanism involving pancreatic hormones

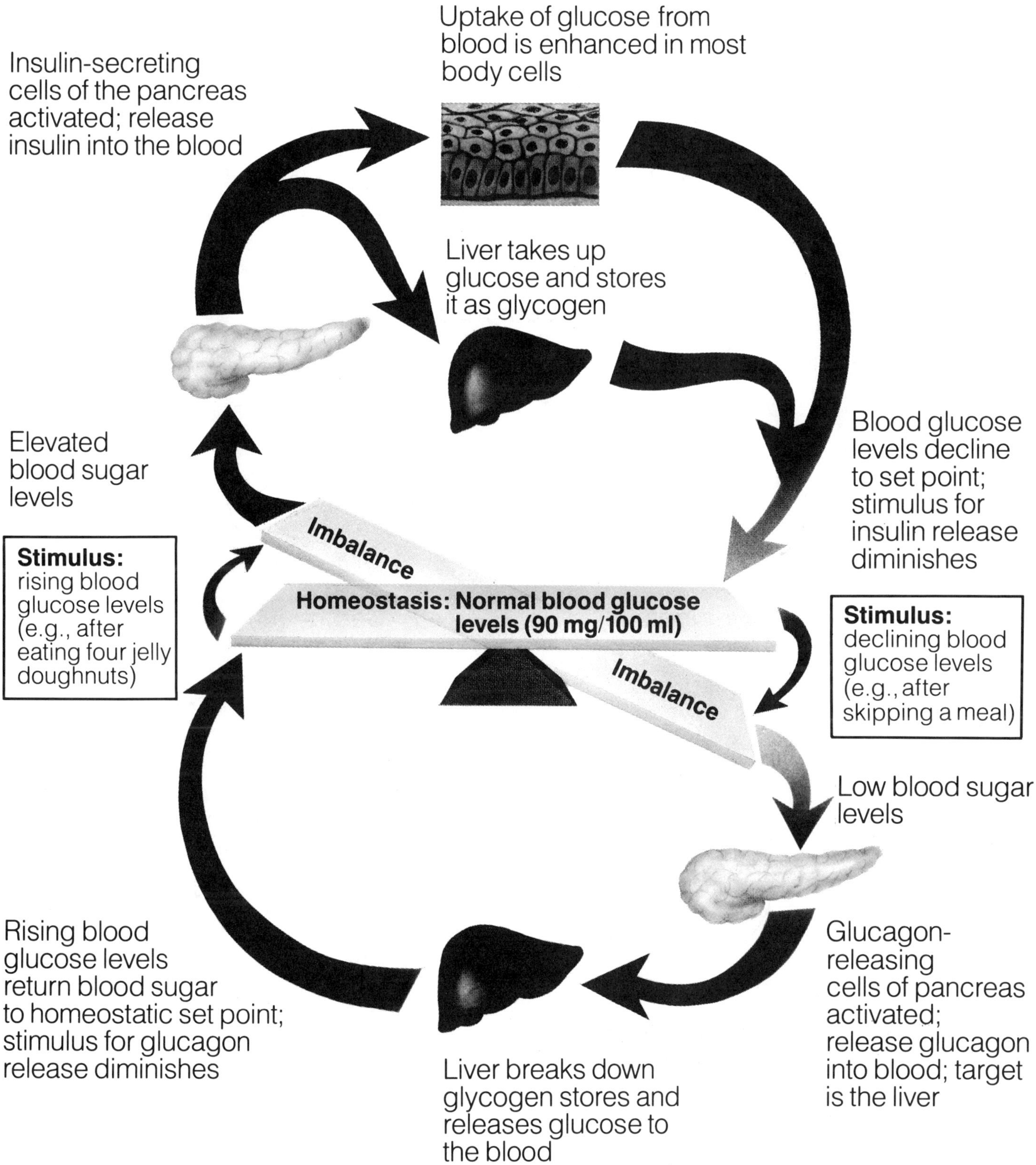

Chemically inert and reactive elements

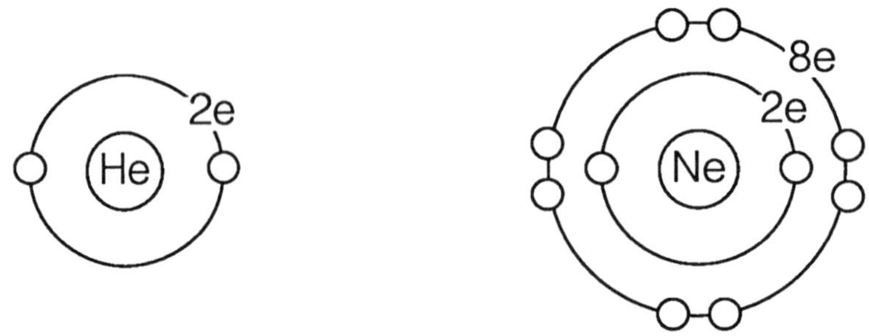

Helium (He)
(2p$^+$; 2n^0; 2e$^-$)

Neon (Ne)
(10p$^+$; 10n^0; 10e$^-$)

(a) Chemically inert elements (valence shell complete)

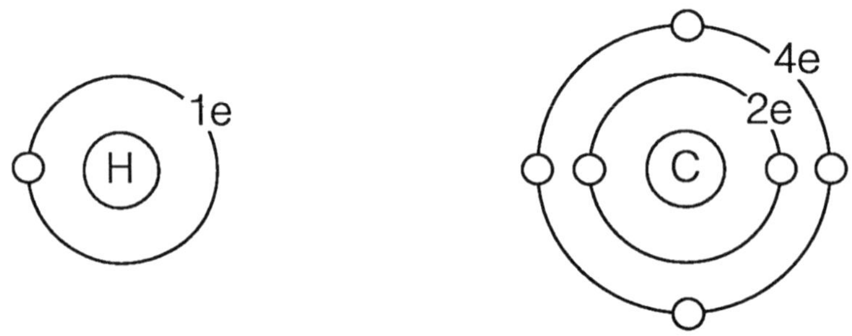

Hydrogen (H)
(1p$^+$; 0n^0; 1e$^-$)

Carbon (C)
(6p$^+$;6n^0;6e$^-$)

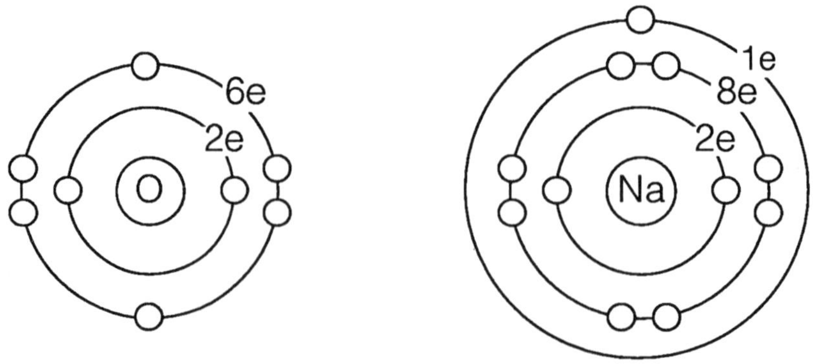

Oxygen (O)
(8p$^+$; 8n^0; 8e$^-$)

Sodium (Na)
(11p$^+$;12n^0;11e$^-$)

(b) Chemically active elements (valence shell incomplete)

Formation of an ionic bond

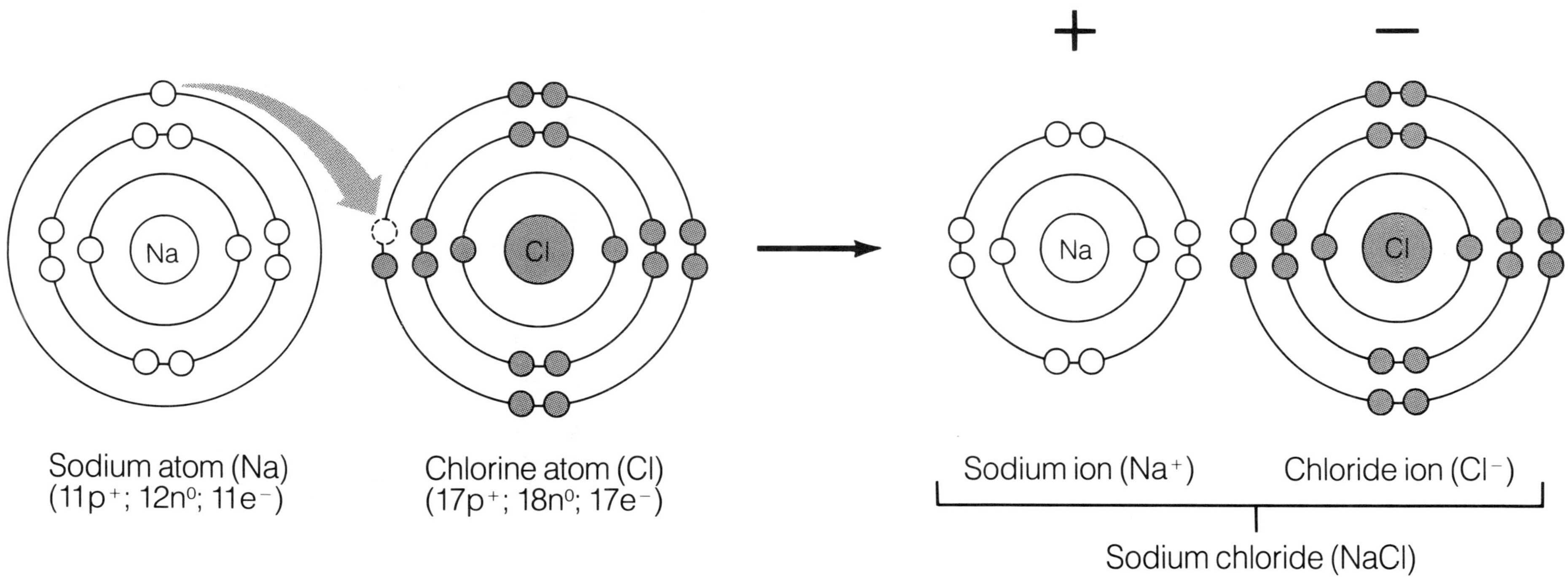

Formation of covalent bonds

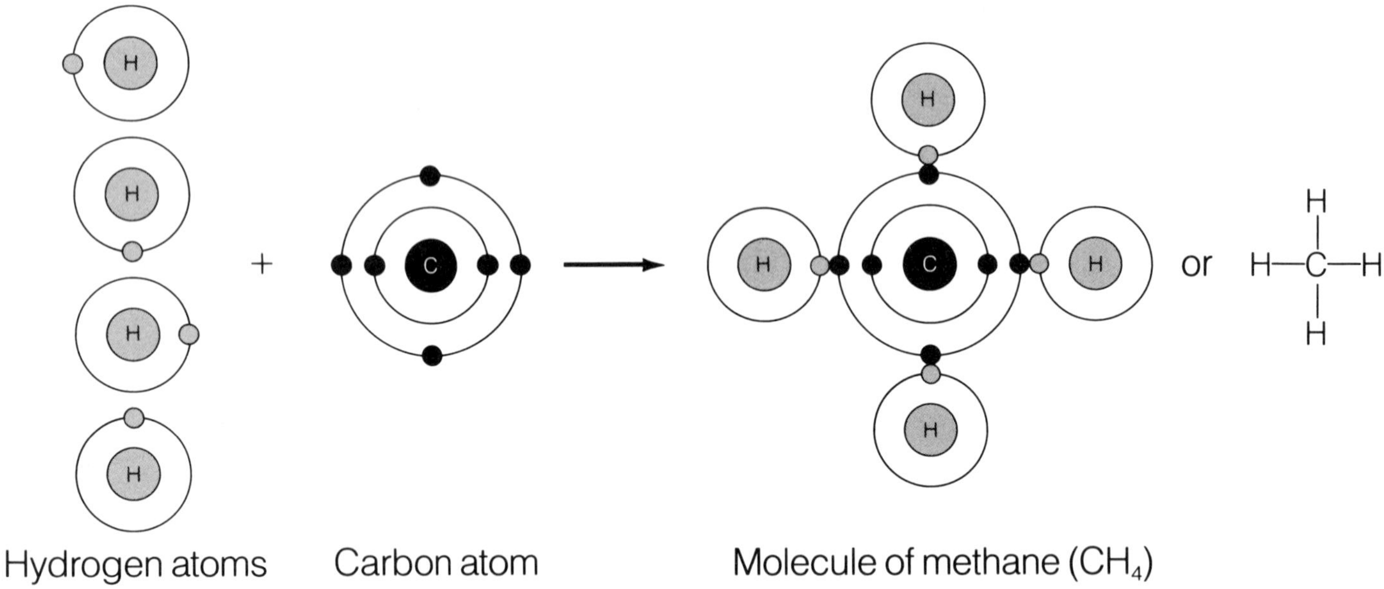

(a) Formation of four single covalent bonds

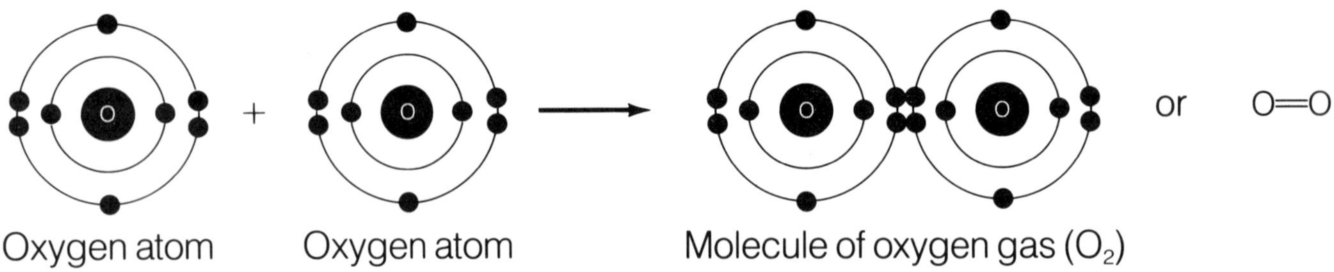

(b) Formation of a double covalent bond

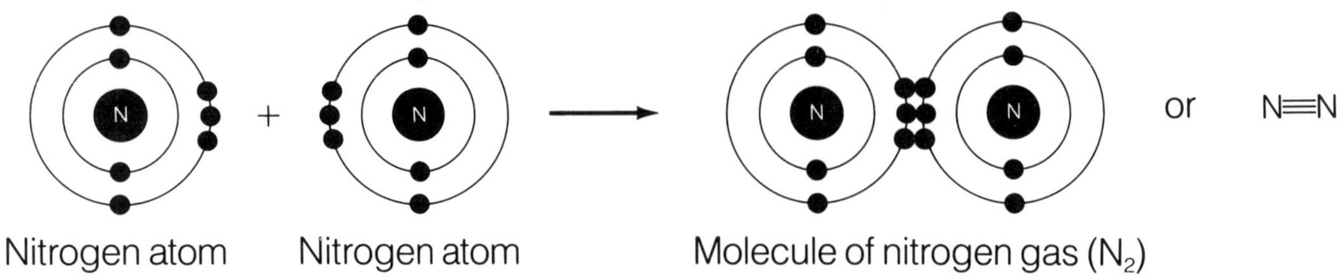

(c) Formation of a triple covalent bond

Comparison of ionic, polar covalent, and nonpolar covalent bonds

Type of bond	Ionic bond	Polar covalent bond	Nonpolar covalent bond
Status of electrons	Complete transfer of electrons	Unequal sharing of electrons	Equal sharing of electrons
Electrical charge distribution	Separate ions (charged particles) form	Slight negative charge (δ^-) at one end of molecule, slight positive charge (δ^+) at other end	Charge balanced among atoms
Example	$Na^+ \; Cl^-$ Sodium chloride	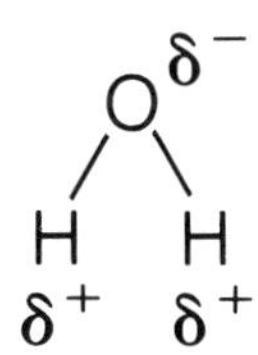 Water	O=C=O Carbon dioxide

Hydrogen bonding between polar water molecules

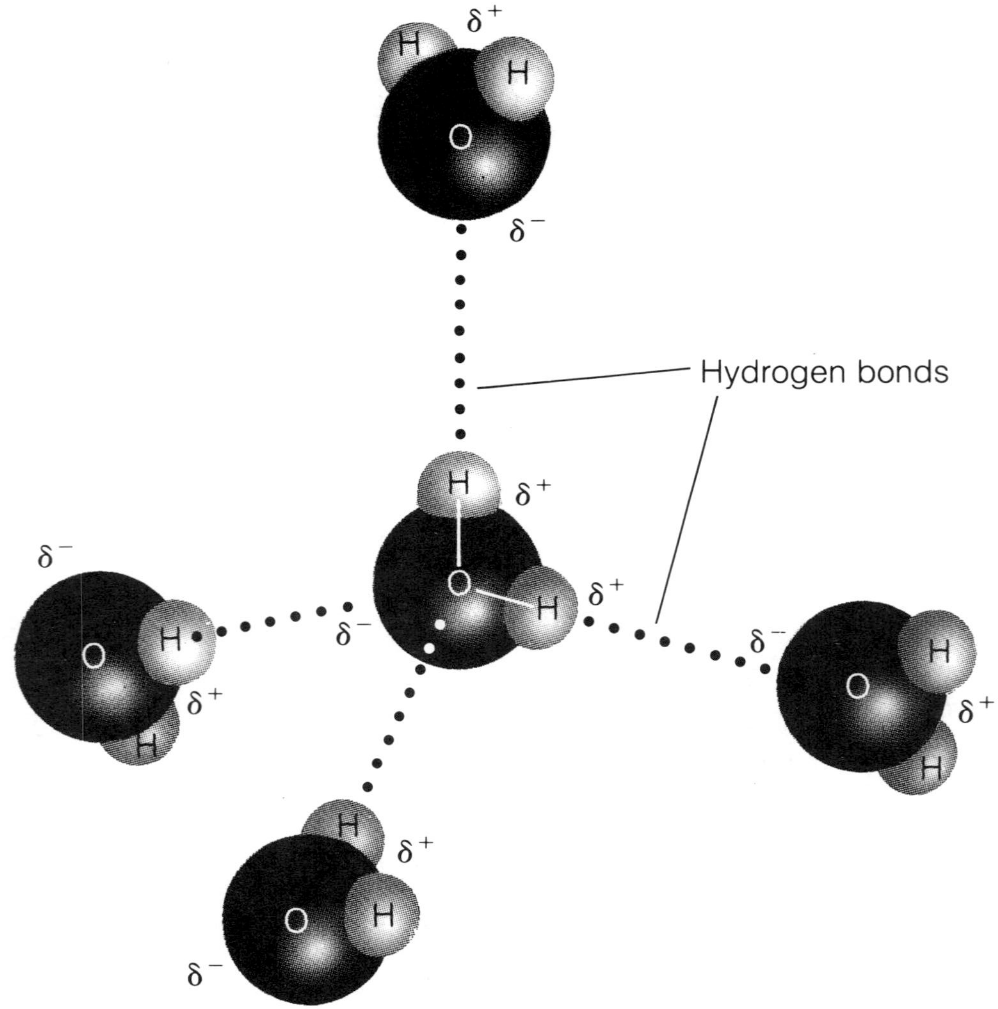

The pH scale

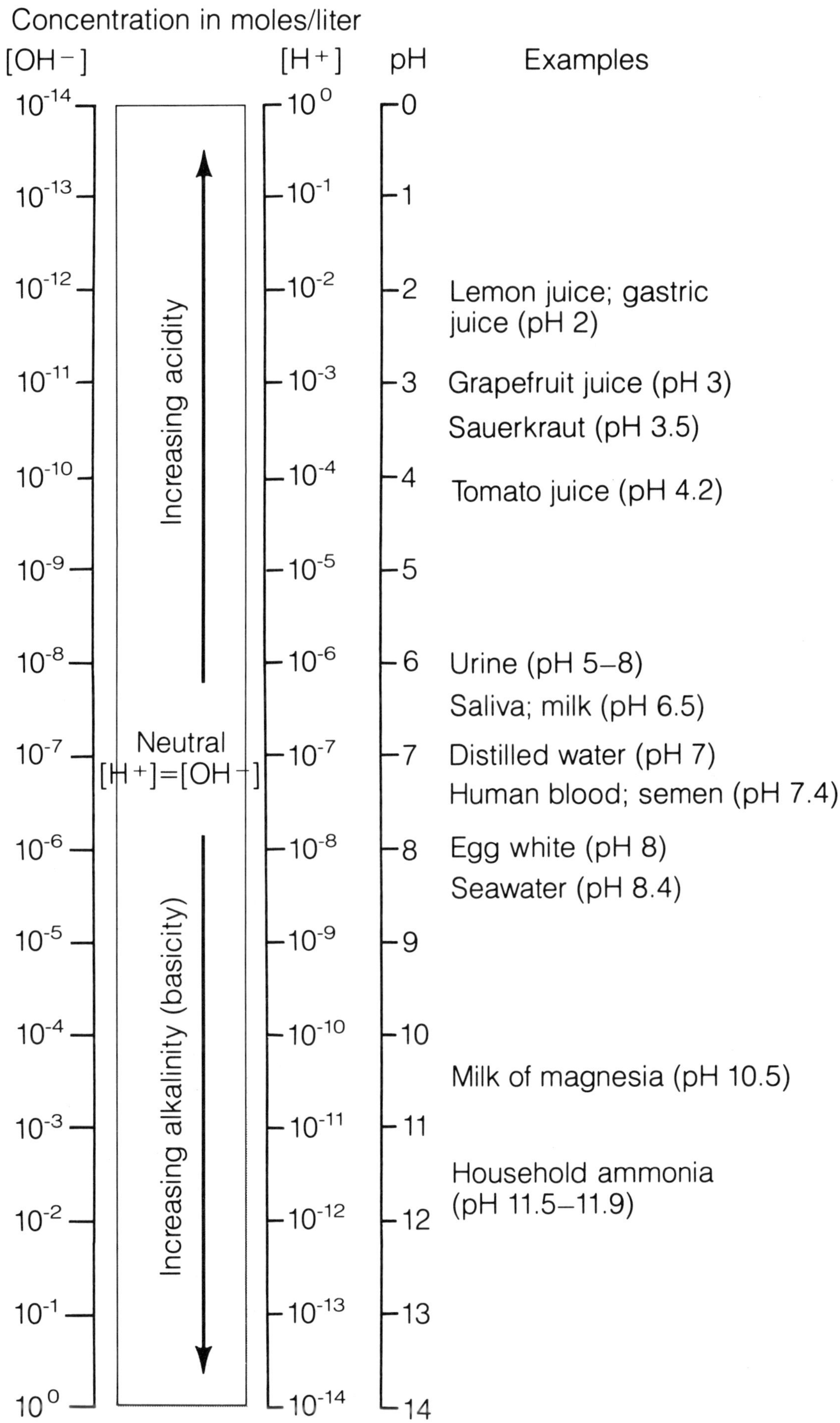
Concentration in moles/liter
[OH−]
[H+]
pH
Examples
10^{-14} 10^{-13} 10^{-12} 10^{-11} 10^{-10} 10^{-9} 10^{-8} 10^{-7} 10^{-6} 10^{-5} 10^{-4} 10^{-3} 10^{-2} 10^{-1} 10^{0}
10^{0} 10^{-1} 10^{-2} 10^{-3} 10^{-4} 10^{-5} 10^{-6} 10^{-7} 10^{-8} 10^{-9} 10^{-10} 10^{-11} 10^{-12} 10^{-13} 10^{-14}
0 1 2 3 4 5 6 7 8 9 10 11 12 13 14
Increasing acidity
Neutral
[H+]=[OH−]
Increasing alkalinity (basicity)
Lemon juice; gastric juice (pH 2)
Grapefruit juice (pH 3)
Sauerkraut (pH 3.5)
Tomato juice (pH 4.2)
Urine (pH 5–8)
Saliva; milk (pH 6.5)
Distilled water (pH 7)
Human blood; semen (pH 7.4)
Egg white (pH 8)
Seawater (pH 8.4)
Milk of magnesia (pH 10.5)
Household ammonia (pH 11.5–11.9)

Carbohydrate molecules

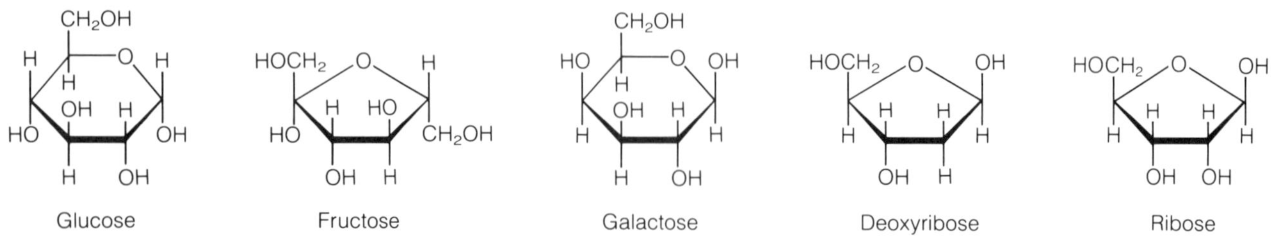

(a) Monosaccharides

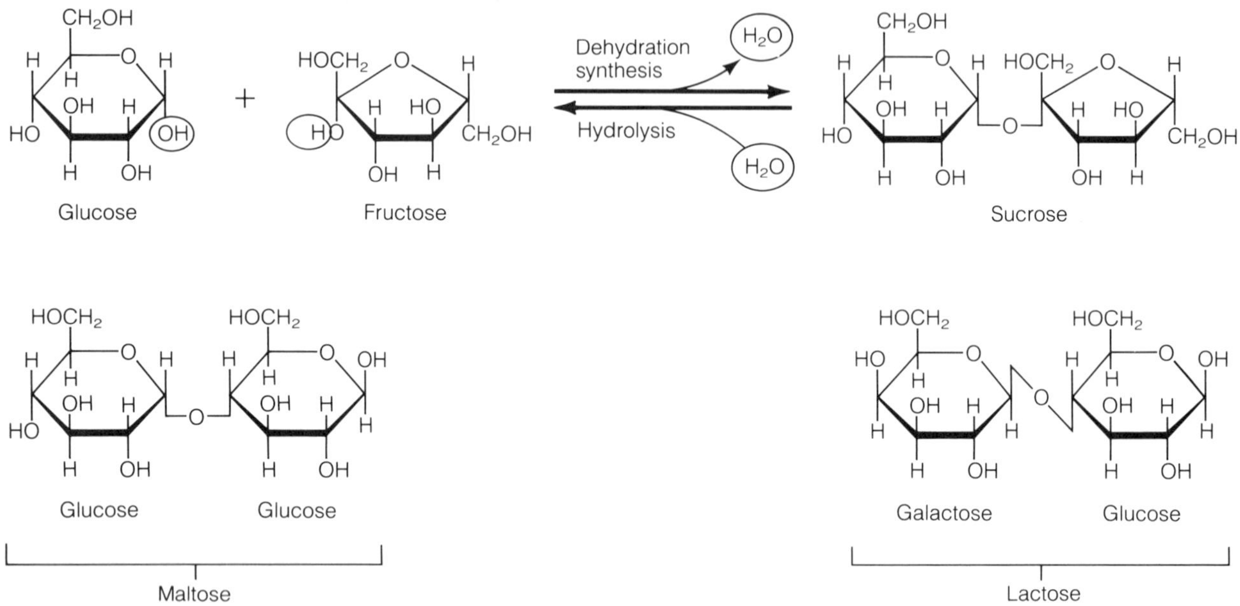

(b) Disaccharides

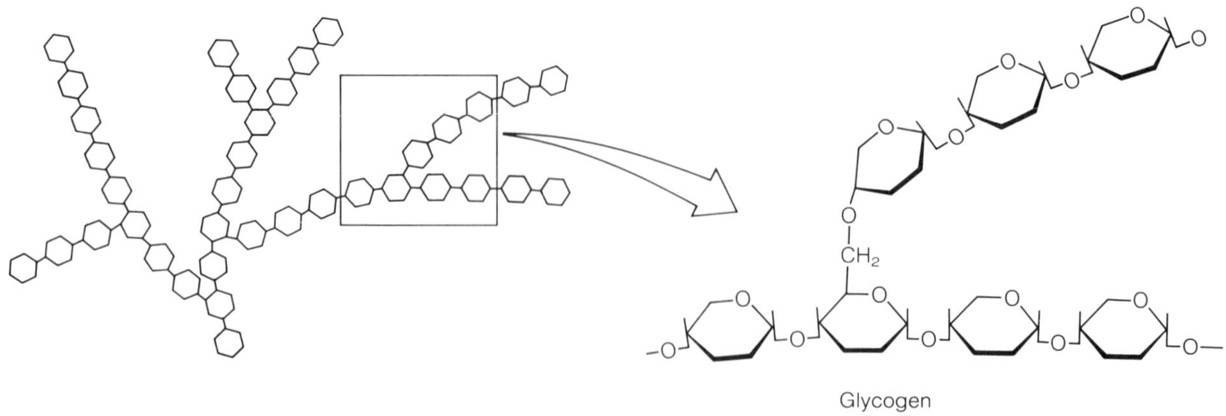

(c) Portion of a polysaccharide molecule (glycogen)

Lipids

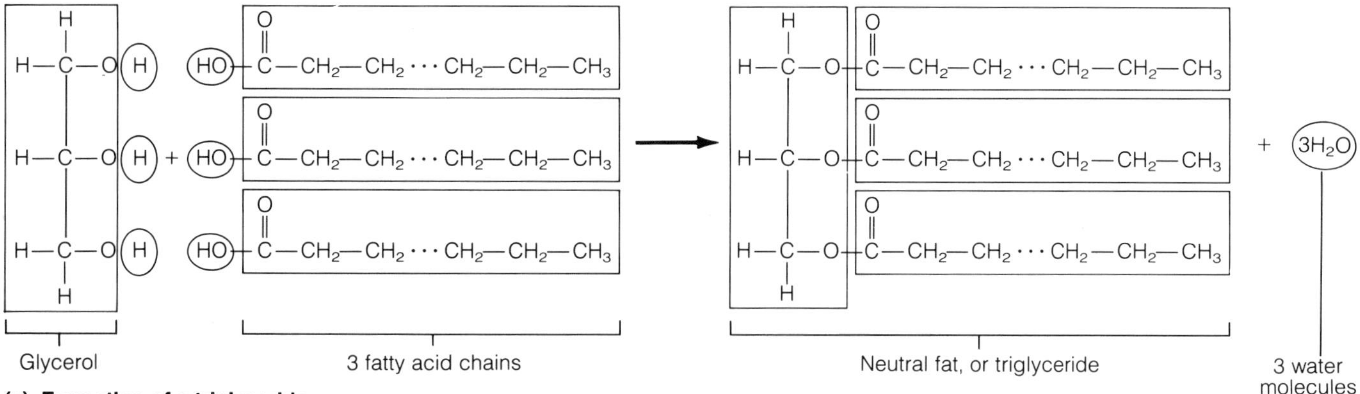

(a) Formation of a triglyceride

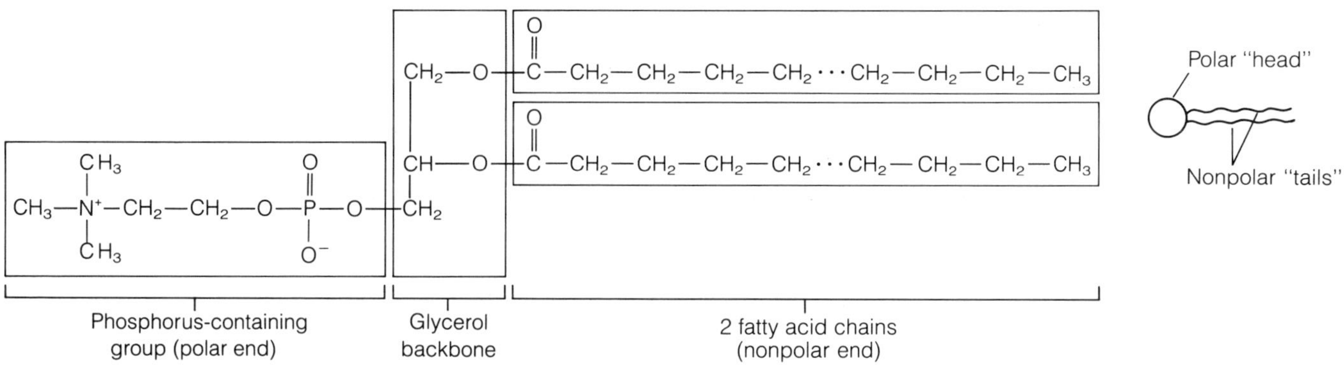

(b) Phospholipid molecule (phosphatidyl choline)

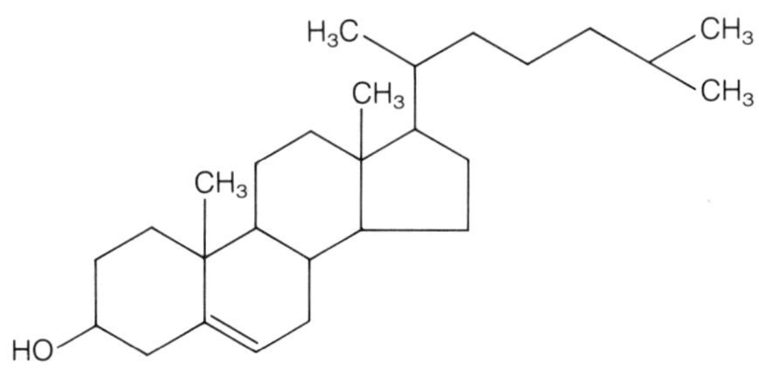

(c) Cholesterol

Amino acid structures

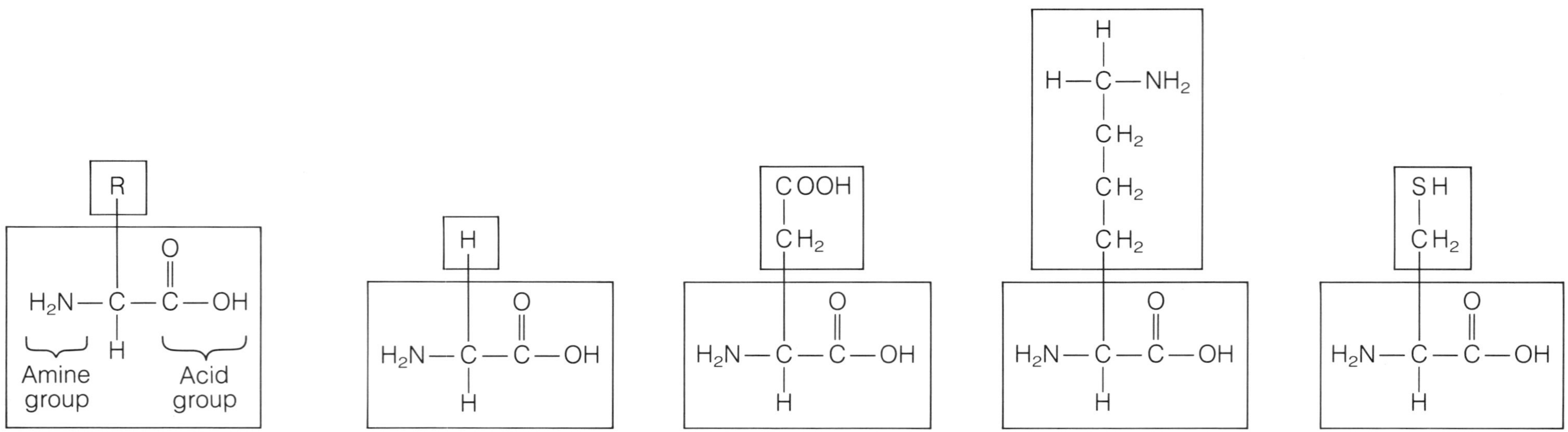

(a) **Generalized structure of all amino acids**

(b) **Glycine (the simplest amino acid)**

(c) **Aspartic acid (an acidic amino acid)**

(d) **Lysine (a basic amino acid)**

(e) **Cysteine (a sulfur-containing amino acid)**

Amino acids are linked together by dehydration synthesis

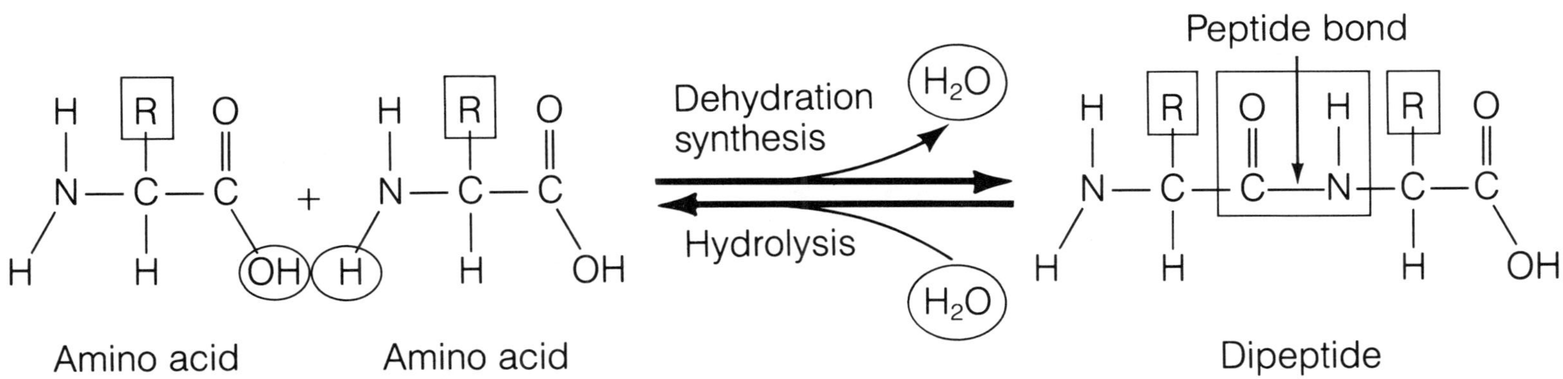

Levels of protein structure

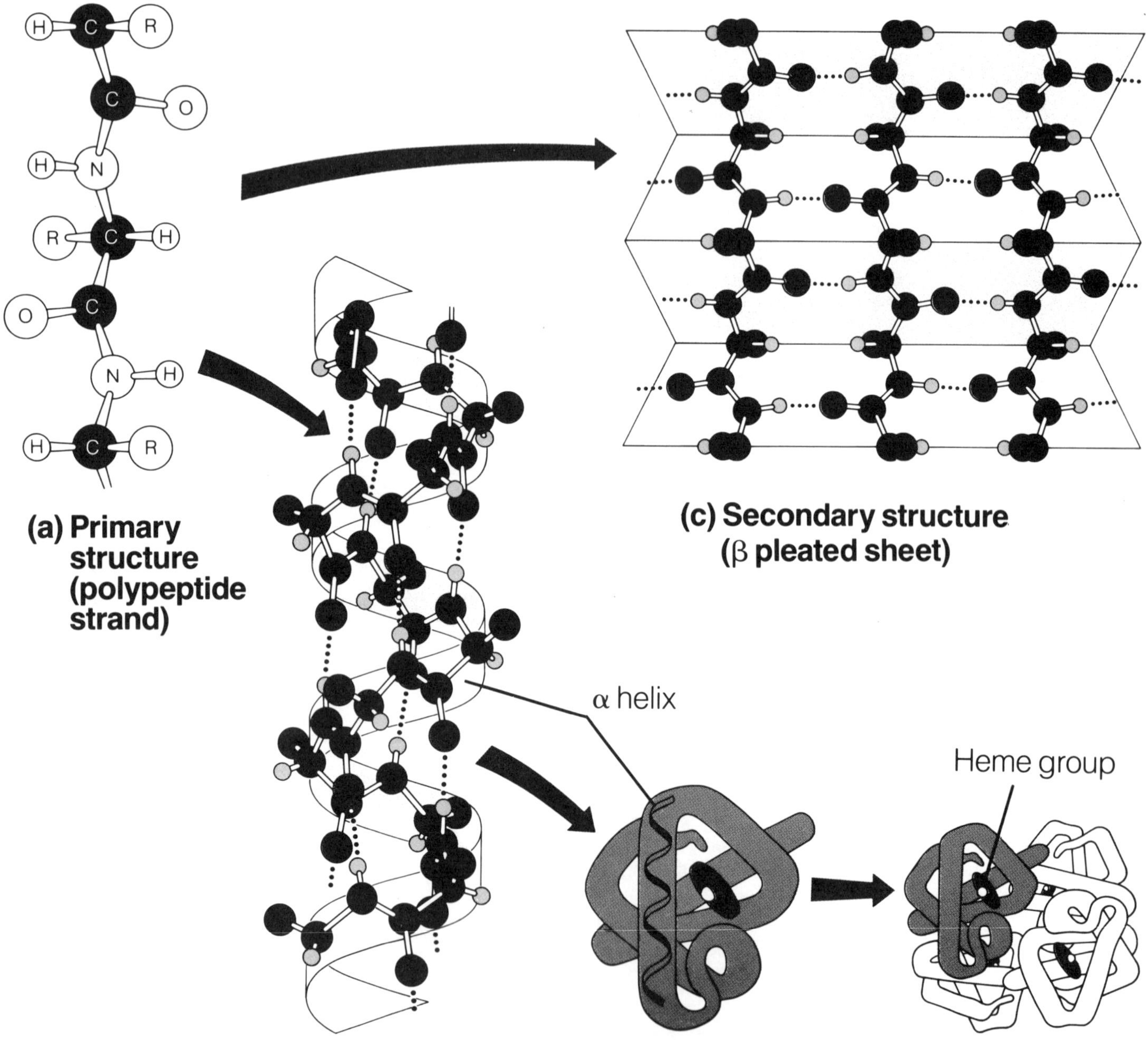

(a) Primary structure (polypeptide strand)

(b) Secondary structure (α helix)

(c) Secondary structure (β pleated sheet)

(d) Tertiary structure (myoglobin molecule)

(e) Quaternary structure (hemoglobin molecule)

Denaturation of a globular protein such as an enzyme

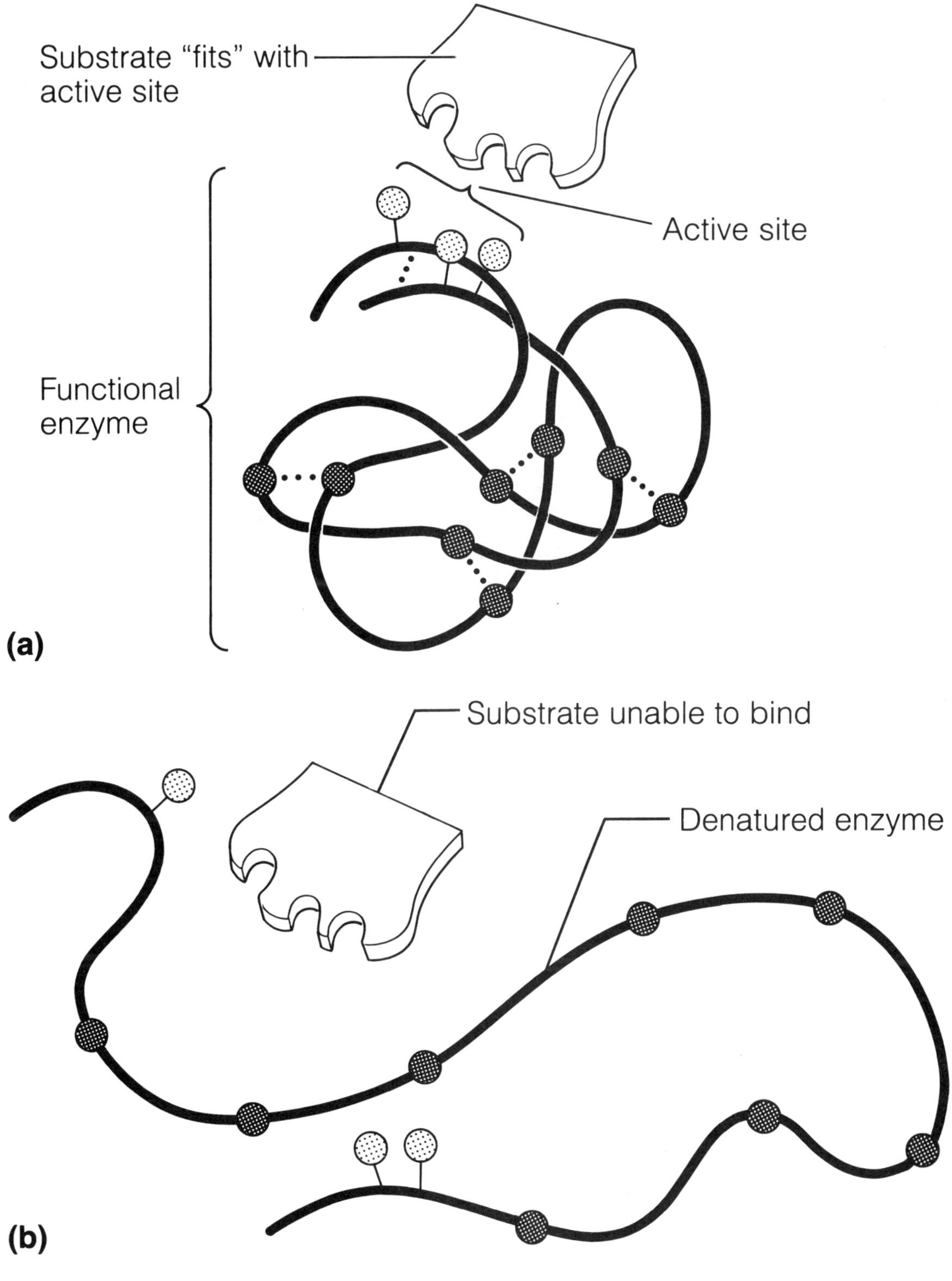

Comparison of the energy barrier in a noncatalyzed reaction and in the same reaction catalyzed by an enzyme

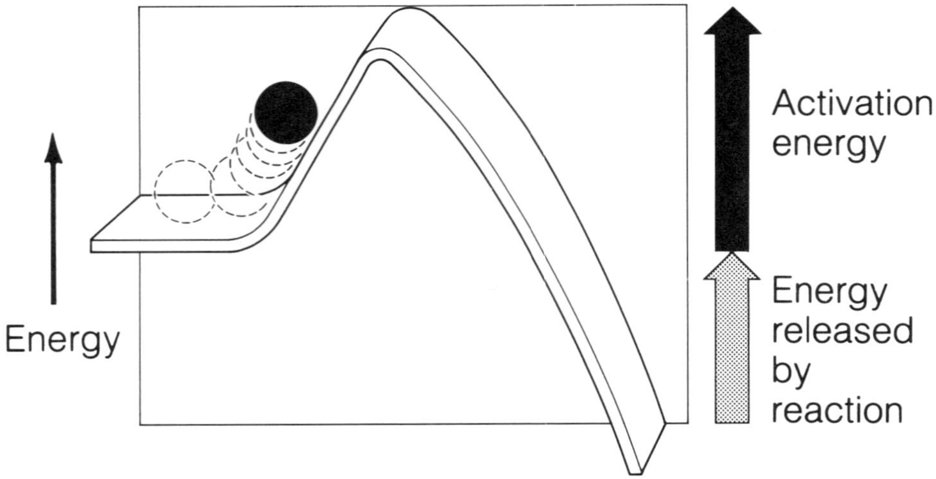

(a) Noncatalyzed reaction

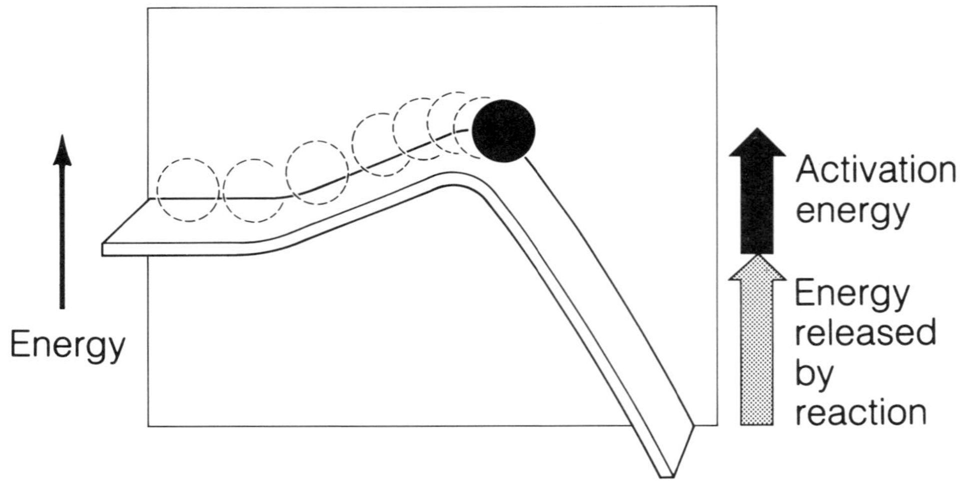

(b) Enzyme-catalyzed reaction

Mechanism of enzyme action

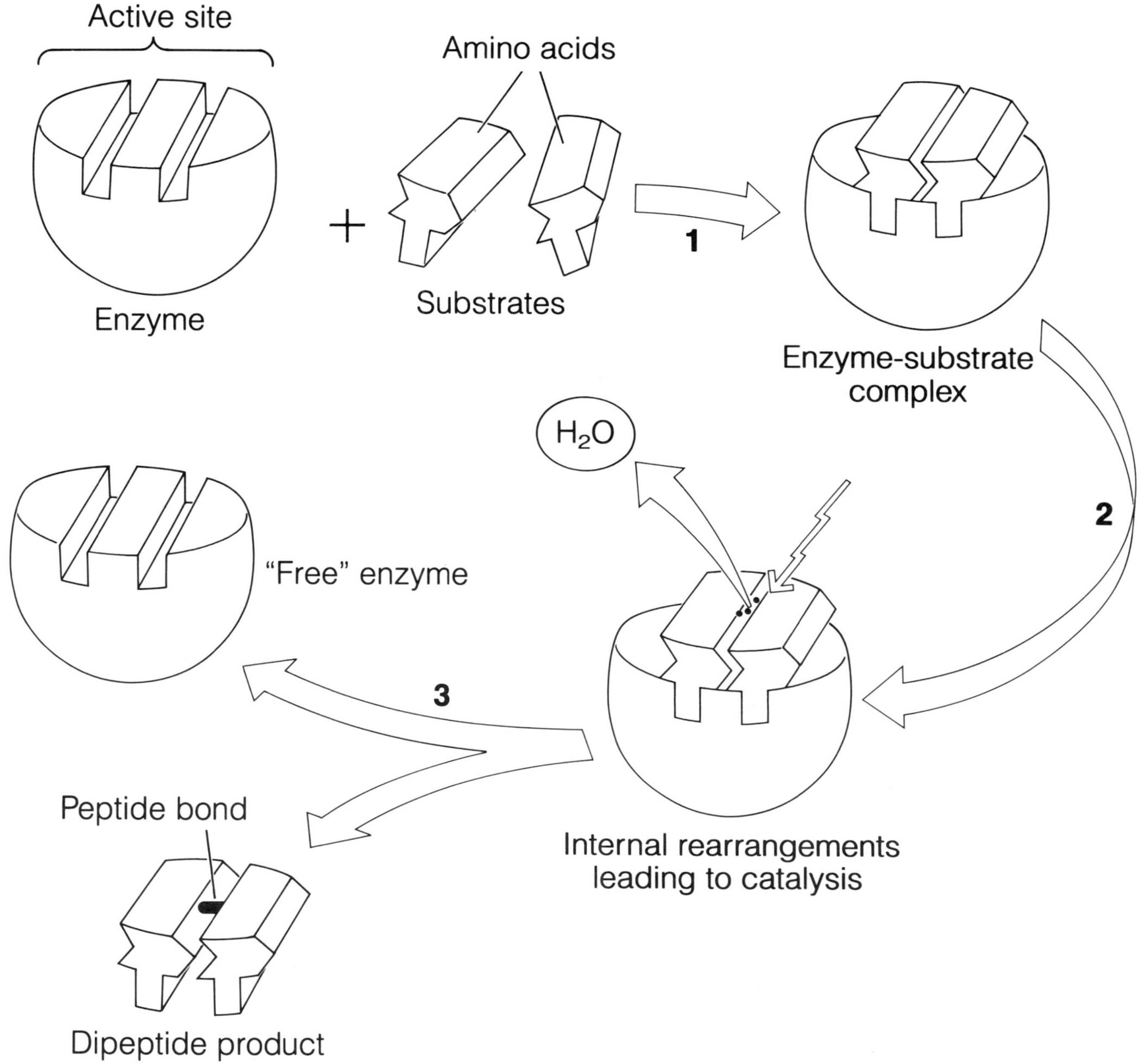

Structure of DNA

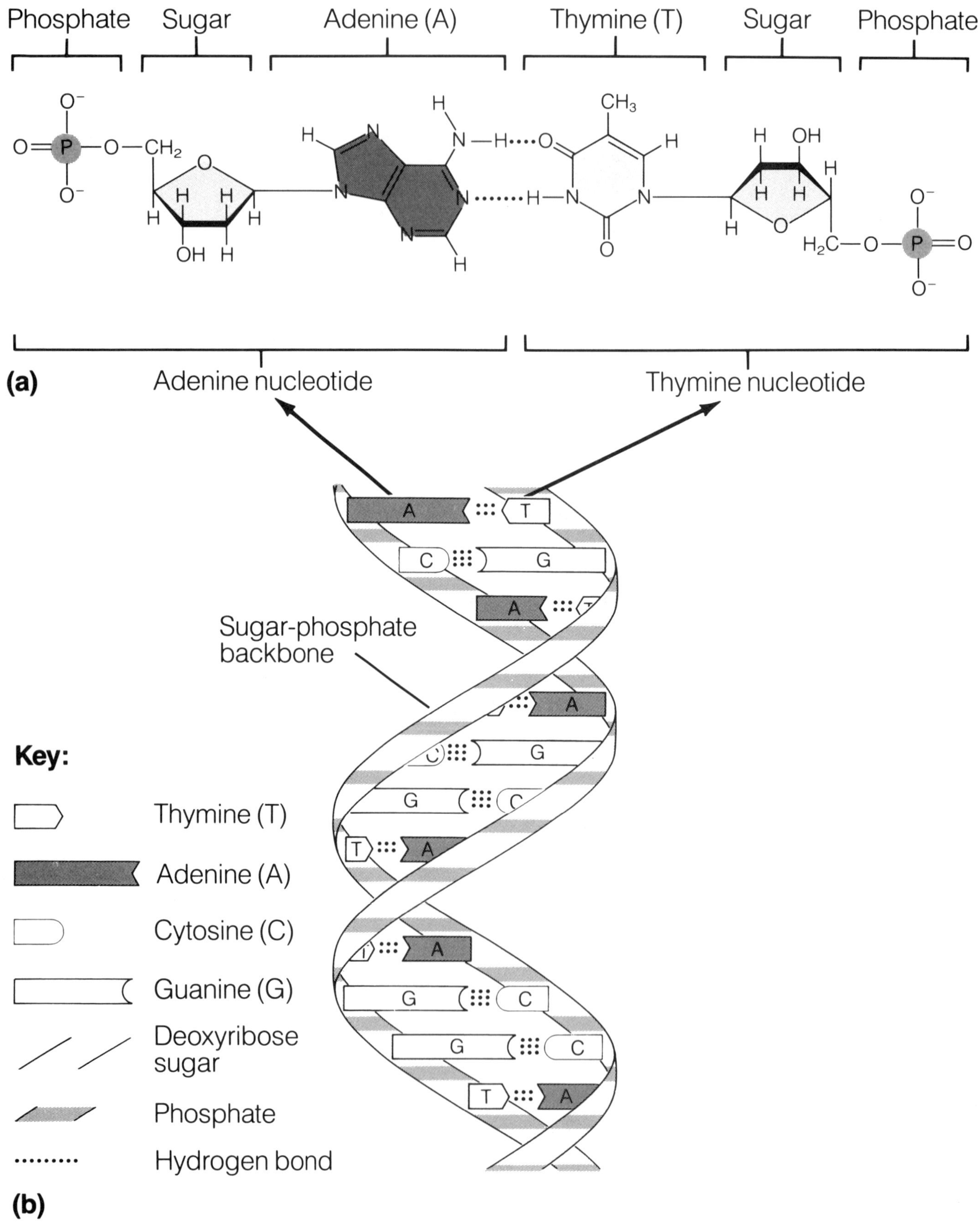

Structure of ATP (adenosine triphosphate)

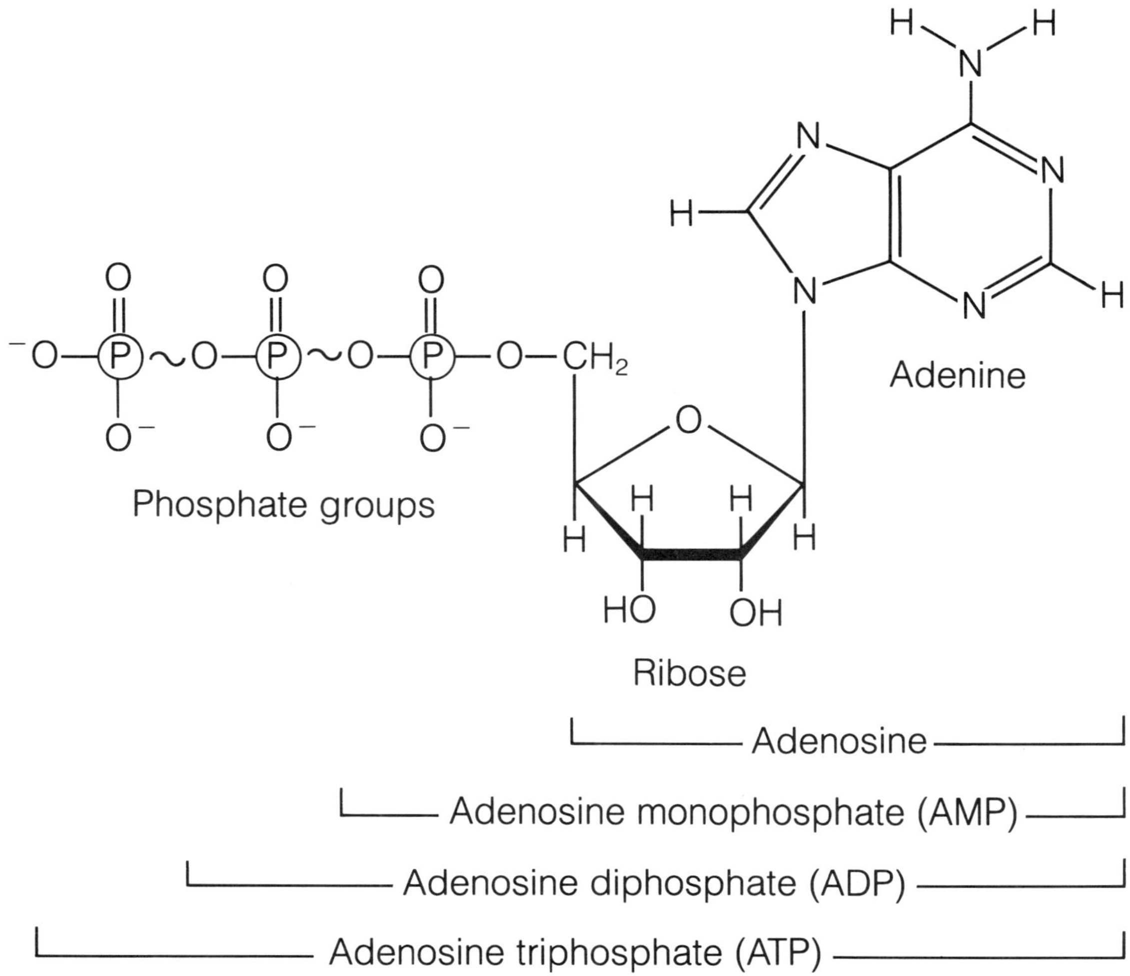

Three examples of how ATP drives cellular work

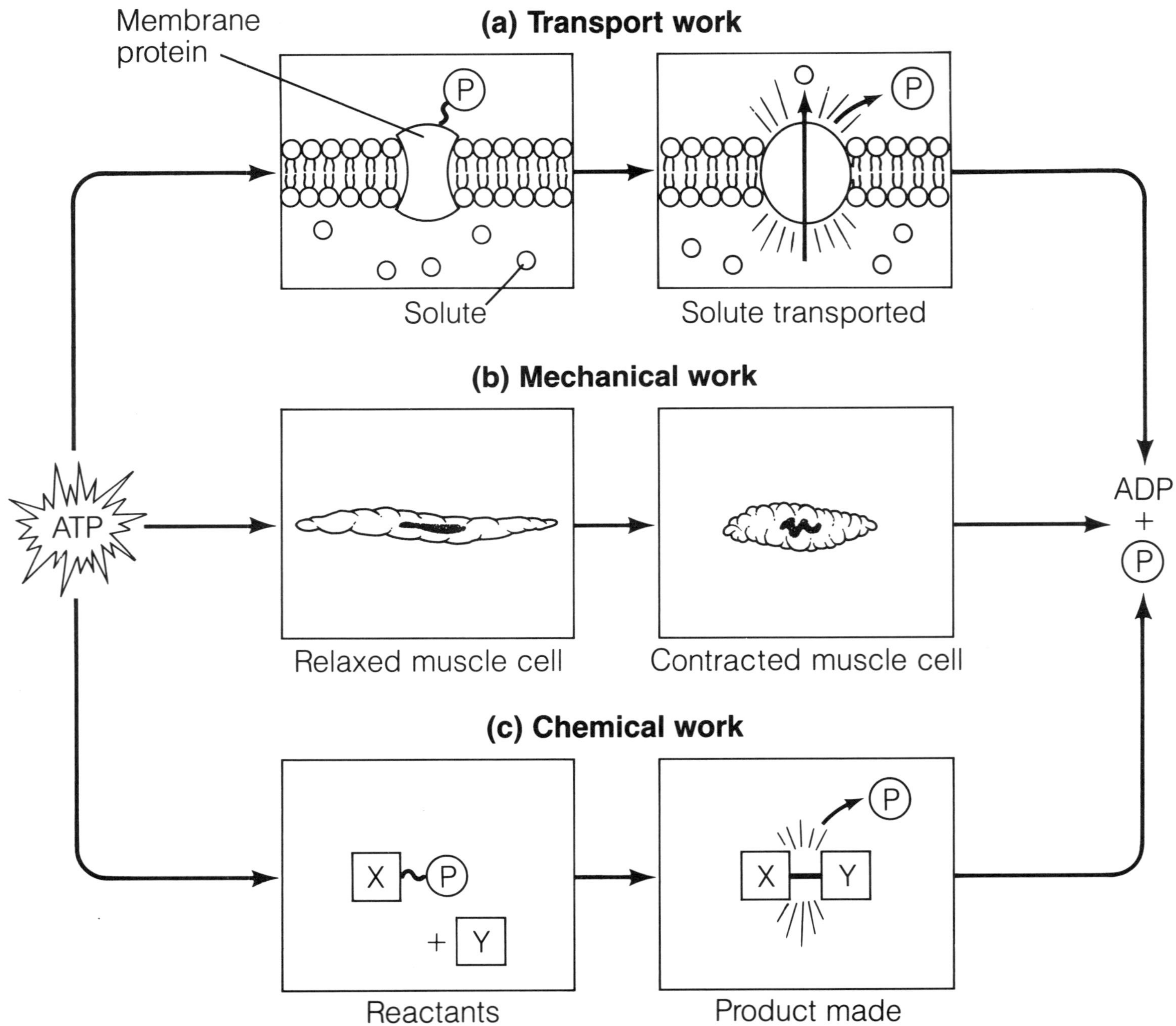

Structure of the plasma membrane

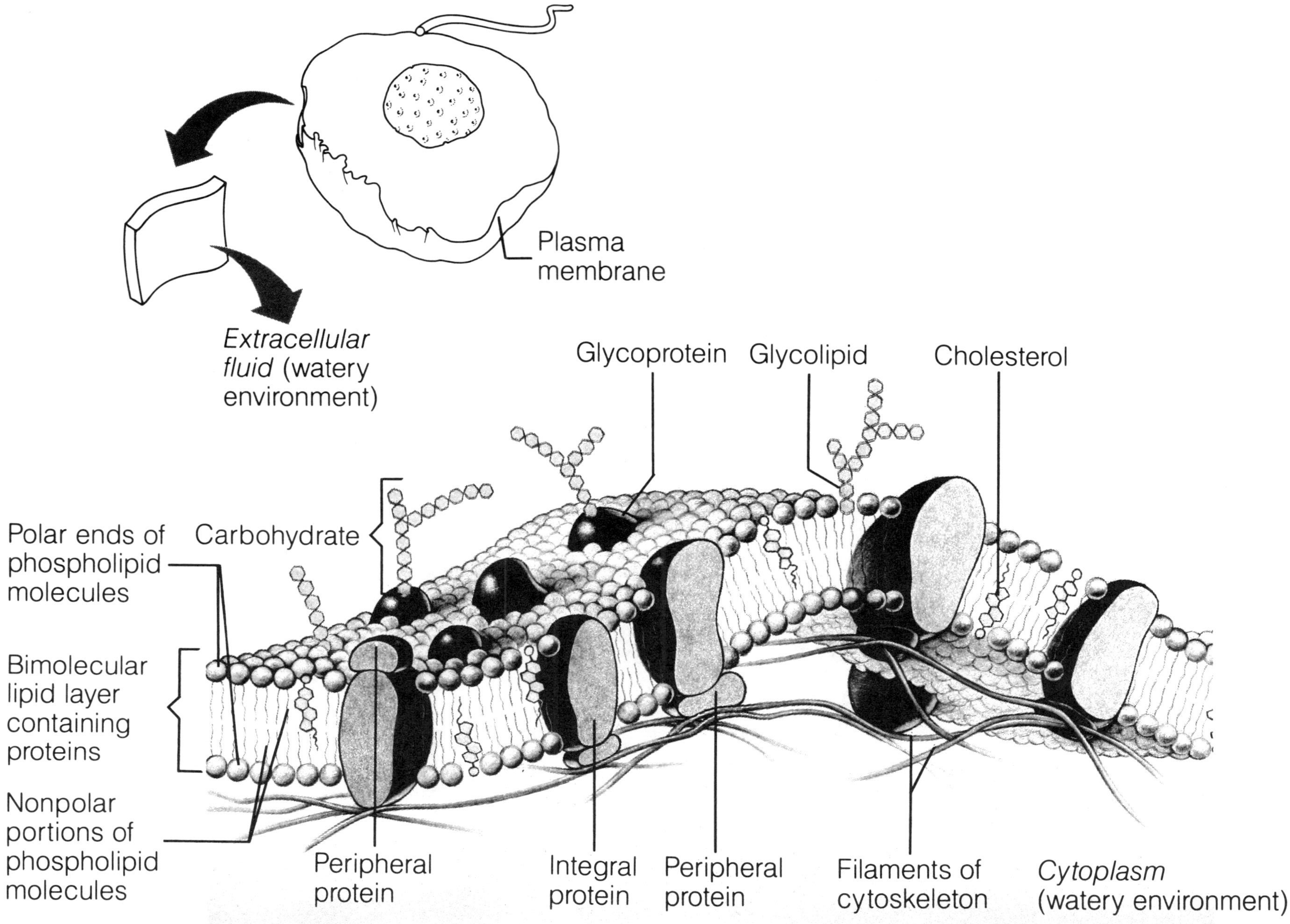

Cell junctions

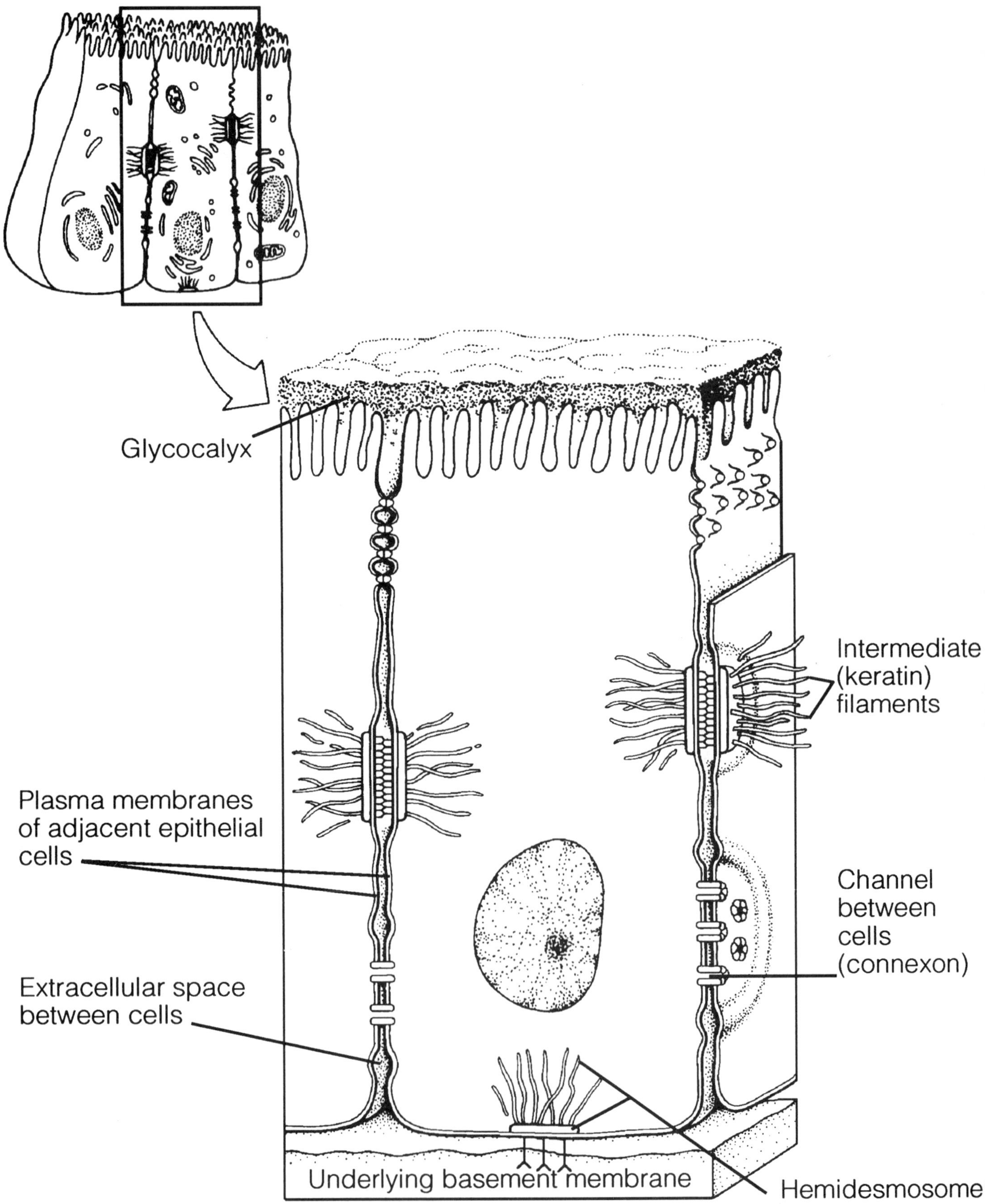

Diffusion through the plasma membrane

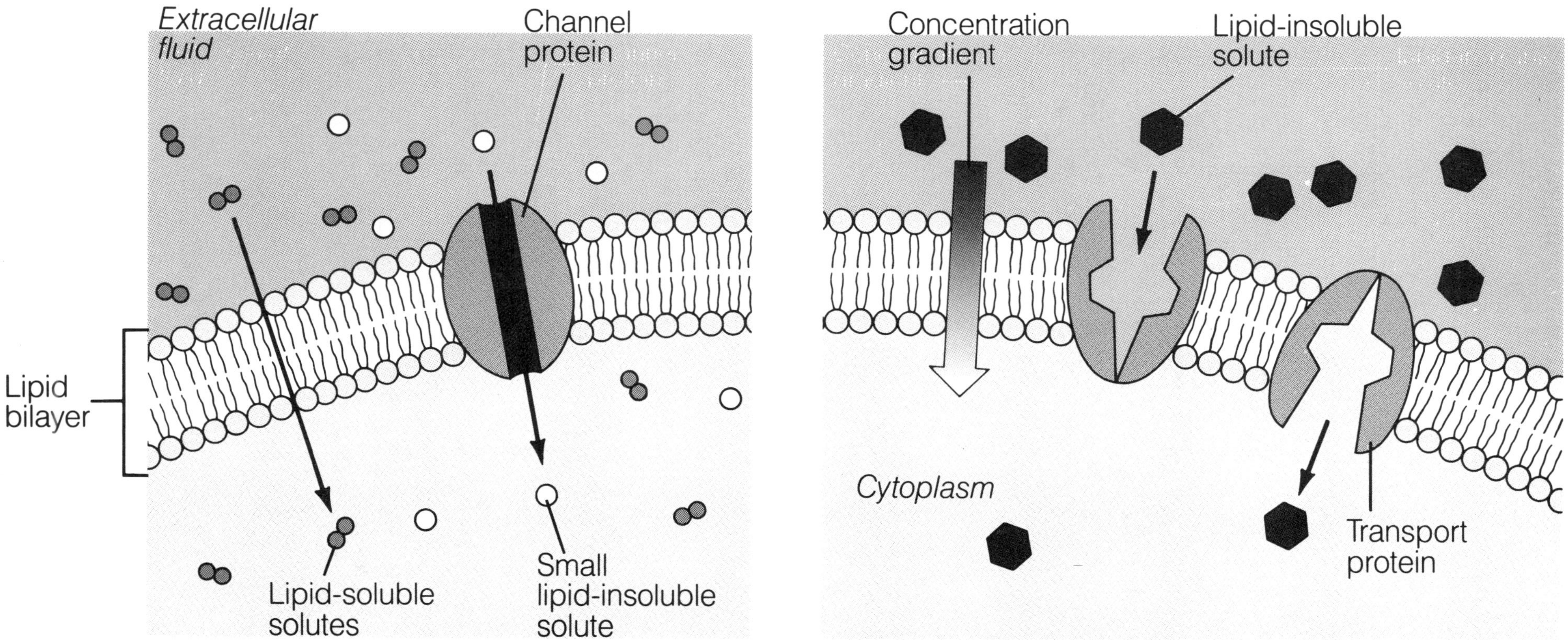

(a) Simple diffusion **(b) Facilitated diffusion**

Influence of membrane permeability on diffusion and osmosis

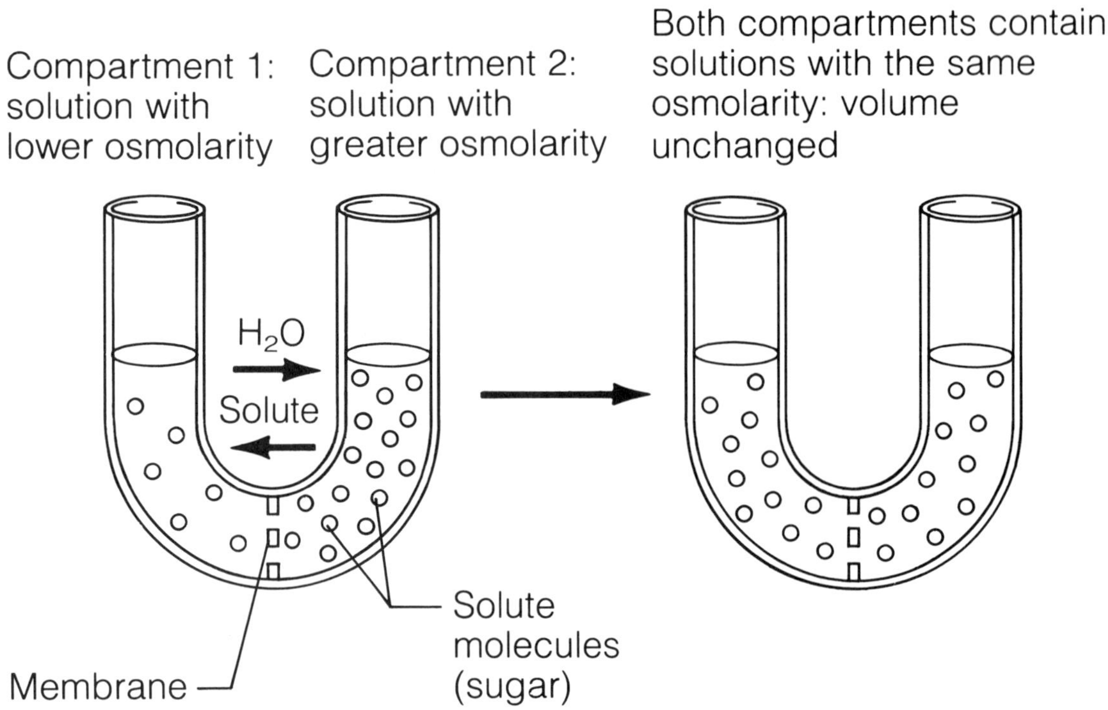

(a) Membrane permeable to both solute molecules and water

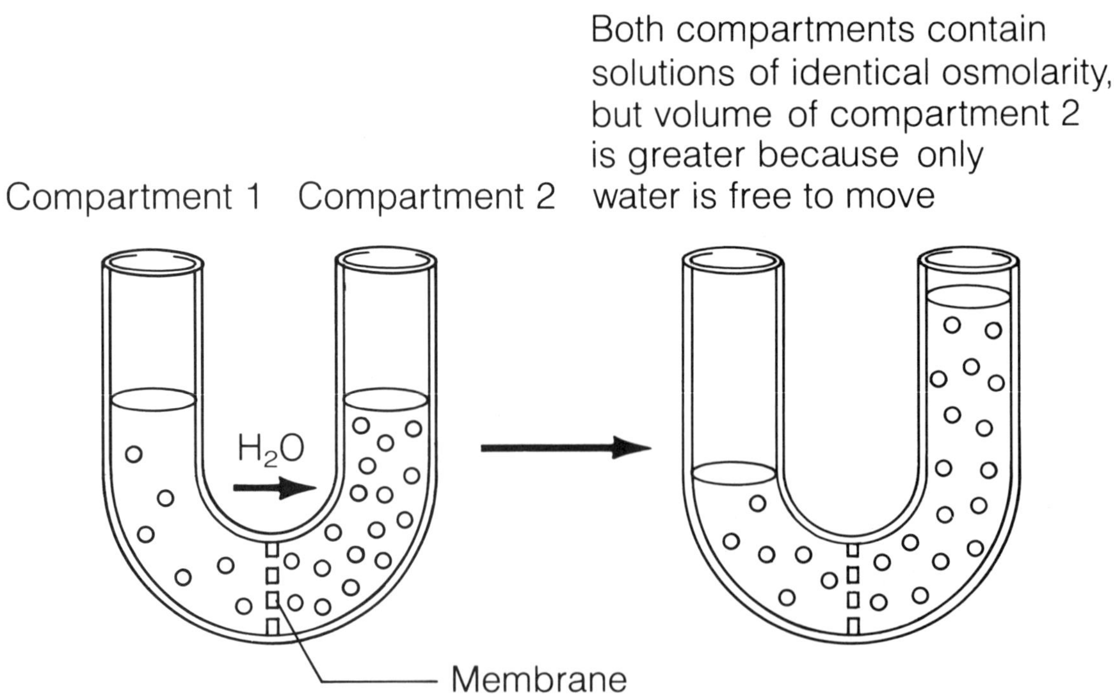

(b) Membrane impermeable to solute molecules, permeable to water

Operation of the sodium-potassium pump

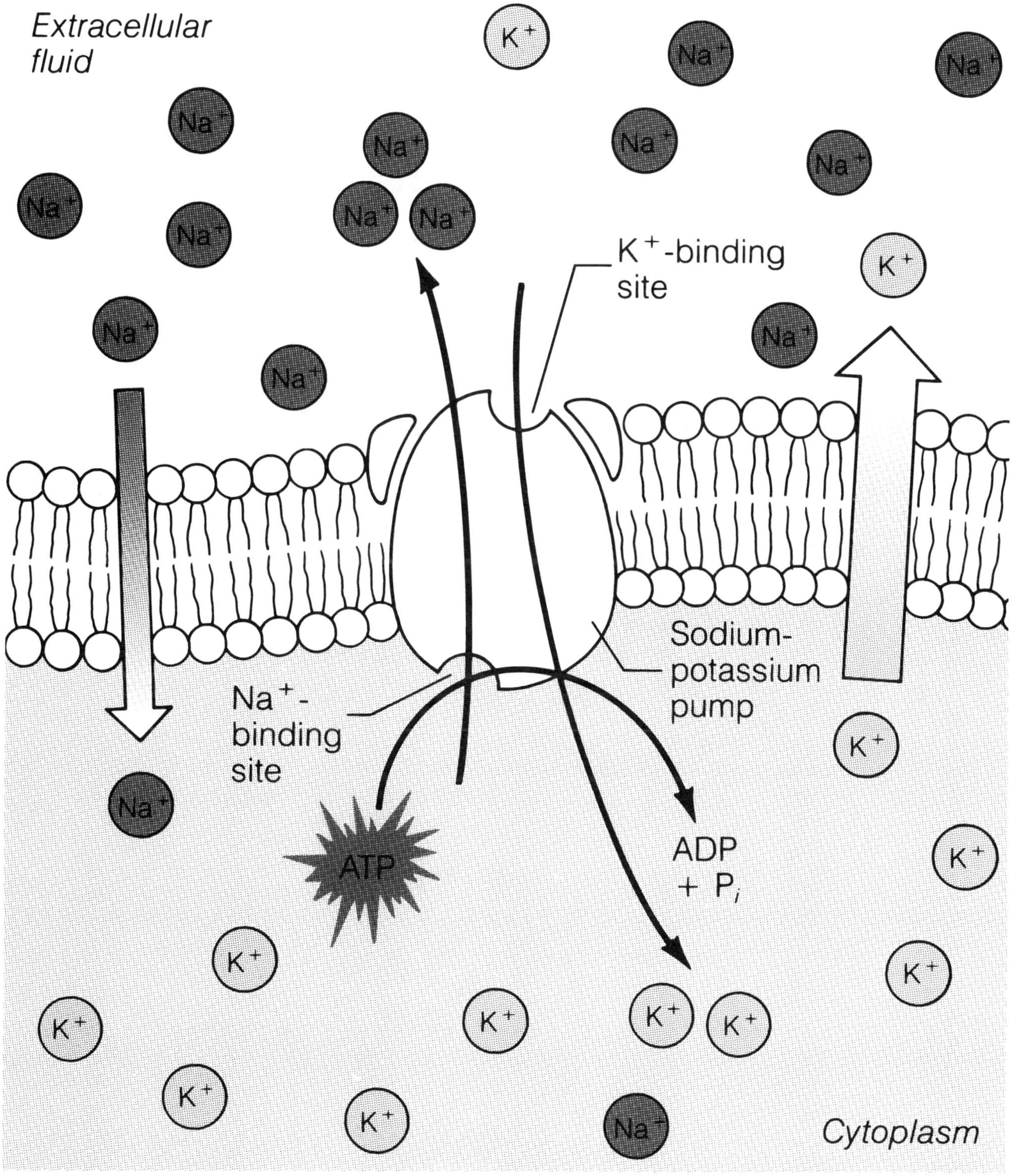

Exocytosis

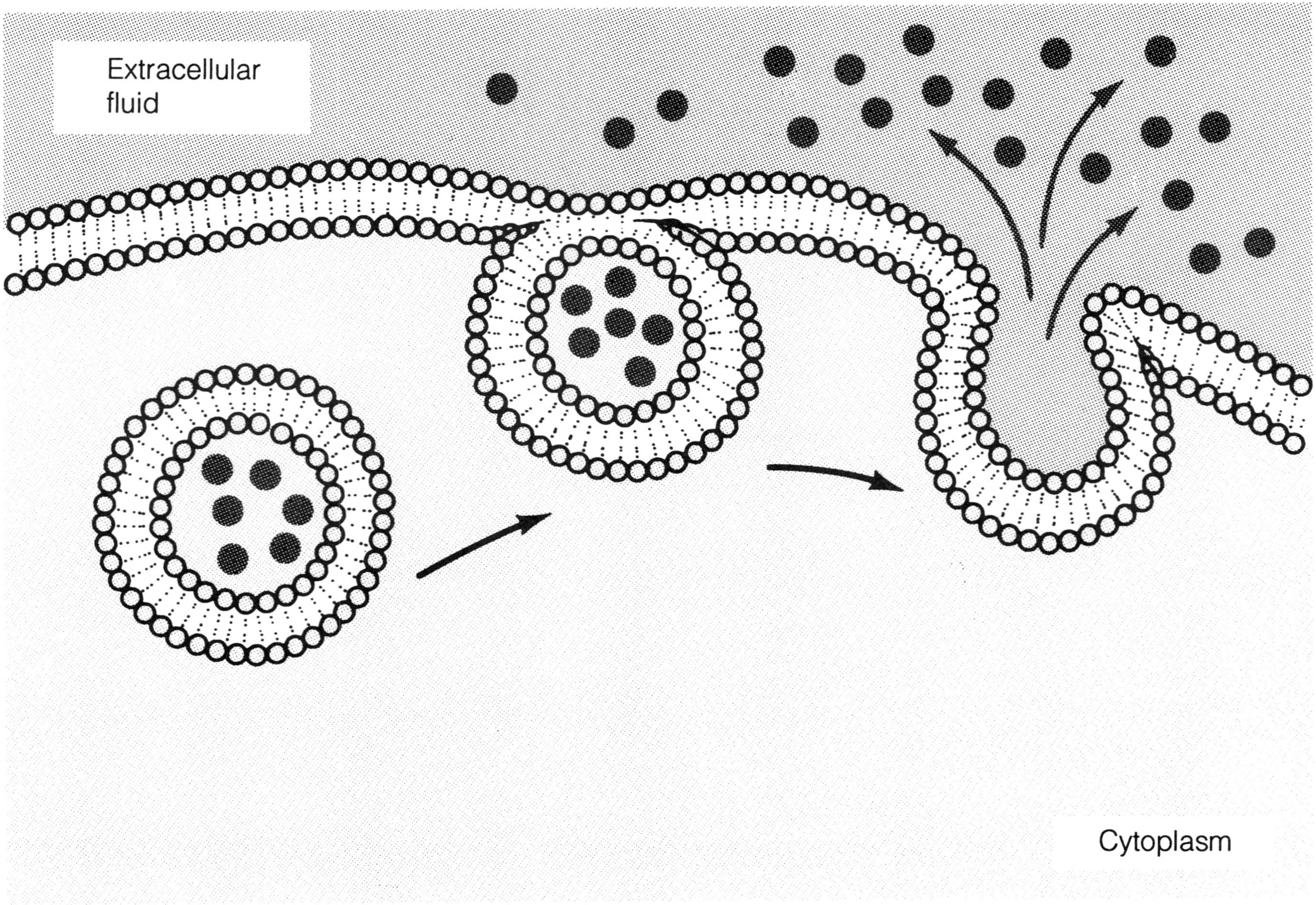
Extracellular
fluid
Cytoplasm

Three types of endocytosis

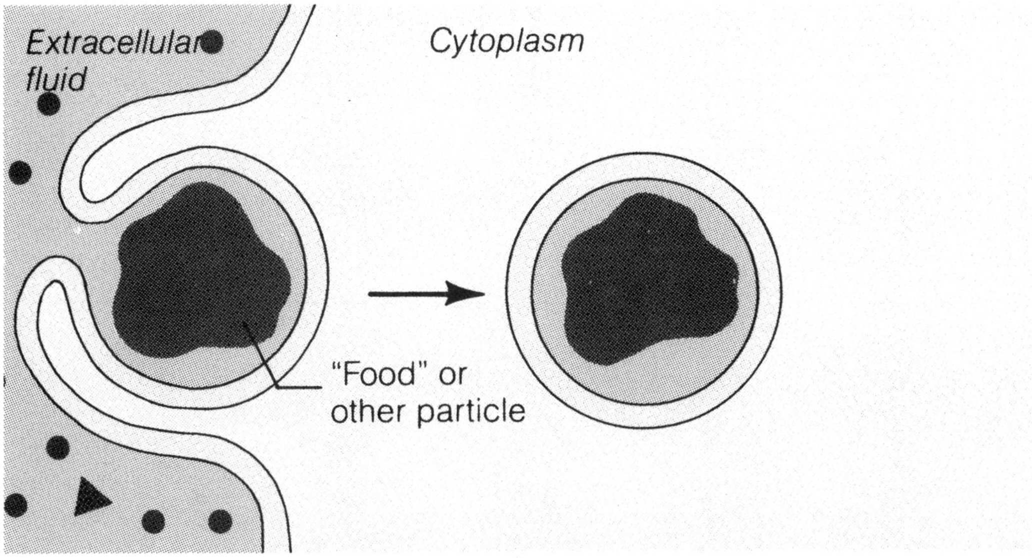

(a) Phagocytosis

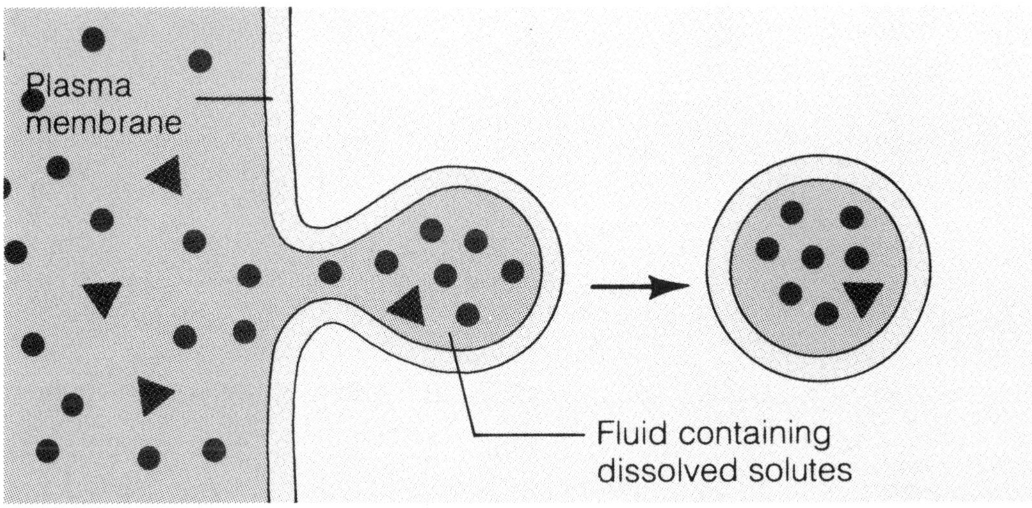

(b) Pinocytosis

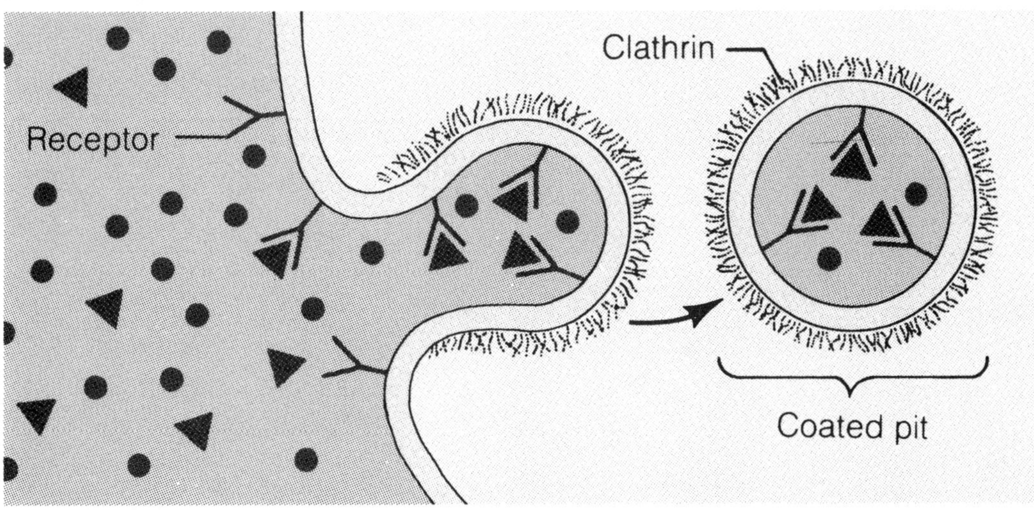

(c) Receptor-mediated endocytosis

Summary of forces that generate and maintain membrane potentials

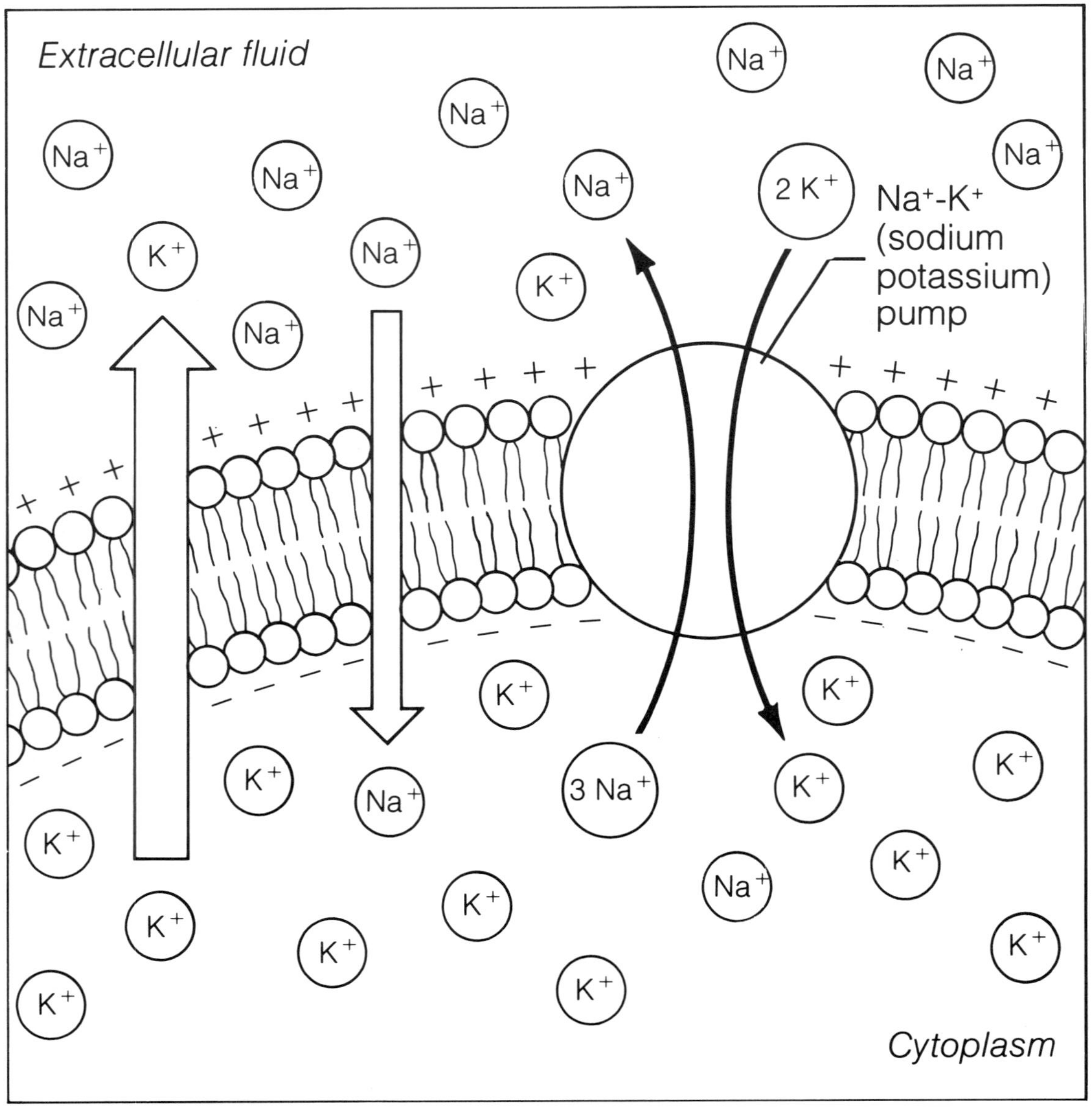

The endoplasmic reticulum

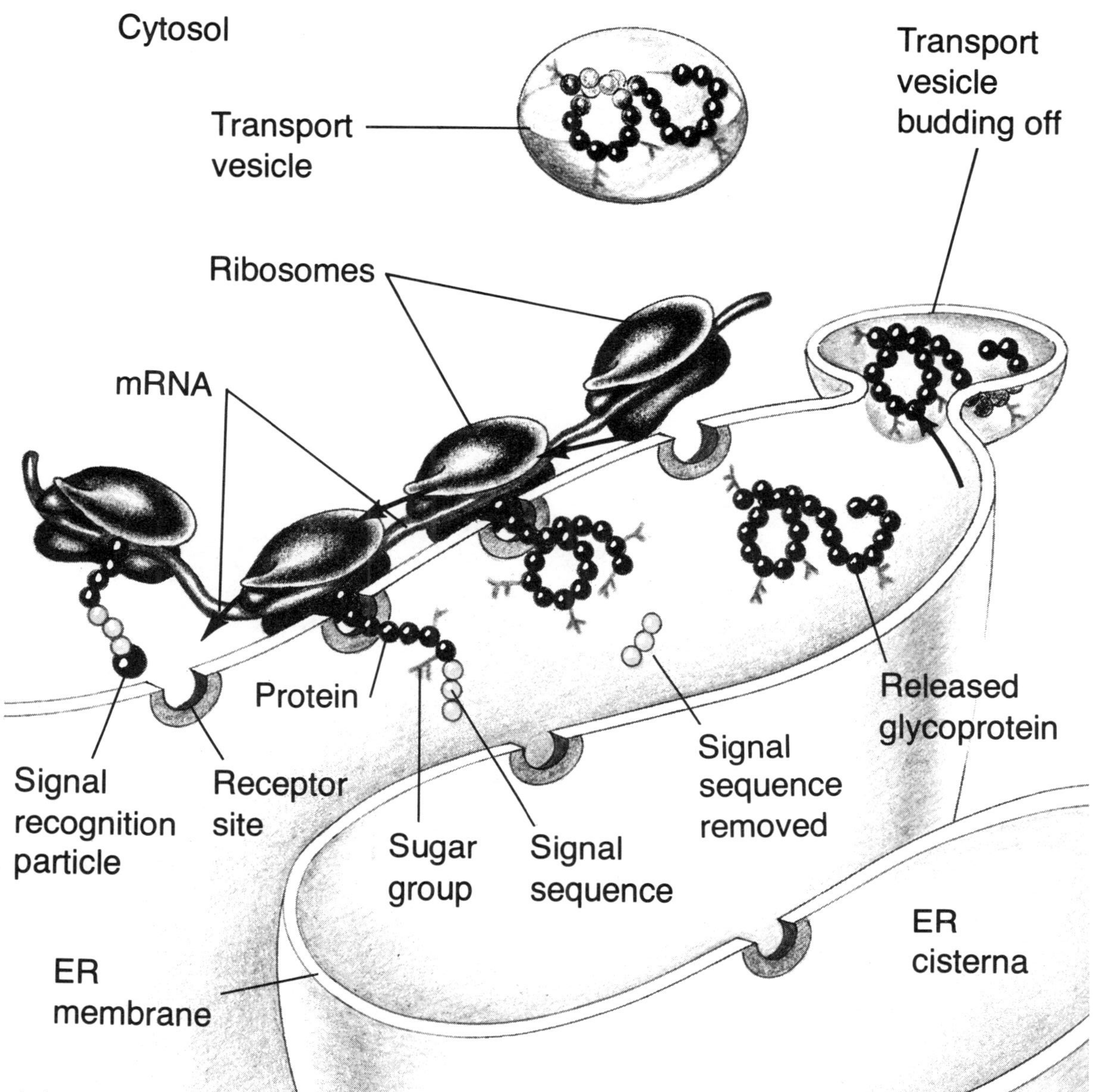

(c)

Role of the Golgi apparatus in packaging proteins for cellular use and for secretion

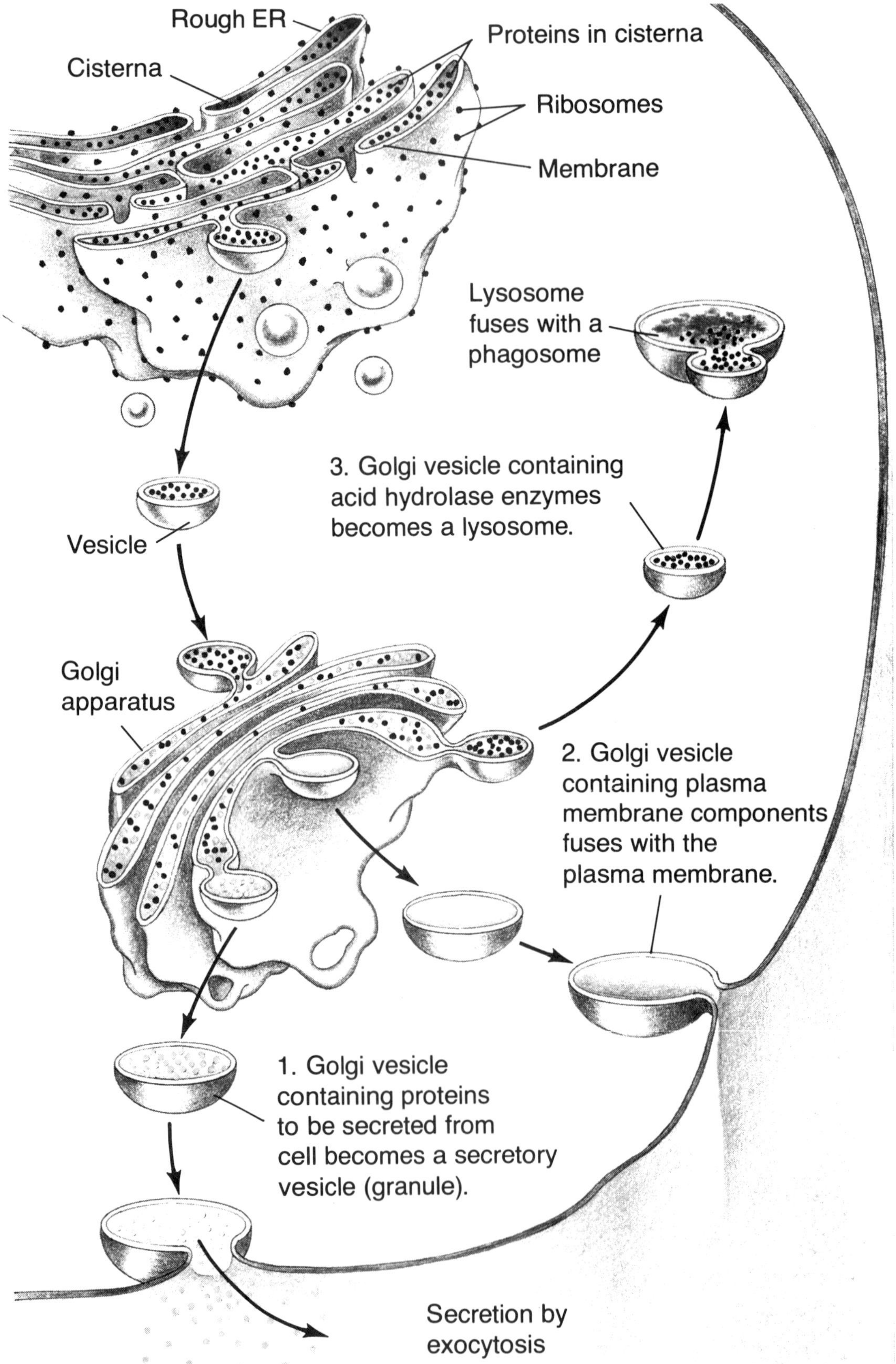

Cilia structure and function

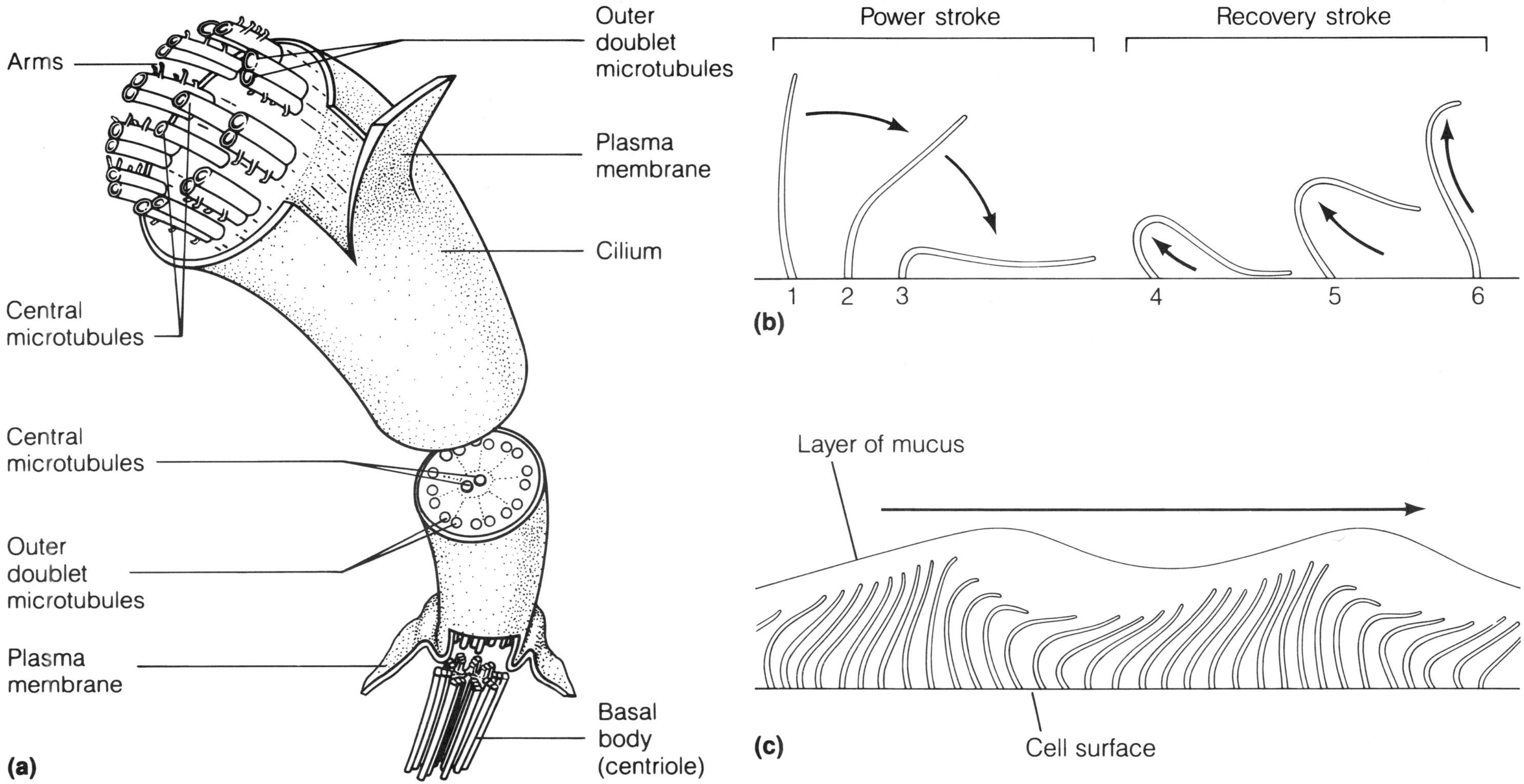

Chromosome structure

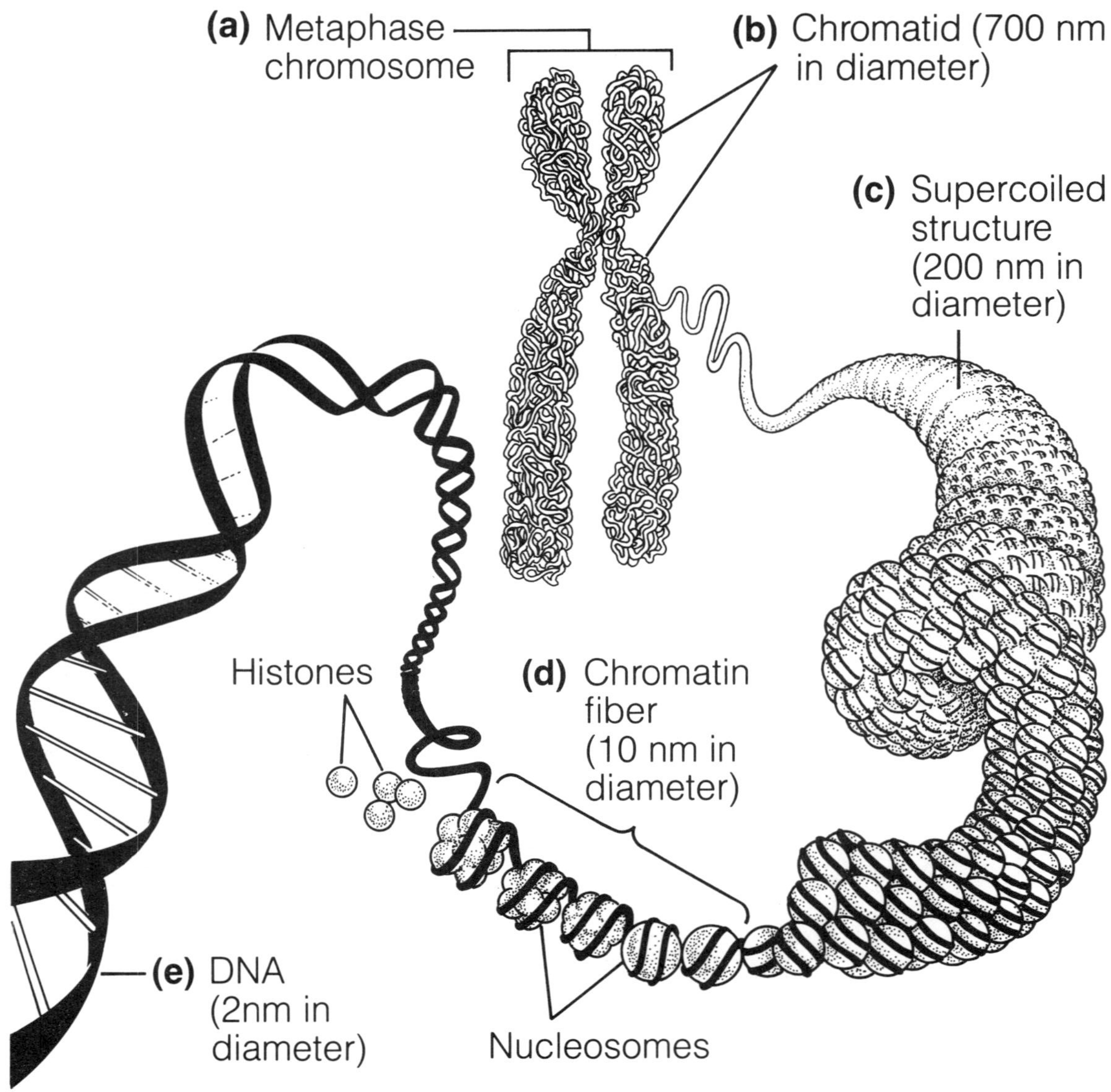

The cell cycle

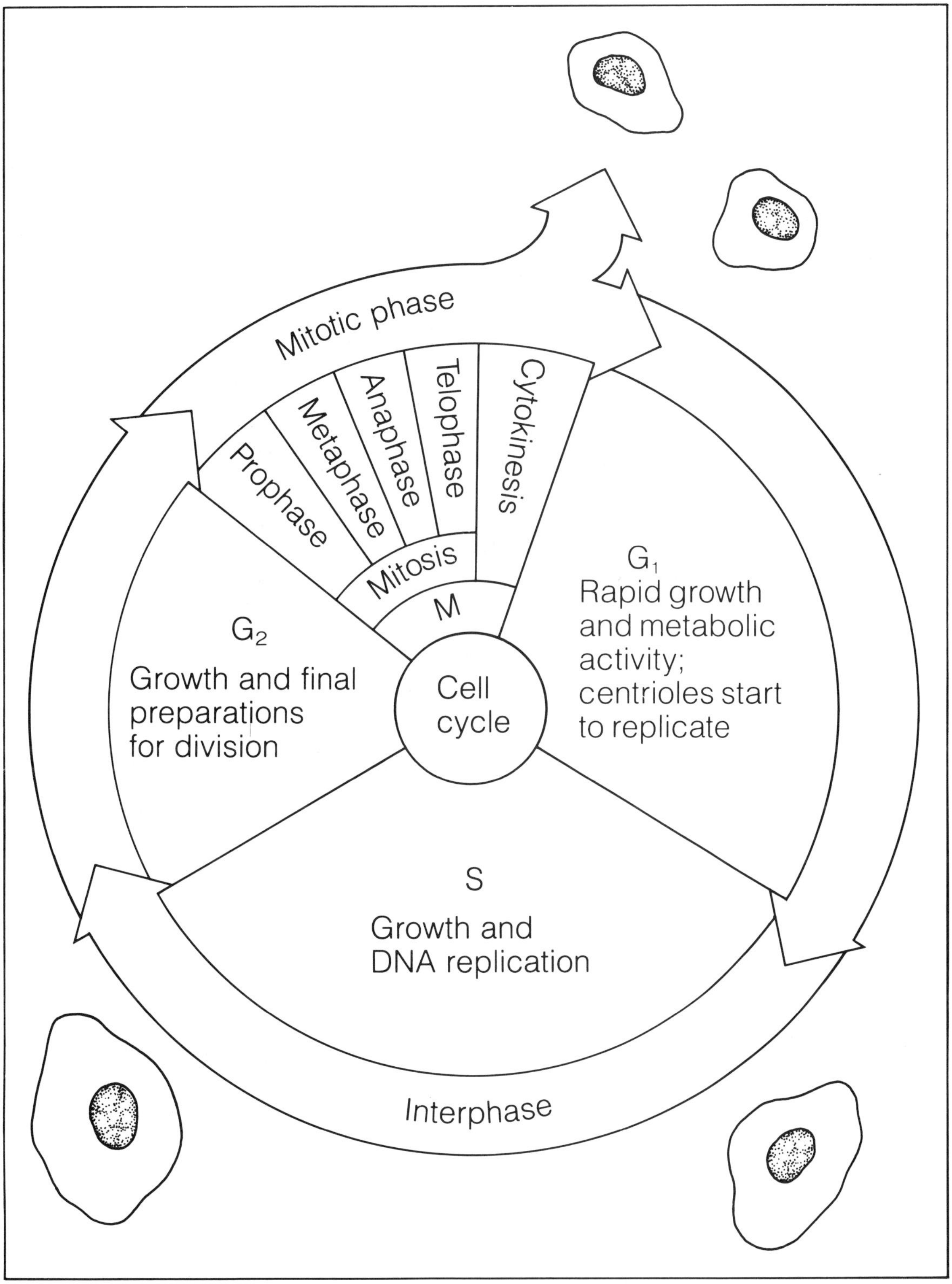

Replication of DNA

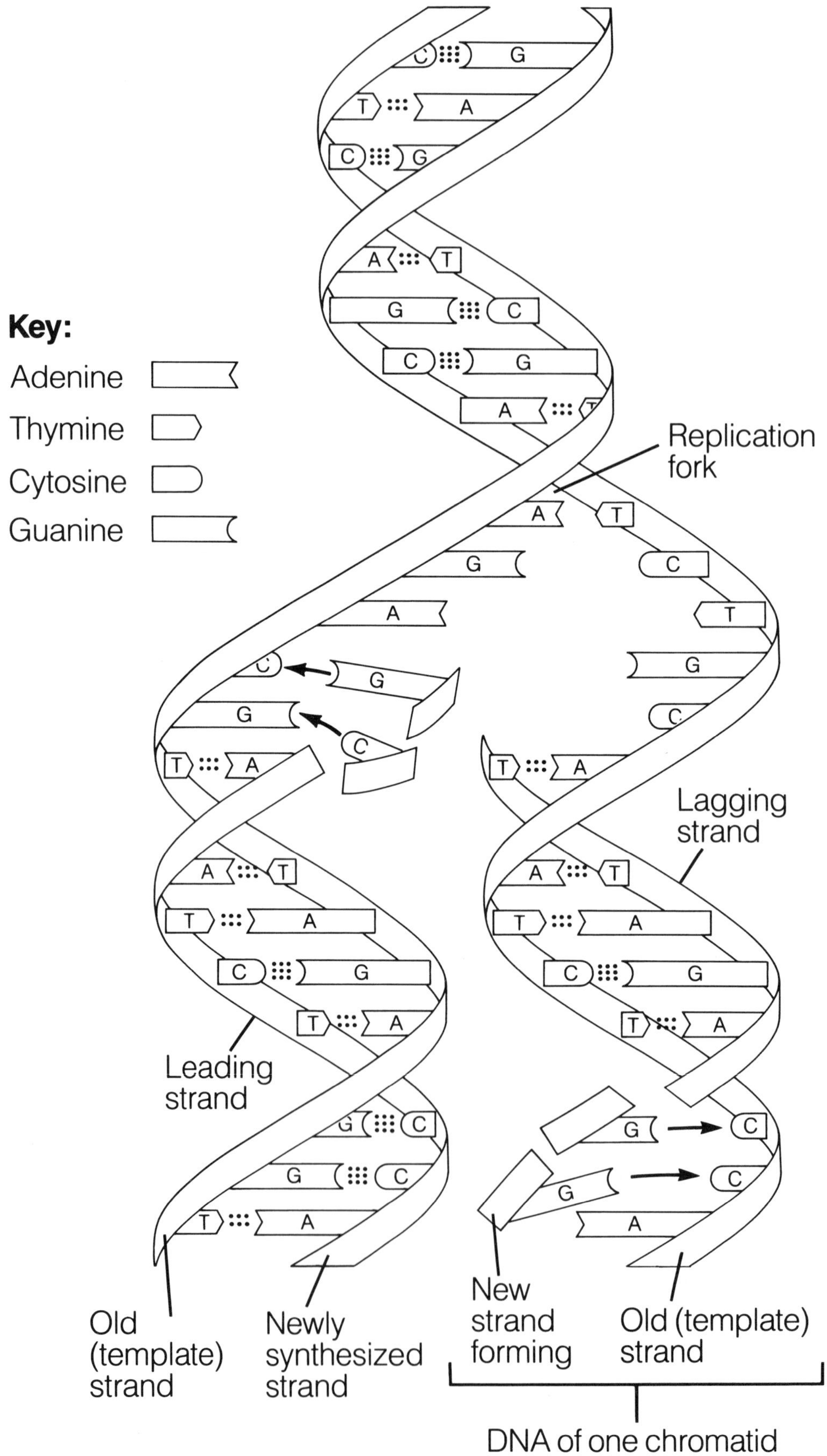

Simplified scheme of information flow from the DNA gene to protein structure

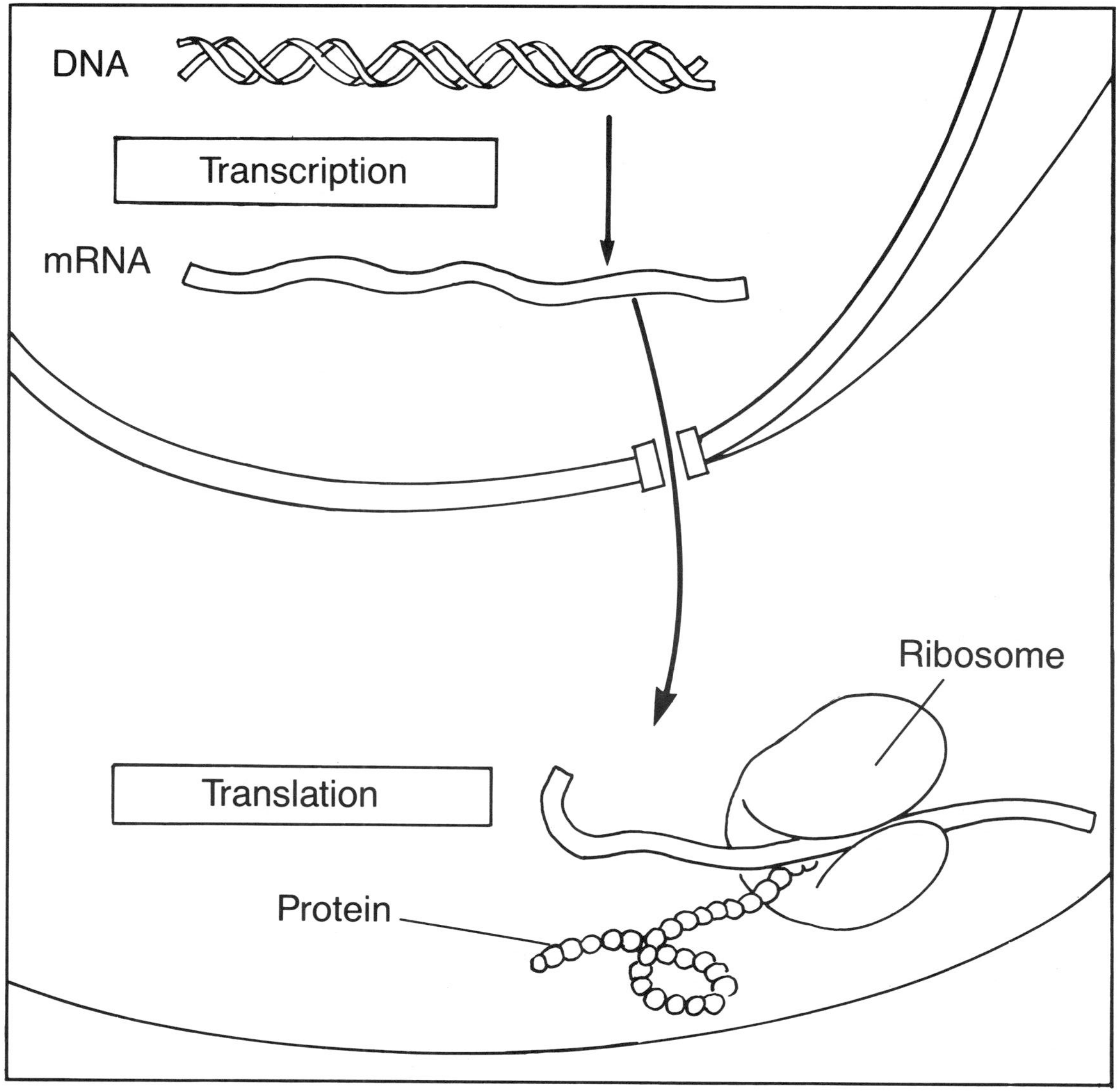

Protein synthesis: From transcription to translation

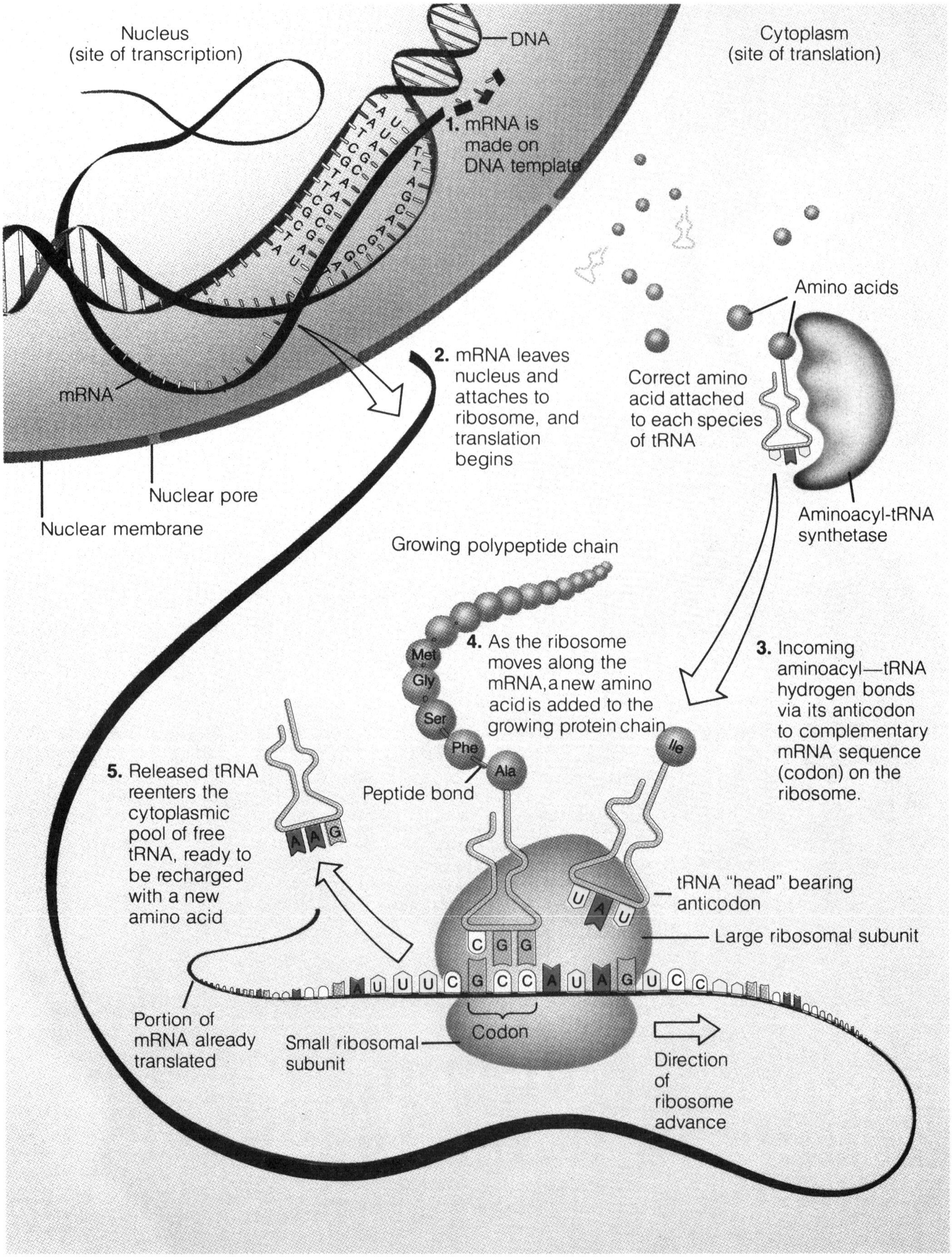

Information transfer from DNA to RNA

DNA base sequence (triplets) of the "gene" coding for the synthesis of a particular polypeptide chain

Base sequence (codons) of the transcribed mRNA

Consecutive base sequences of tRNA *anticodons* capable of recognizing the mRNA codons "calling for" the amino acids they transport

Amino acid sequence of the polypeptide chain

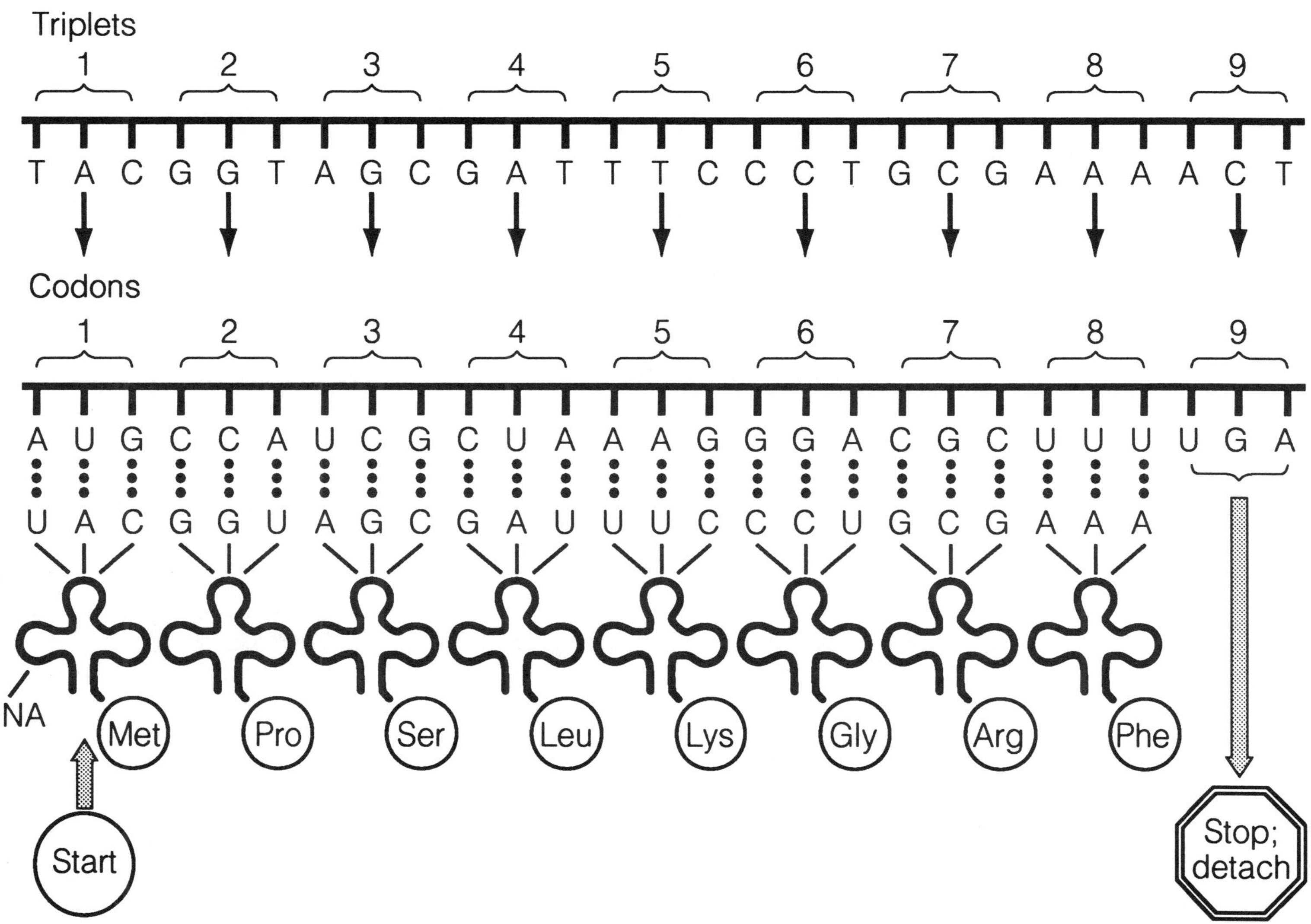

Classification of epithelia

(a)

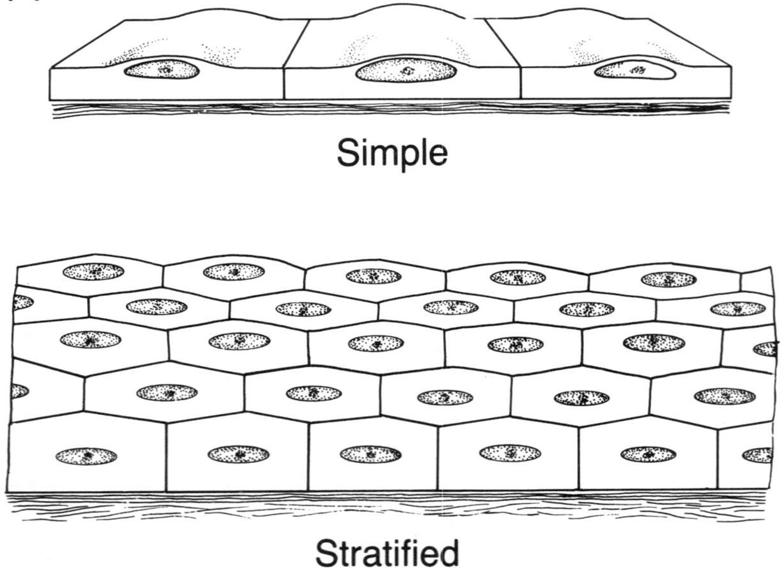

(b)

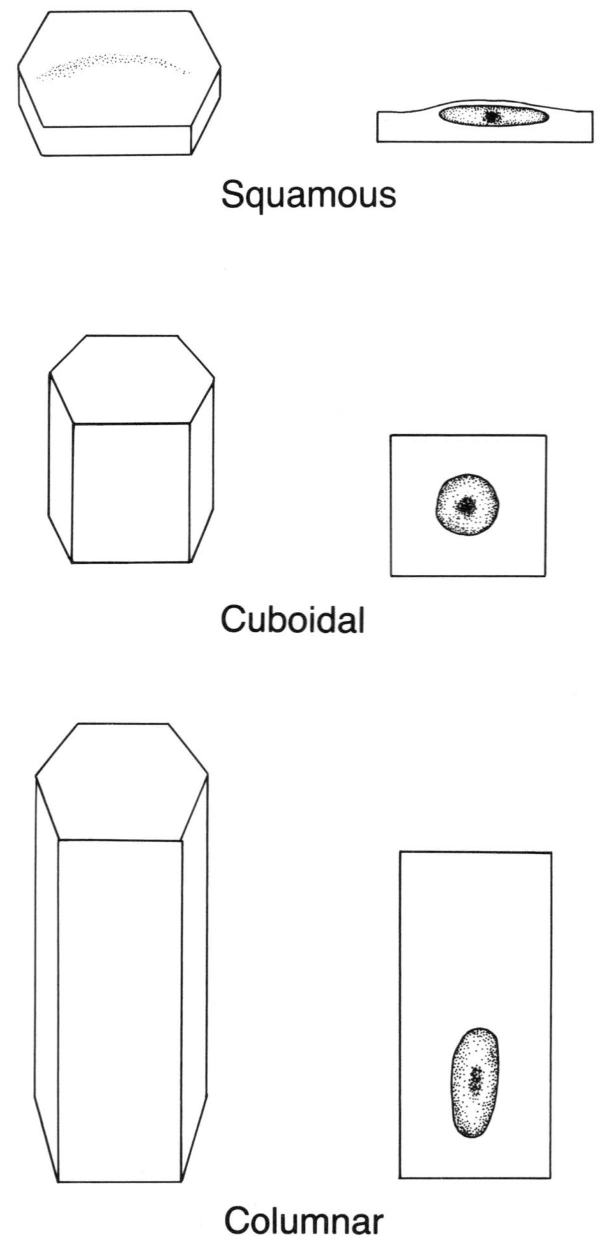

Goblet cells, examples of unicellular exocrine glands

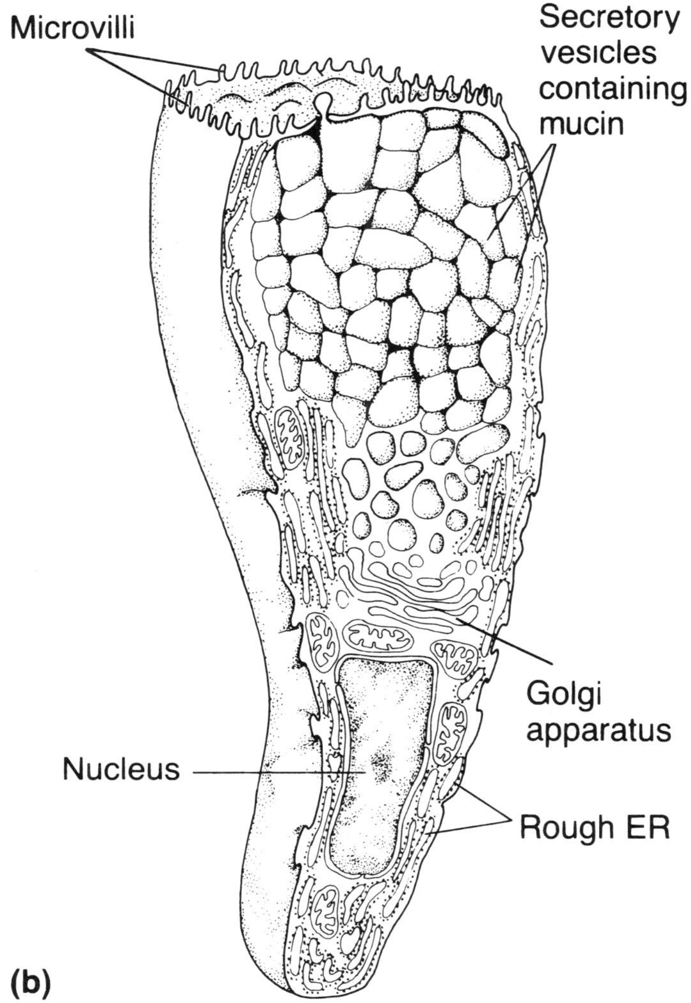

(b)

Types of multicellular exocrine glands

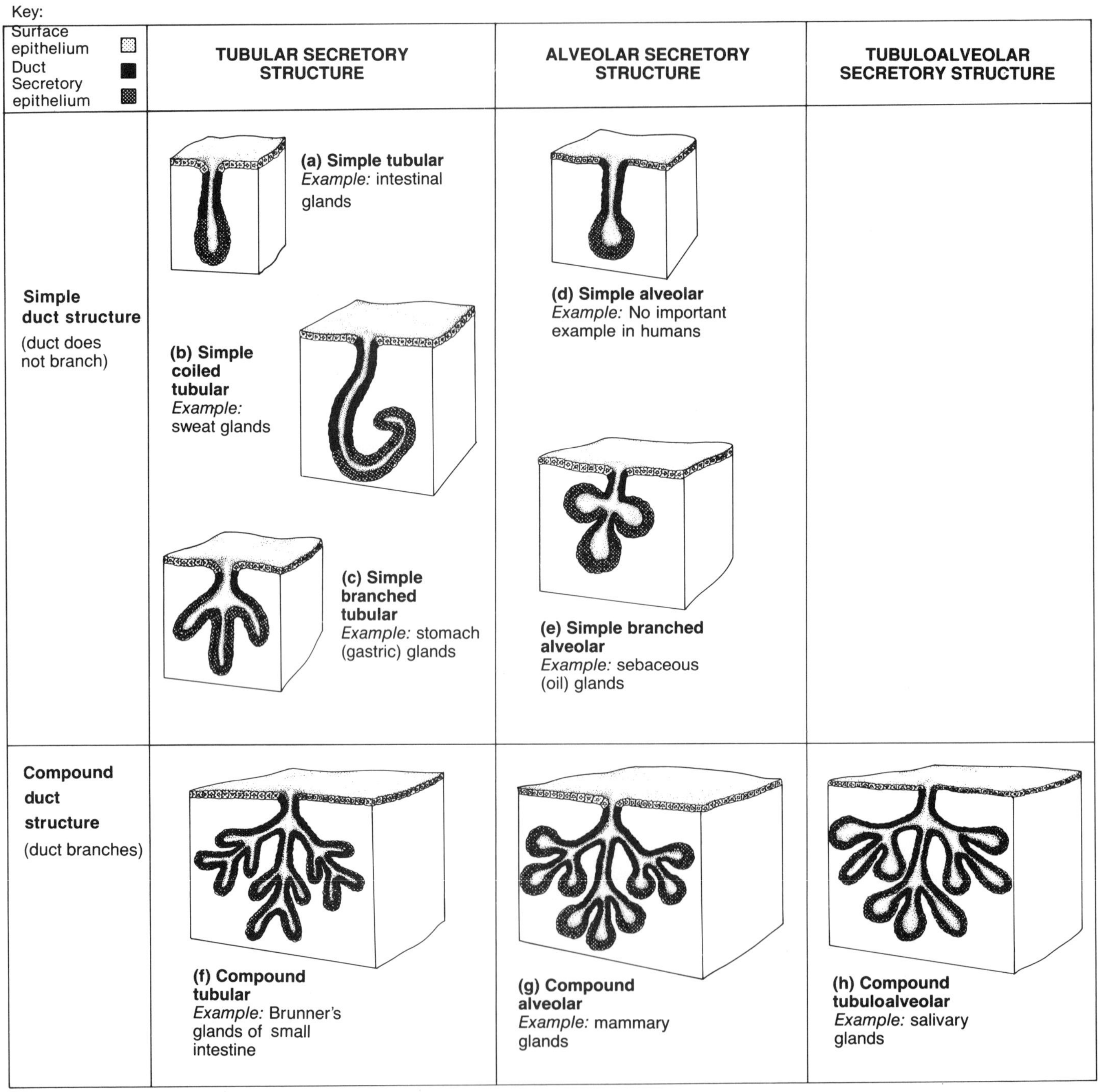

Major classes of connective tissue

Common embryonic origin:	**Mesenchyme**			
Cellular descendants:	Fibroblast	Chondroblast	Osteoblast	Hemocytoblast
	Fibrocyte	Chondrocyte	Osteocyte	Blood cells* (and macrophages)
Class of connective tissue resulting:	Connective tissue proper	Cartilage	Osseous (bone)	Blood
Subclasses:	**1.** Loose connective tissue Types: Areolar, Adipose, Reticular **2.** Dense connective tissue Types: Regular, Irregular, Elastic	**1.** Hyaline cartilage **2.** Fibrocartilage **3.** Elastic cartilage	**1.** Compact bone **2.** Spongy (cancellous) bone	**Blood cell formation and differentiation are quite complex. Details are provided in Chapter 18.*

Embryonic germ layers and the primary tissue types they produce

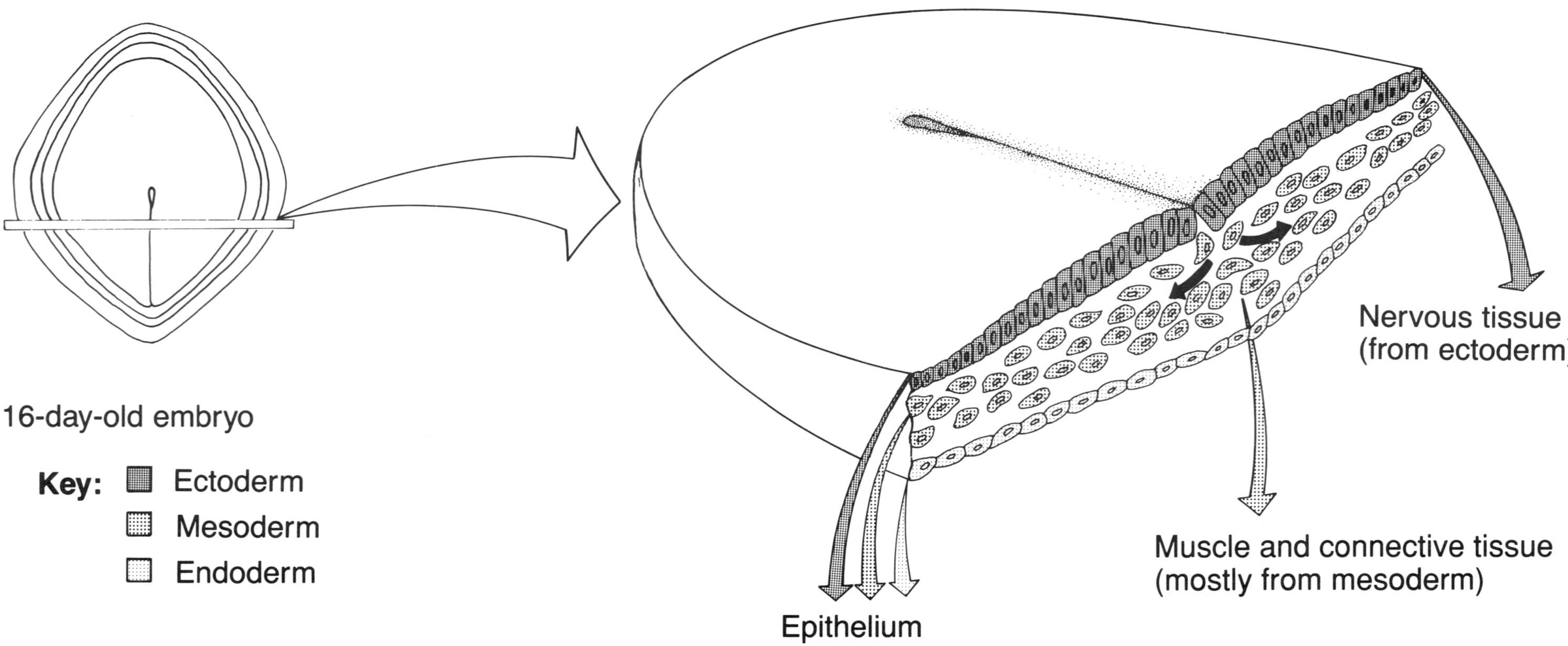

Diagram of a single osteon

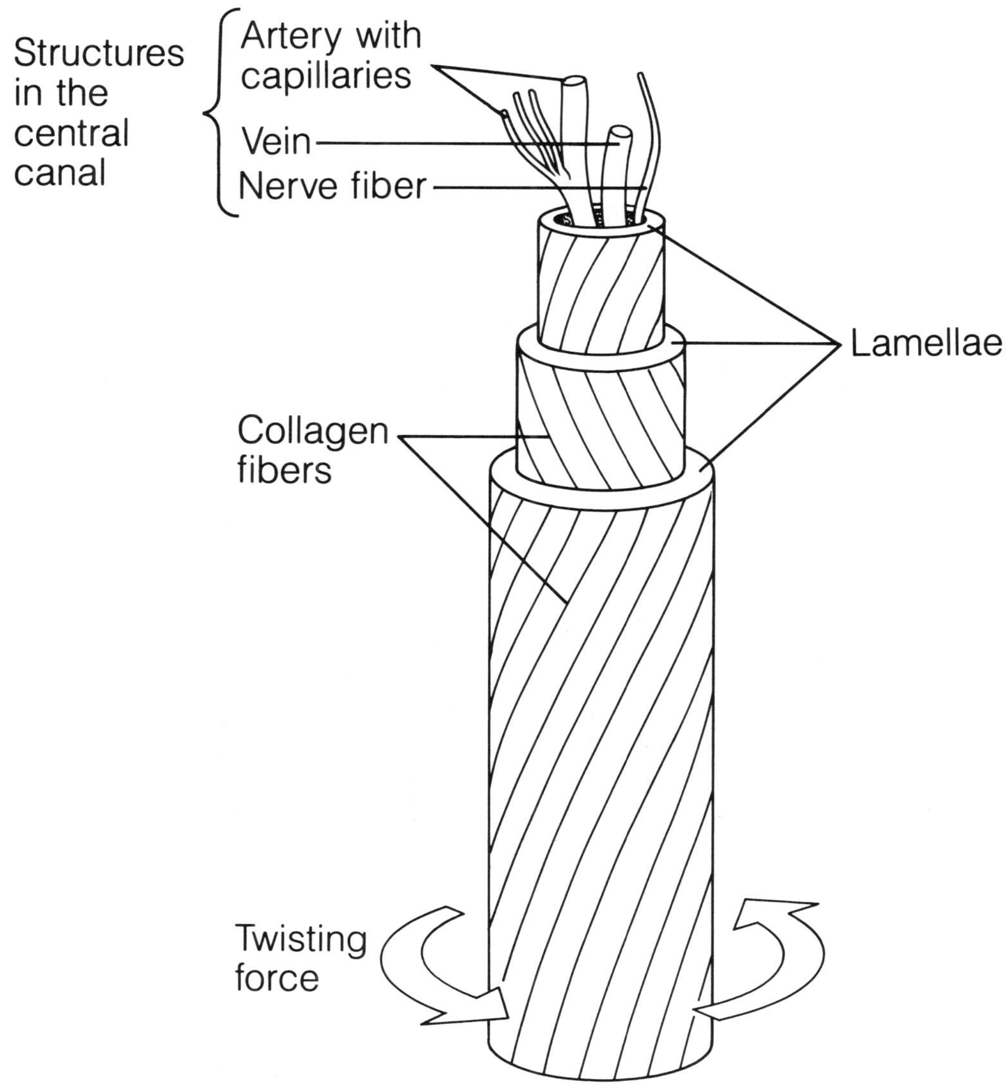

The stages in intramembranous ossification

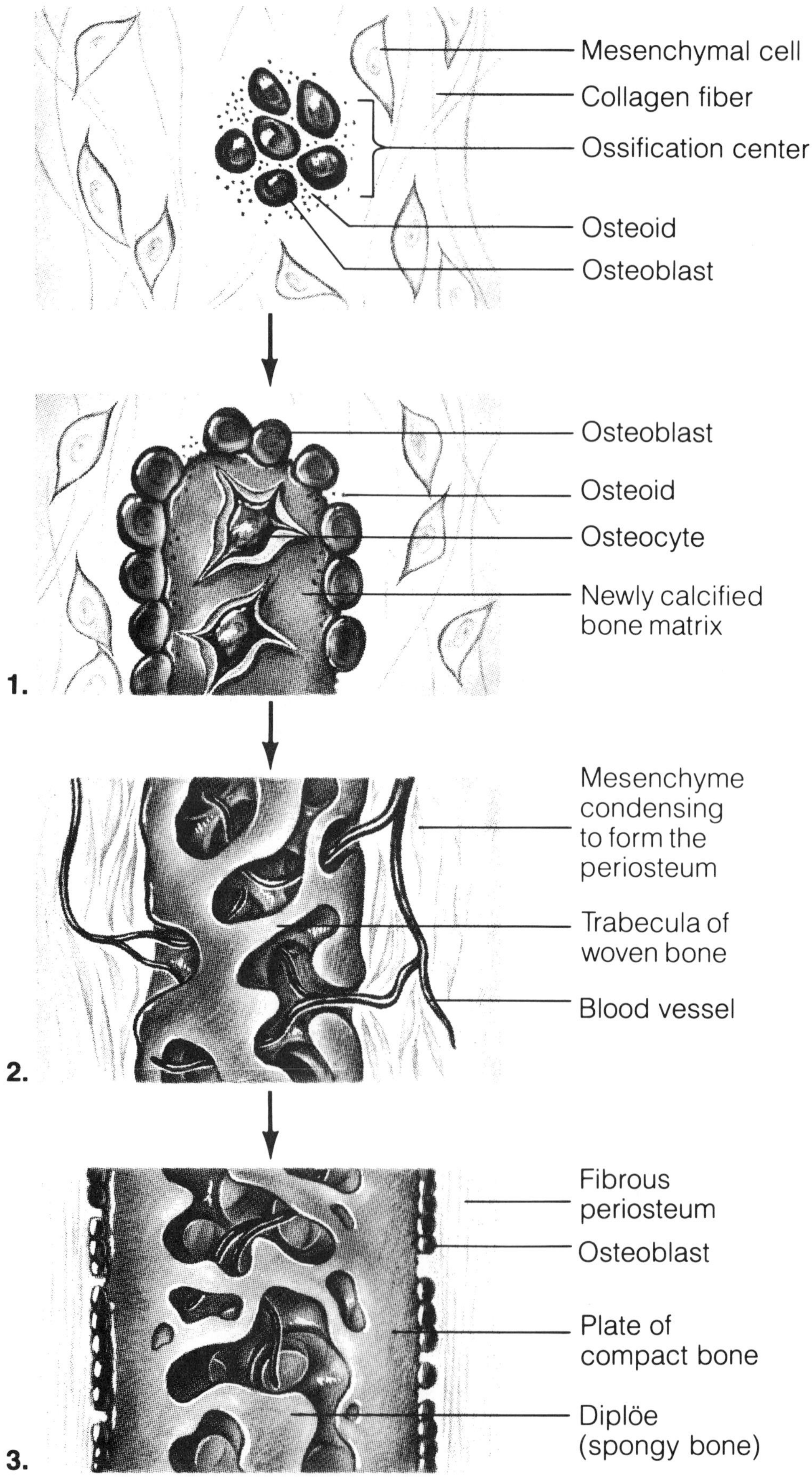

Stages in endochondral ossification occurring in a long bone

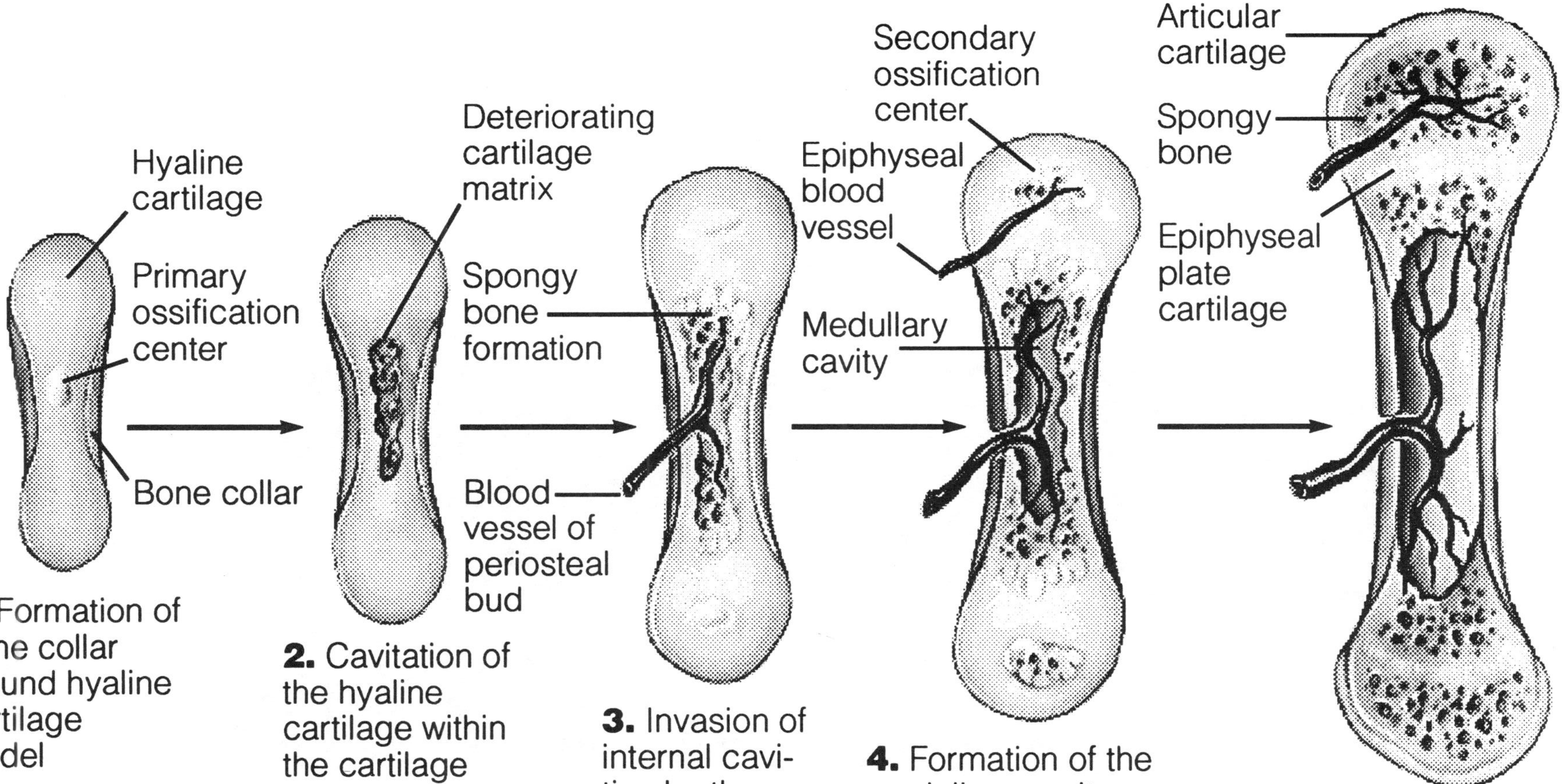

1. Formation of bone collar around hyaline cartilage model

2. Cavitation of the hyaline cartilage within the cartilage model

3. Invasion of internal cavities by the periosteal bud and spongy bone formation

4. Formation of the medullary cavity as ossification continues; appearance of secondary ossification centers in the epiphyses in preparation for stage 5.

5. Ossification of the epiphyses; when completed, hyaline cartilage remains only in the epiphyseal plates and articular cartilages. (The epiphyseal plates promote longitudinal growth until young adulthood.)

Growth in length of a long bone

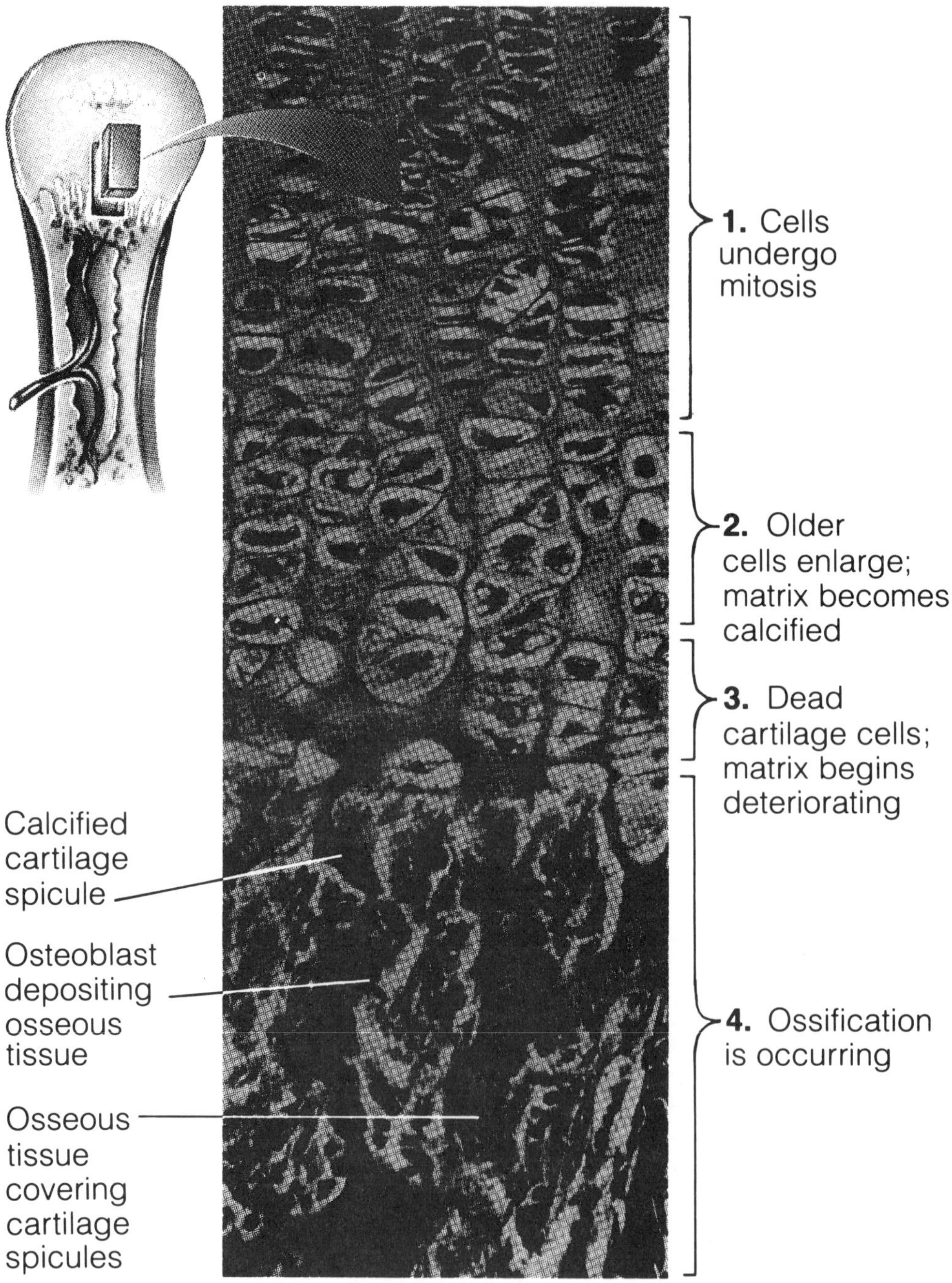

Long bone growth and remodeling during youth

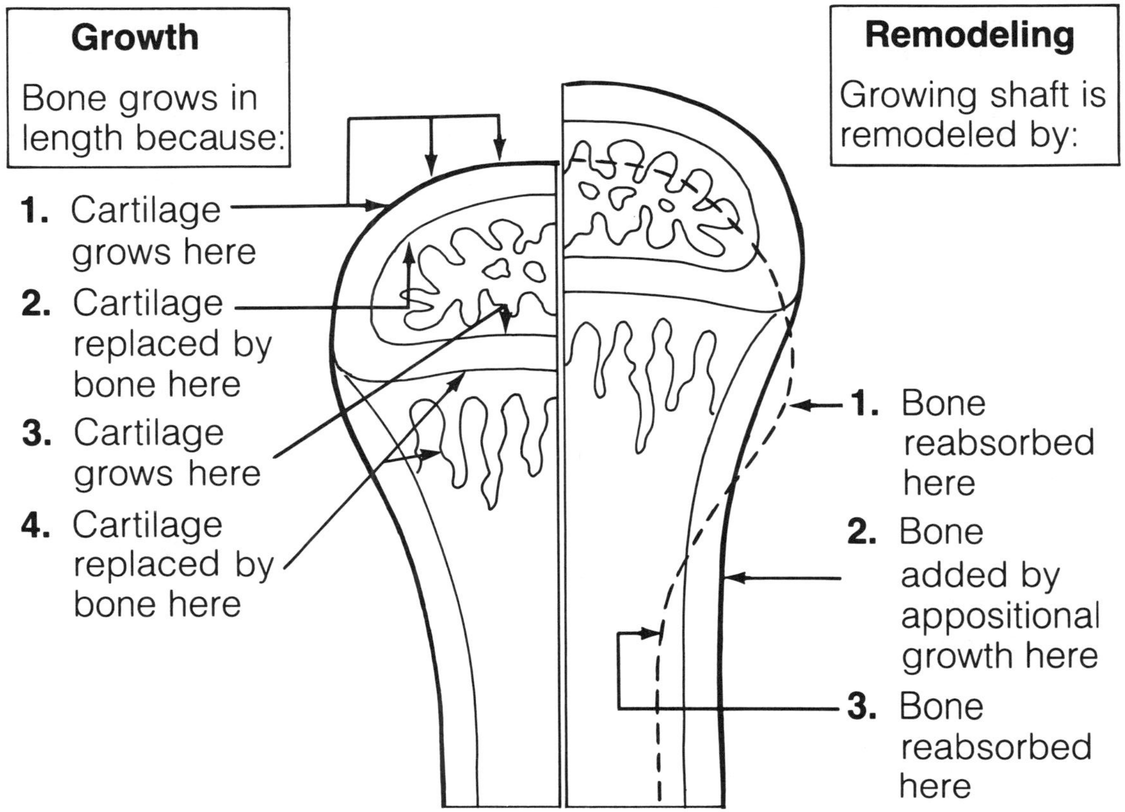

Hormonal controls of ionic calcium levels in the blood

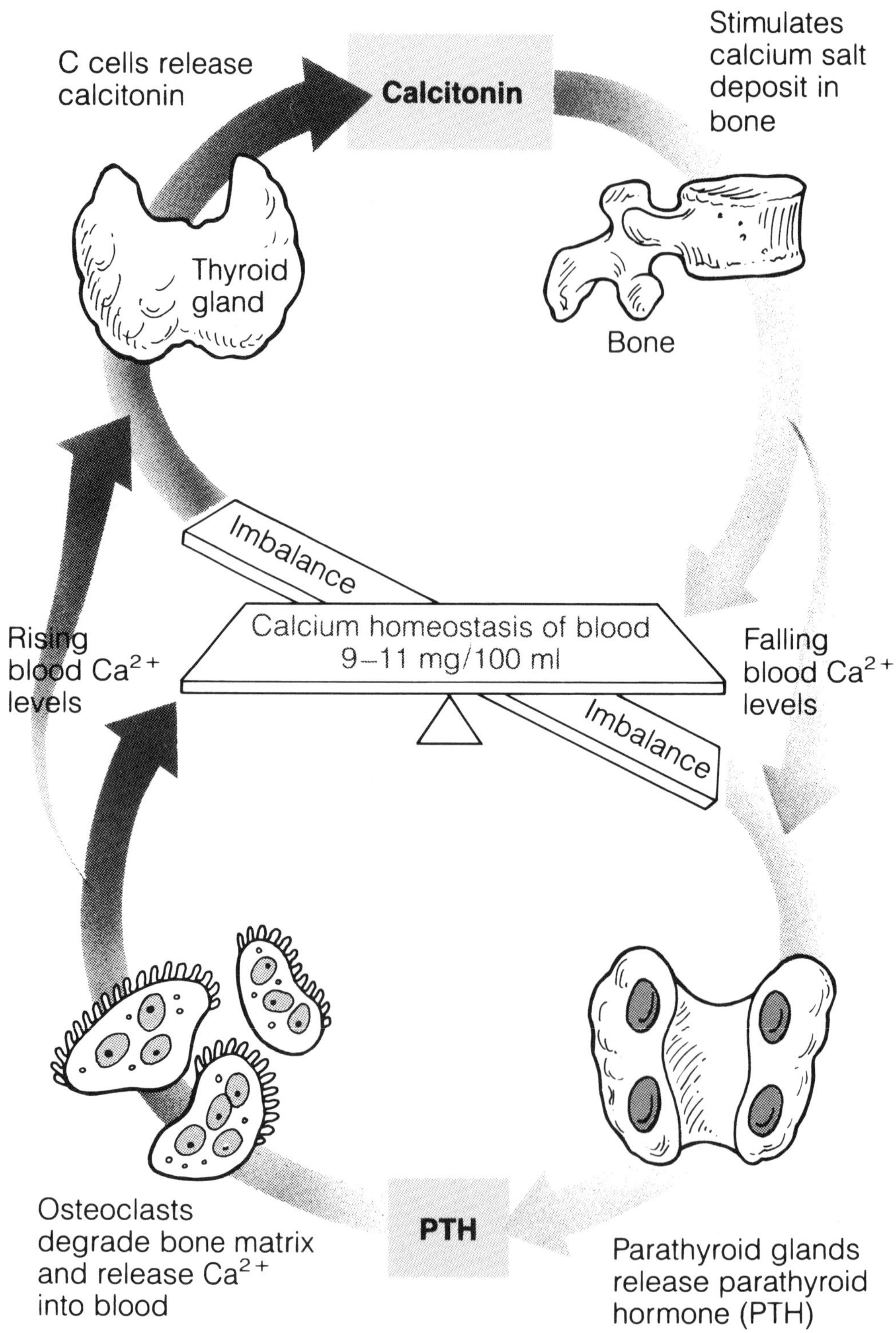

Effect of bending on a long bone

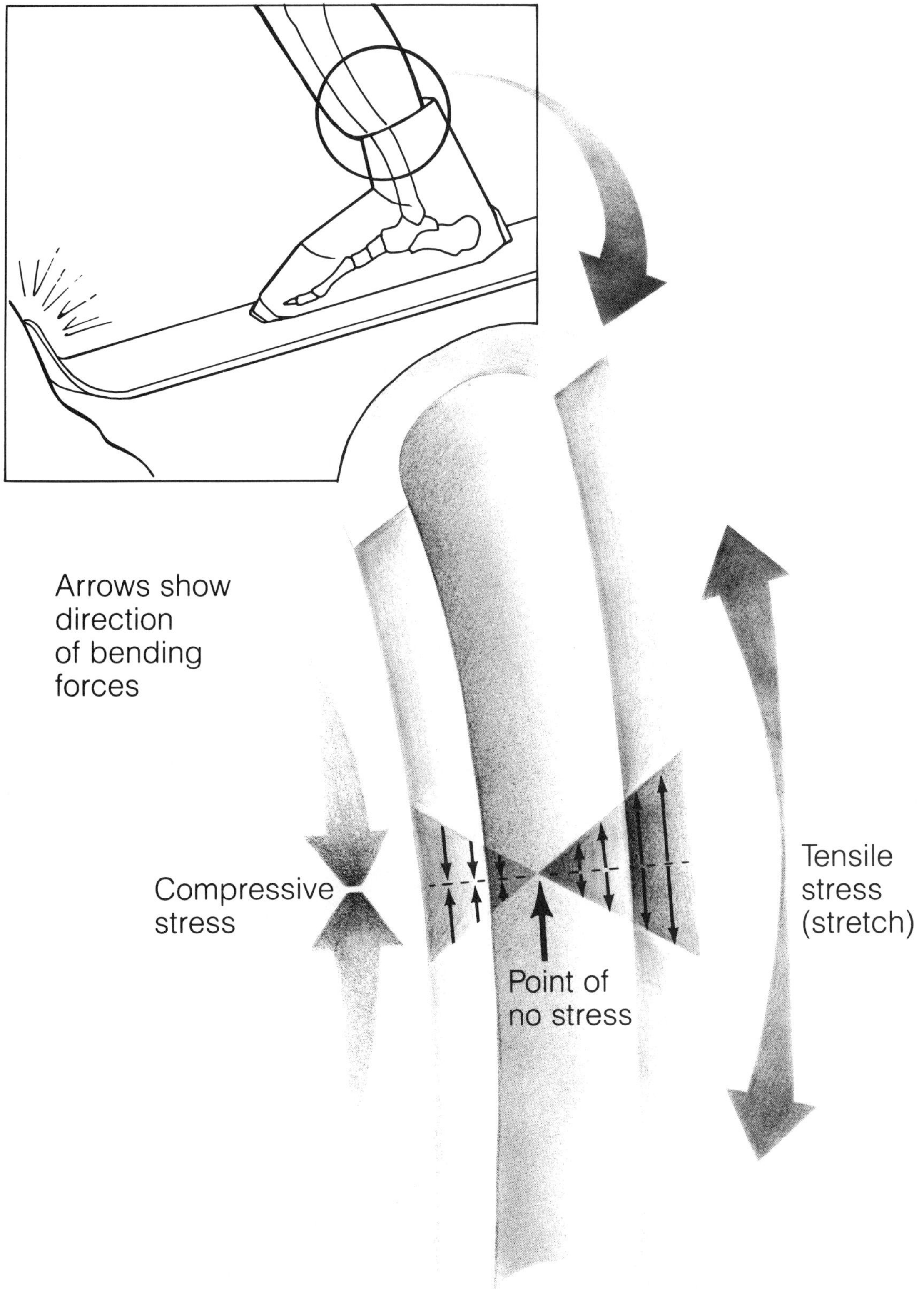

Phases of fracture healing

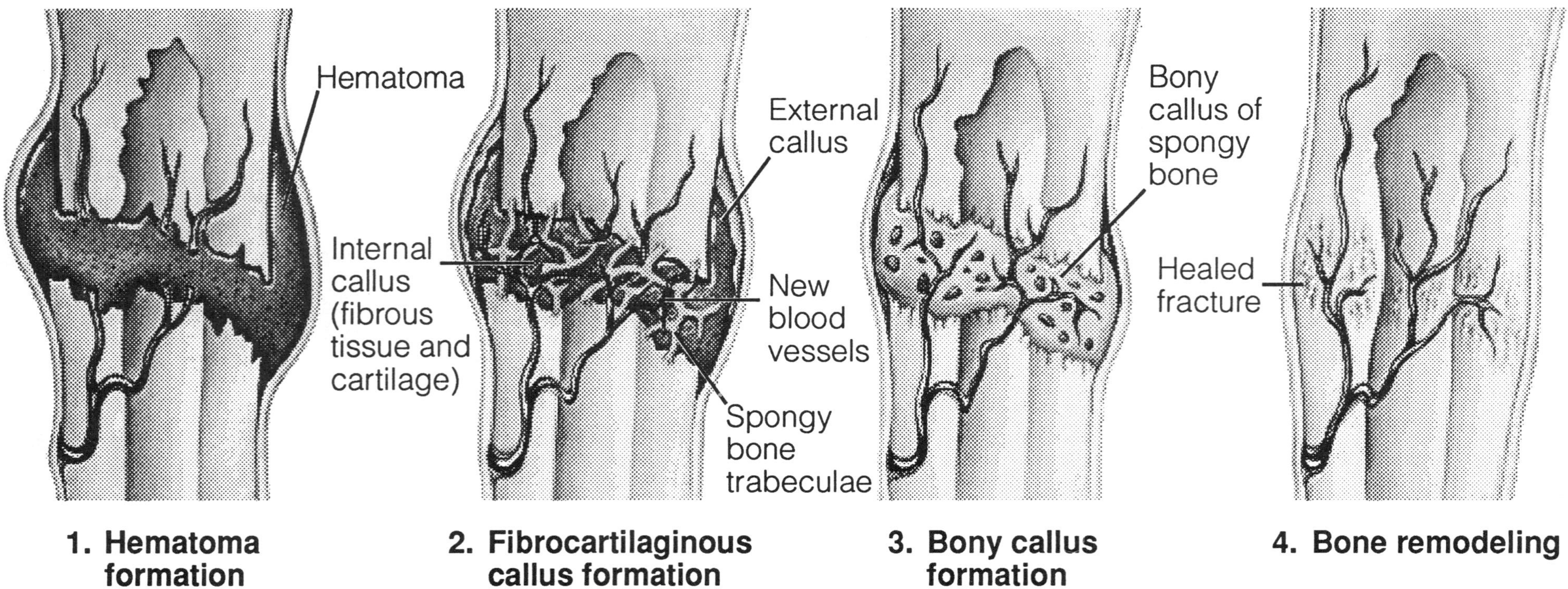

Microscopic anatomy of a skeletal muscle fiber

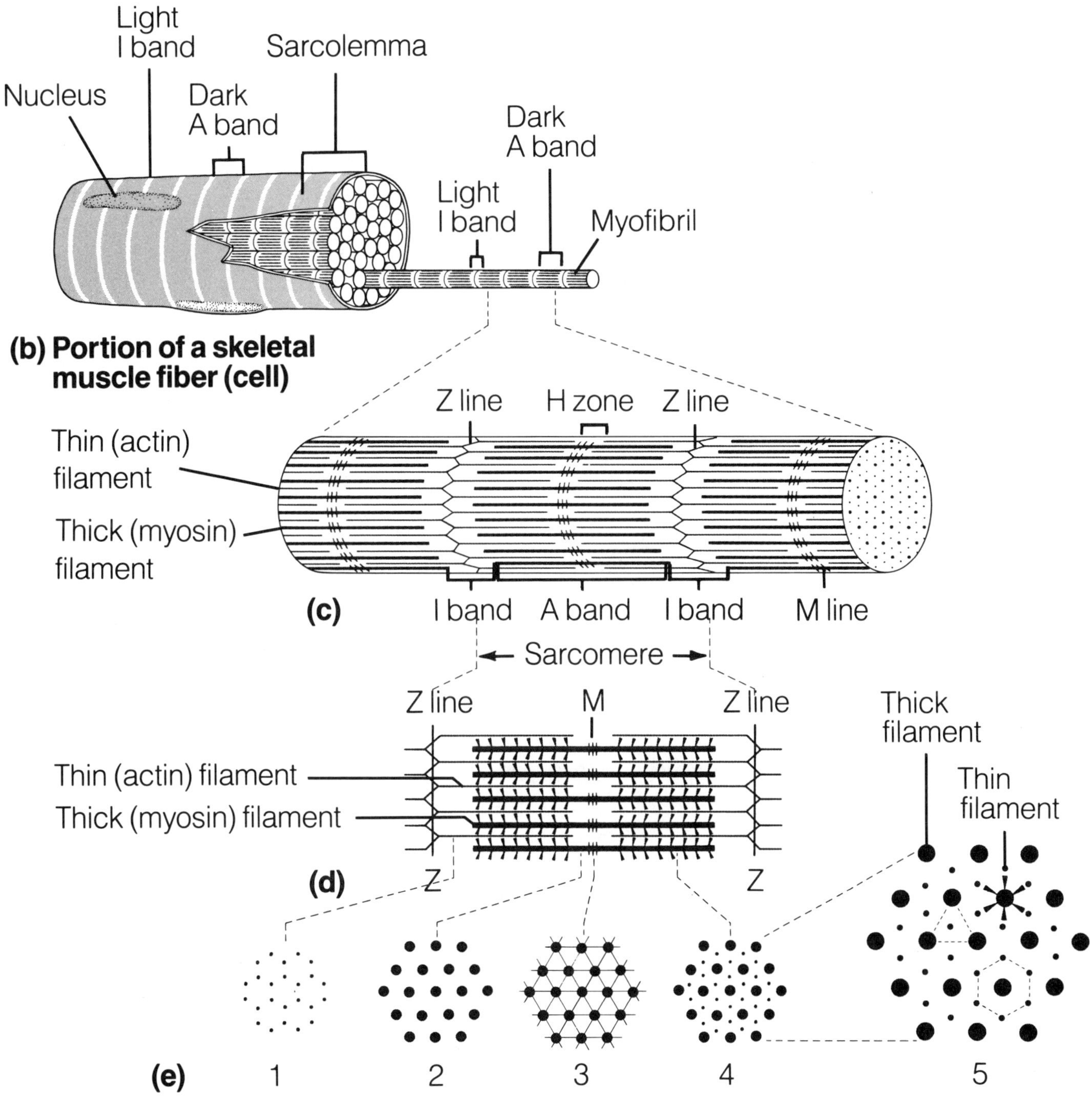

(b) Portion of a skeletal muscle fiber (cell)

(c)

(d)

(e)

Myofilament composition in skeletal muscle

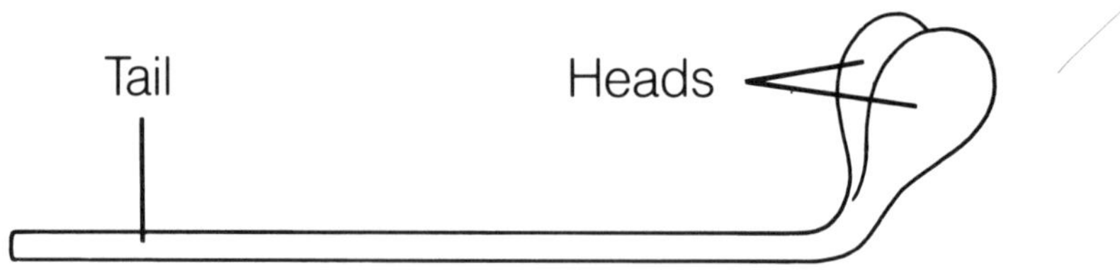

(a) Myosin molecule

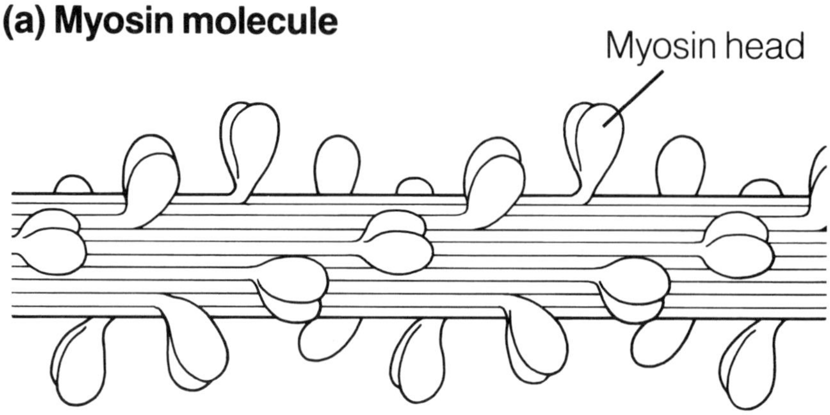

(b) Portion of a thick filament

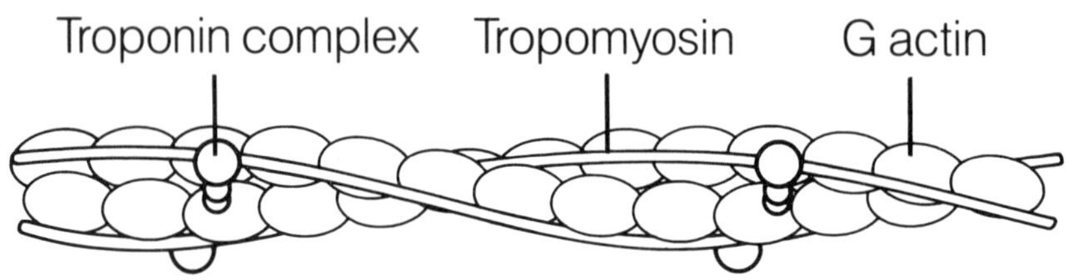

(c) Portion of a thin filament

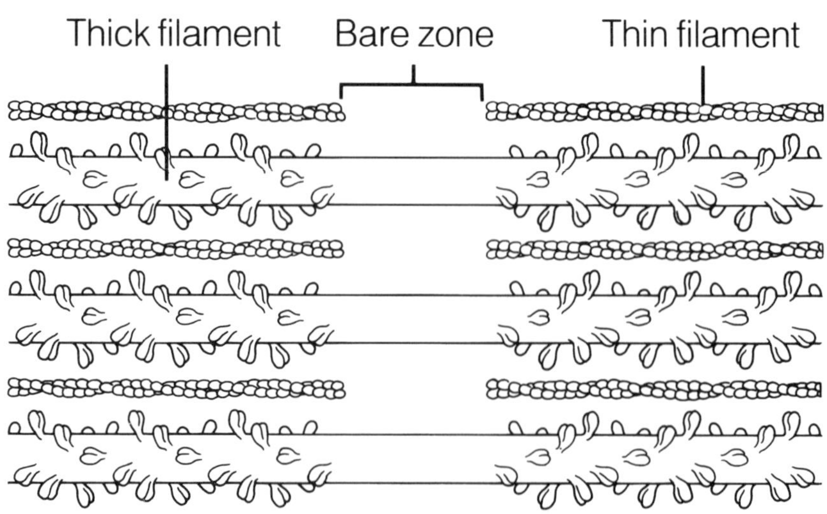

(d) Longitudinal section of filaments within one sarcomere of a myofibril

Relationship of the sarcoplasmic reticulum and T tubules to the myofibrils of skeletal muscle

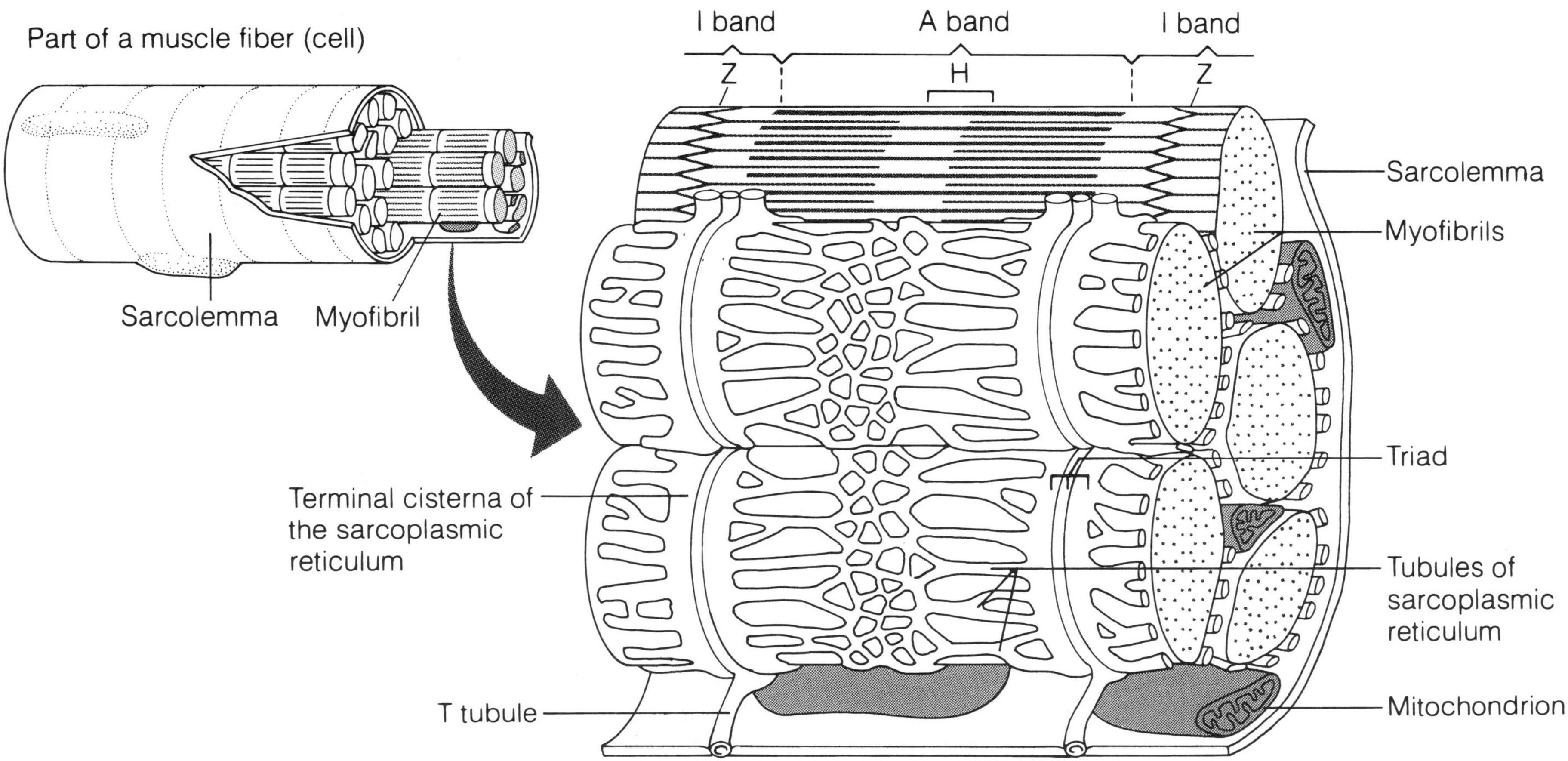

Role of ionic calcium in the contraction mechanism

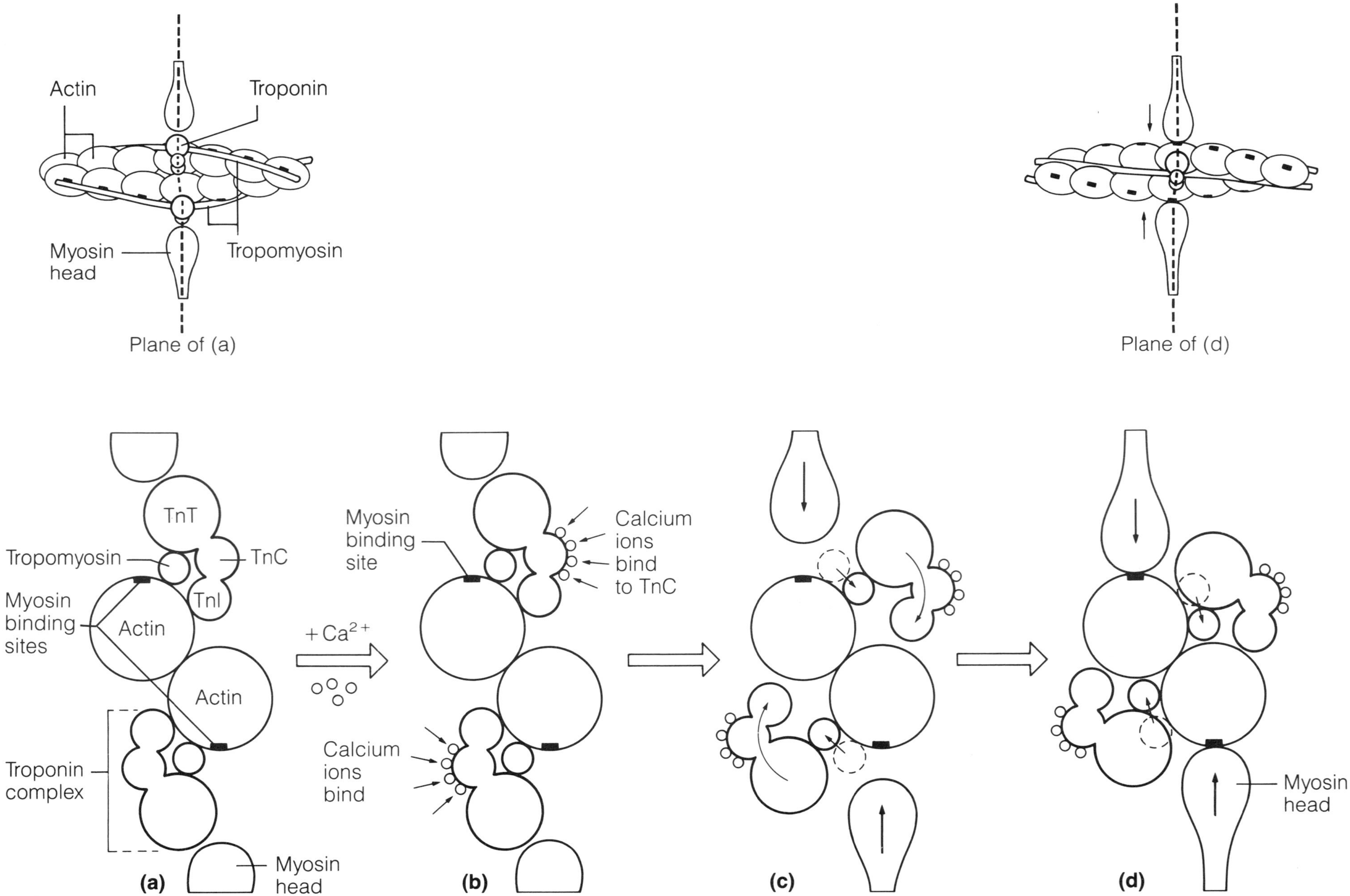

Sequence of events involved in the sliding of the actin filaments during contraction

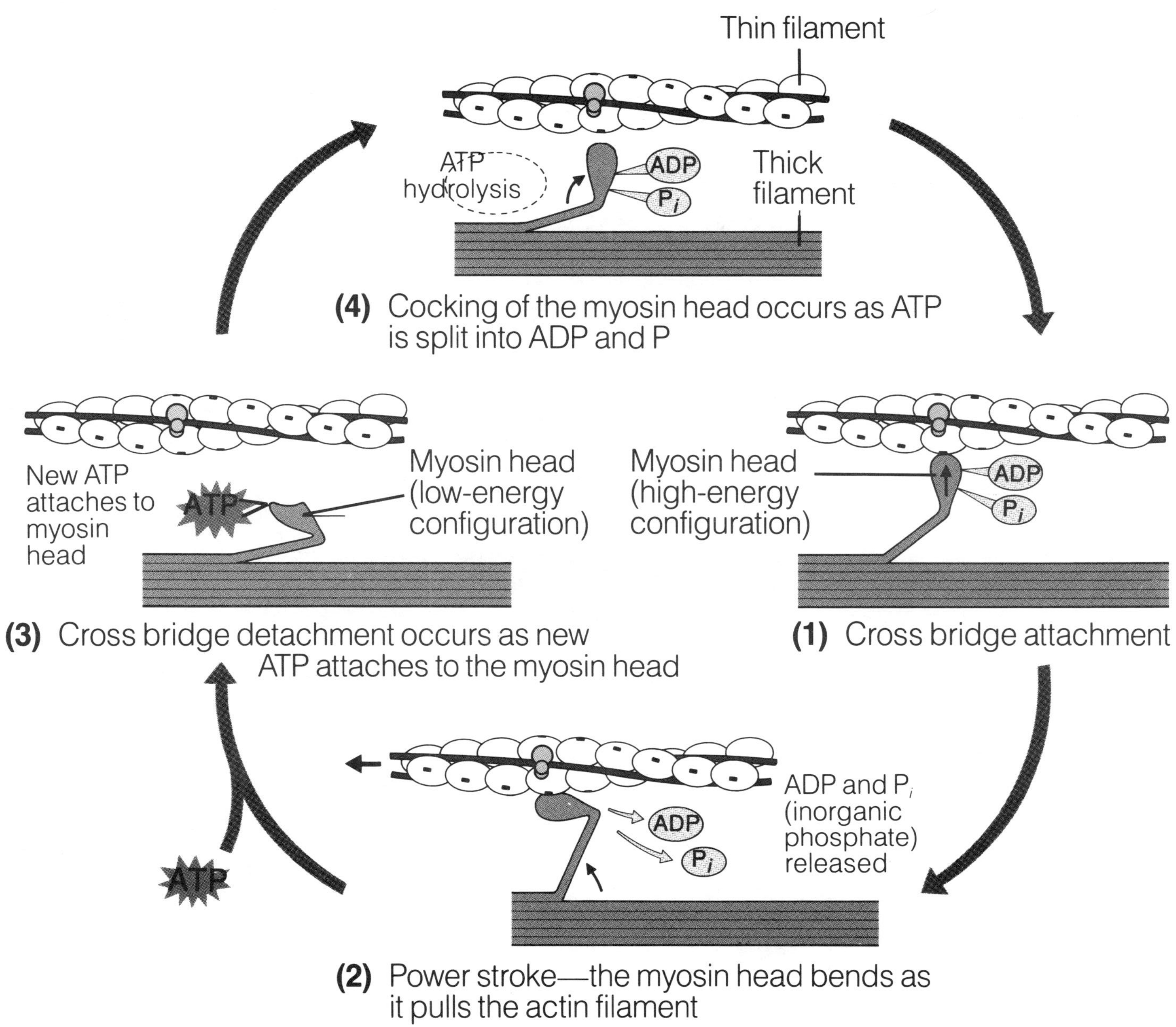

The neuromuscular junction

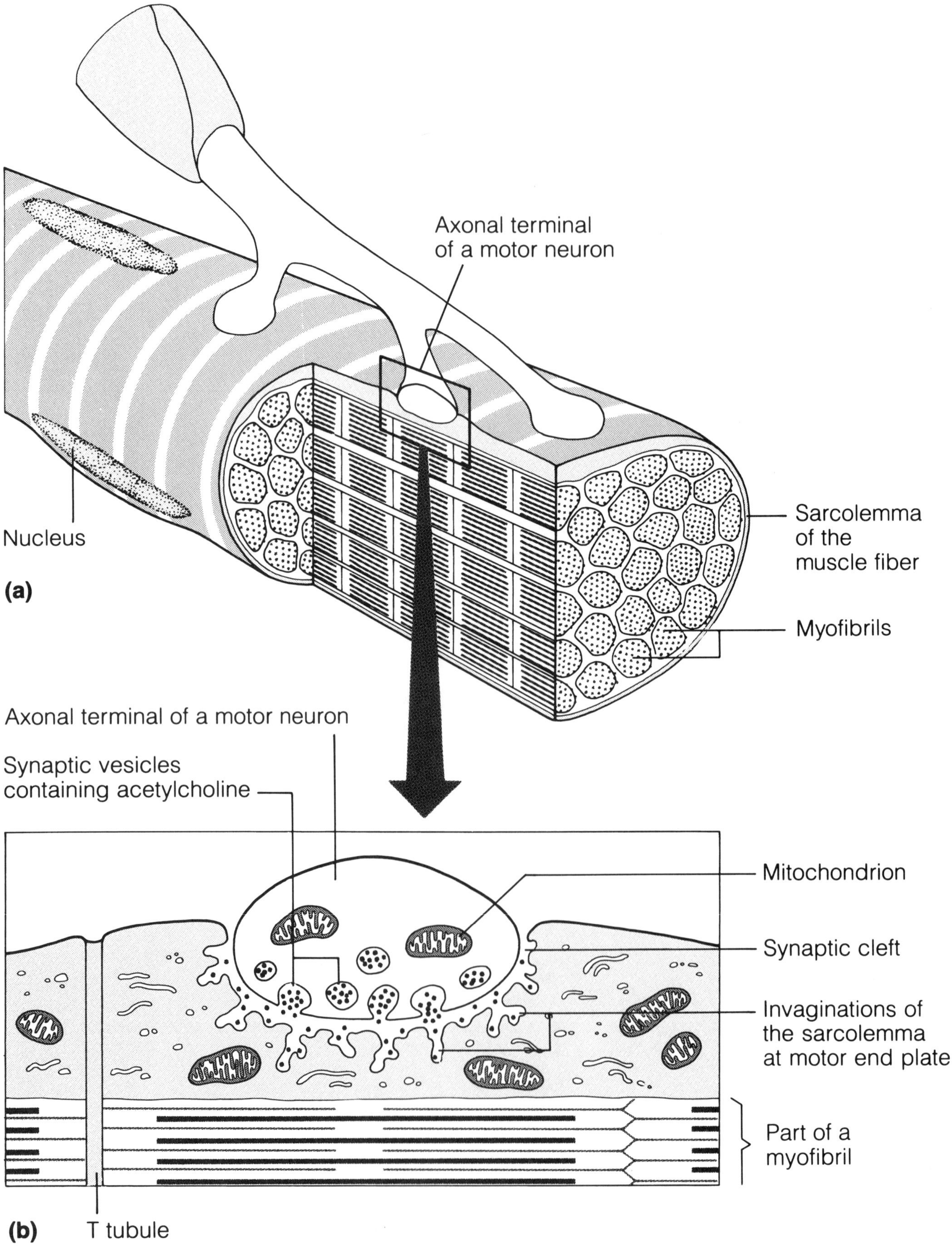

Summary of events in the generation and propagation of an action potential in a skeletal muscle fiber

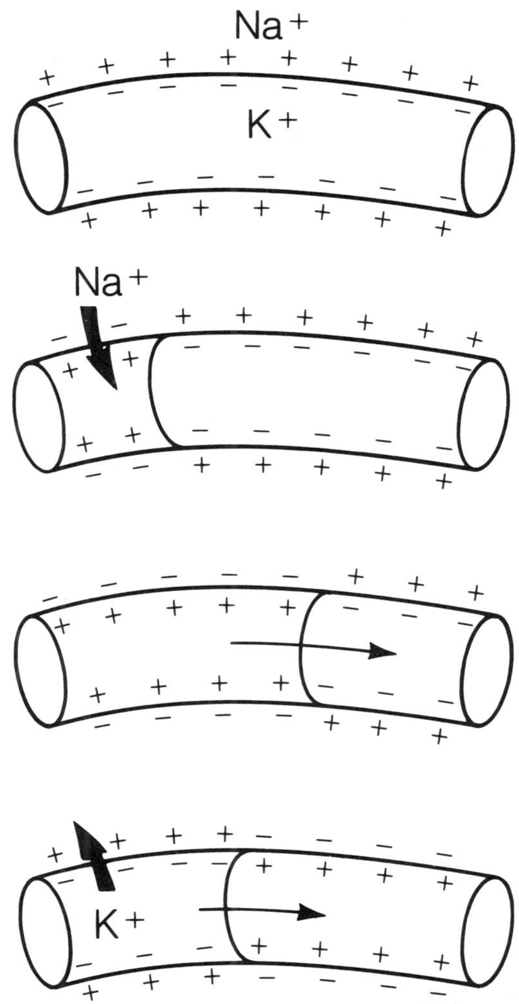

(a)

(b)

(c)

(d)

Sequence of events in excitation-contraction coupling

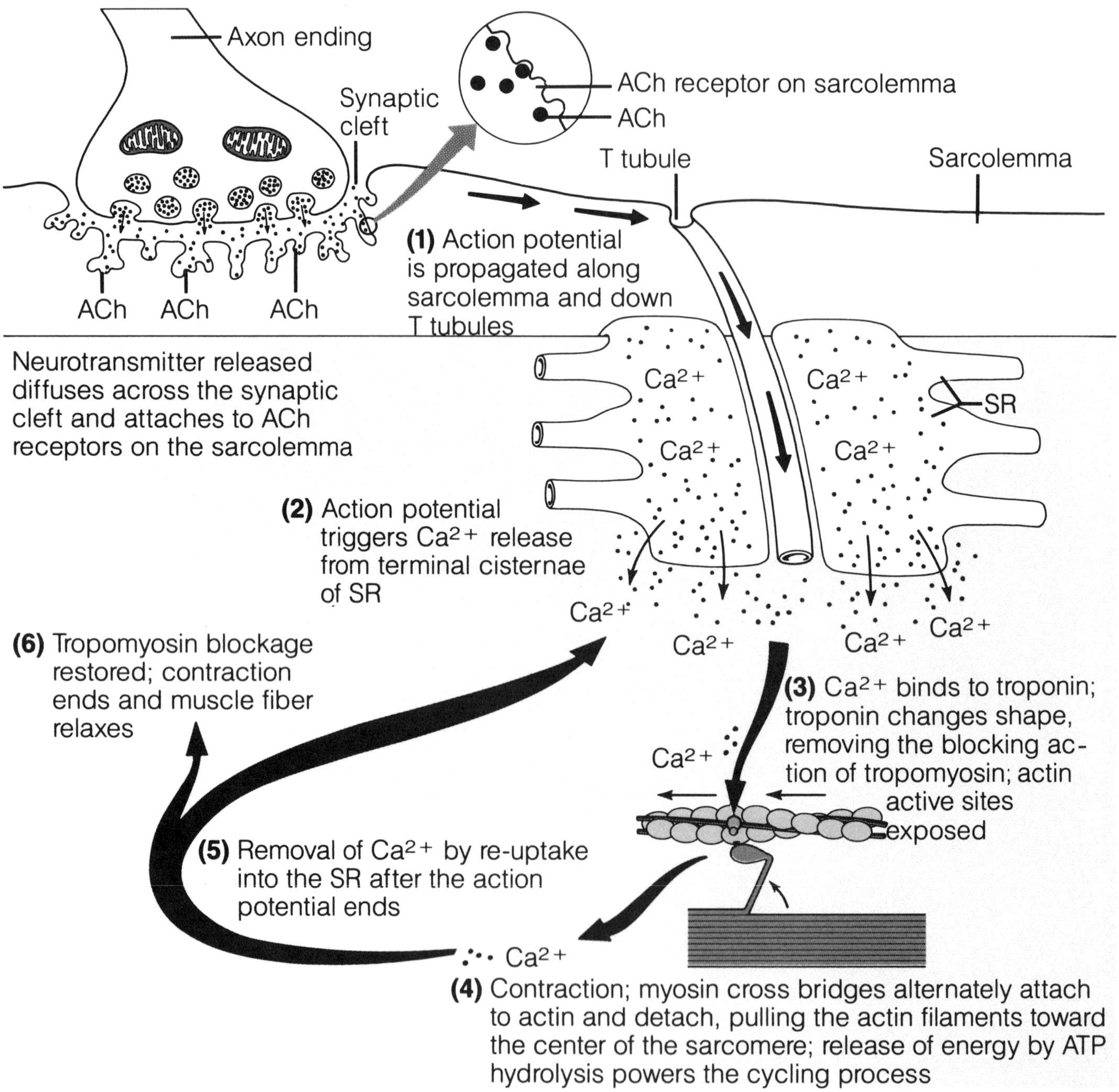

Motor units

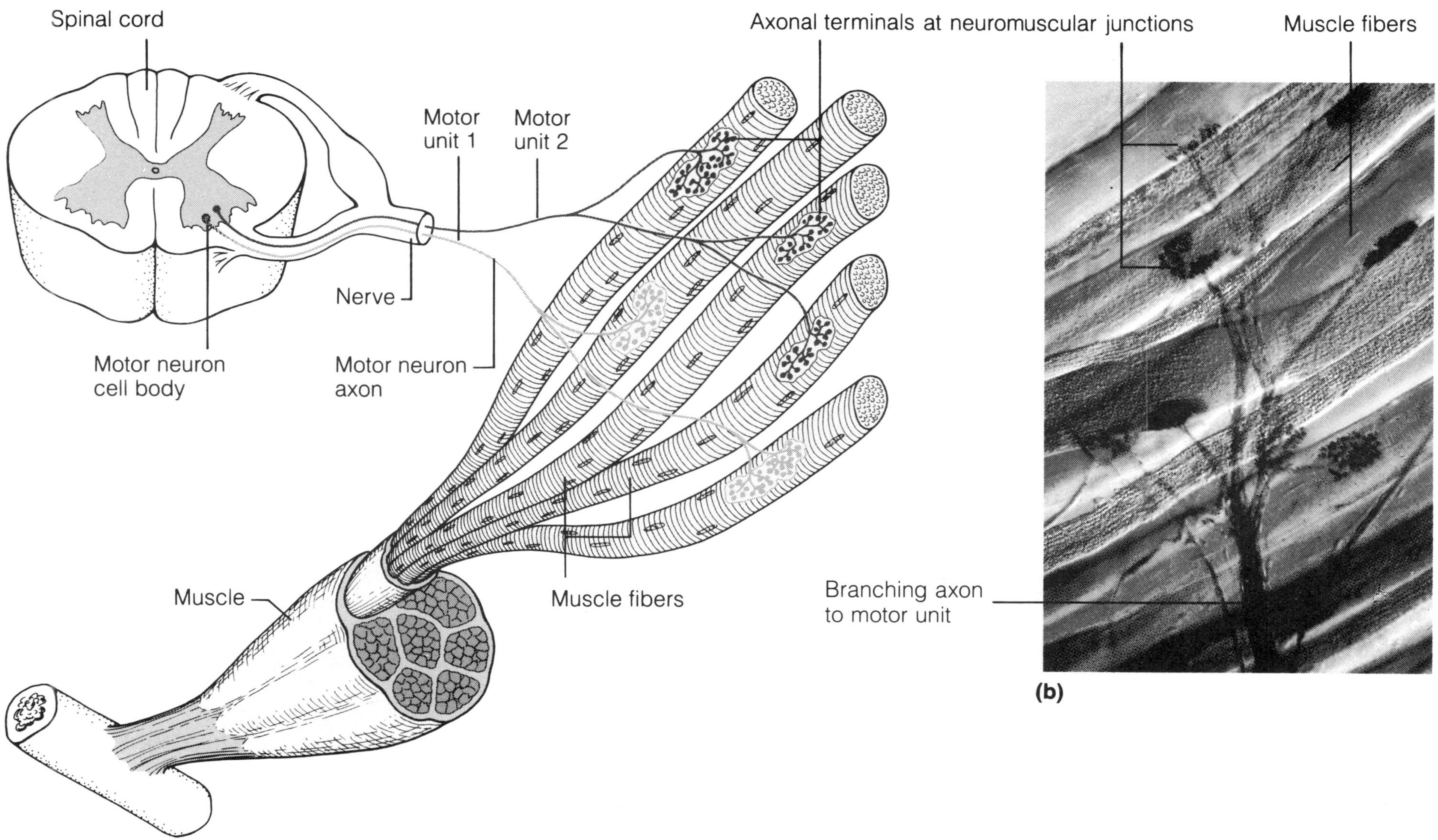
Spinal cord
Motor neuron cell body
Motor unit 1
Motor unit 2
Nerve
Motor neuron axon
Axonal terminals at neuromuscular junctions
Muscle fibers
Muscle
Muscle fibers
Branching axon to motor unit
(b)

The muscle twitch

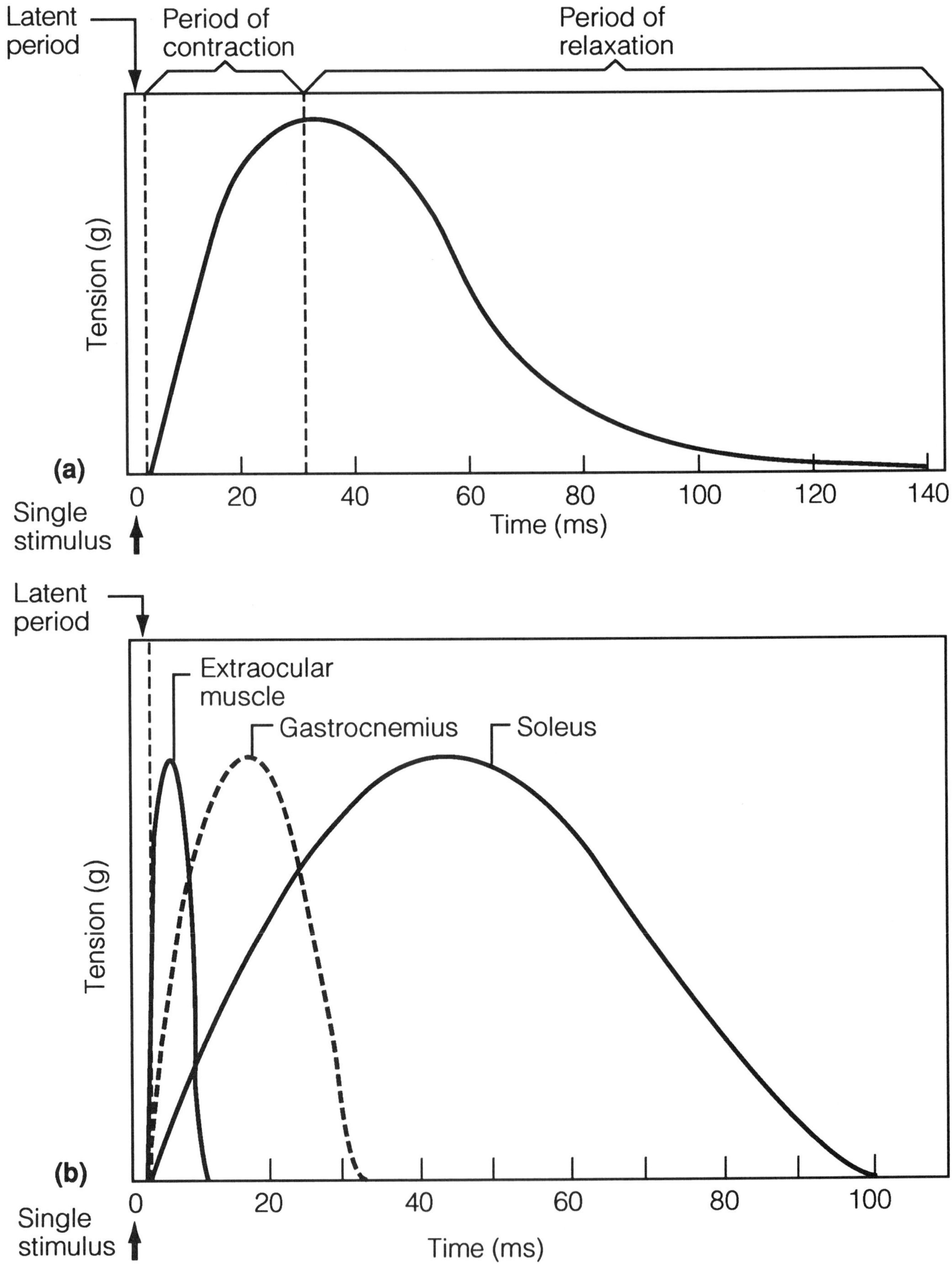

Wave summation and tetanus

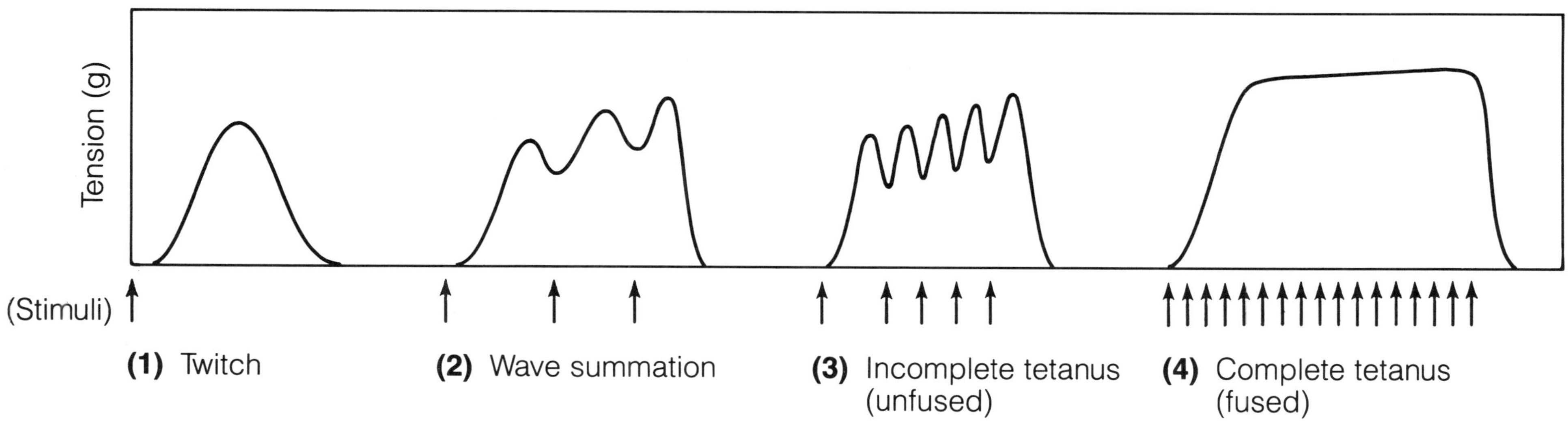

Asynchronous motor unit summation

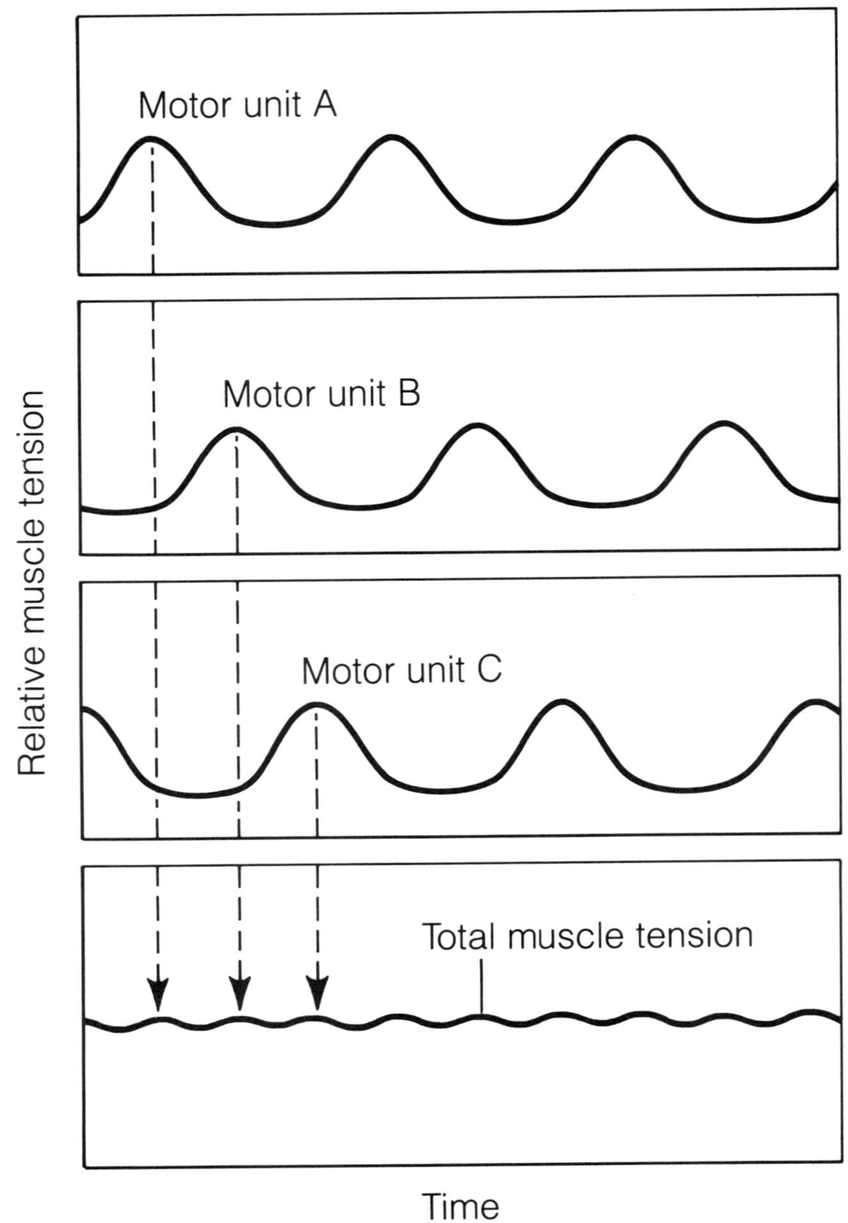

Myogram of treppe (the staircase effect)

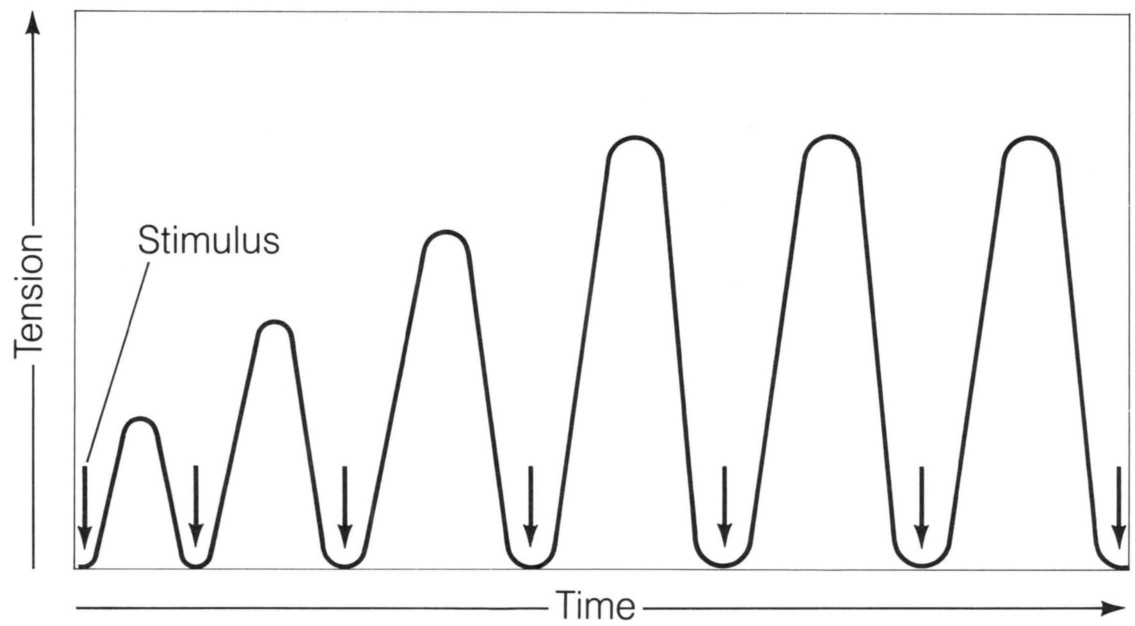

Skeletal muscle energy metabolism

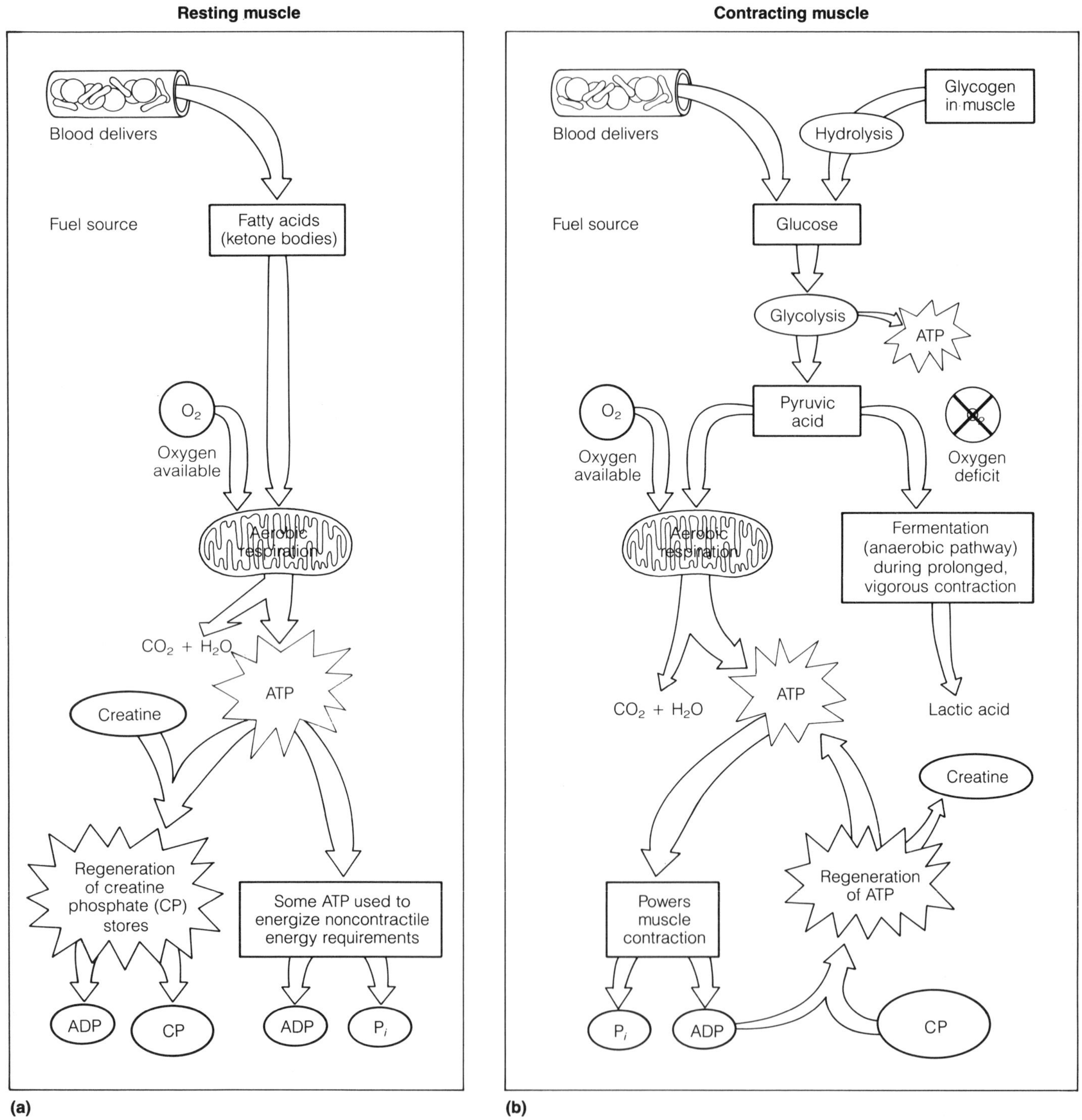

Factors influencing force, velocity, and duration of skeletal muscle contraction

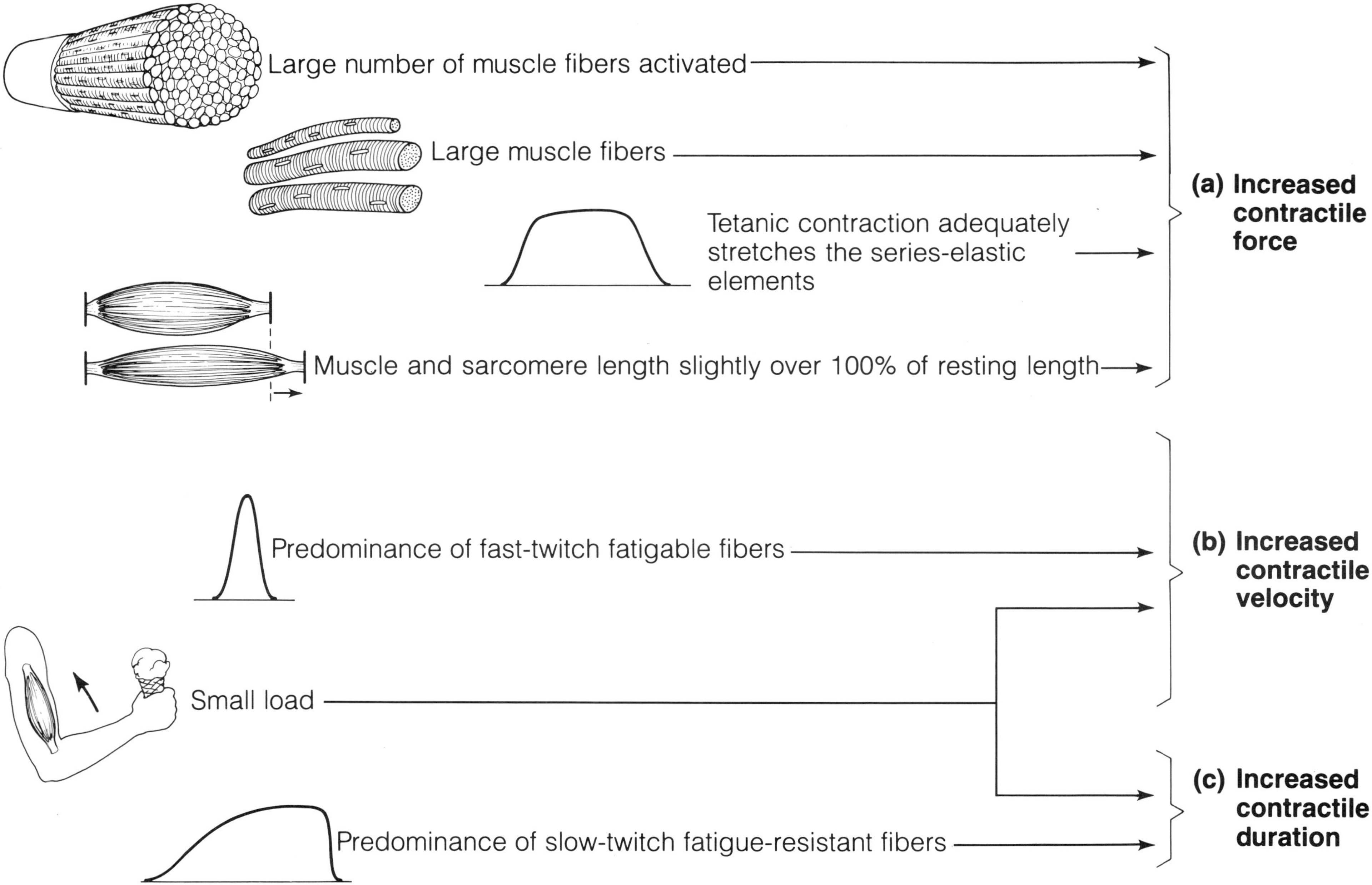

Relationship of stimulus frequency to external tension exerted on the load

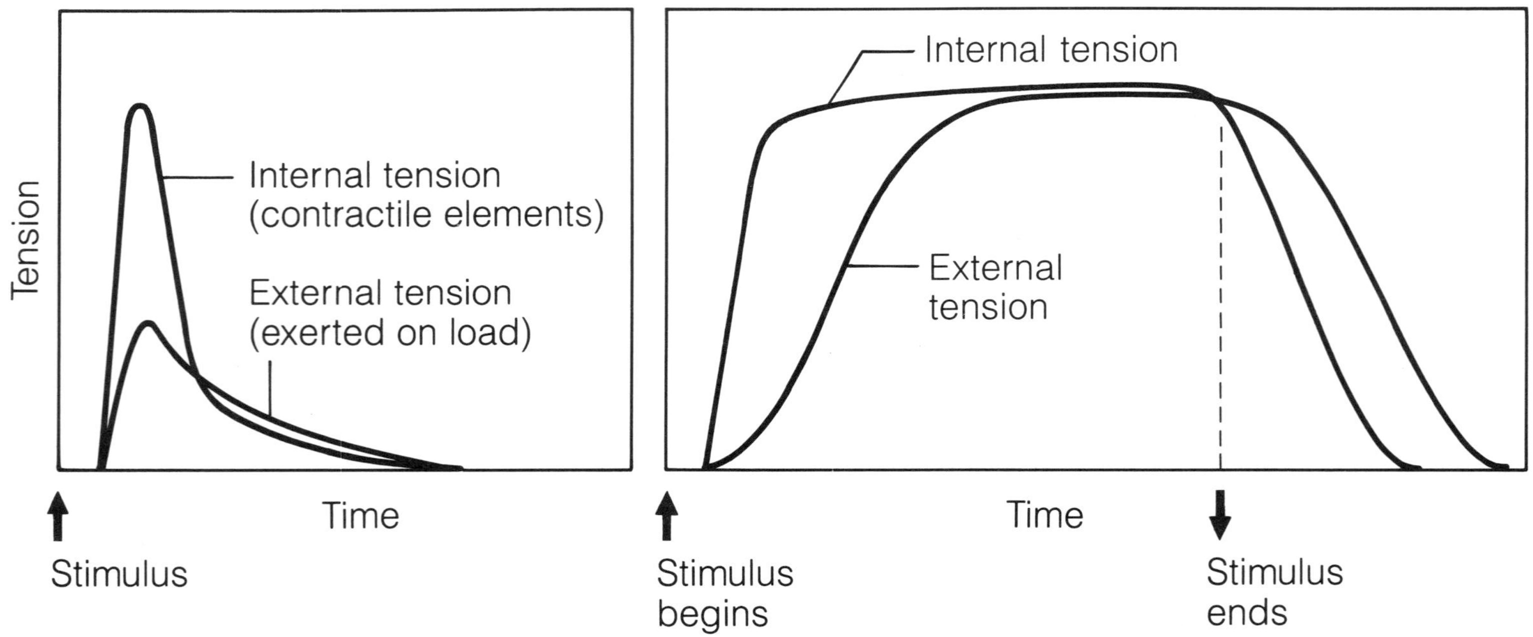

(a) Twitch

(b) Tetanic contraction

Length-tension relationships in skeletal muscles

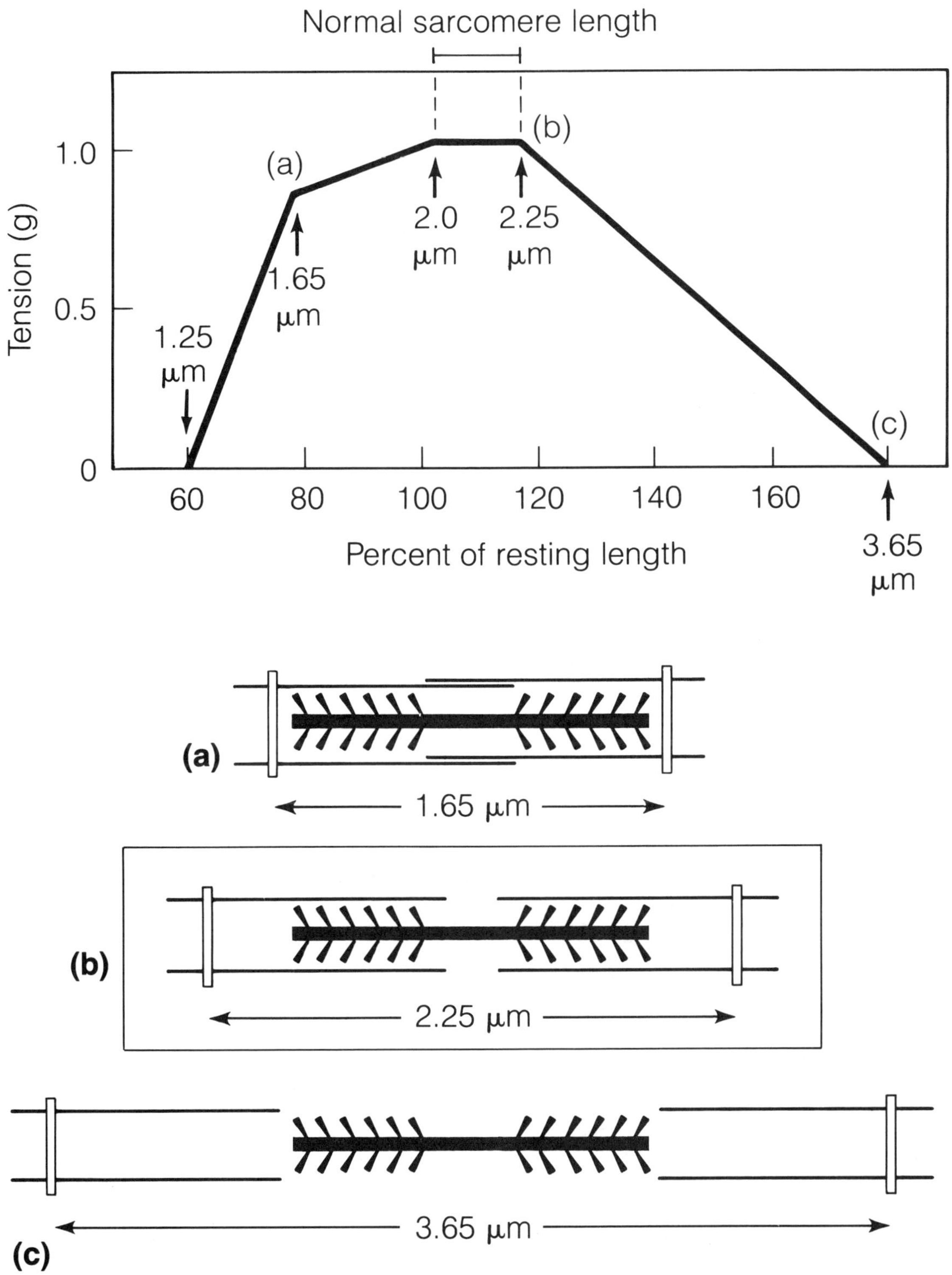

Influence of load on velocity and duration of contraction

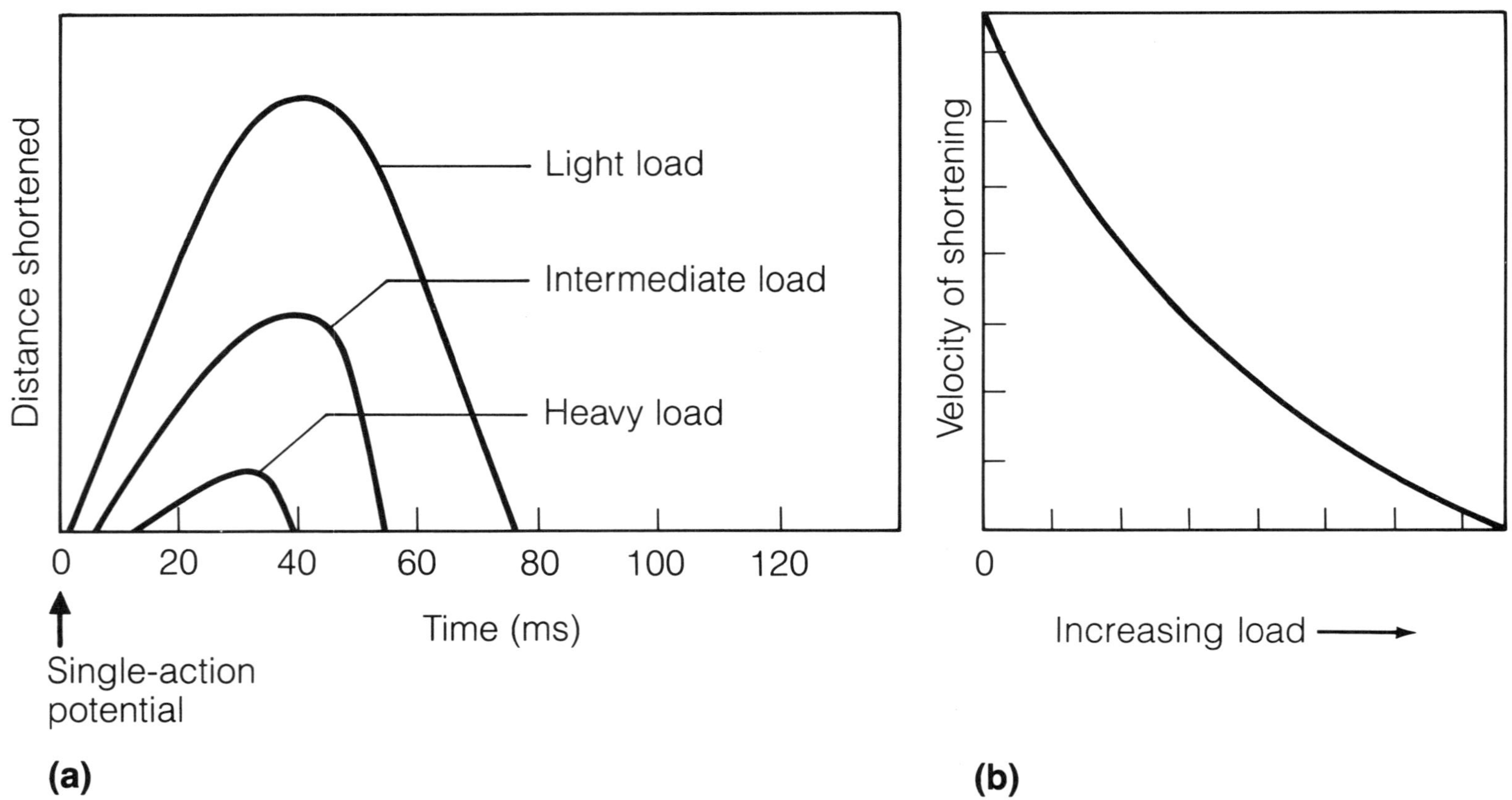

Intermediate filaments and dense bodies of smooth muscle fibers

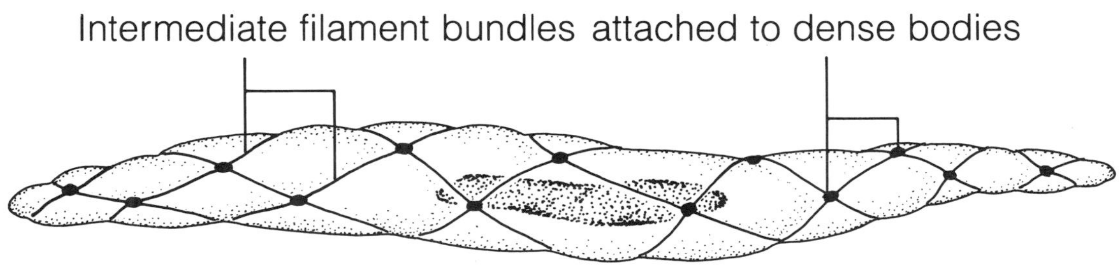

(a) Relaxed smooth muscle cell

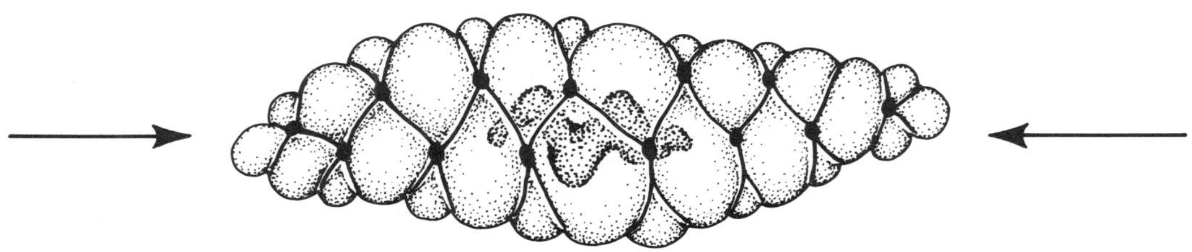

(b) Contracted smooth muscle cell

Structure and organizational levels of skeletal muscle

Structure and organizational level	Description	Connective tissue wrappings
Muscle (organ) 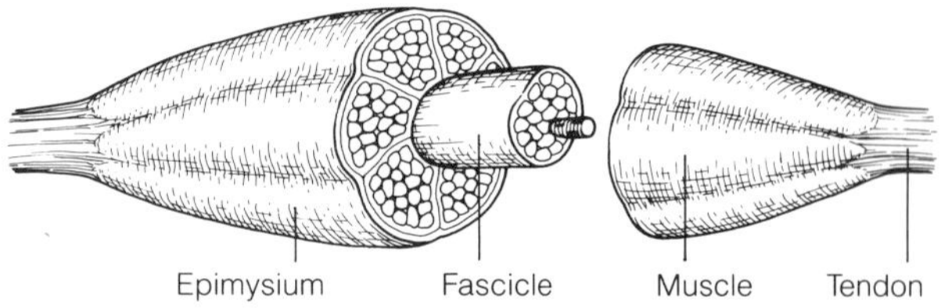	Consists of hundreds to thousands of muscle cells, plus connective tissue wrappings, blood vessels, and nerve fibers	Covered externally by the epimysium
Fascicle (a portion of the muscle) 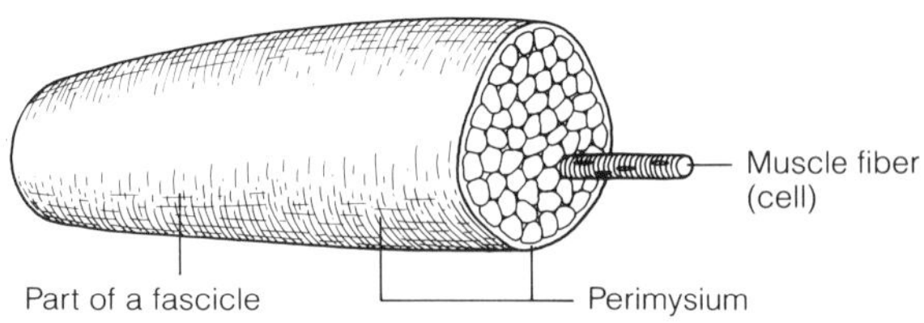	Discrete bundle of muscle cells, segregated from the rest of the muscle by a connective tissue sheath	Surrounded by a perimysium
Muscle fiber (cell) 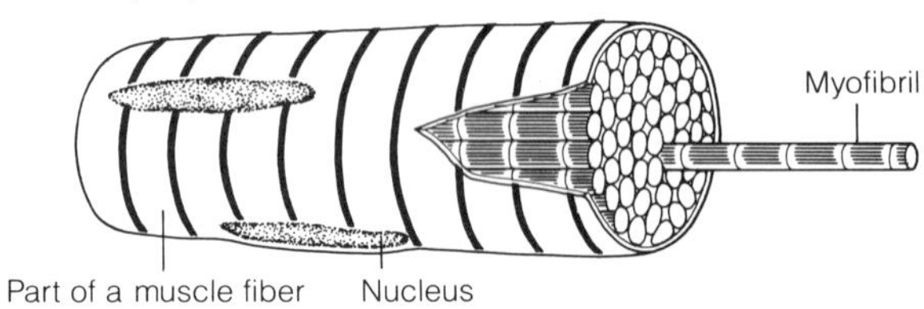	Elongated multinucleate cell; has a banded (striated) appearance	Surrounded by the endomysium
Myofibril or fibril (complex organelle composed of bundles of myofilaments) 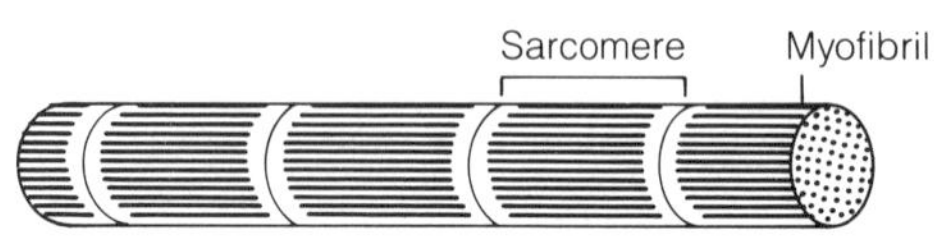	Rodlike contractile element; myofibrils occupy most of the muscle cell volume; appear banded, and bands of adjacent myofibrils are aligned; composed of sarcomeres, end to end	
Sarcomere (a segment of a myofibril) 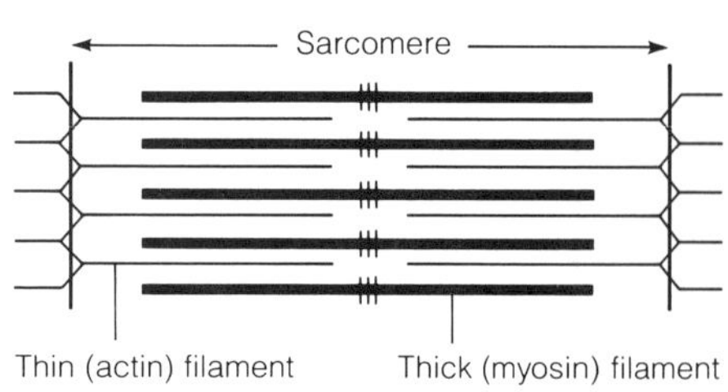	The contractile unit, composed of myofilaments made up of contractile proteins	
Myofilament or filament (extended macromolecular structure) 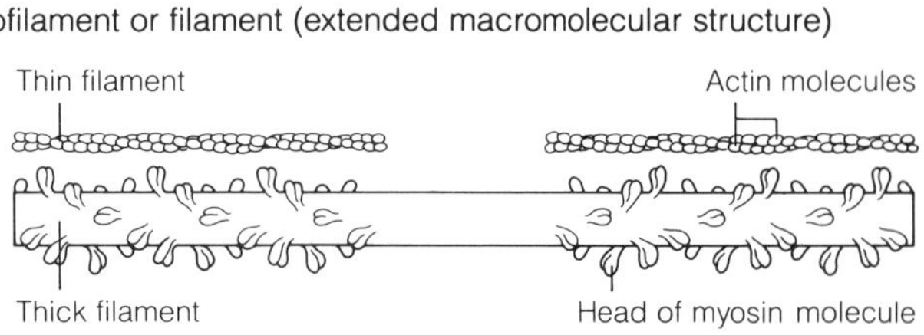	Myofilaments are of two types—thick and thin—composed of contractile proteins; the thick filaments contain bundled myosin molecules; the thin filaments contain actin molecules (plus other proteins); the sliding of the thin filaments past the thick filaments produces muscle shortening	

Structural and functional characteristics of the three types of skeletal muscle fibers

Feature	Red slow-twitch (fatigue-resistant)	White fast-twitch (fatigable)	Intermediate fast-twitch (fatigue-resistant)
Metabolic characteristics:			
Twitch rate	Slow	Fast	Fast
Myosin ATPase activity	Slow	Fast	Fast
Primary pathway for ATP synthesis	Aerobic	Anaerobic	Aerobic
Myoglobin content	High	Low	High
Glycogen stores	Low	High	Intermediate
Rate of fatigue	Slow	Fast	Intermediate
Structural characteristics:			
Color	Red	White (pale)	Red (pink)
Fiber diameter	Small	Large	Intermediate
Mitochondria	Many	Few	Many
Capillaries	Many	Few	Many

Levels of organization in the nervous system

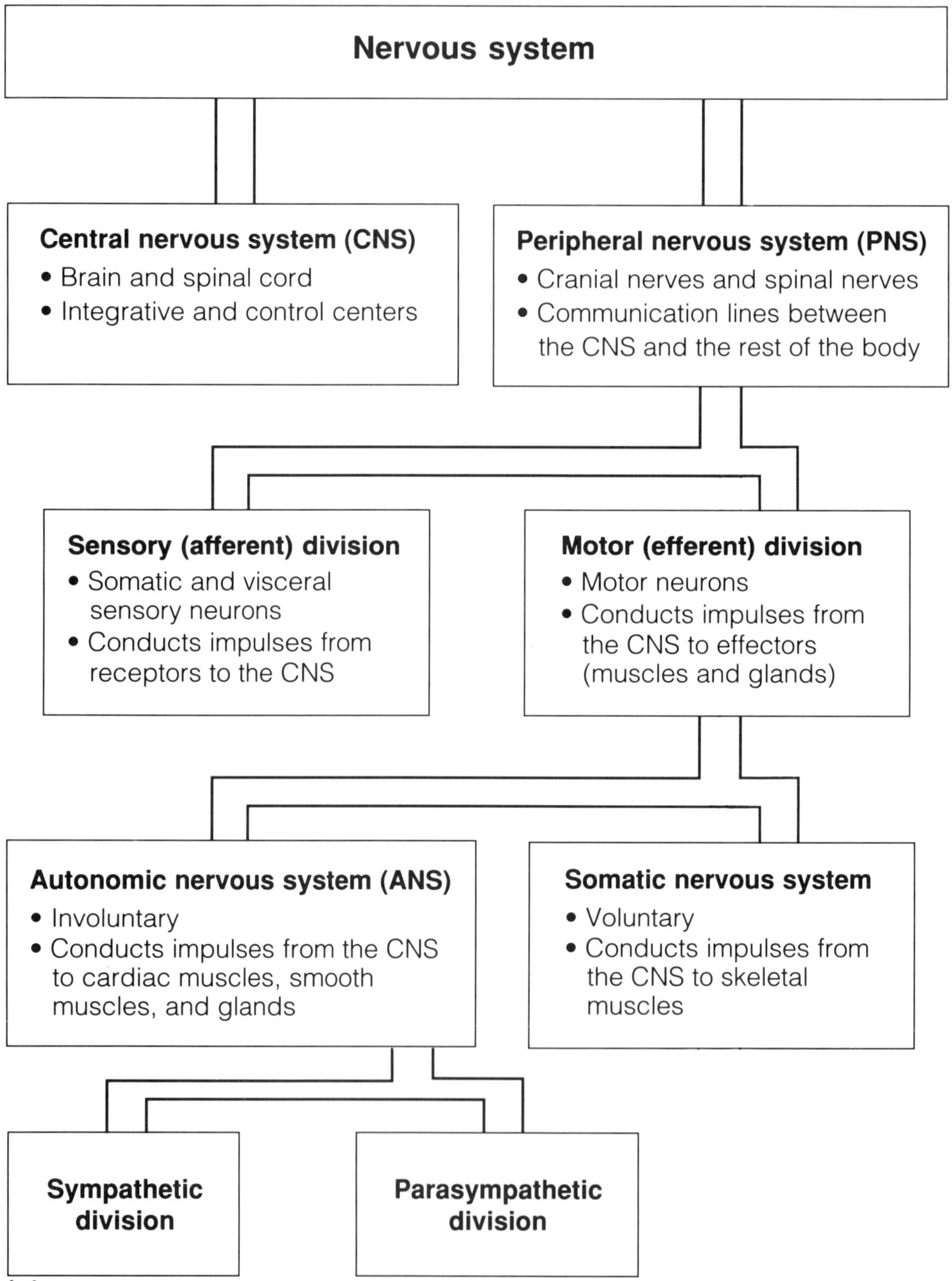

(a)

Visceral organs

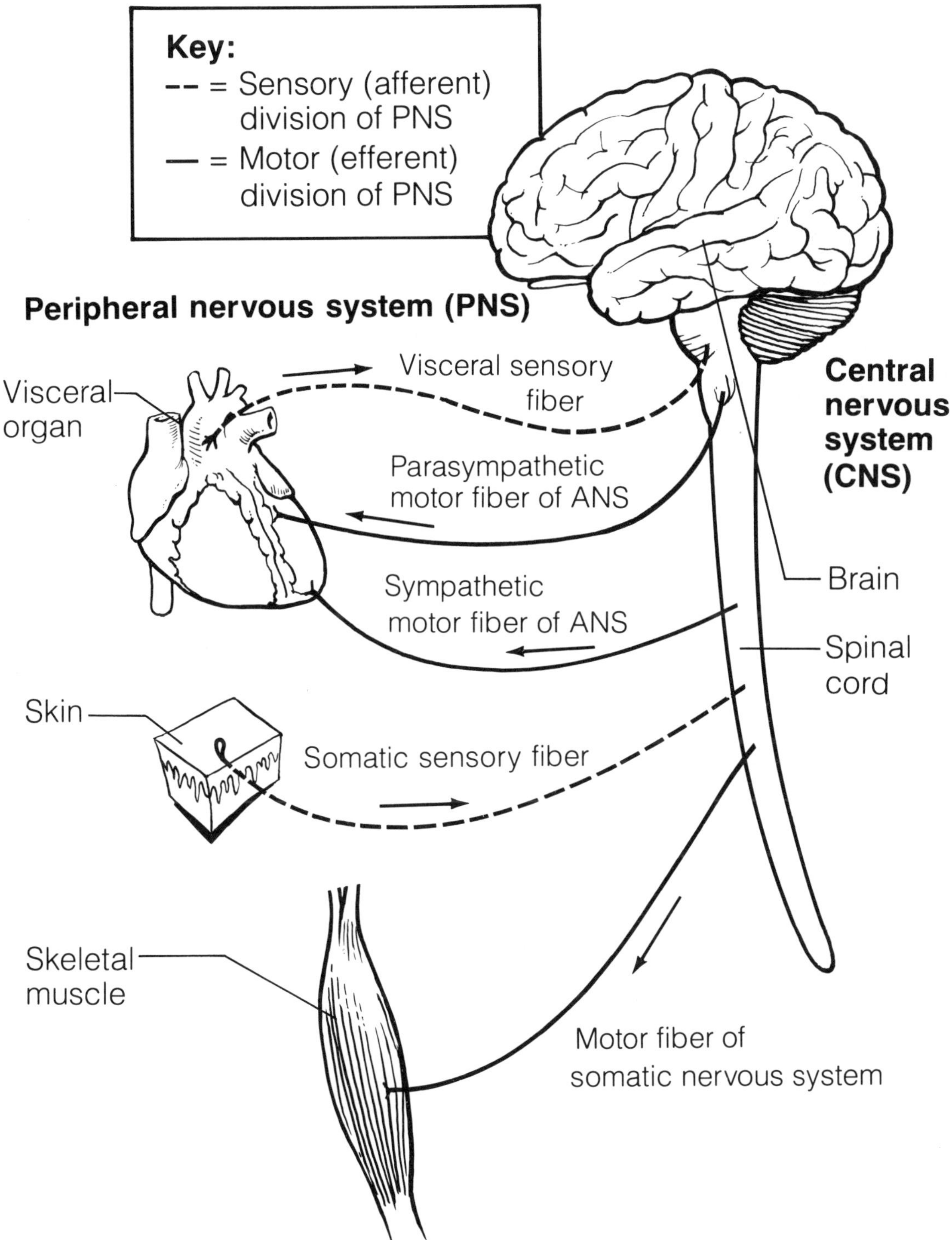

(b)

Supporting cells of nervous tissue

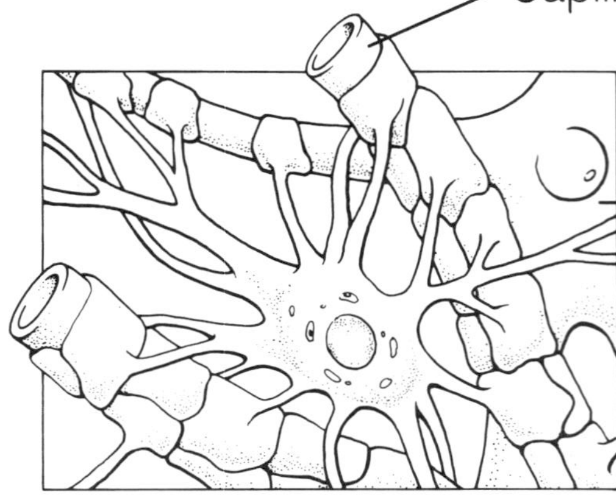

(a) Astrocyte

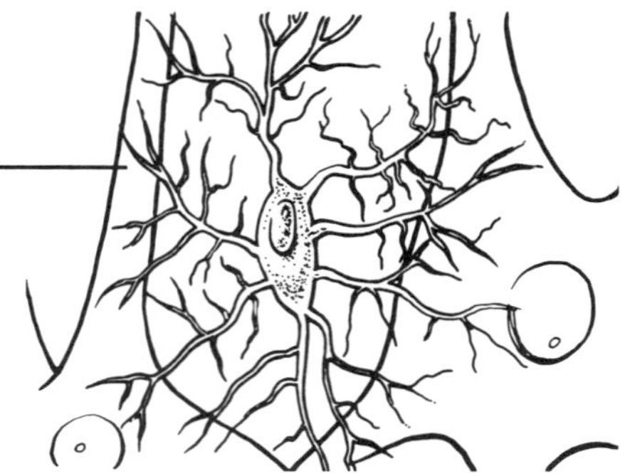

(b) Microglial cell

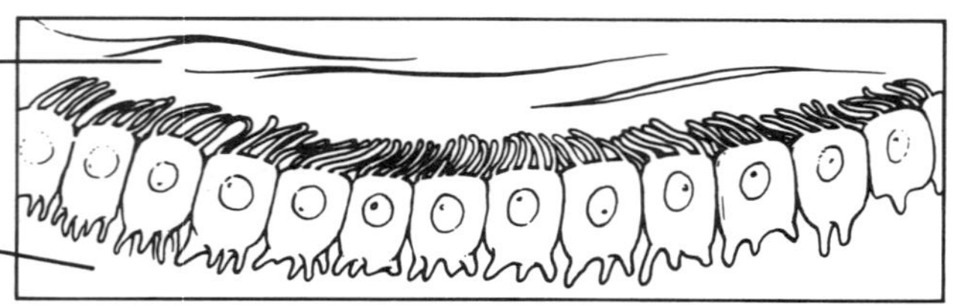

(c) Ependymal cells

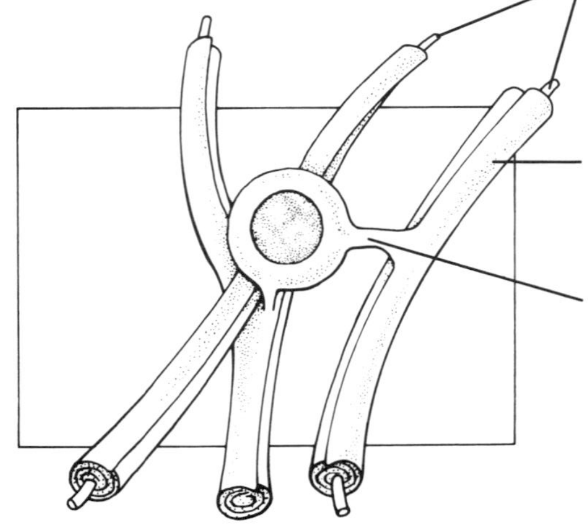

(d) Oligodendrocyte

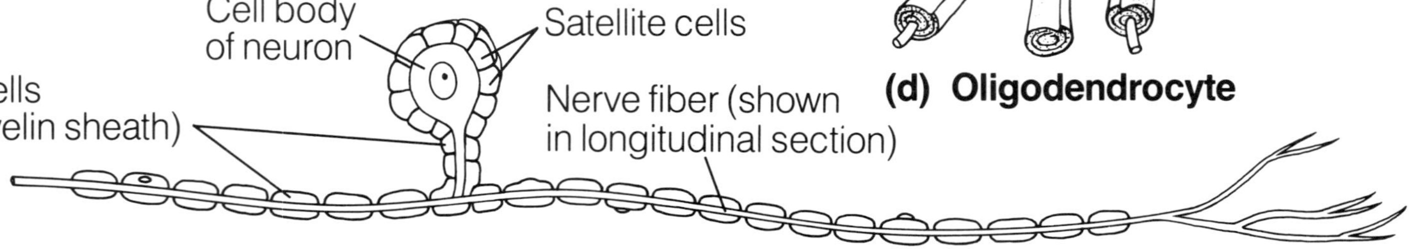

(e) Sensory neuron with Schwann cells and satellite cells

Relationship of Schwann cells to axons in the peripheral nervous system

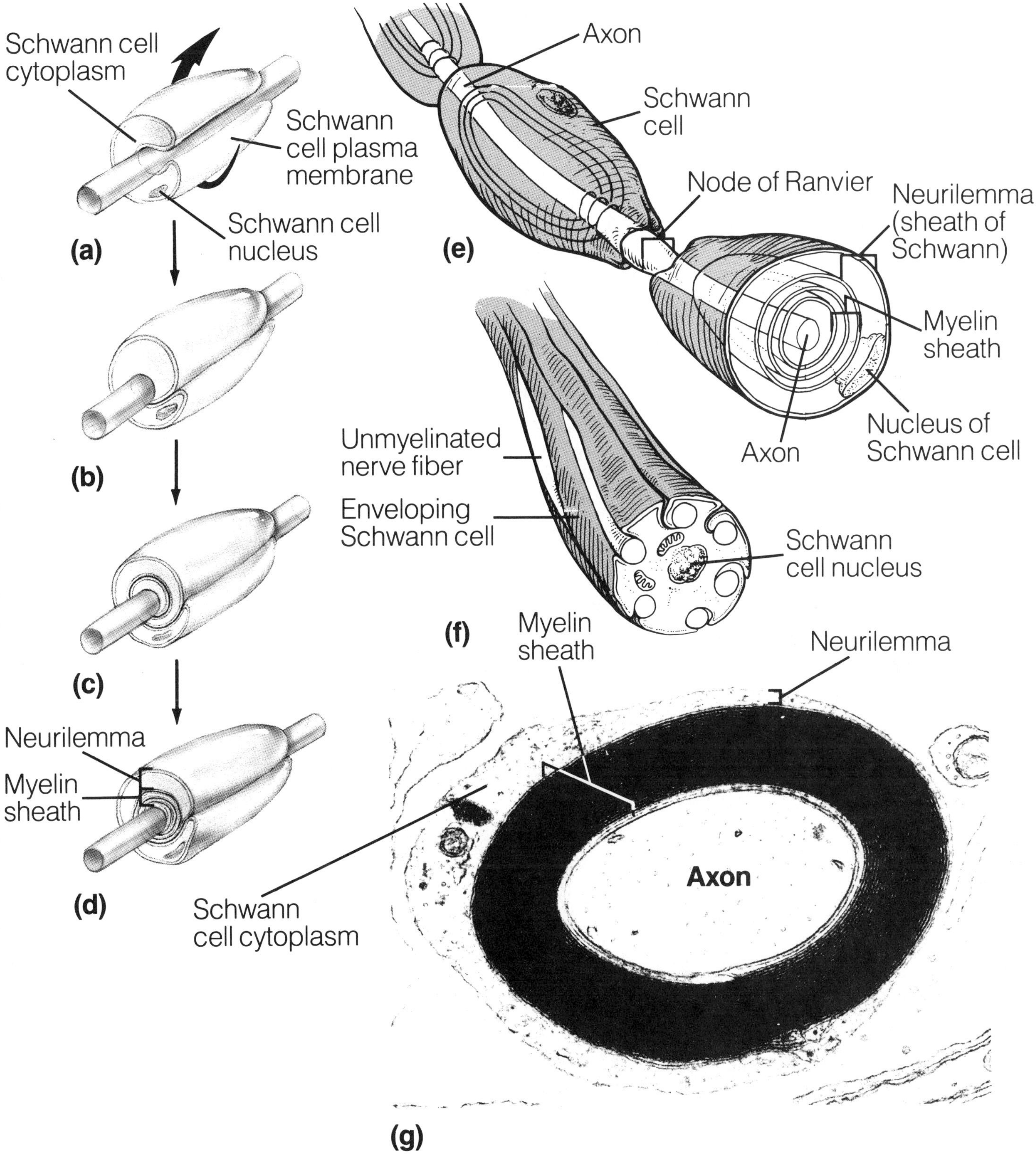

Types of plasma membrane ion channels

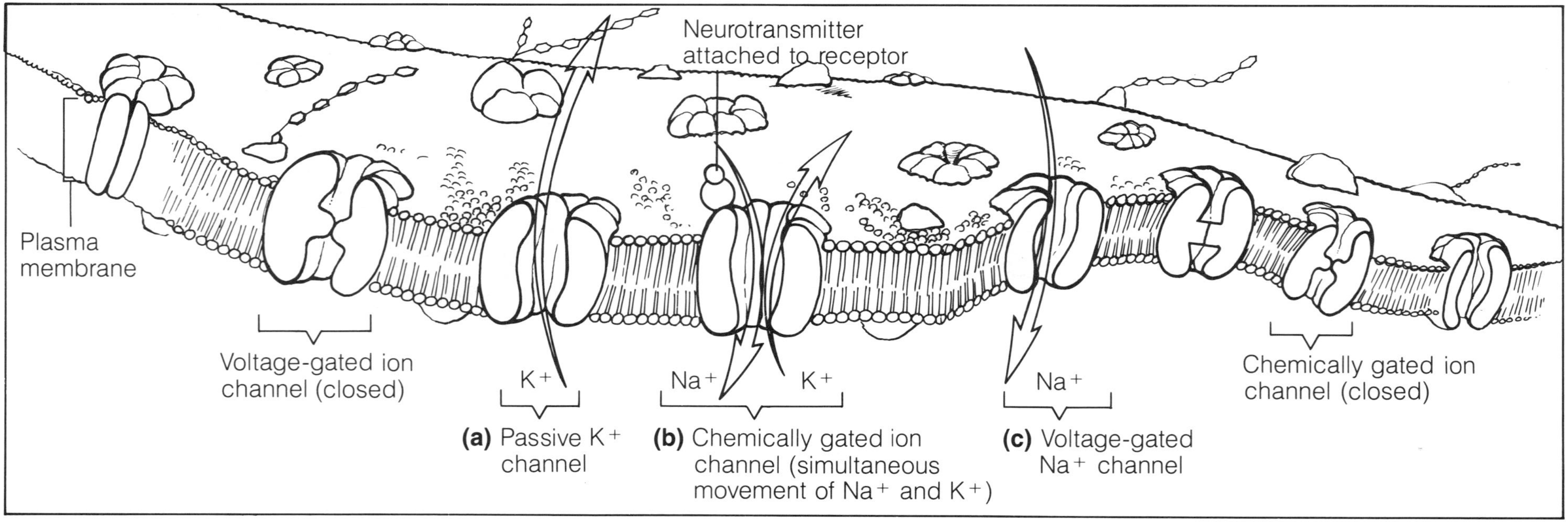

Measuring potential difference between two points in neurons

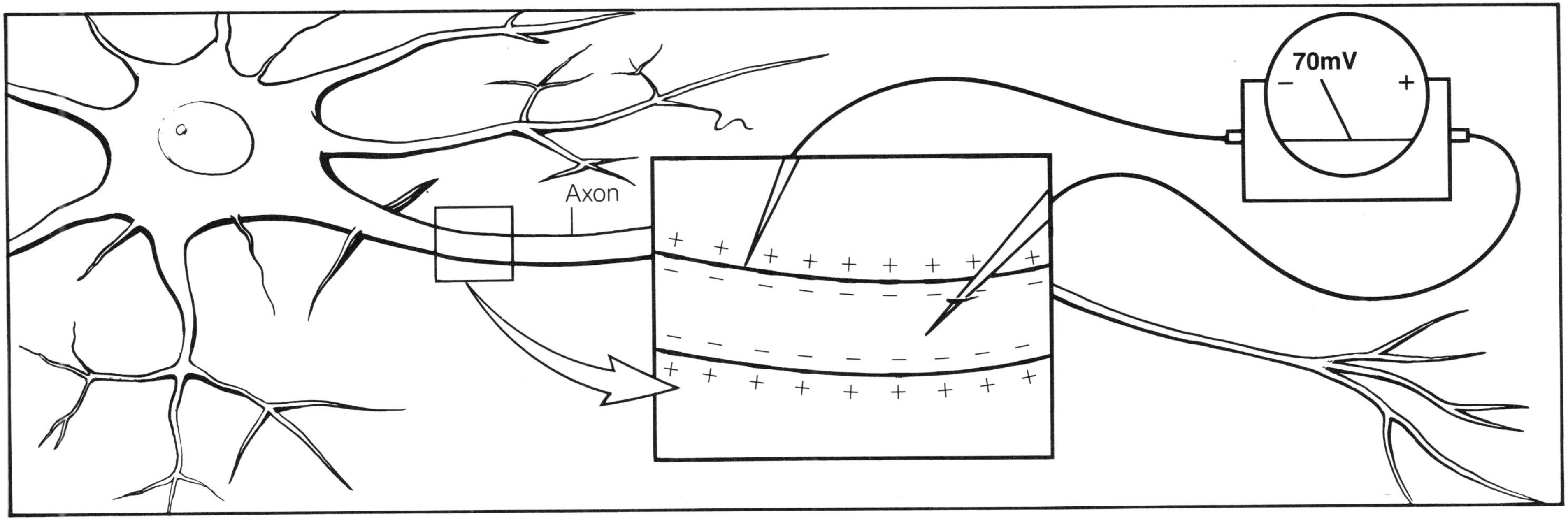

Passive and active forces that establish and maintain the resting membrane potential

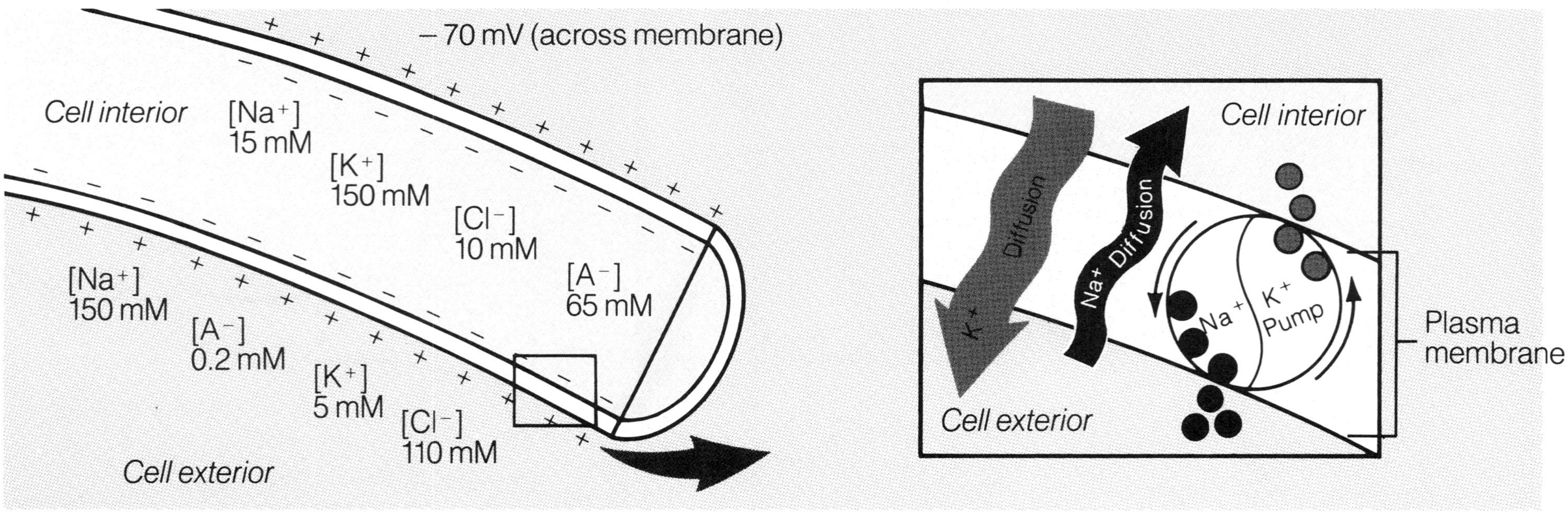

Depolarization and hyperpolarization of the resting membrane potential

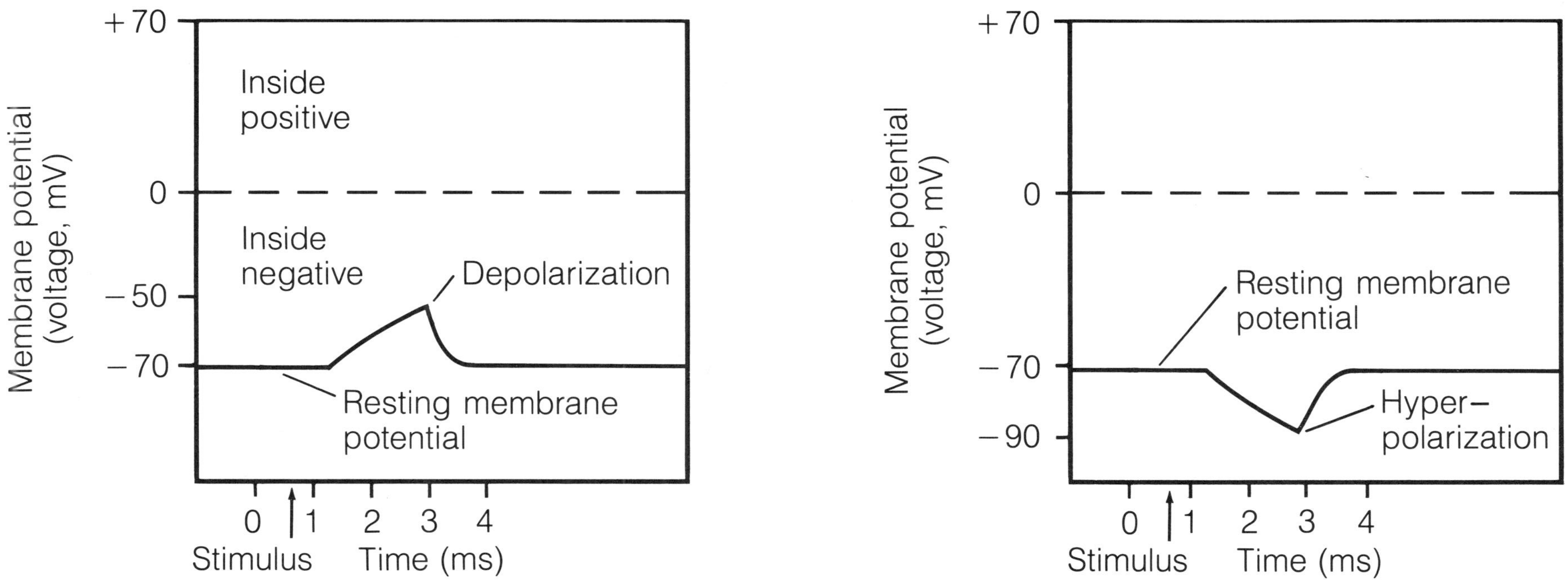

(a) Depolarization **(b) Hyperpolarization**

The mechanism of a graded potential

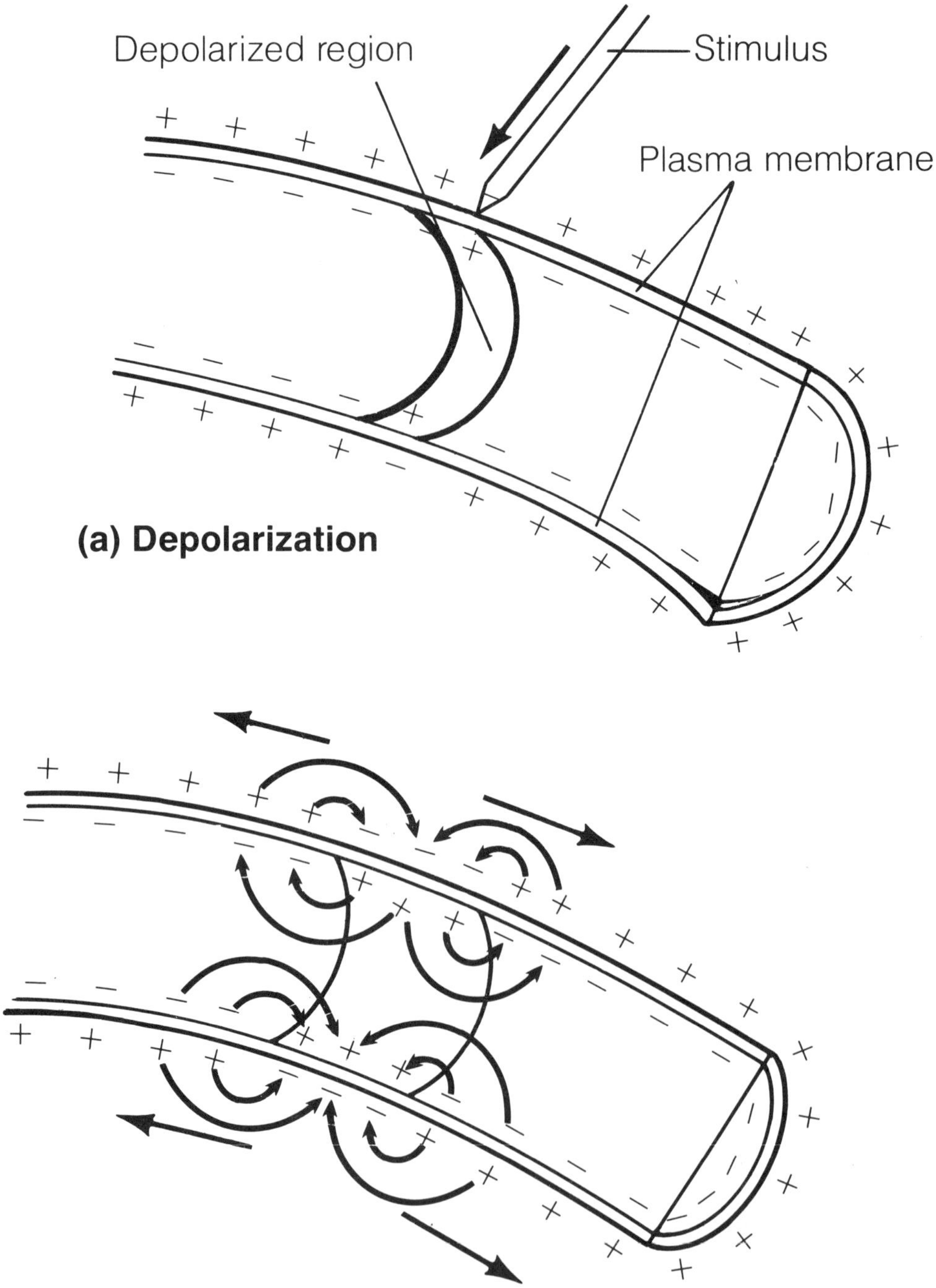

(a) Depolarization

(b) Spread of depolarization

Changes in membrane voltage produced by a depolarizing graded potential

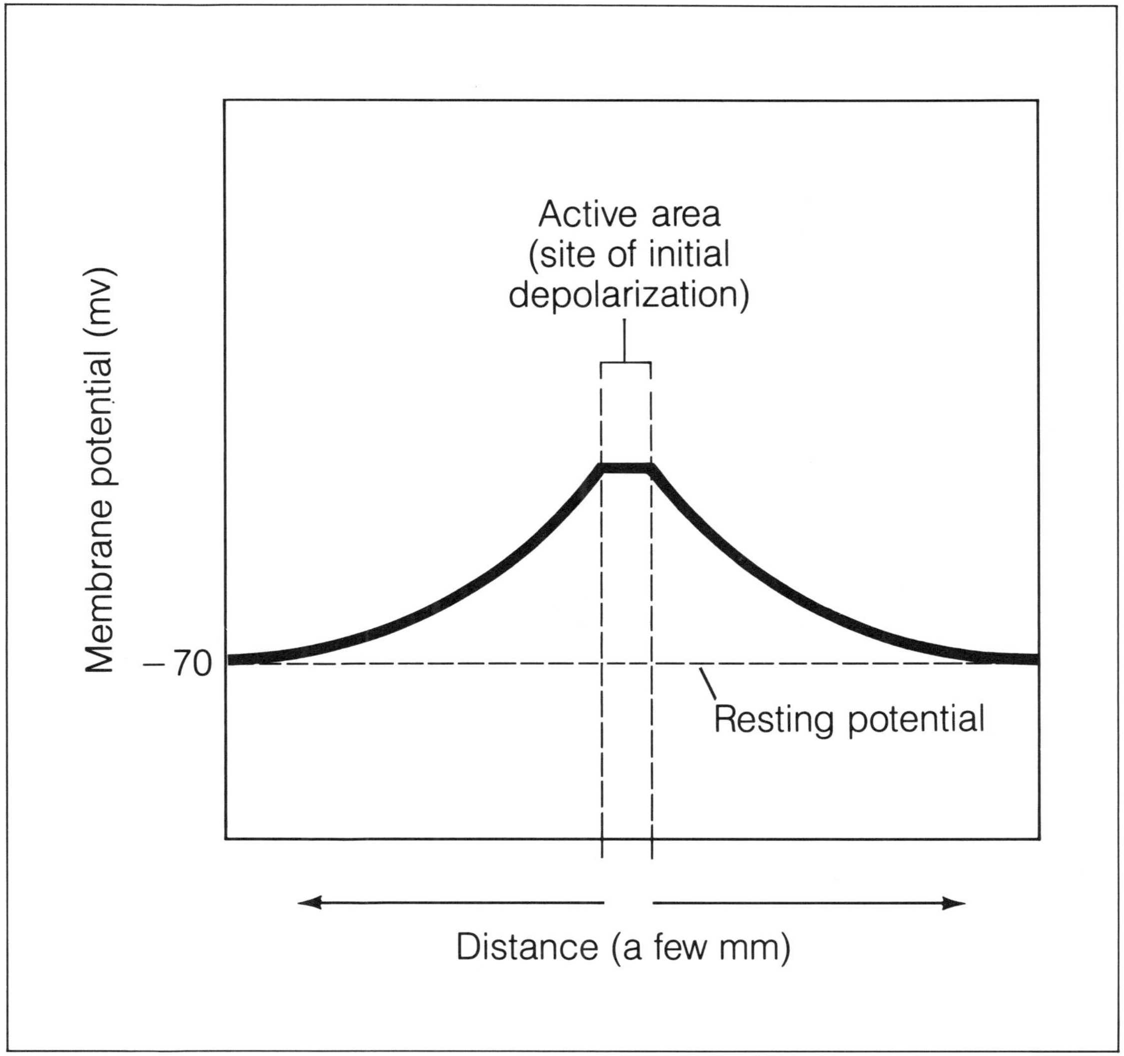

Voltage and membrane permeability changes occurring during action potential generation

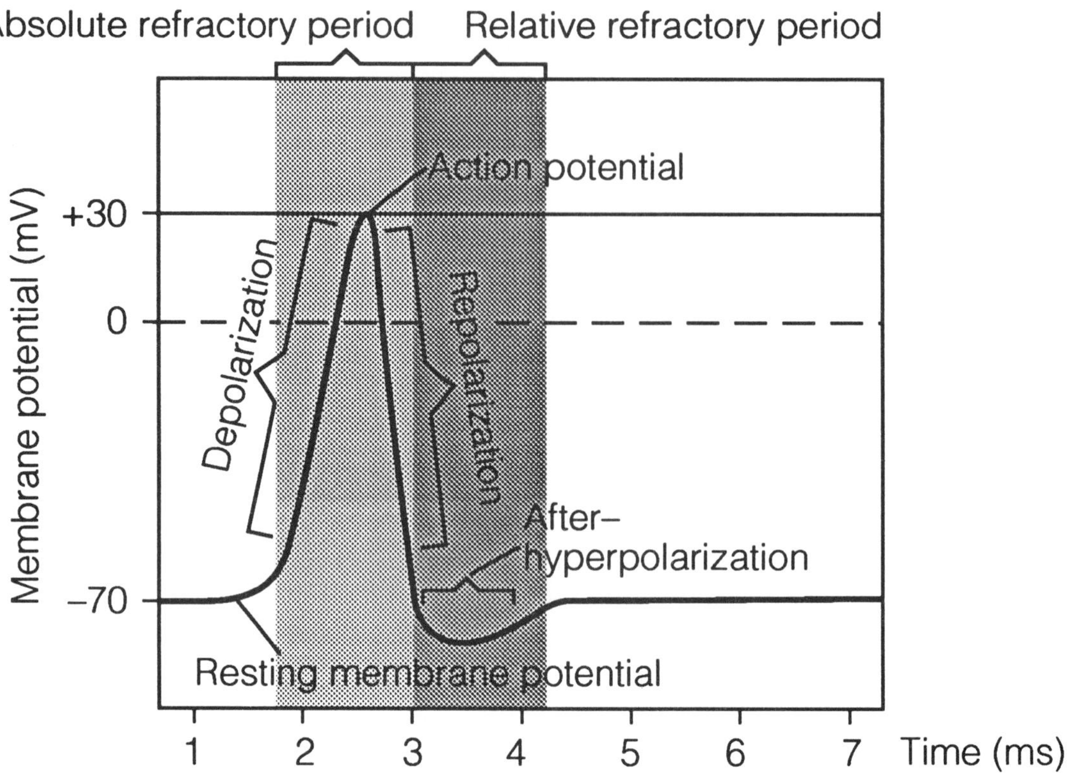

(a) Recording of an action potential

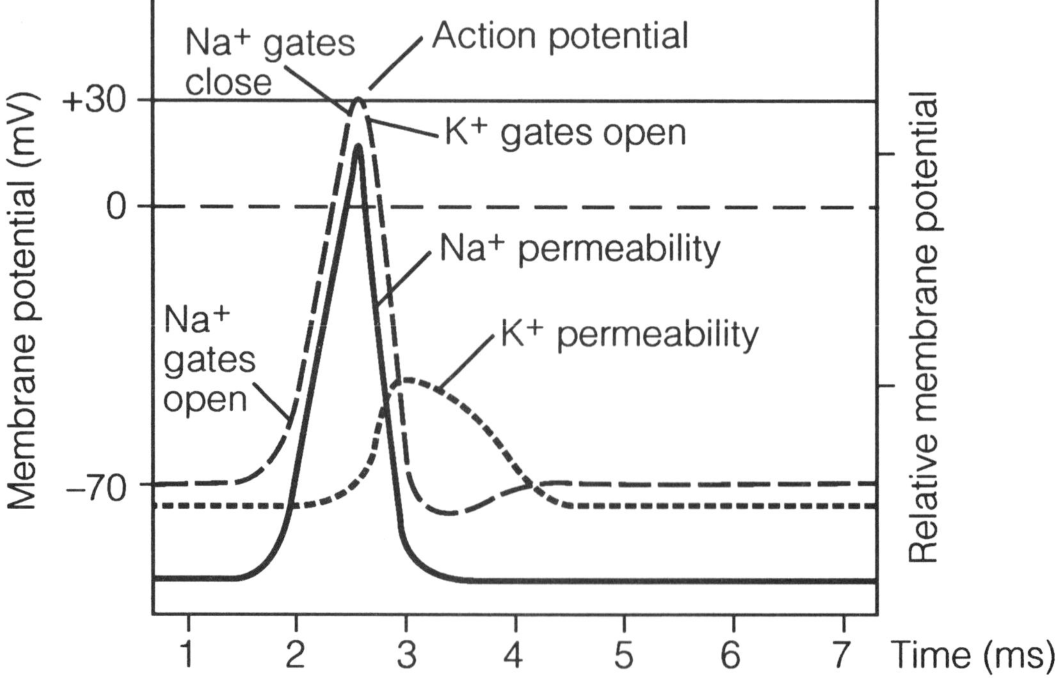

(b) Relationship of membrane permeability changes to voltage changes during the action potential

The Hodgkin cycle, which produces the rising phase of the action potential

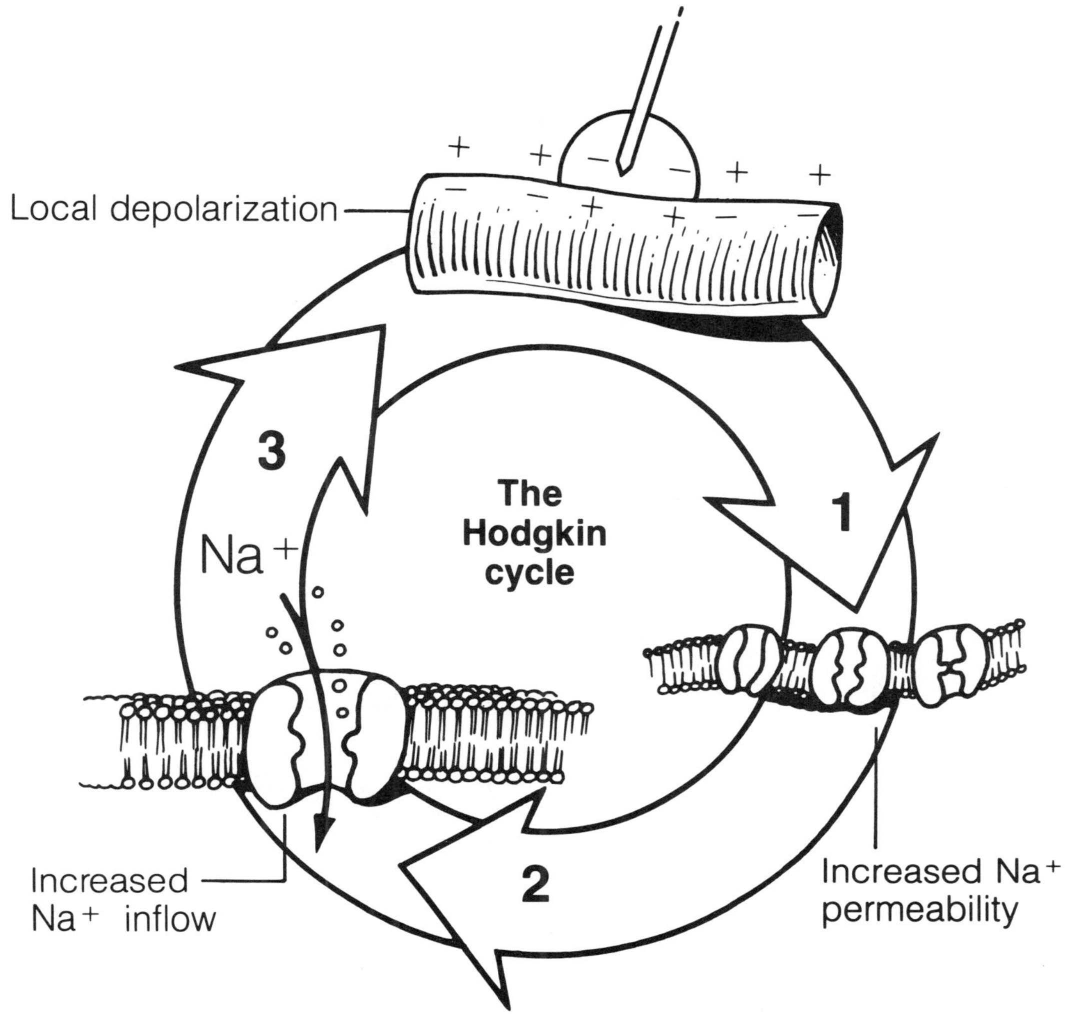

Propagation of an action potential

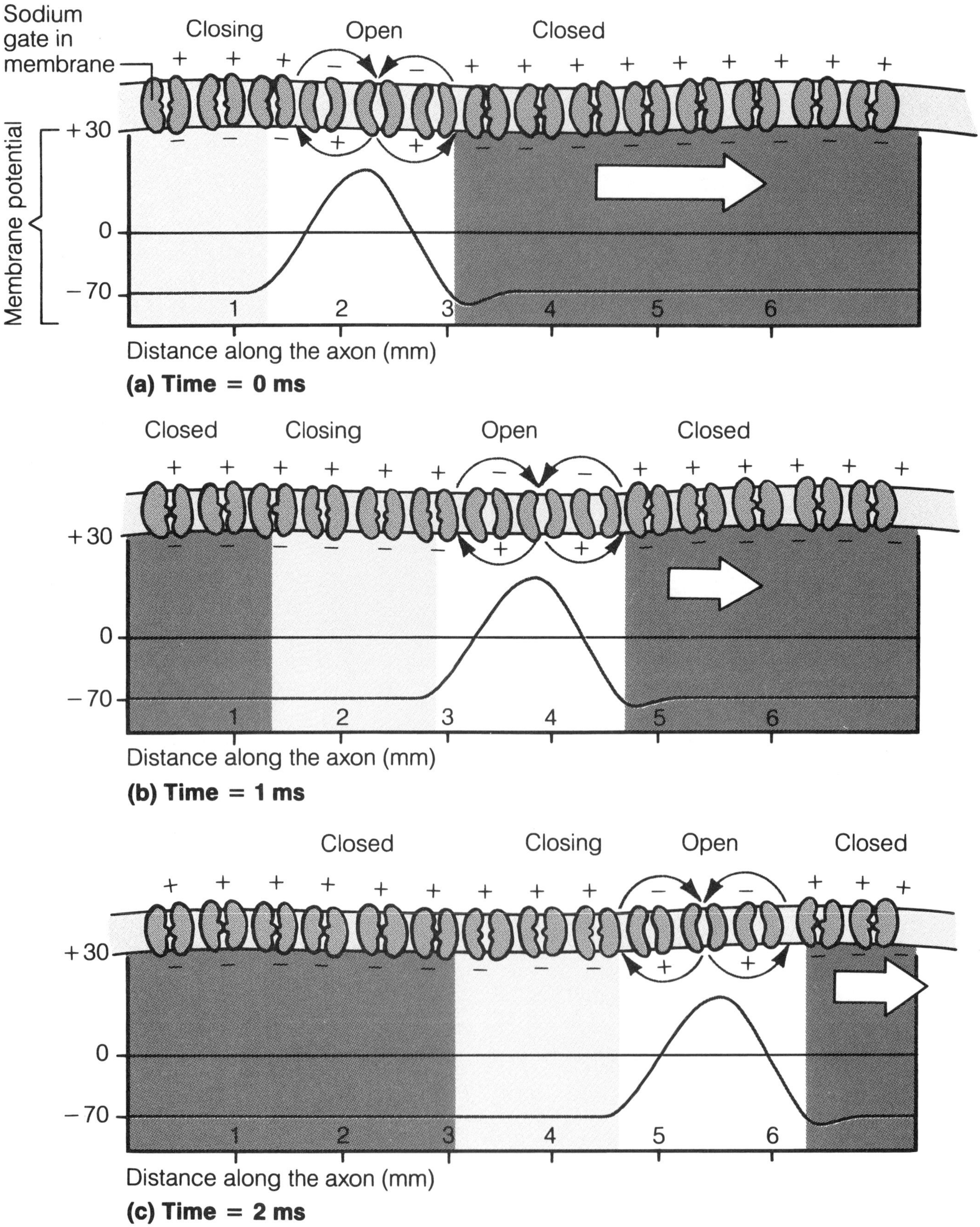

Relationship between stimulus strength, the local potential produced, and action potential frequency

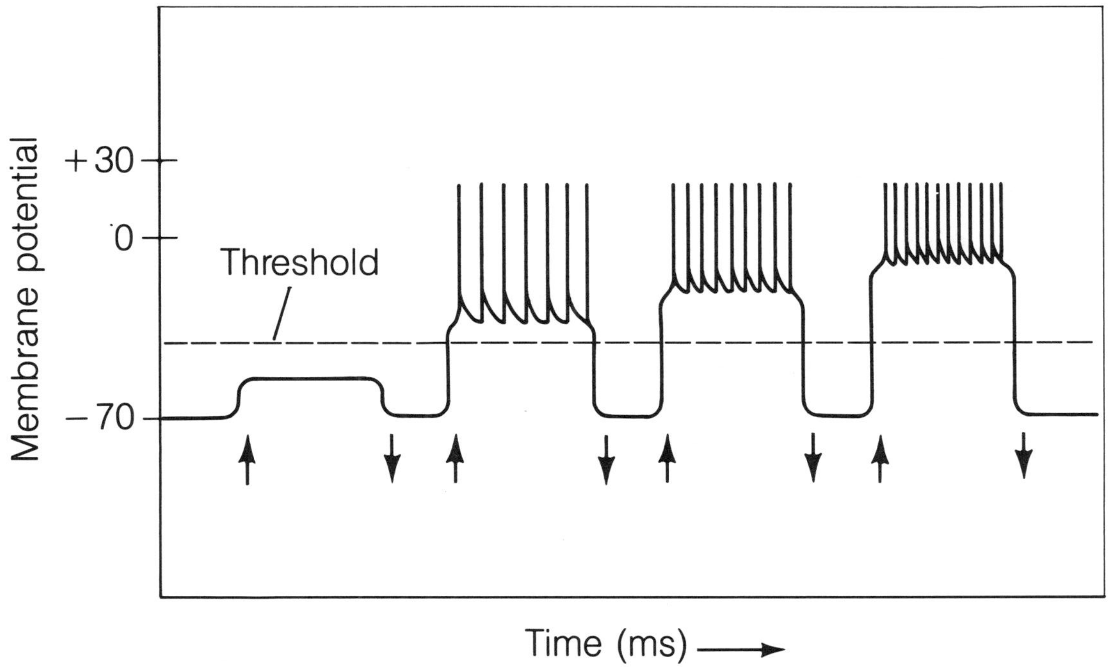

Saltatory conduction in a myelinated axon

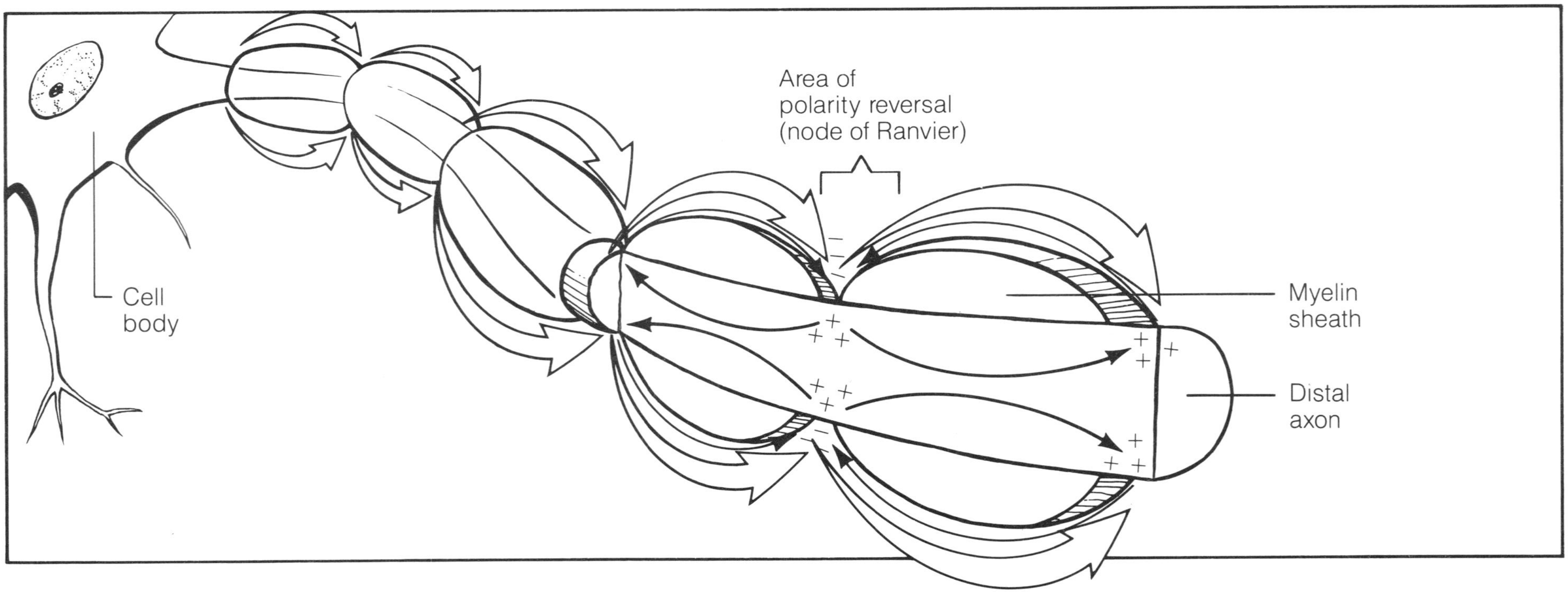

Postsynaptic potentials

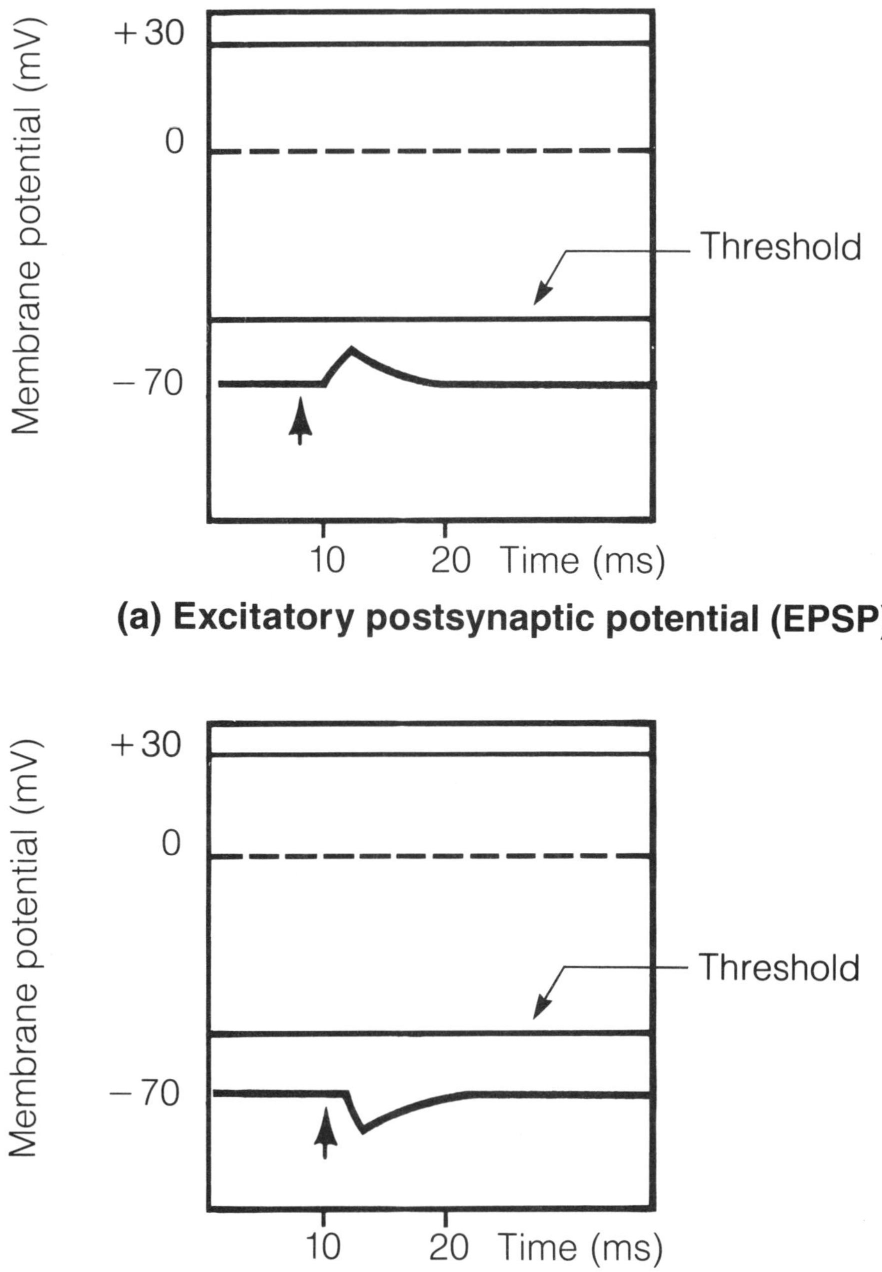

(a) Excitatory postsynaptic potential (EPSP)

(b) Inhibitory postsynaptic potential (IPSP)

Neural integration of EPSPs and IPSPs at the axonal membrane of the postsynaptic cell

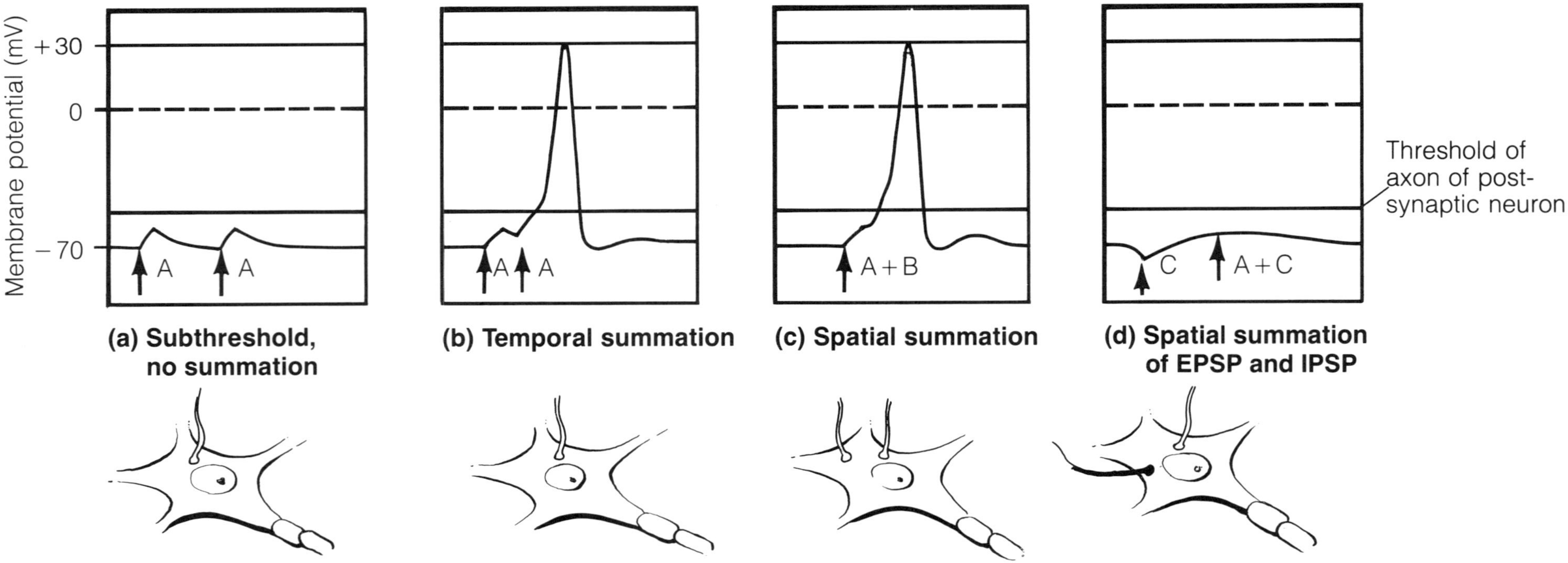

(a) Subthreshold, no summation
(b) Temporal summation
(c) Spatial summation
(d) Spatial summation of EPSP and IPSP

Presynaptic inhibition

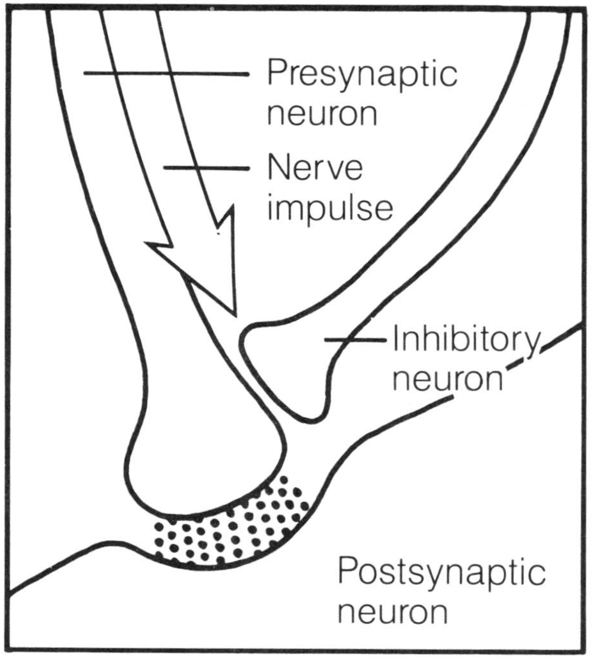

(a) Inhibitory neuron inactive

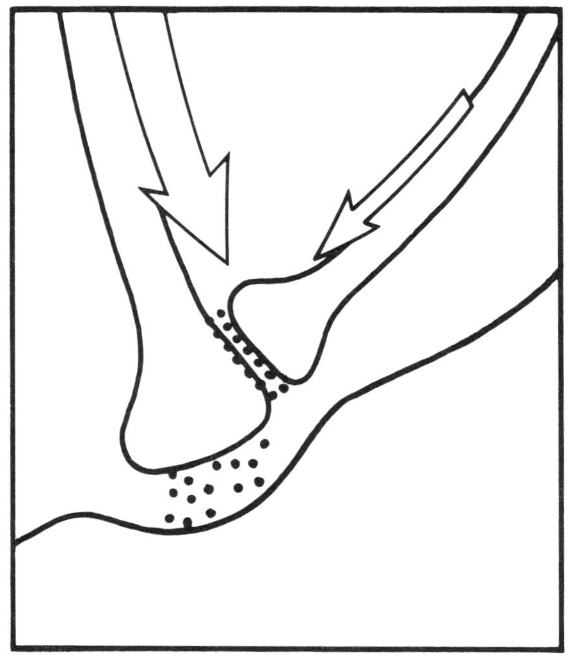

(b) Inhibitory neuron active

Common pathway for synthesis of dopamine, norepinephrine, and epinephrine

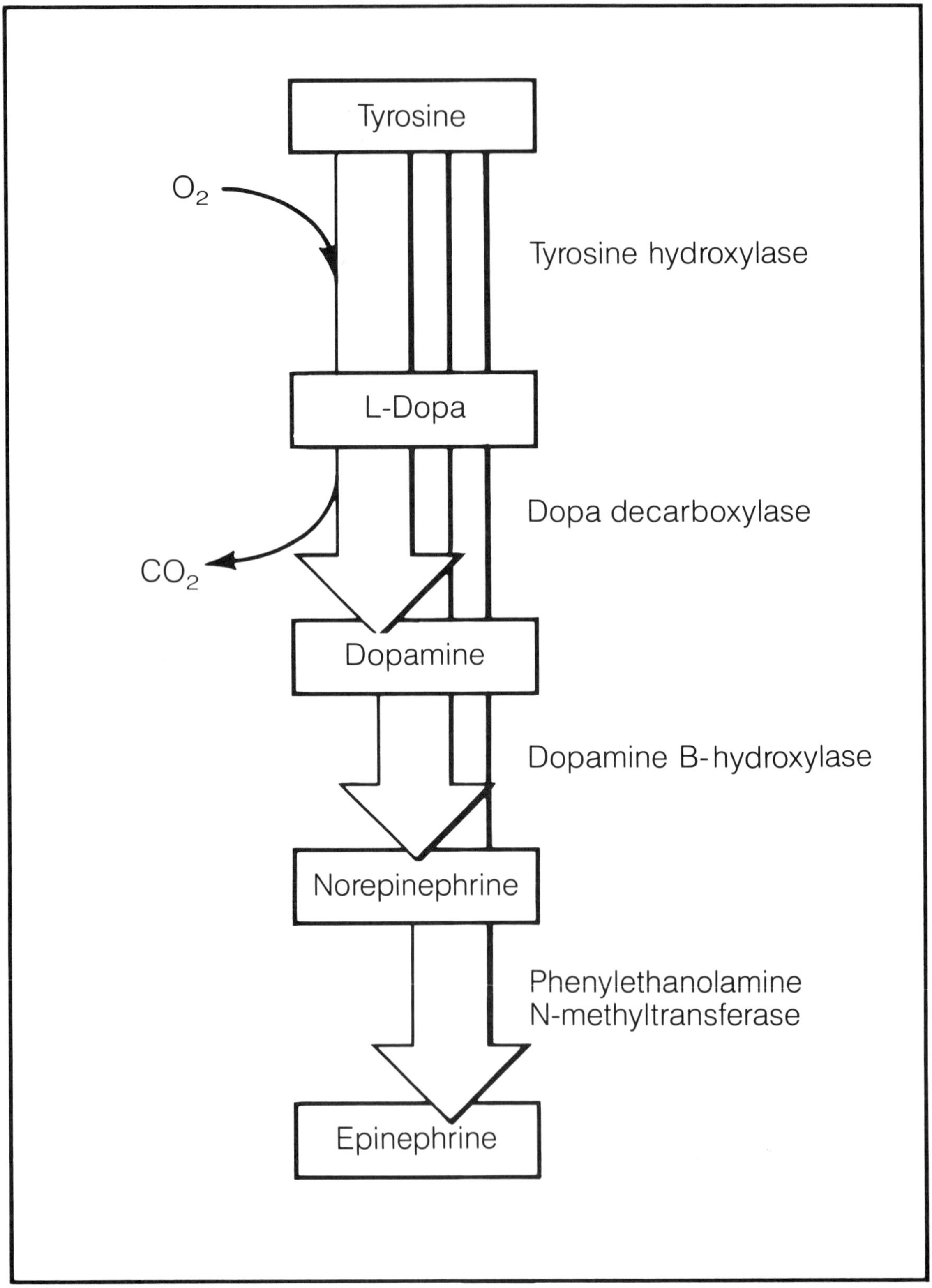

Mechanism of a metabotropic neurotransmitter

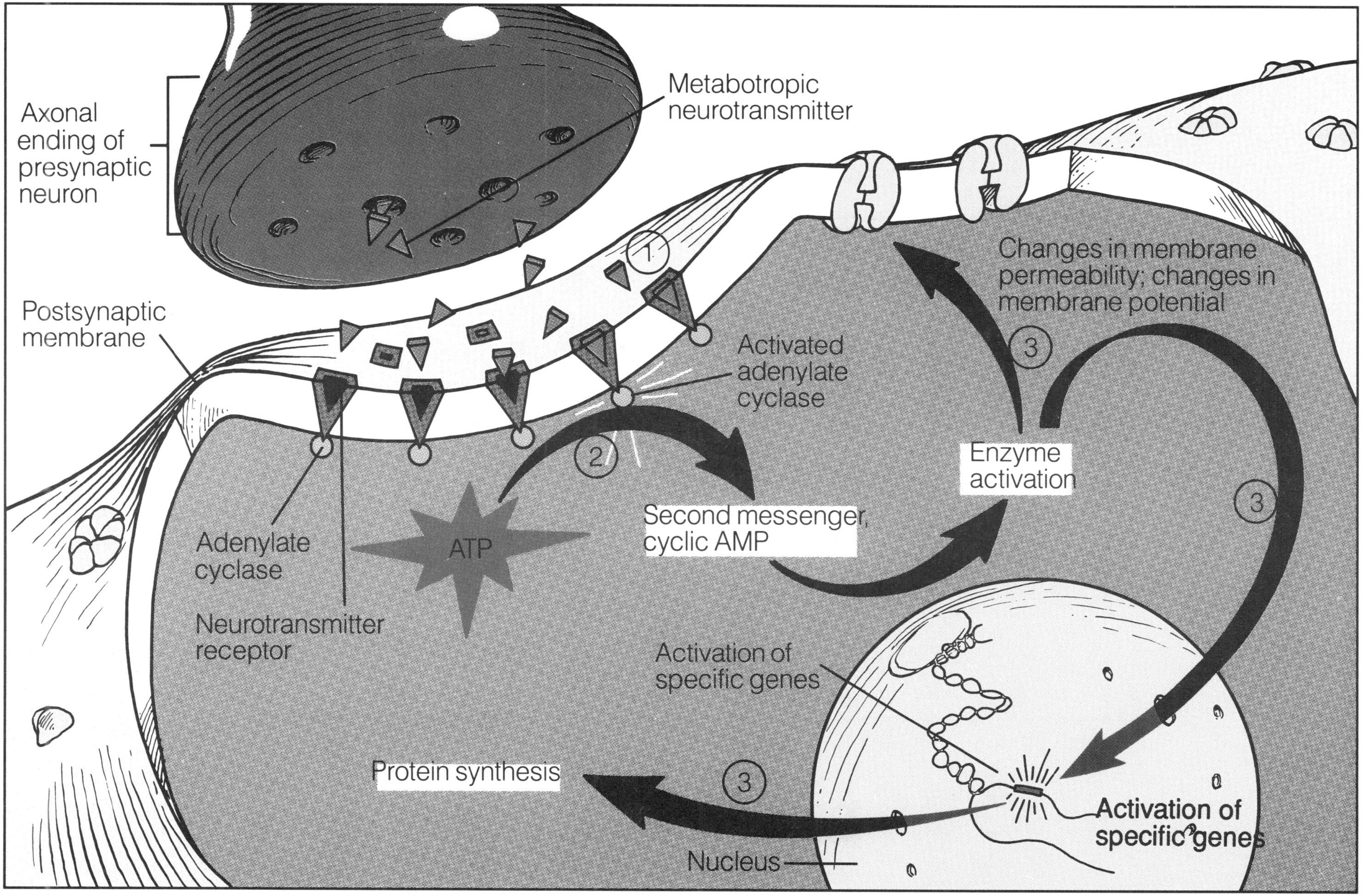

Simple neuronal pool

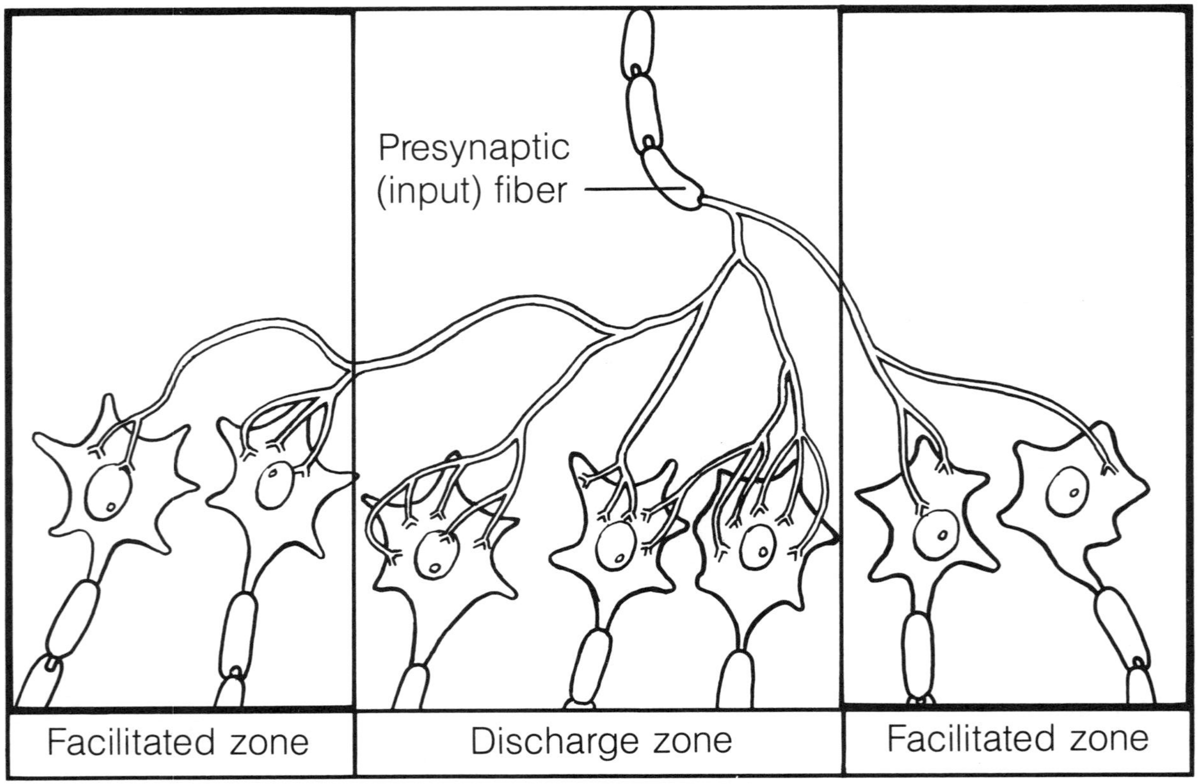

Types of circuits in neuronal pools

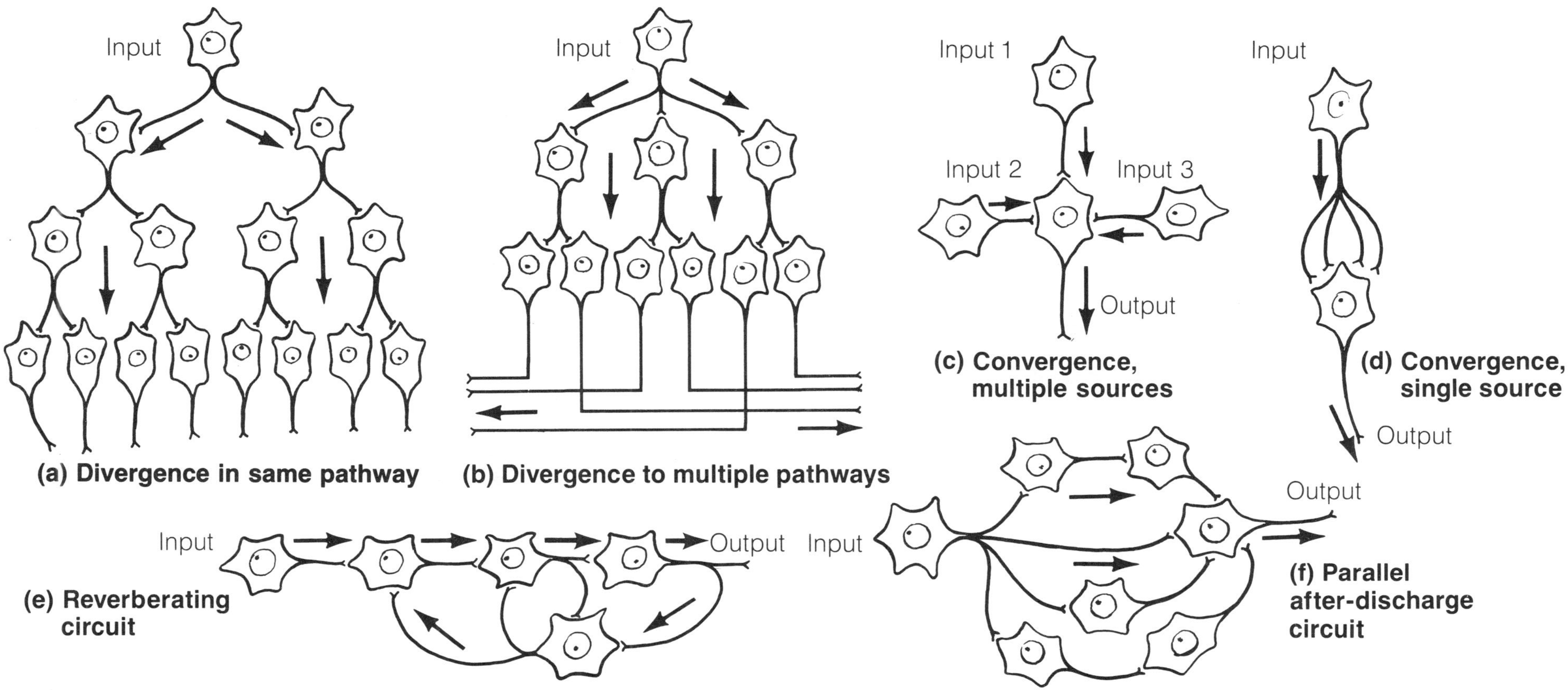

A simple reflex arc

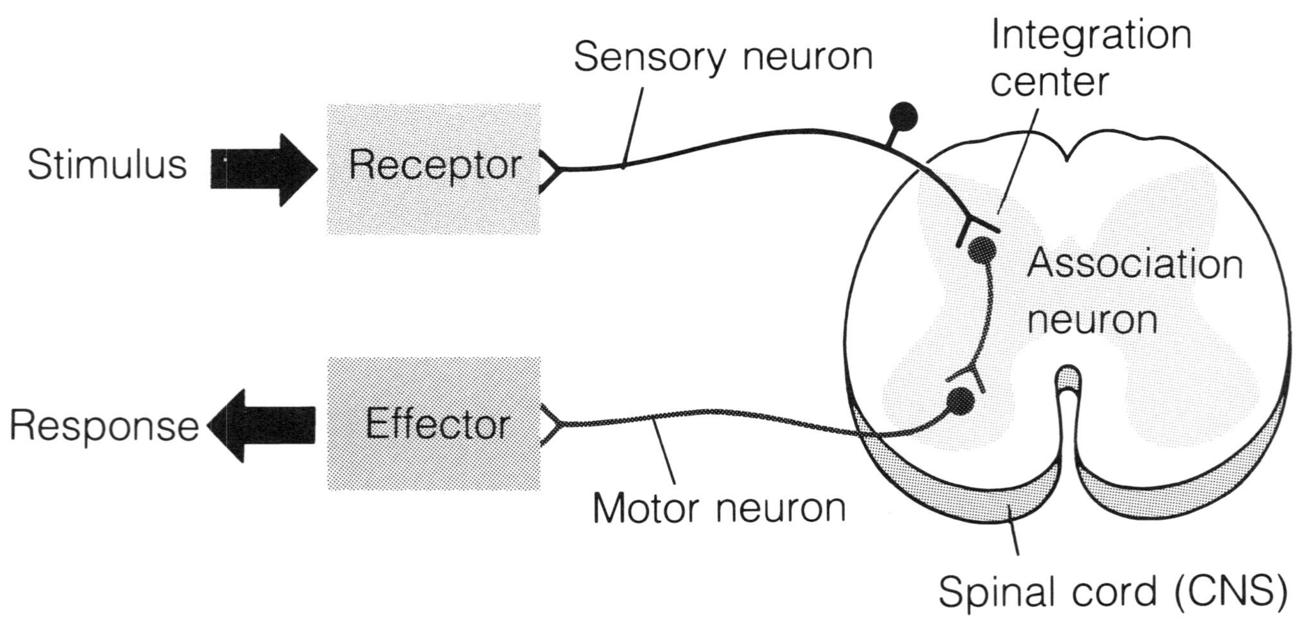

Lobes and fissures of the cerebral hemispheres: (a) left lateral view of the brain; (b) medial surface of the right hemisphere

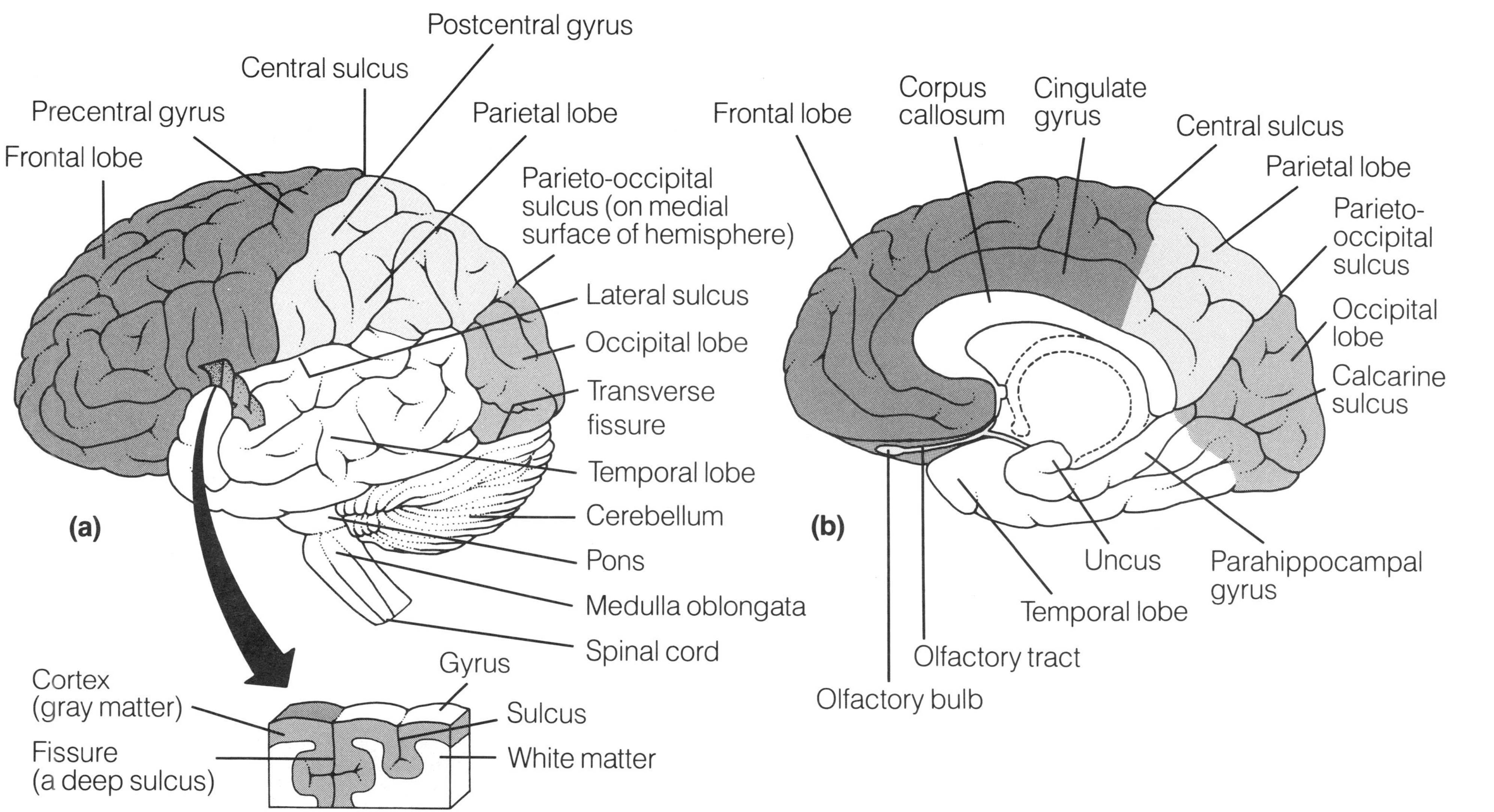

Lobes and fissures of the cerebral hemispheres: Superior view

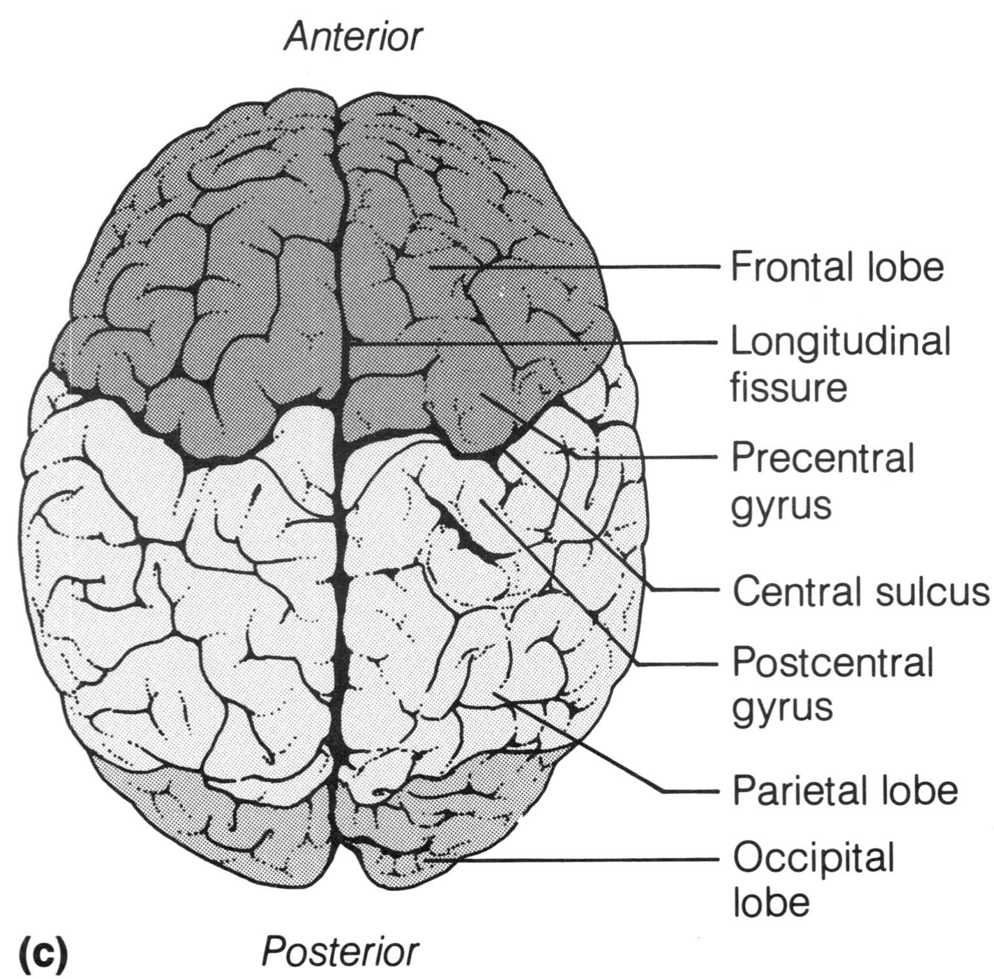

(c)

Cerebral hemispheres: Coronal section

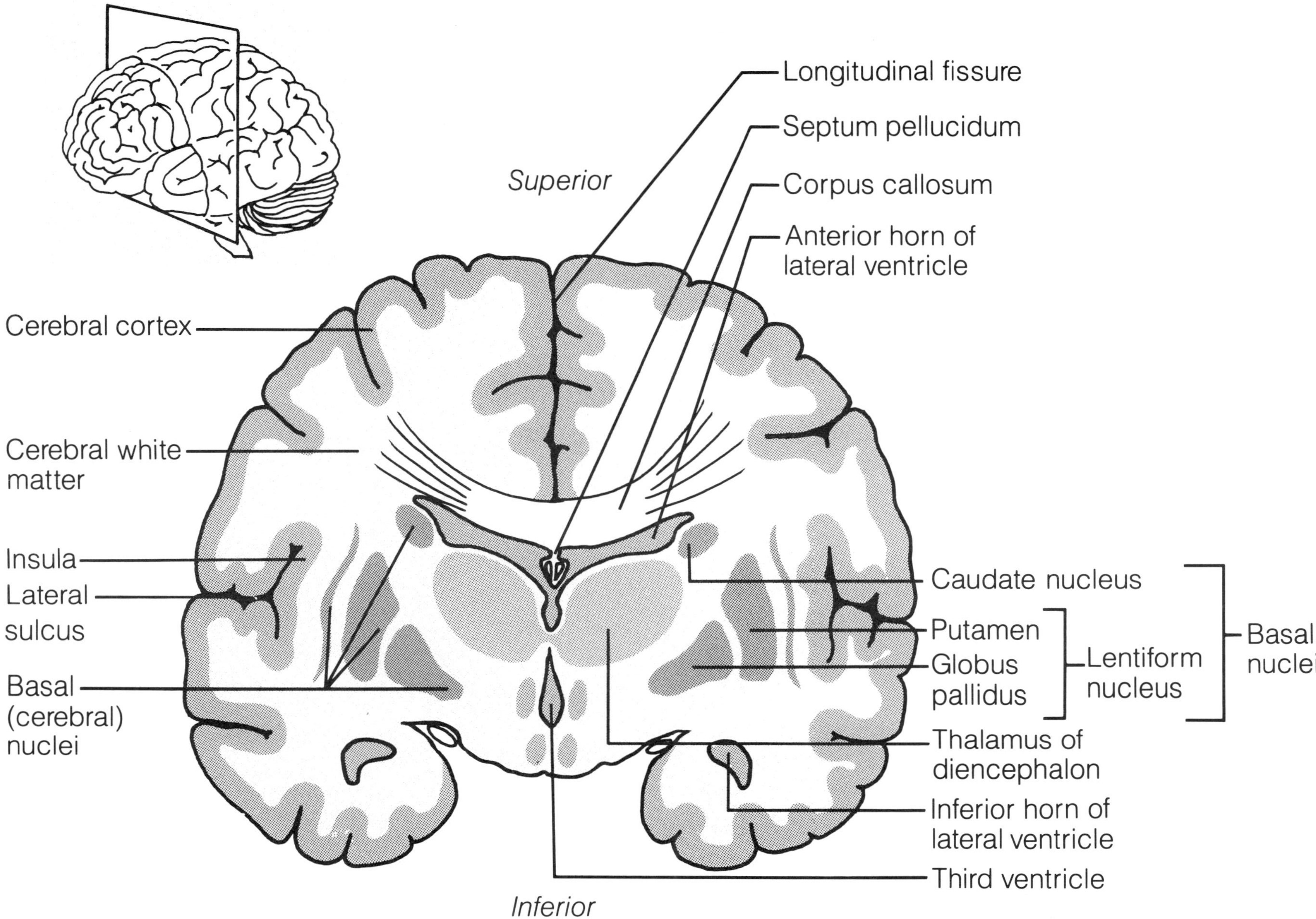

Functional areas of the left cerebral cortex

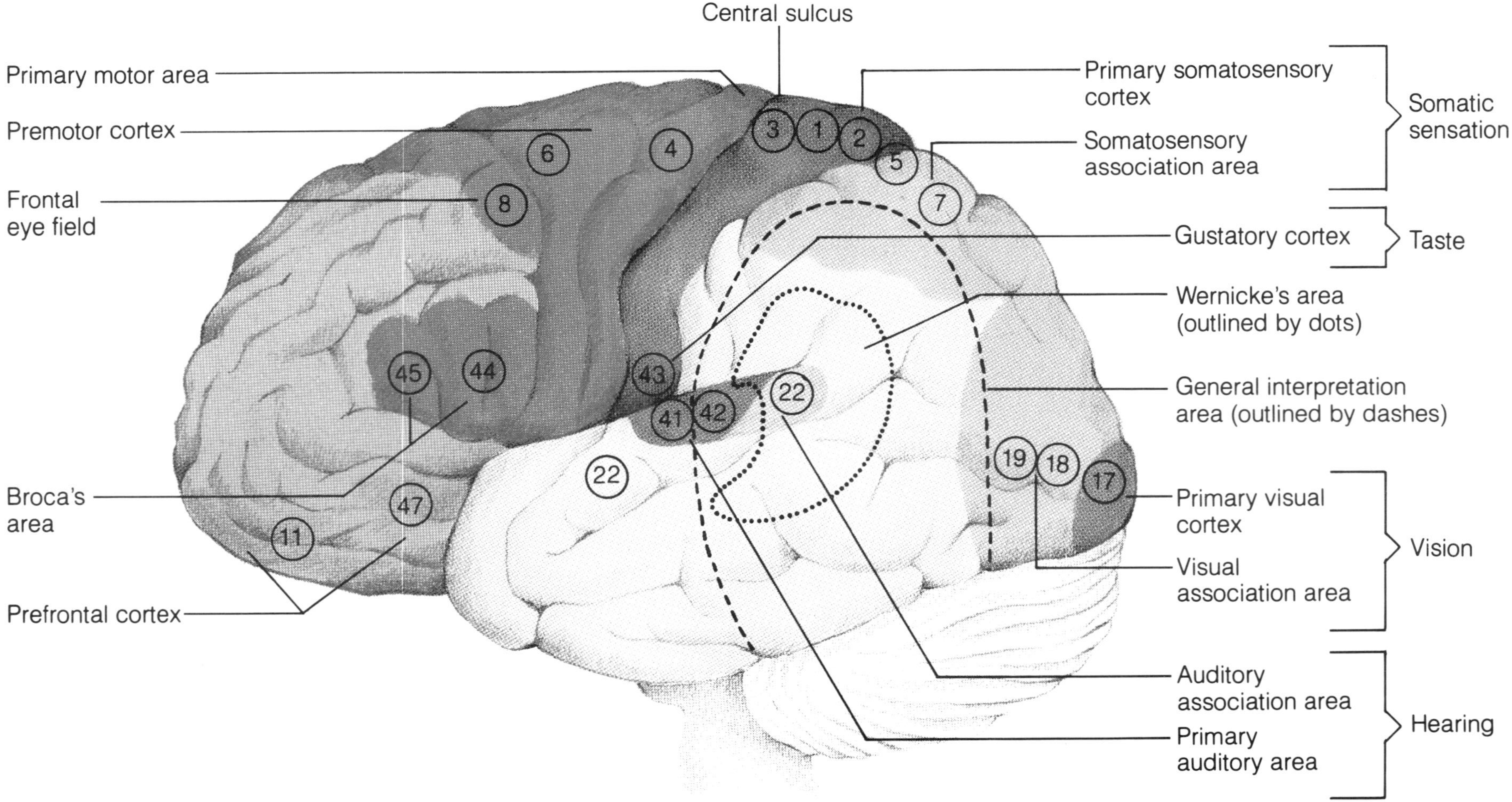

Sensory and motor areas of the cerebral cortex

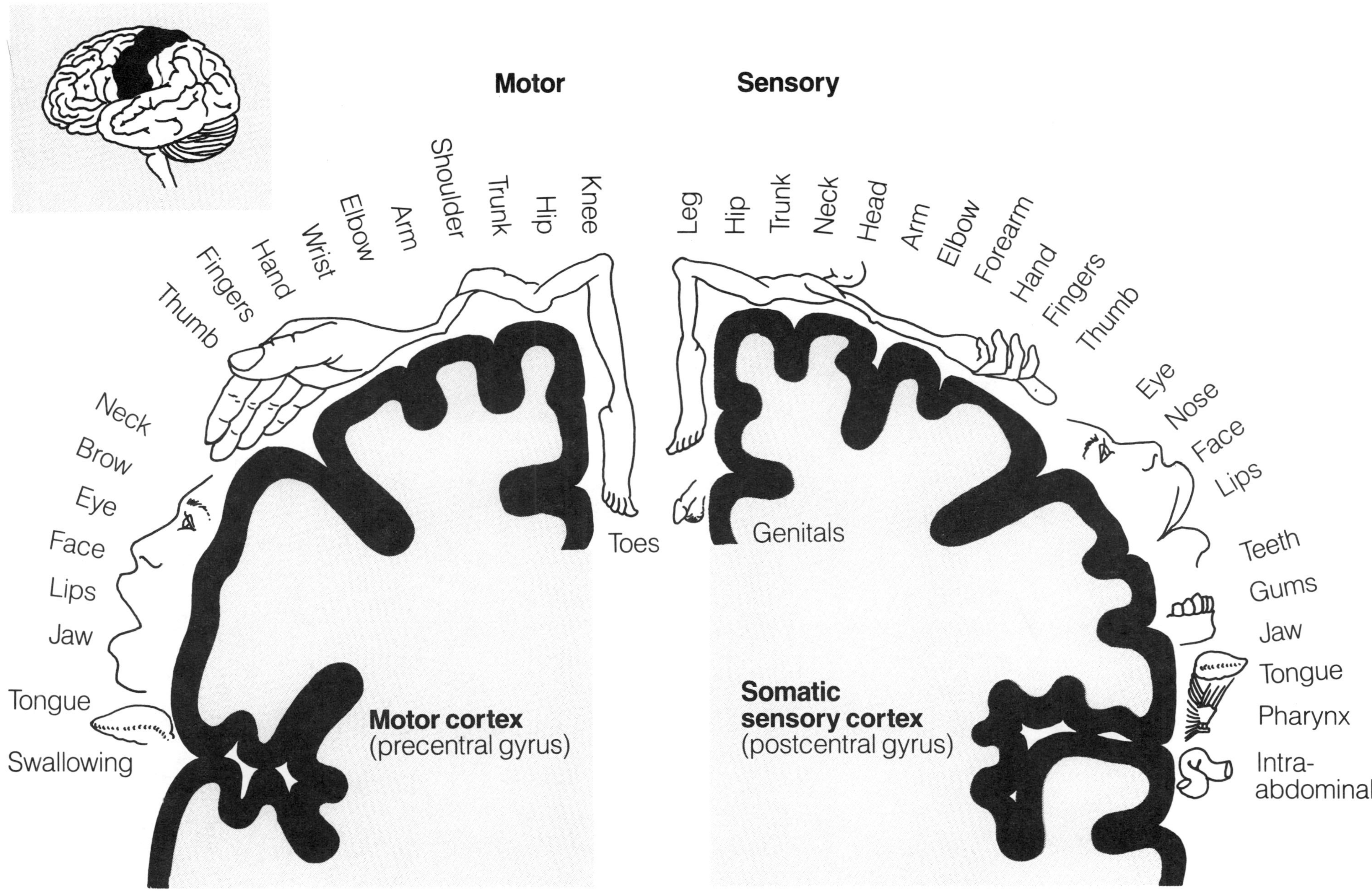

Types of fiber tracts composing the white matter of the cerebral medulla

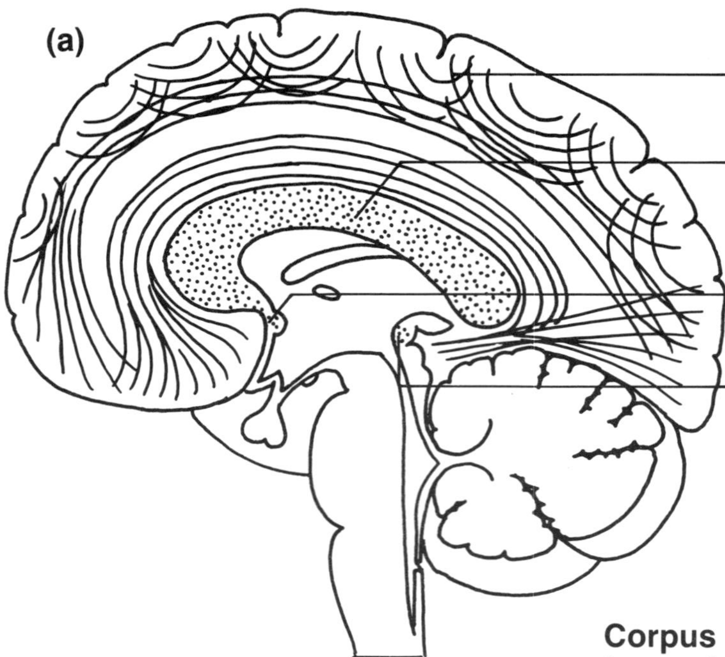

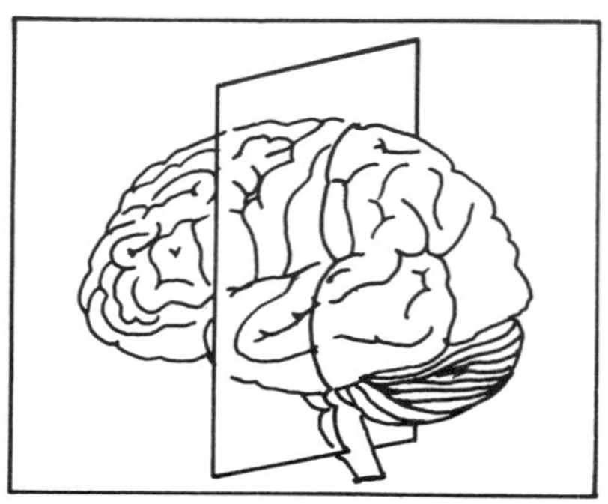

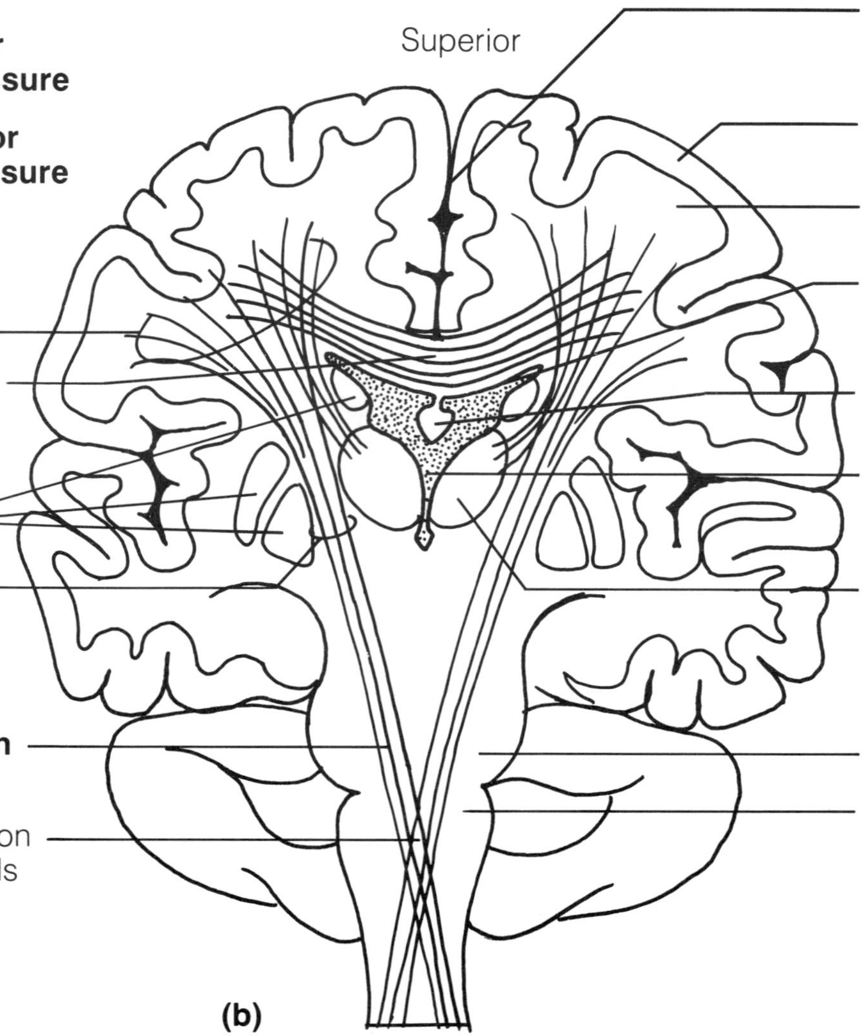

Basal (cerebral) nuclei: (a) 3D view; (b) Transverse section

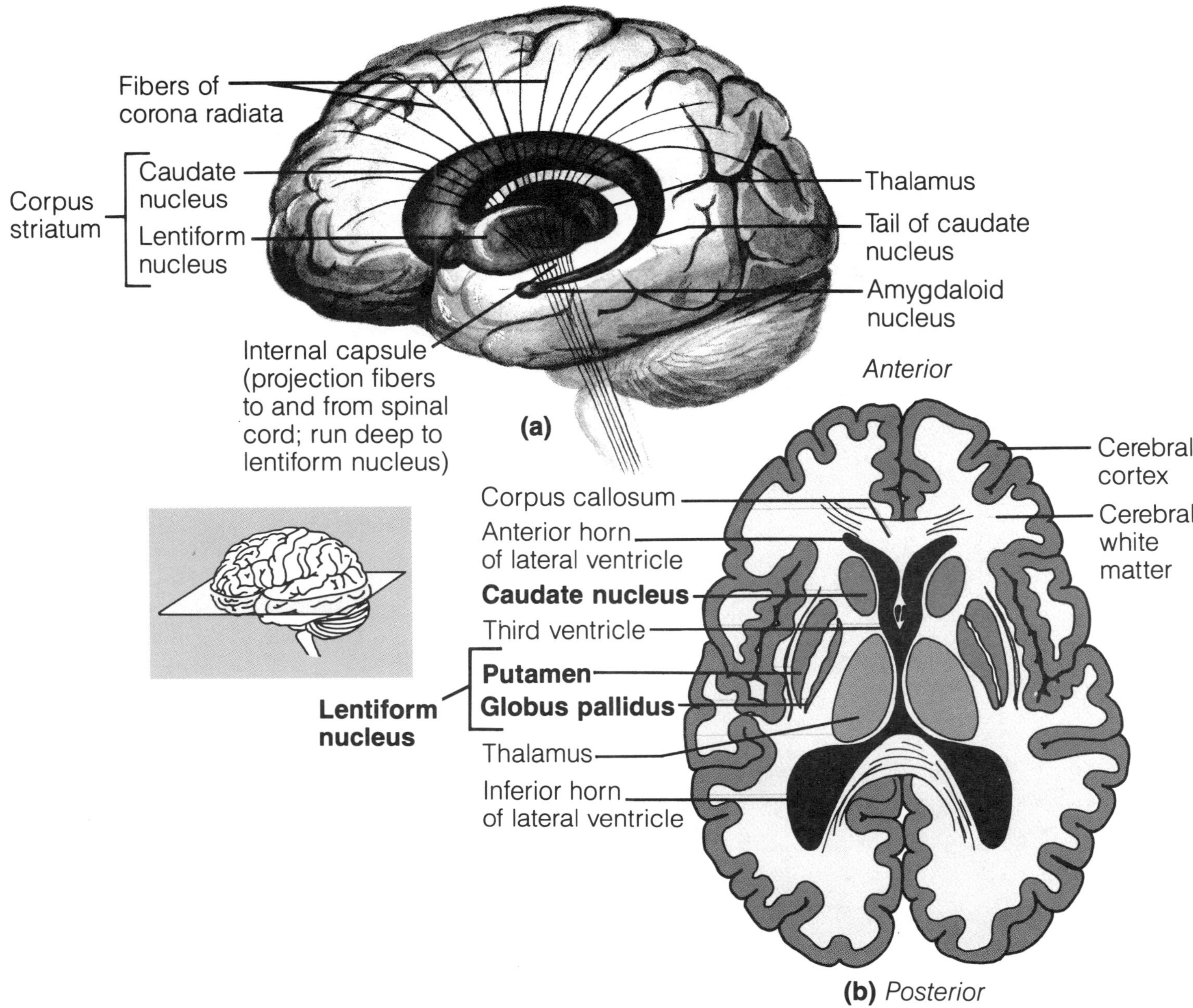

Diencephalon and brain stem structures

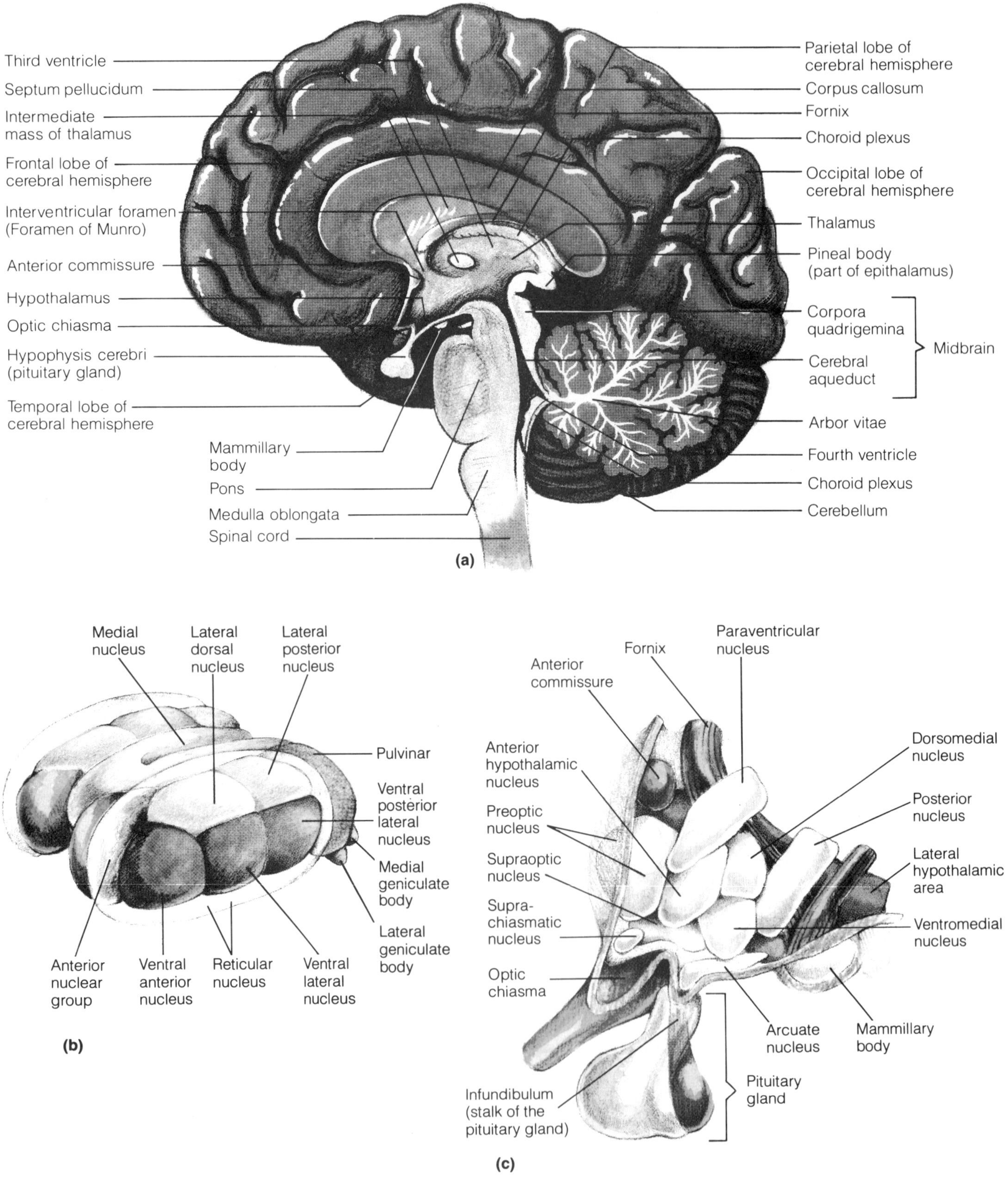

Relationship of the brain stem to the diencephalon and basal nuclei

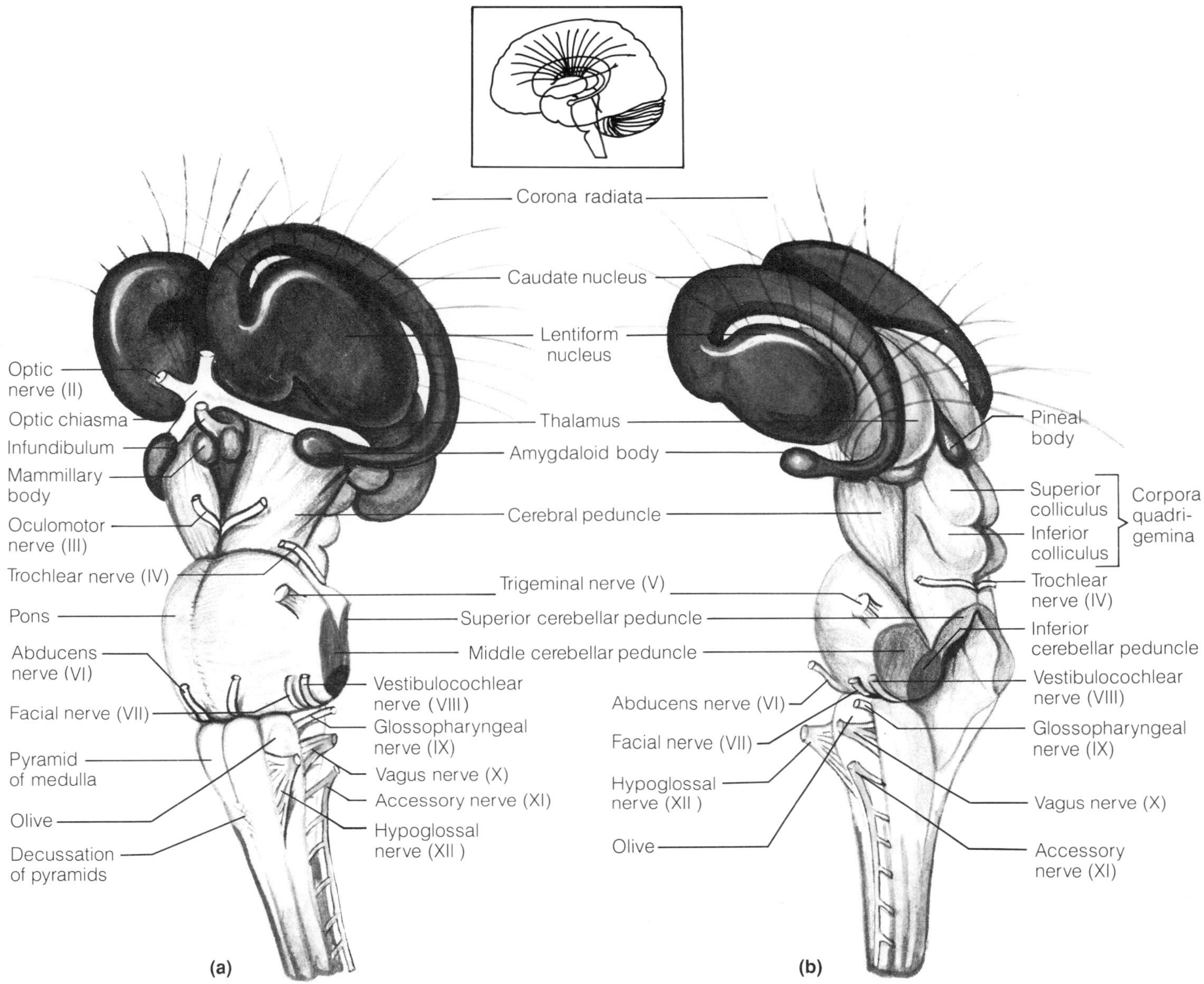

Cerebellum

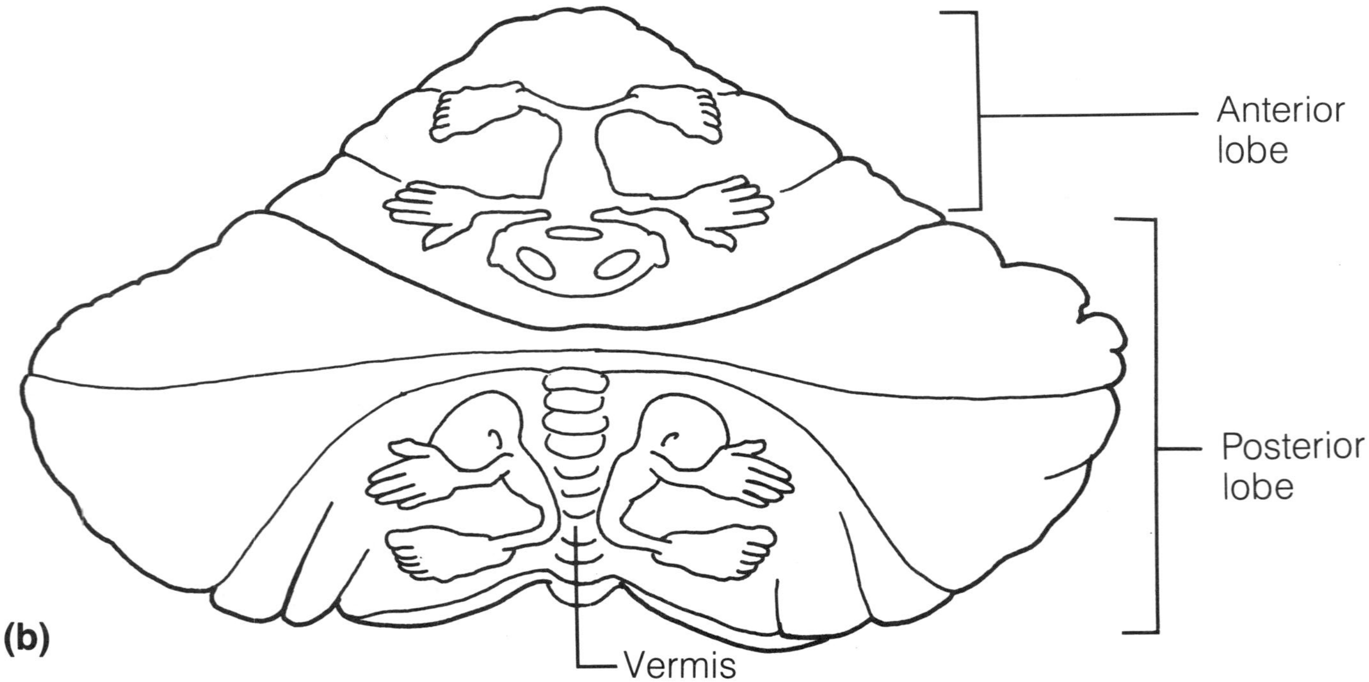

(b)

The limbic system

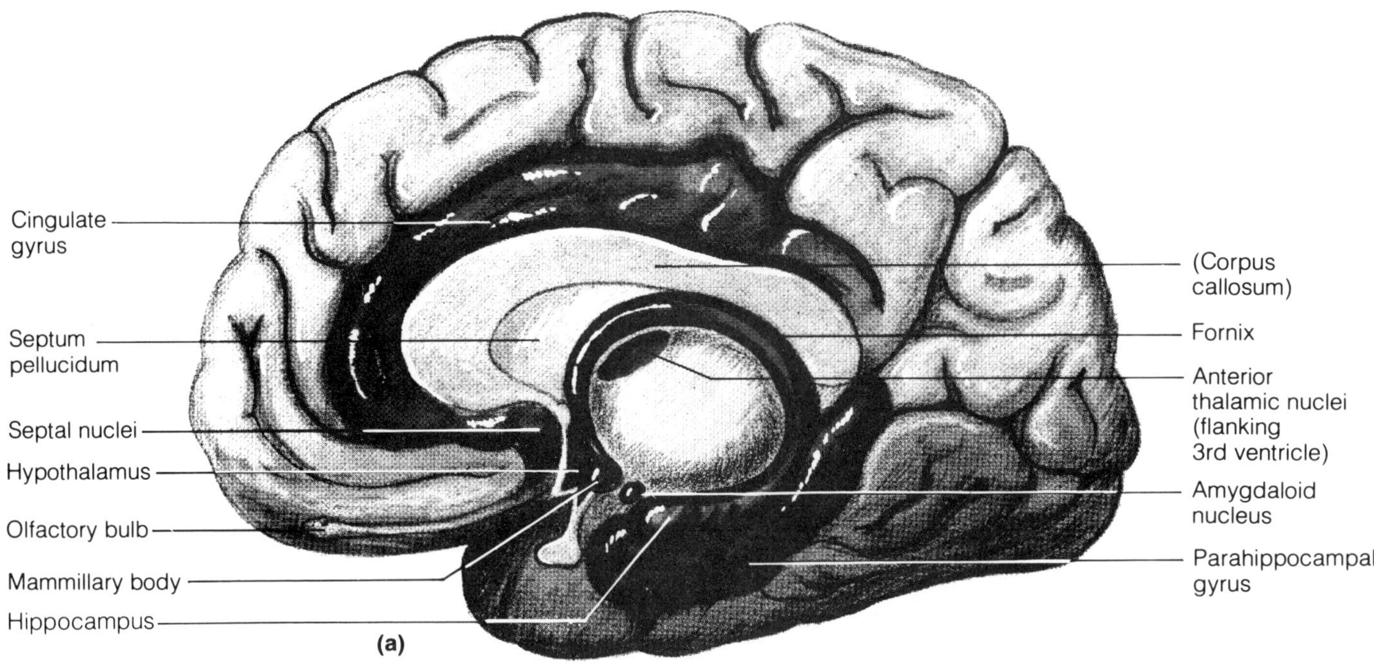

(a)

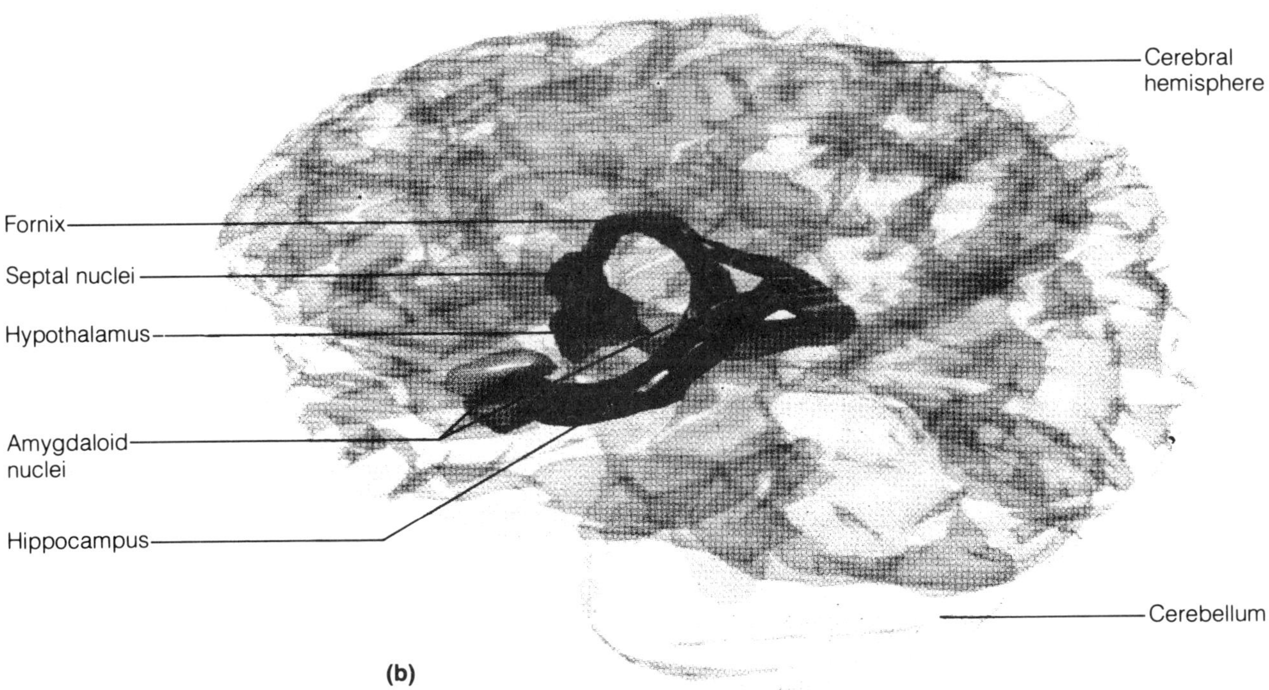

(b)

The reticular formation

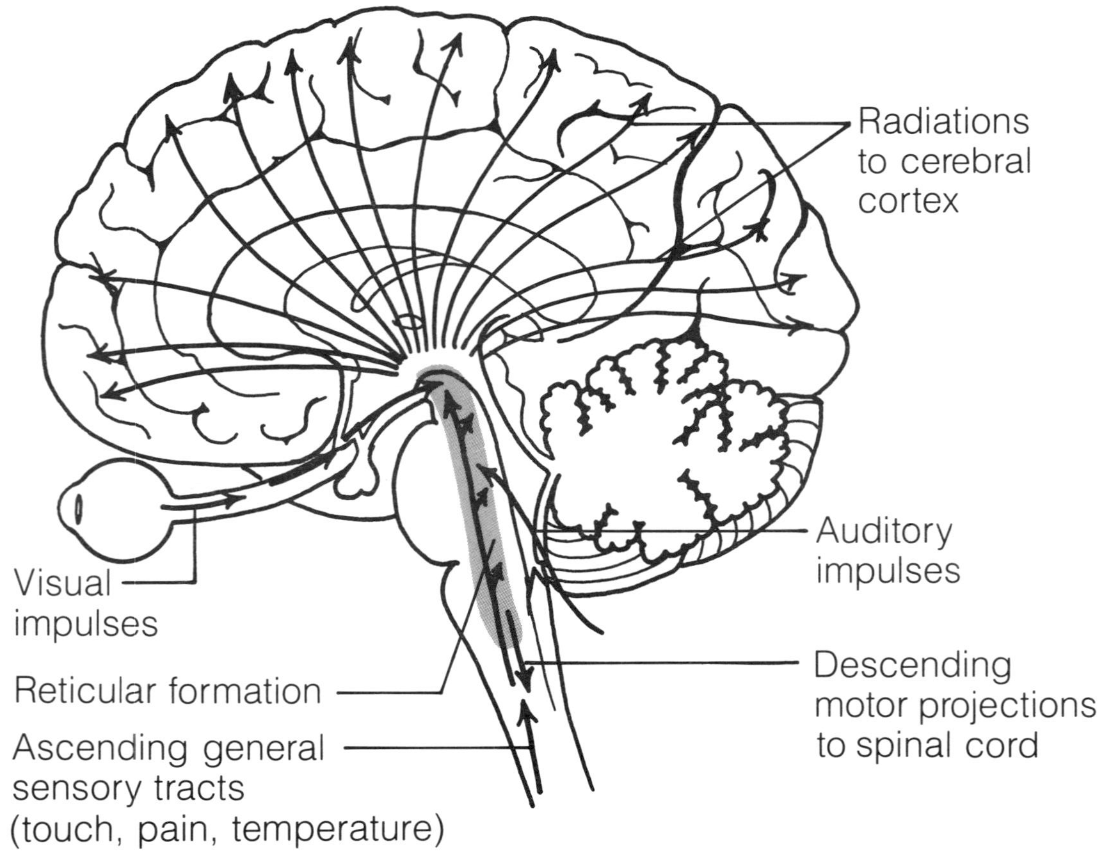

Protection of the brain

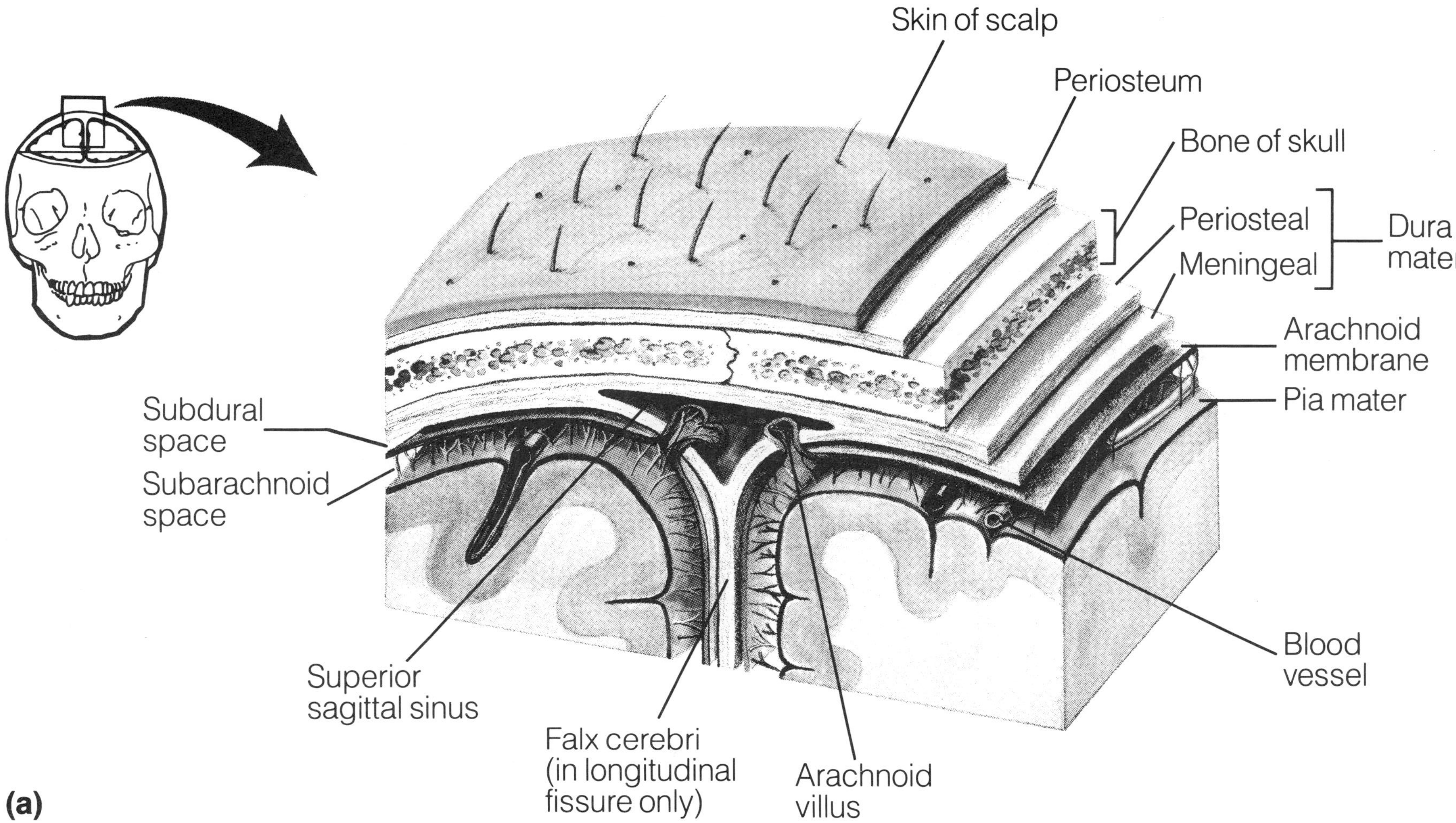

(a)

Location and circulatory pattern of cerebrospinal fluid

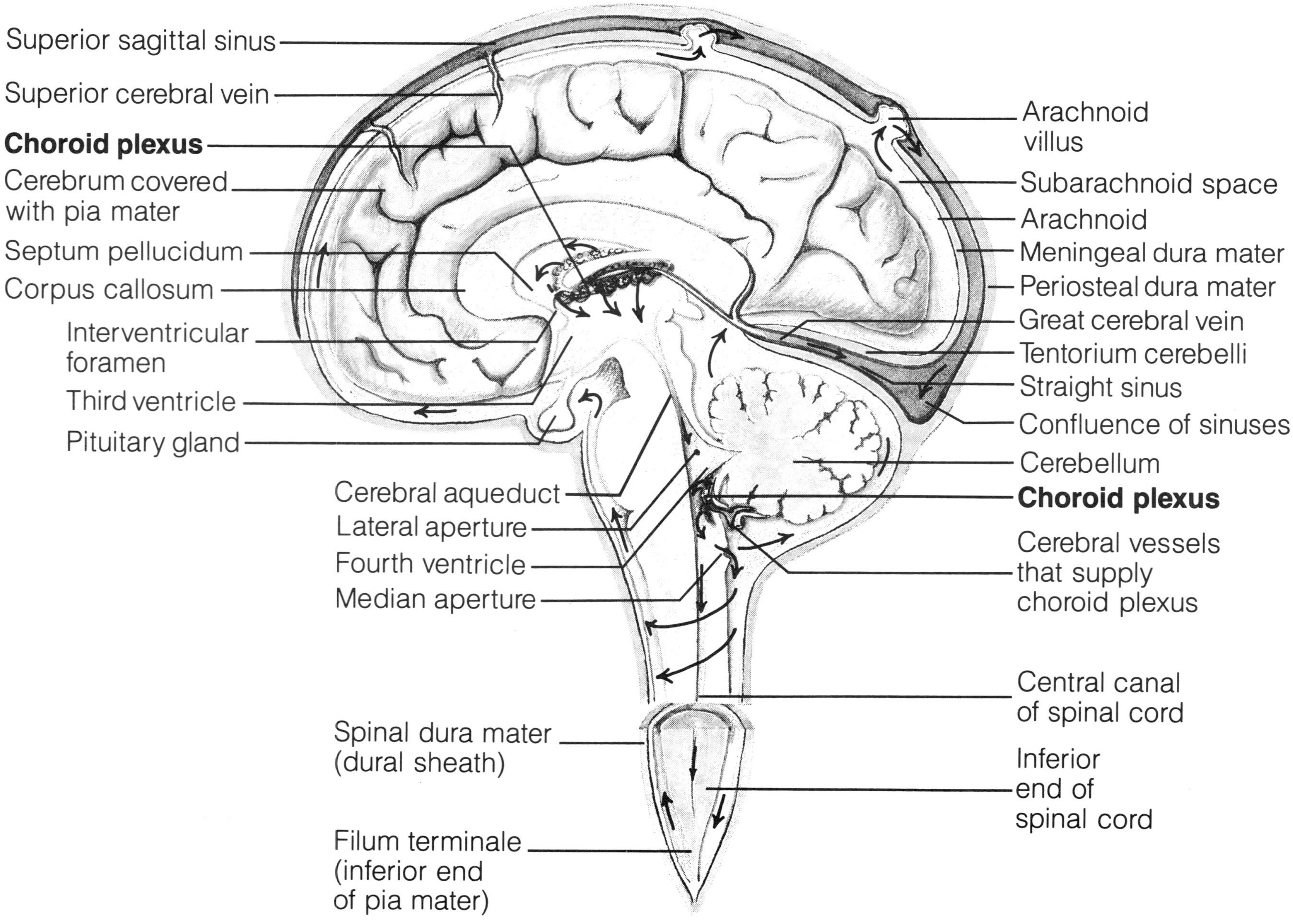

Structure of the spinal cord

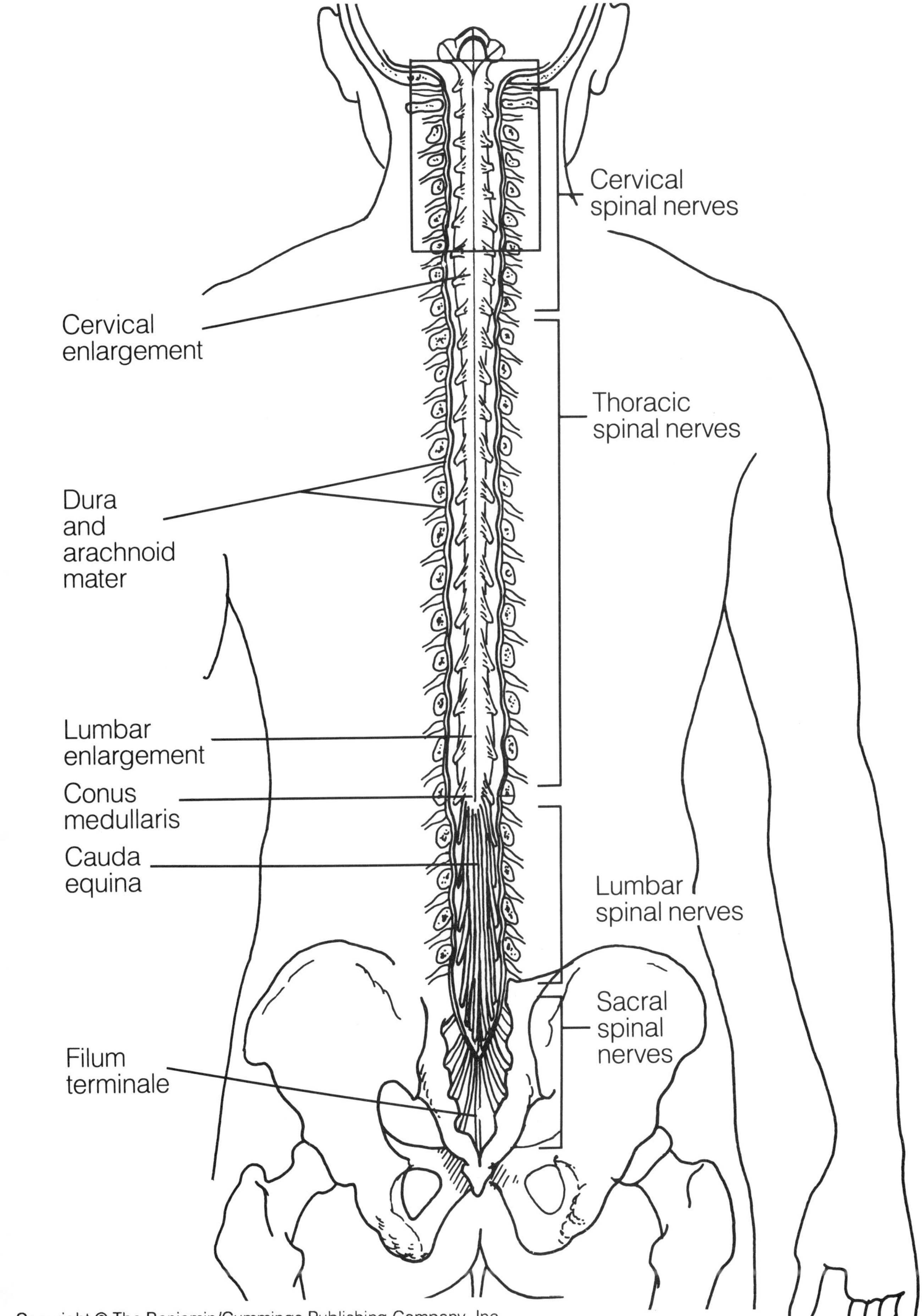

Major ascending (sensory) and descending (motor) tracts of the spinal cord, cross-sectional view

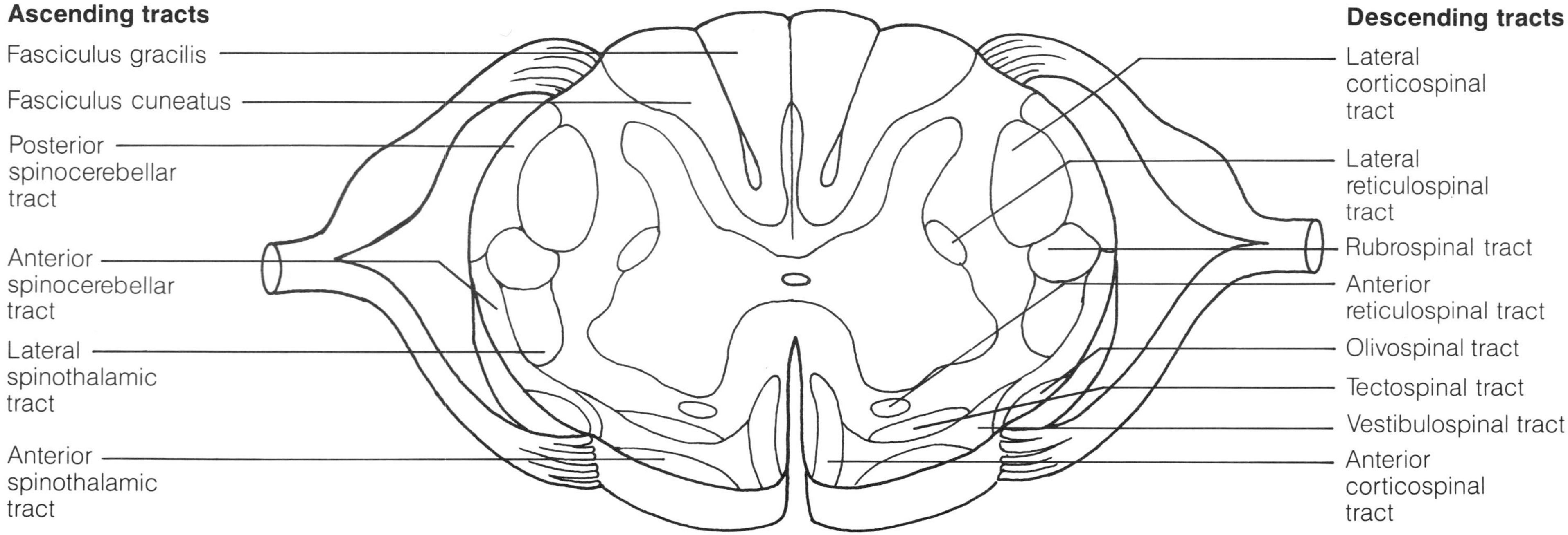

Pathways of selected ascending spinal cord tracts

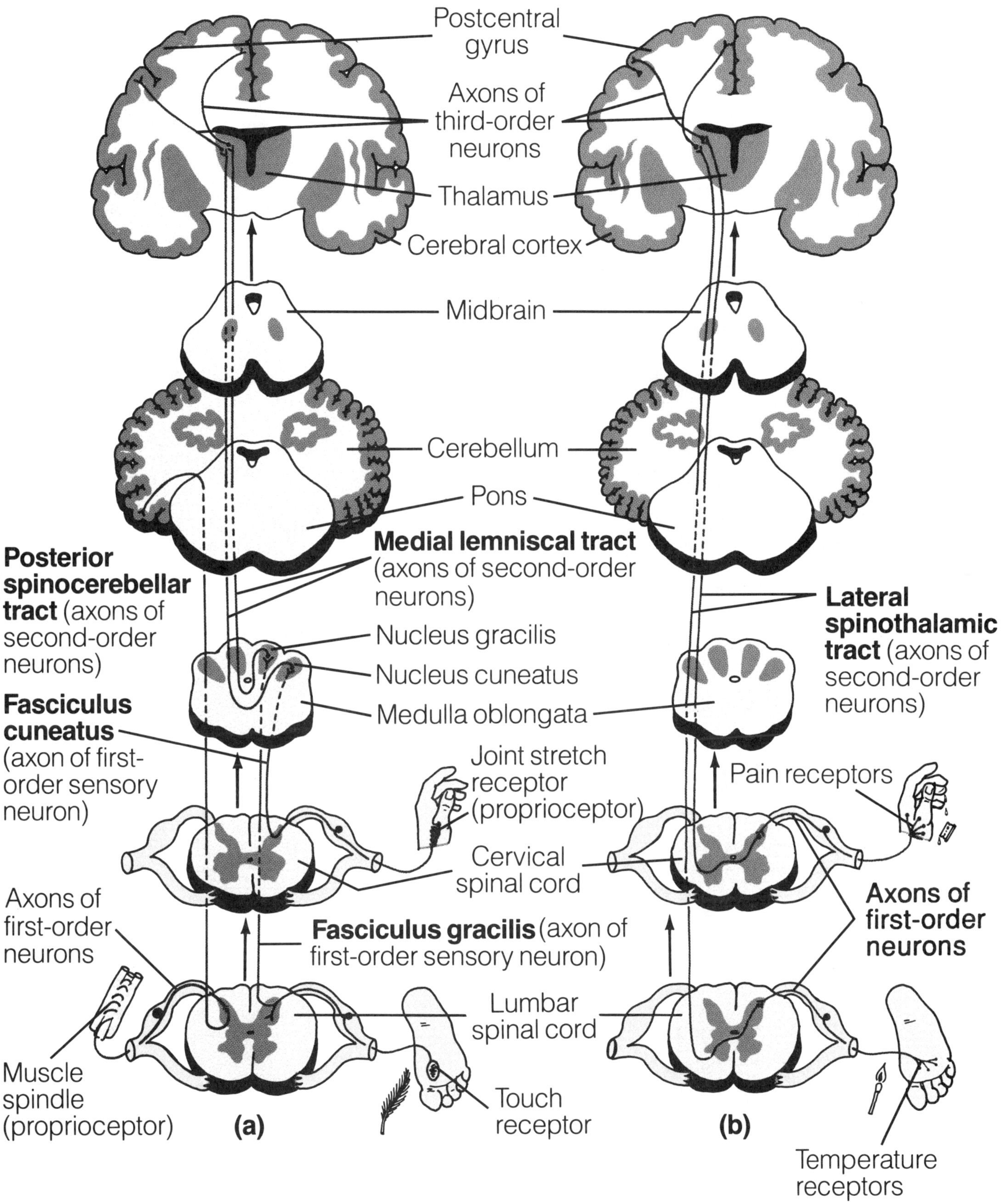

Pathways of selected descending spinal cord tracts

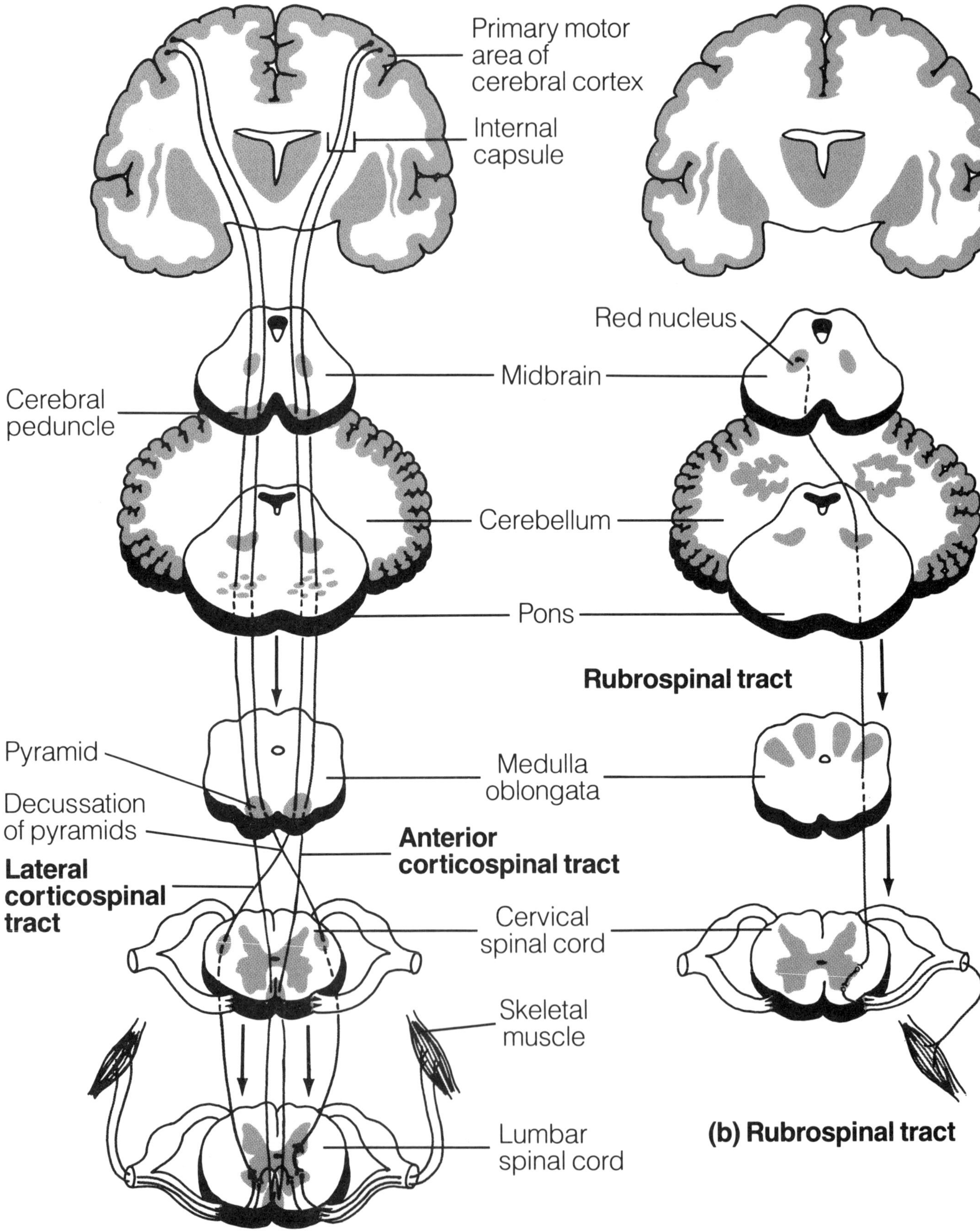

(a) Lateral and anterior corticospinal tracts

(b) Rubrospinal tract

Events leading to regeneration of nerve fibers in the peripheral nerves

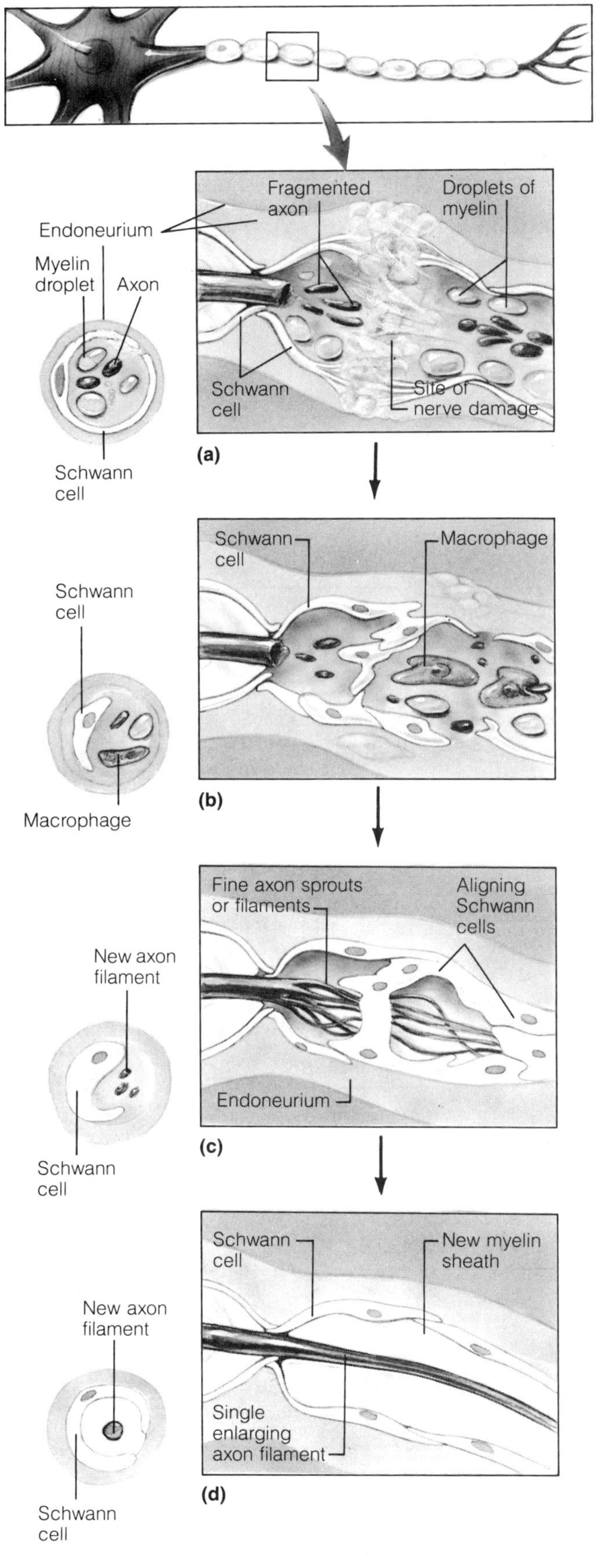

The cervical plexus

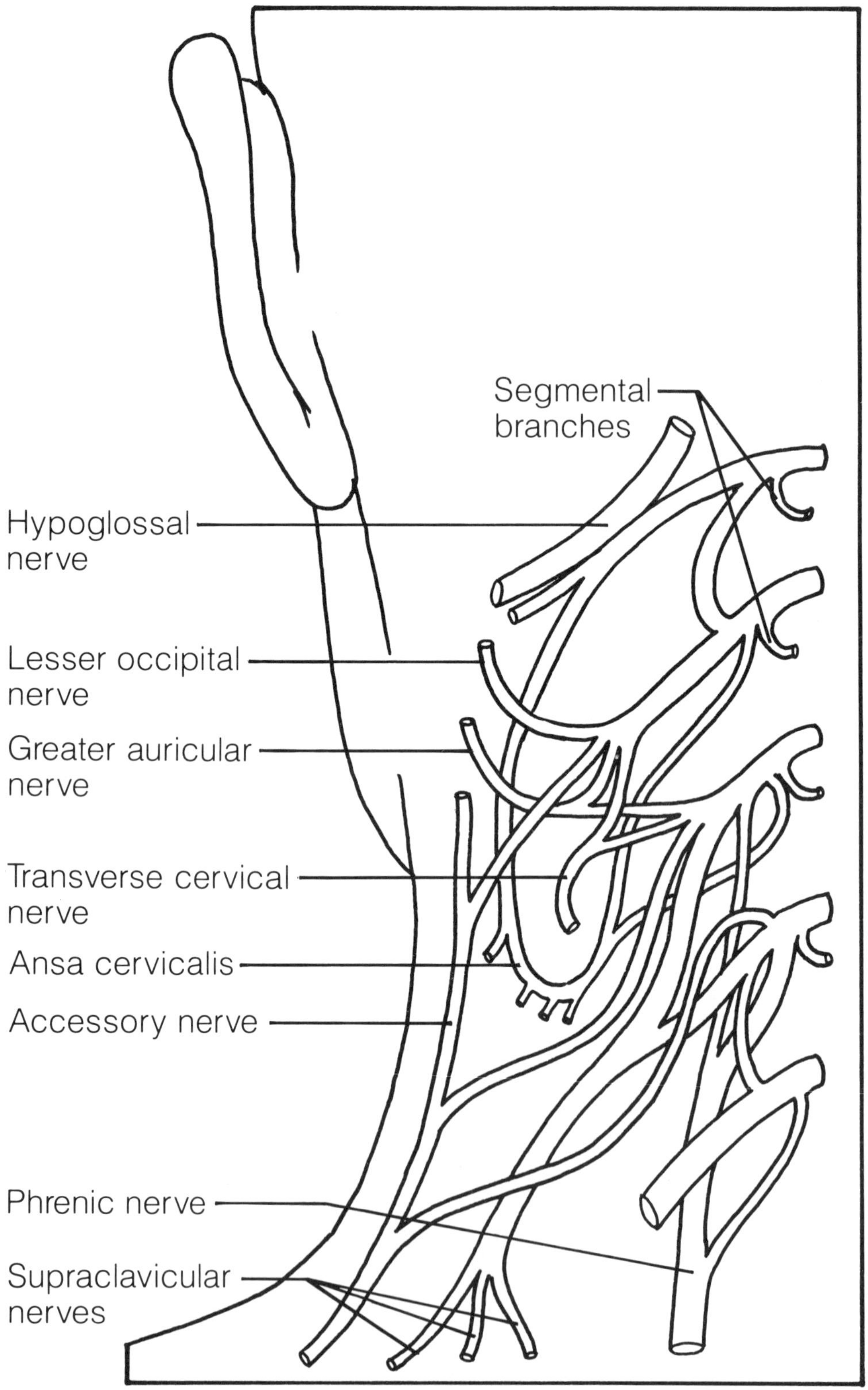

Dermatomes

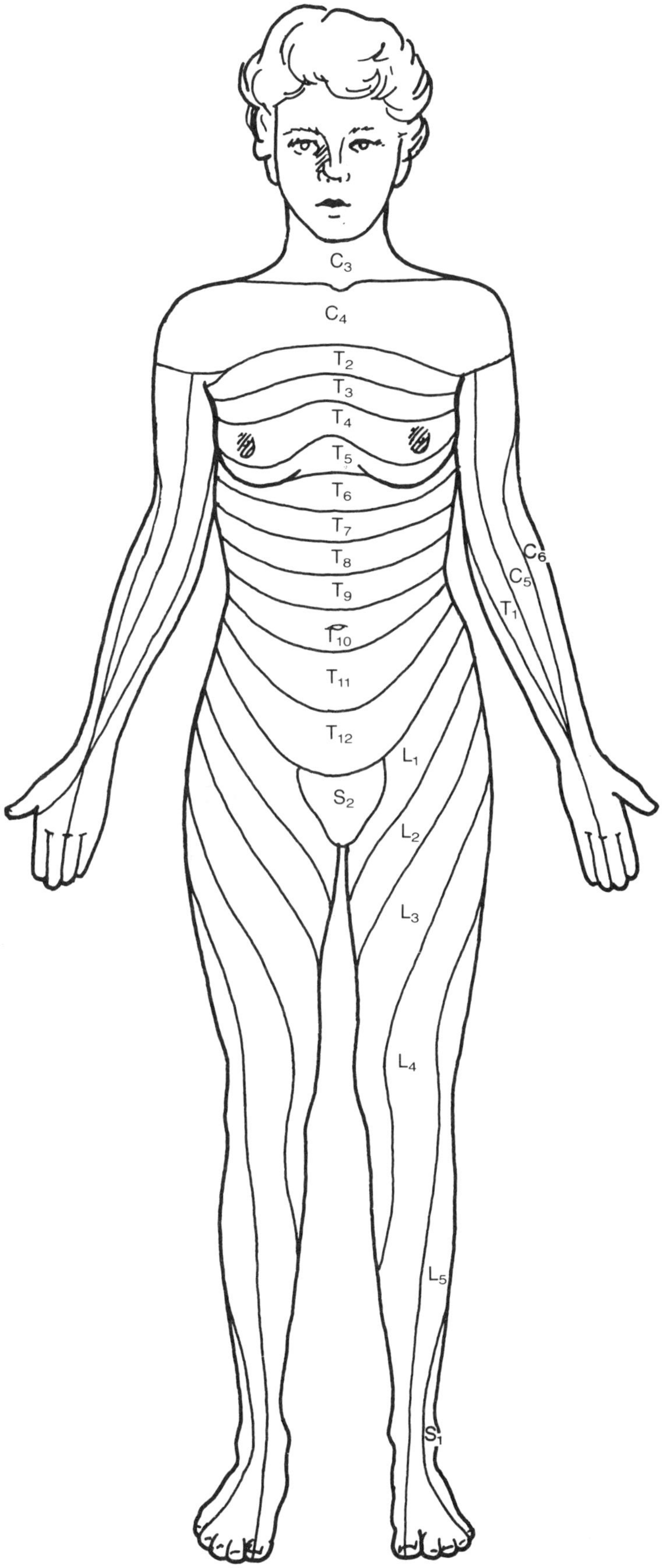
C3
C4
T2
T3
T4
T5
T6
T7
T8
T9
T10
T11
T12
C6
C5
T1
L1
S2
L2
L3
L4
L5
S1

The basic components of all human reflex arcs—receptor, sensory neuron, integration center (one or more synapses in the CNS), motor neuron, and effector

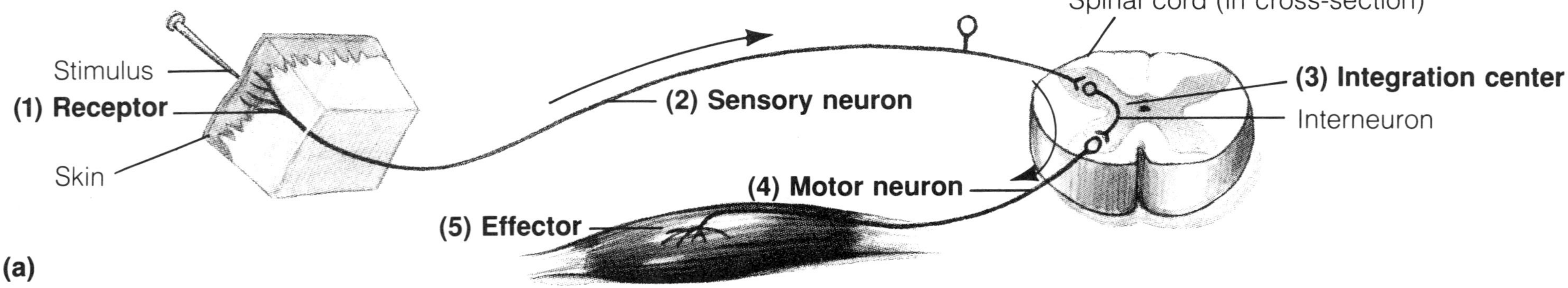

(a)

Anatomy of the muscle spindle and Golgi tendon organ

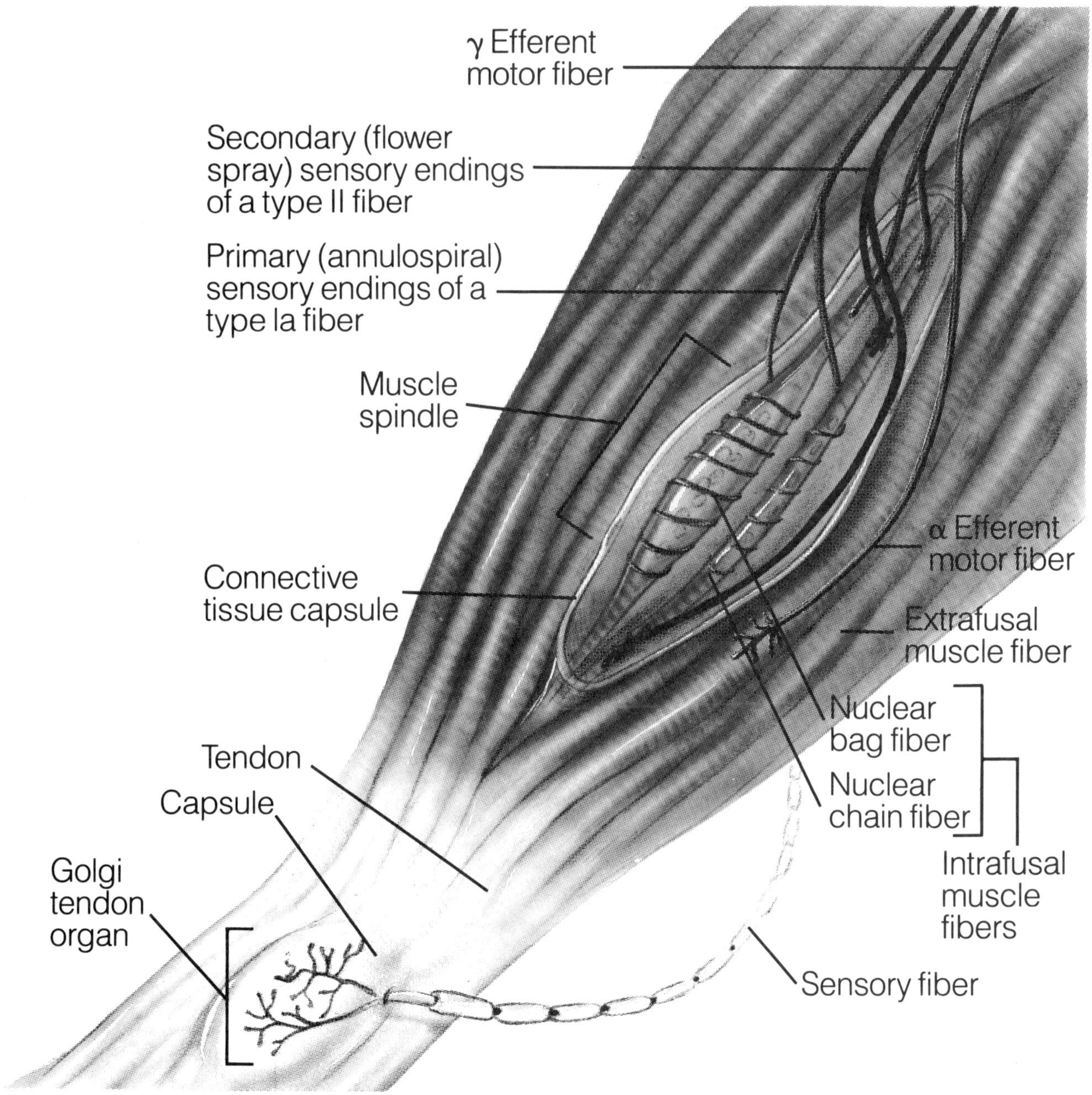

Events of the stretch reflex

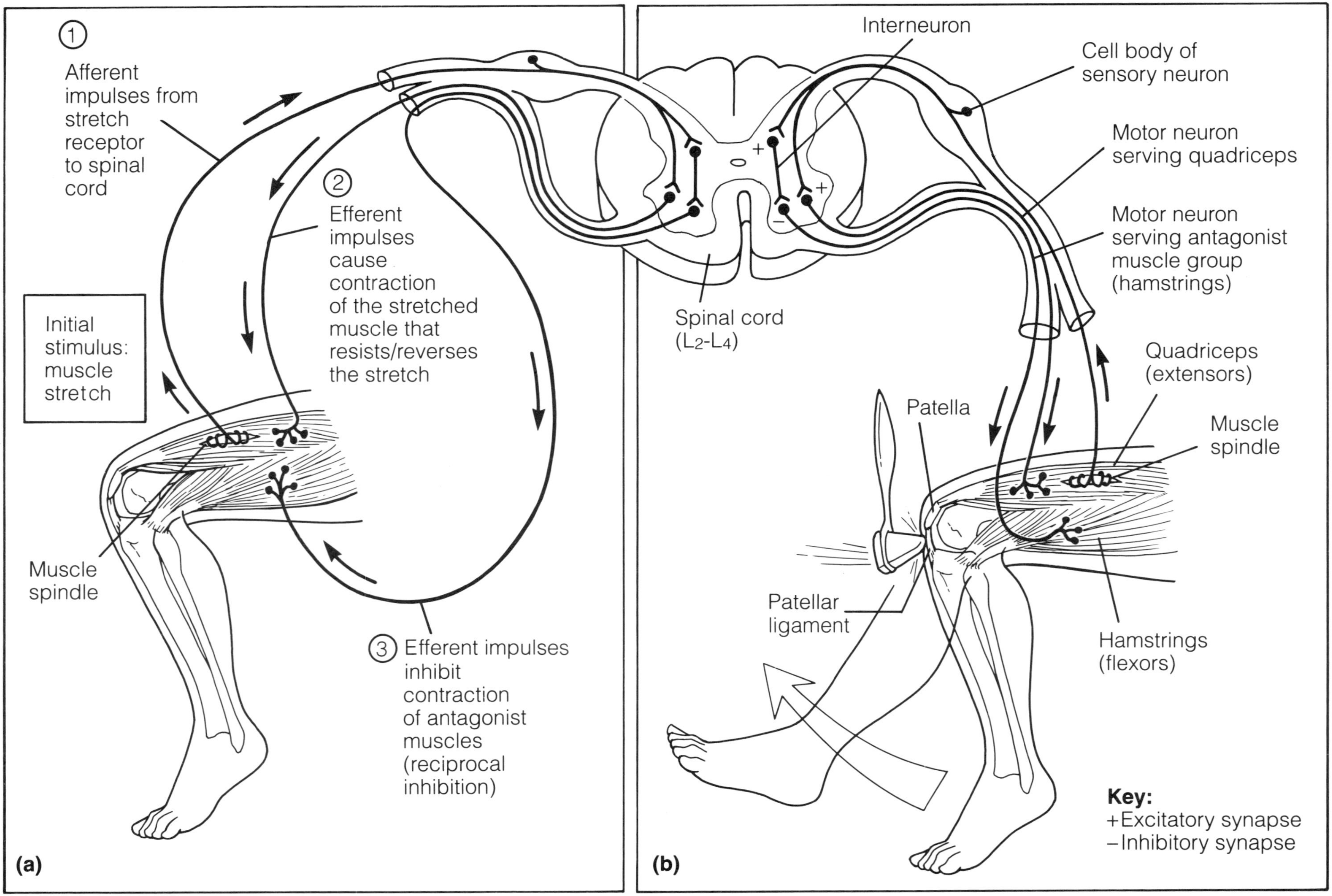

The deep tendon reflex

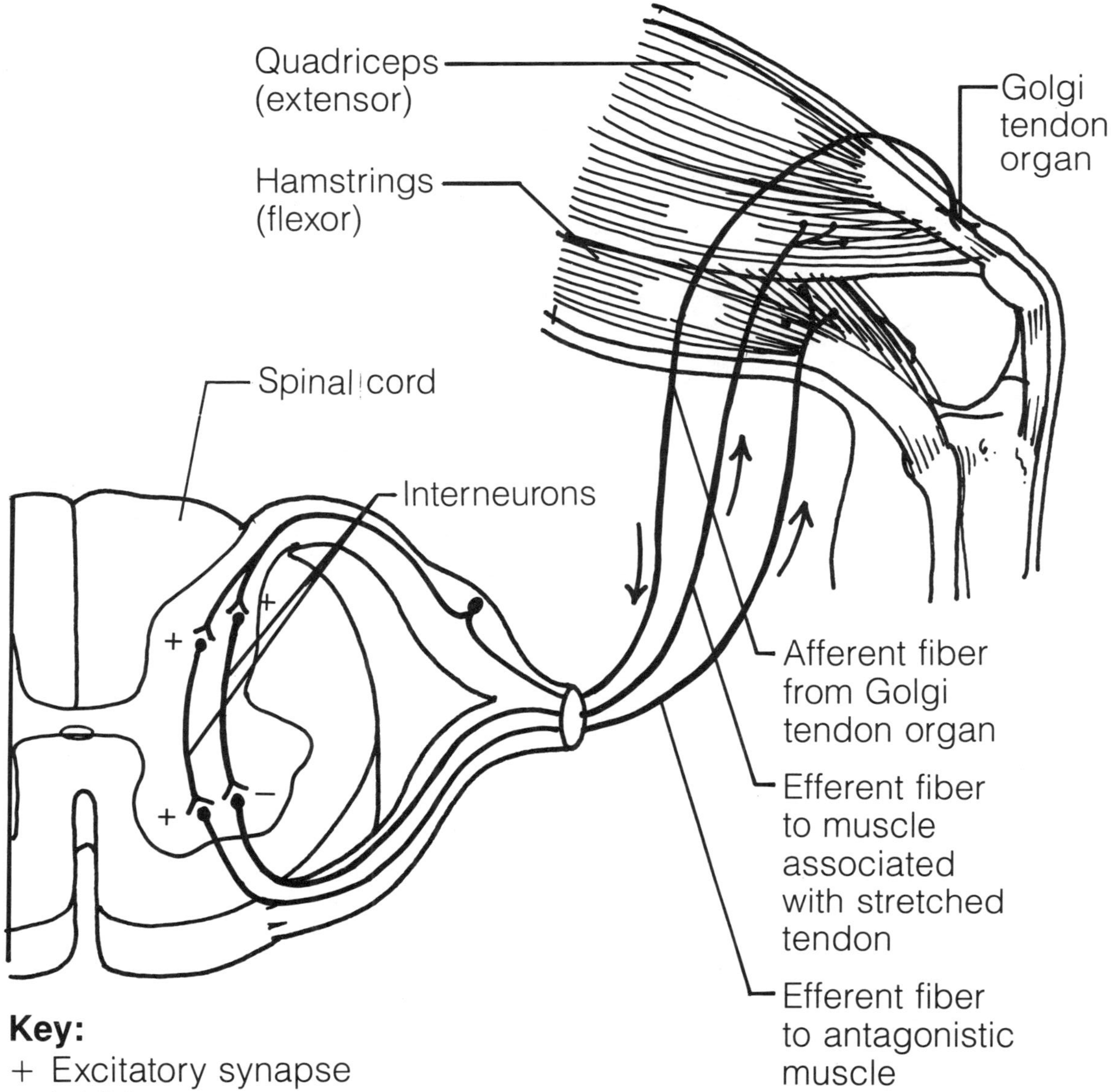

Key:
+ Excitatory synapse
– Inhibitory synapse

The crossed extensor reflex

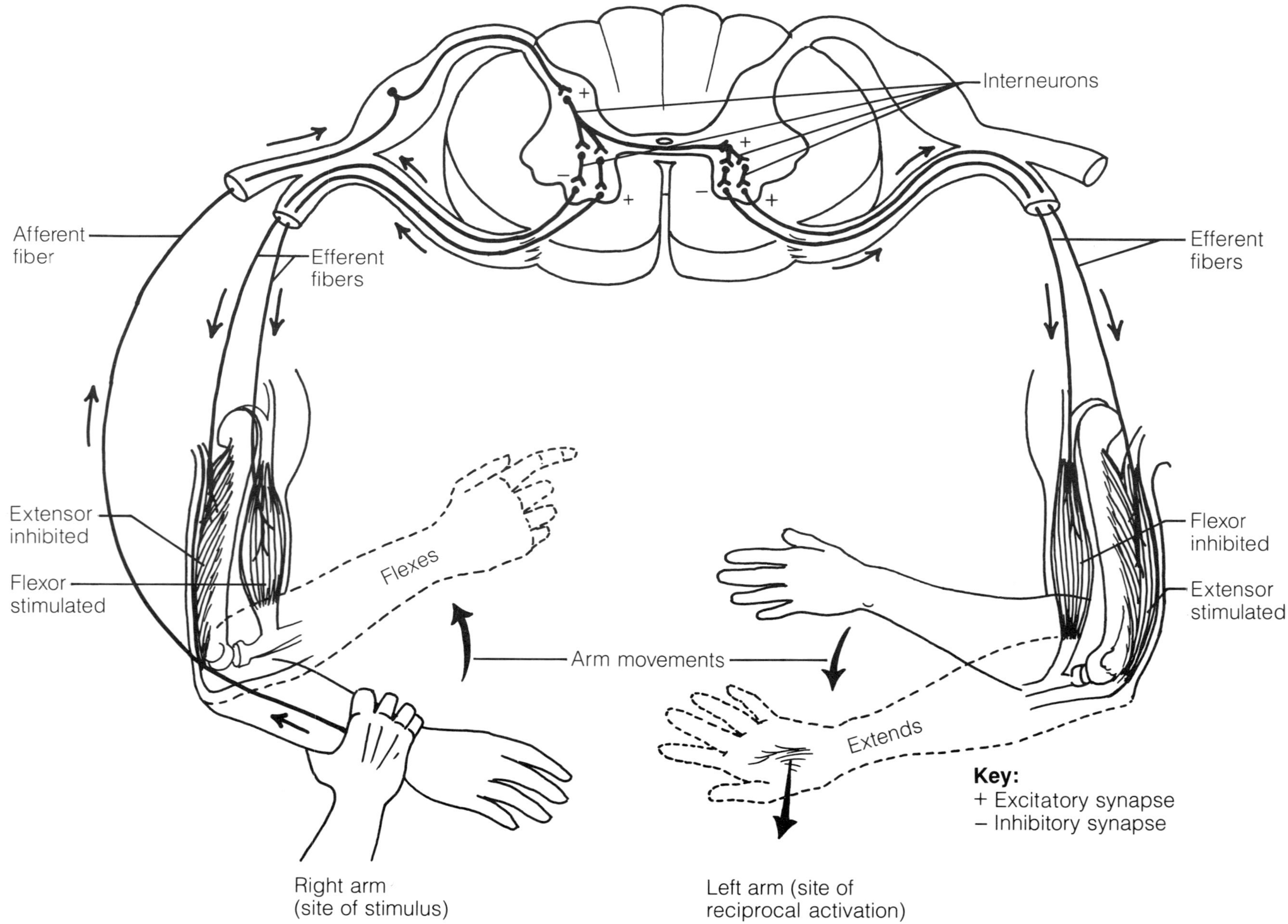

Comparison of the somatic and autonomic nervous systems

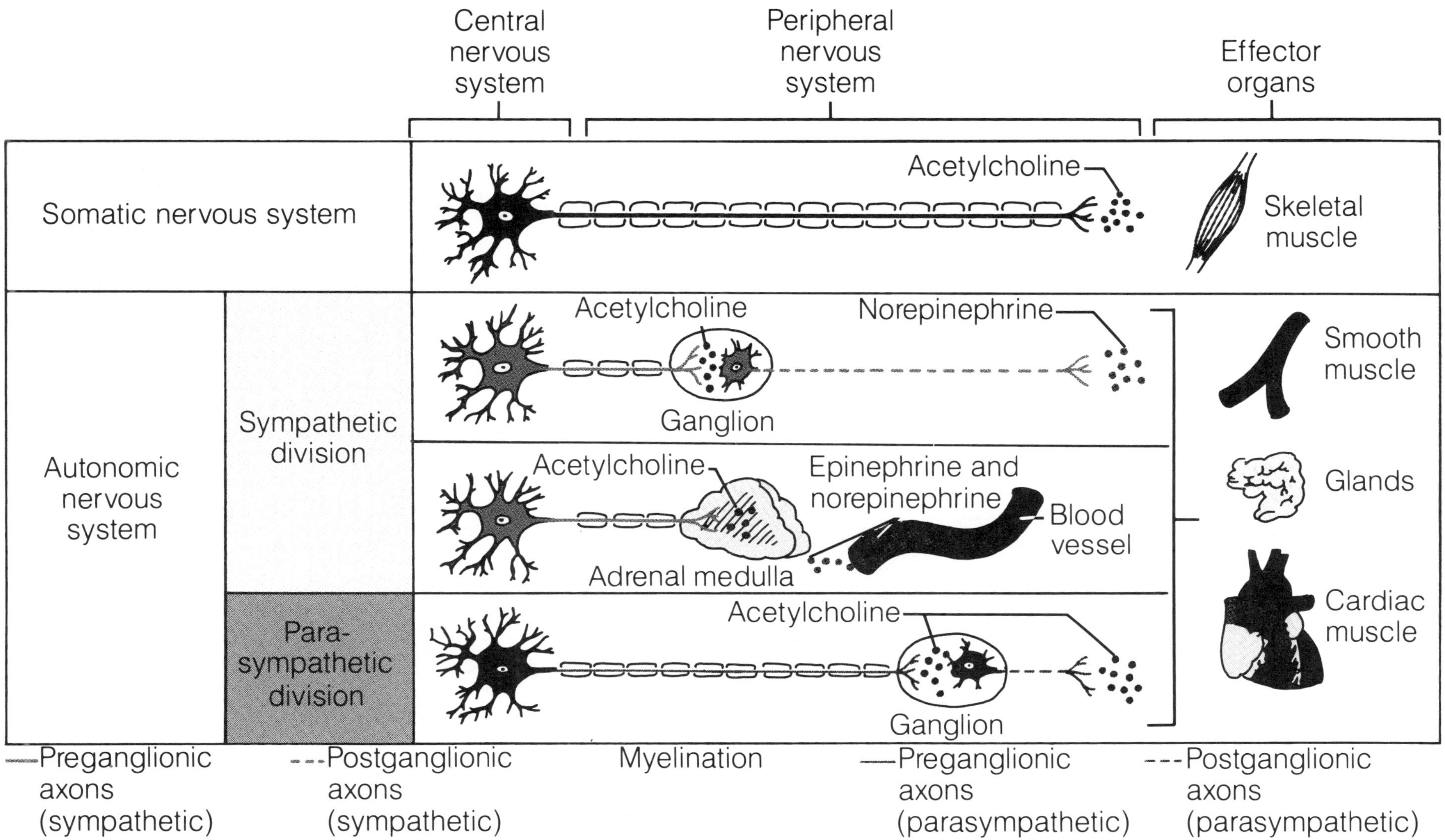

The autonomic nervous system

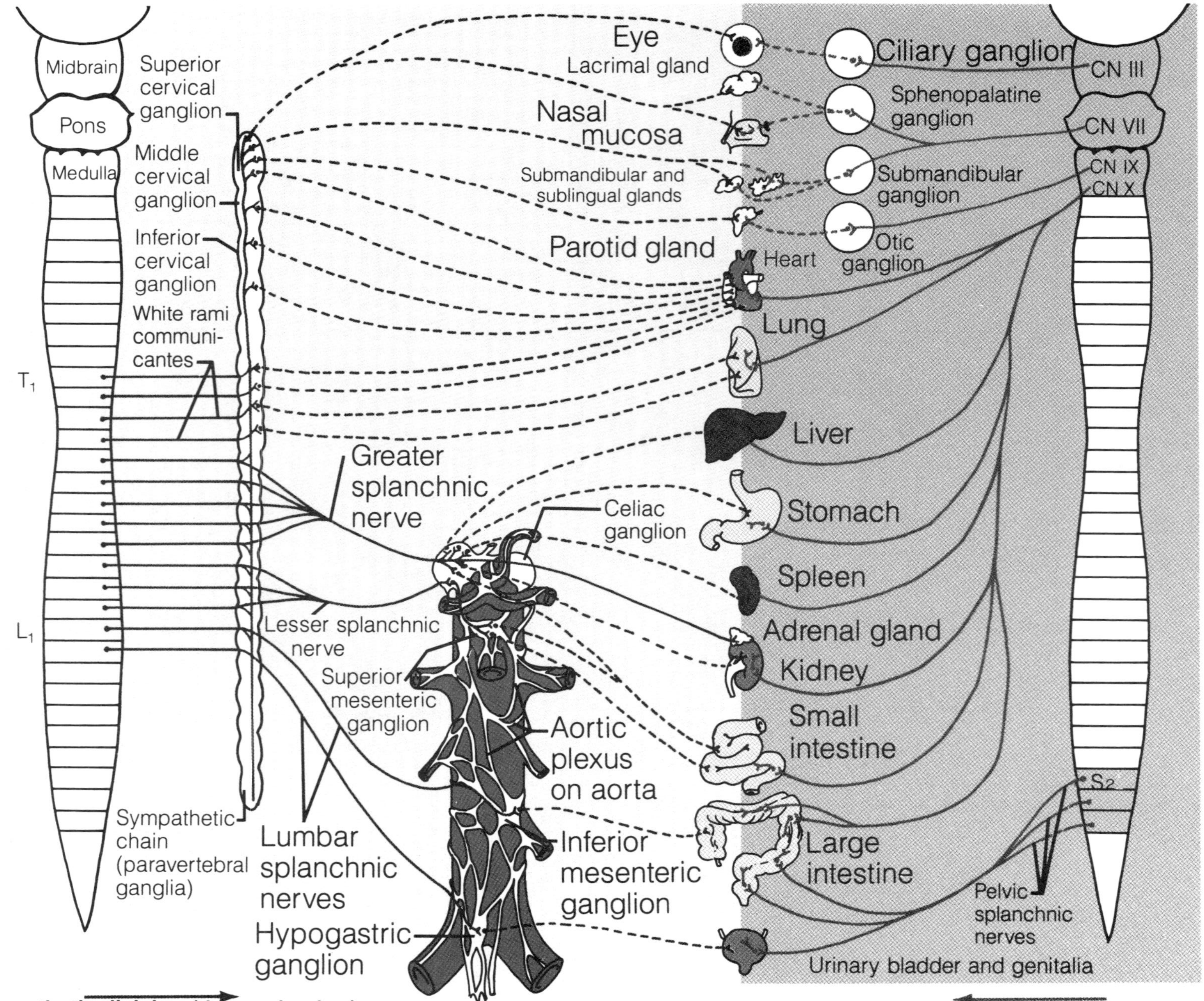

Sympathetic pathways

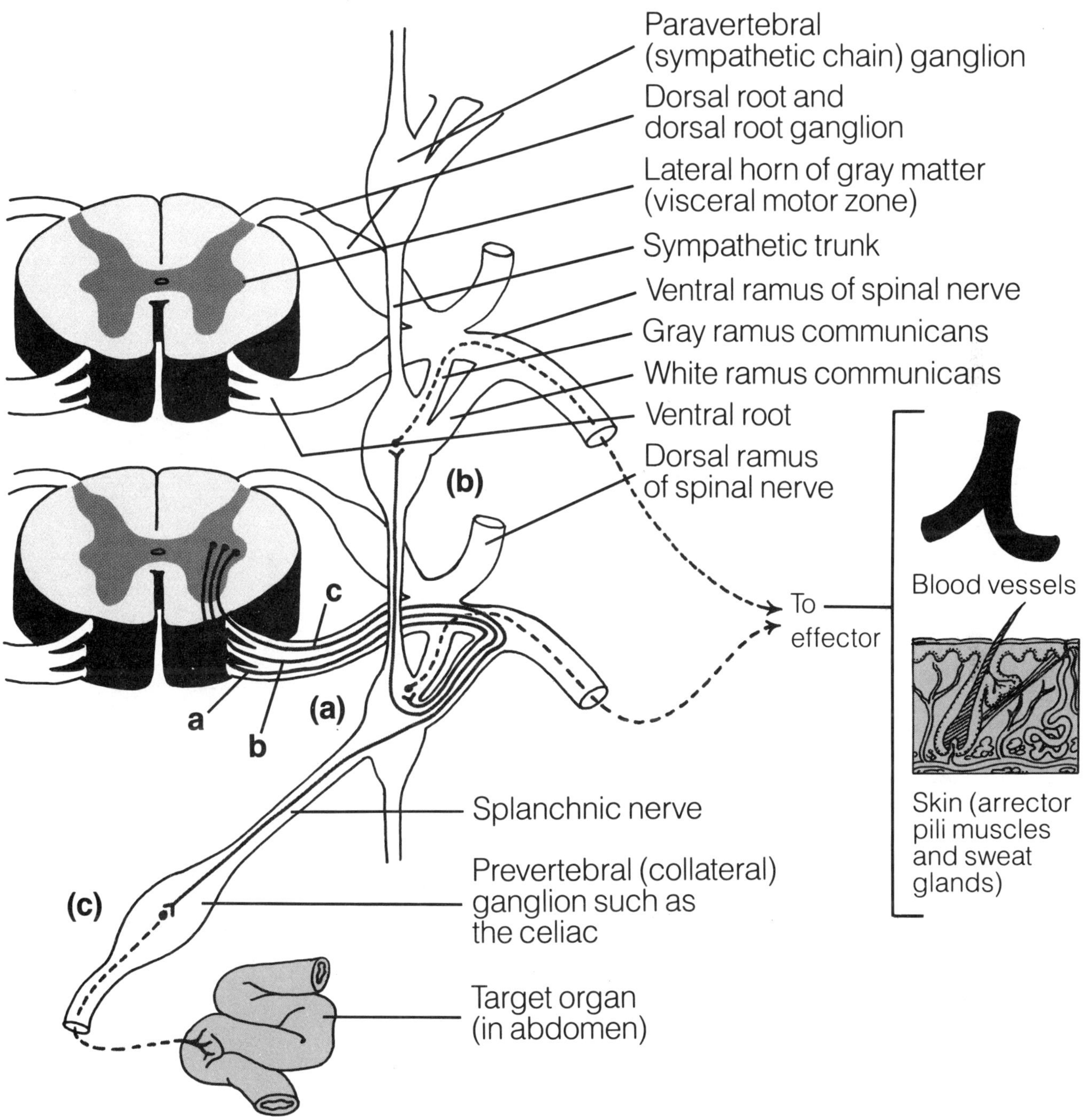

Referred pain

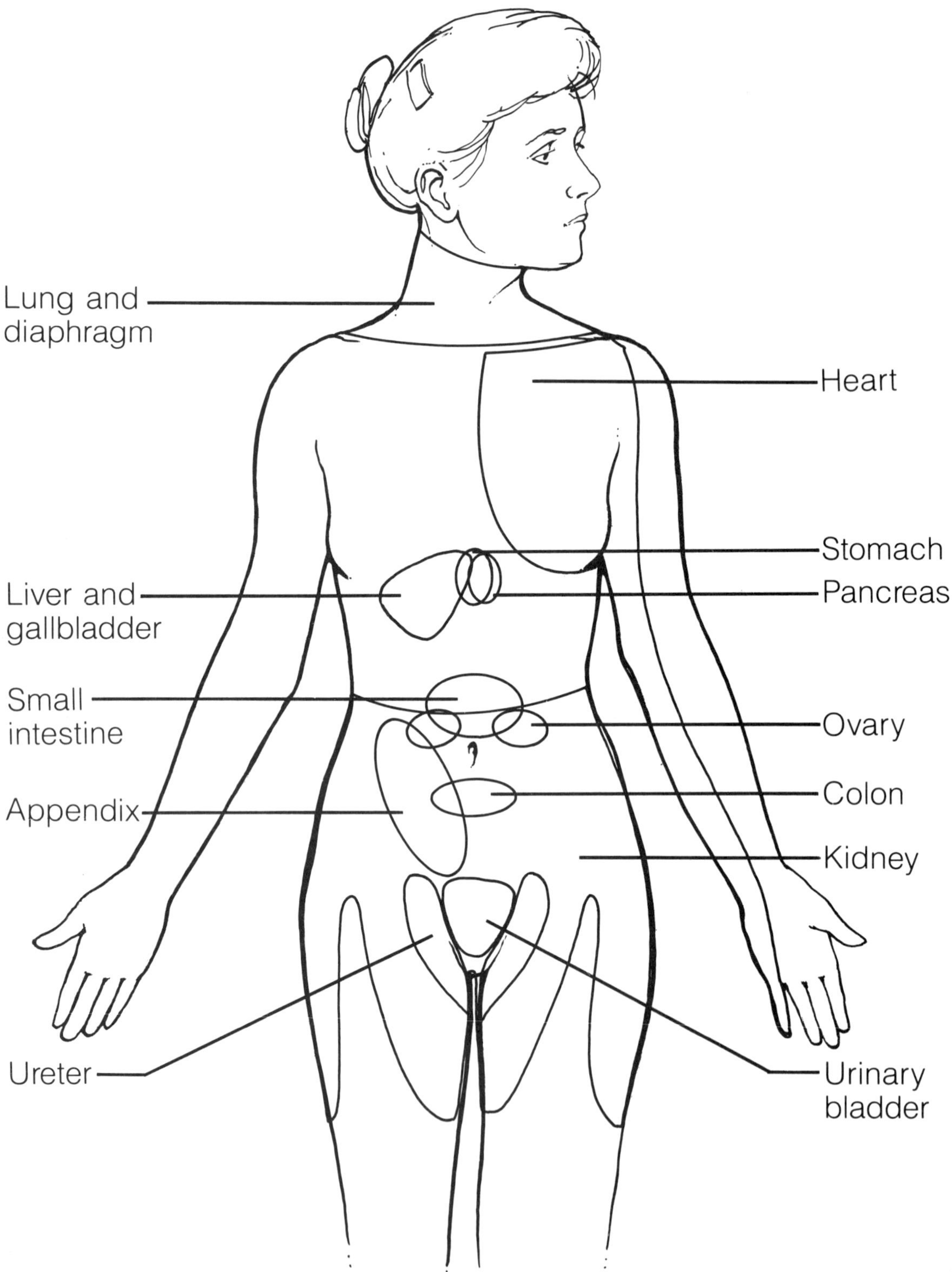

General organization of the somatosensory system

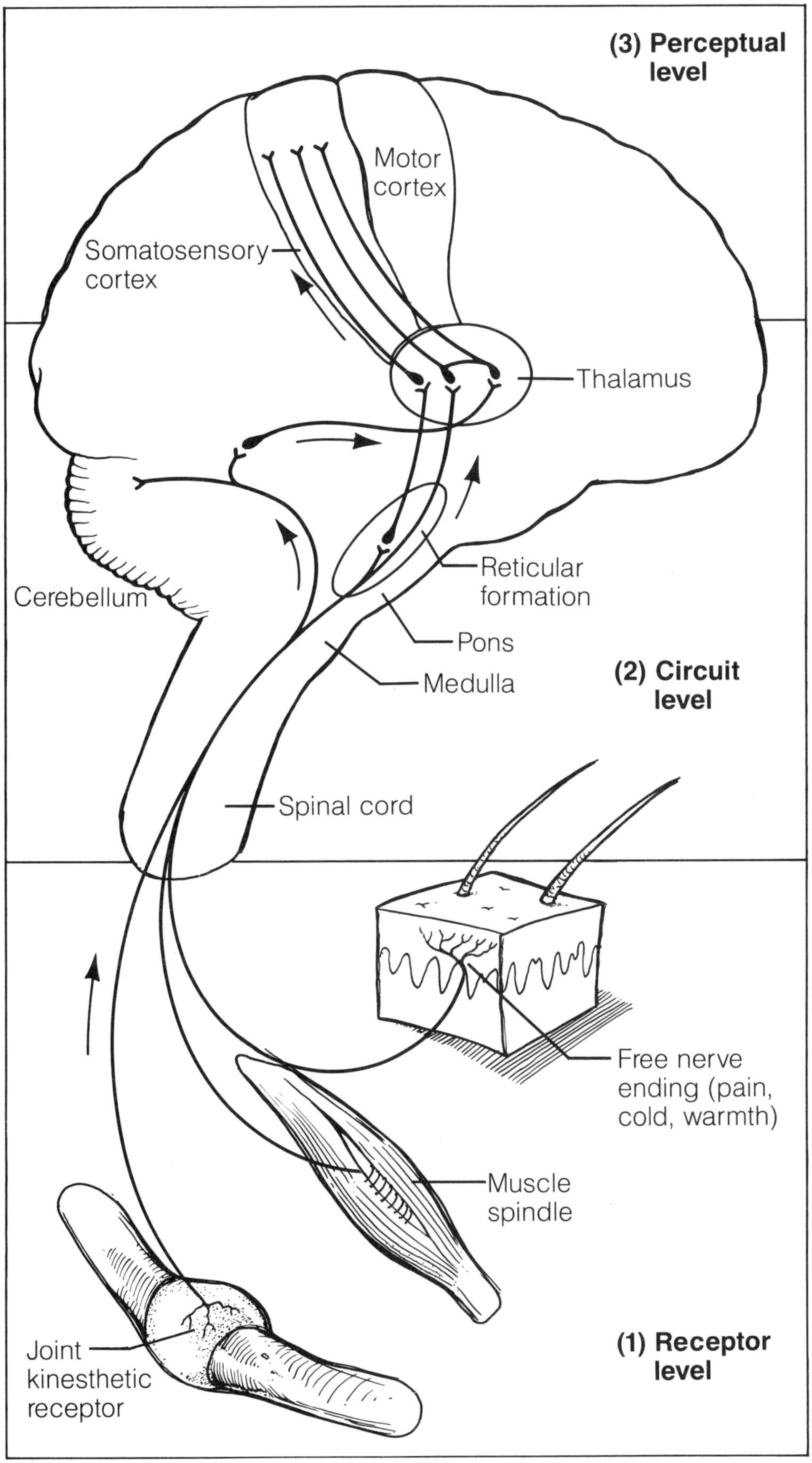

Two-point thresholds in the adult

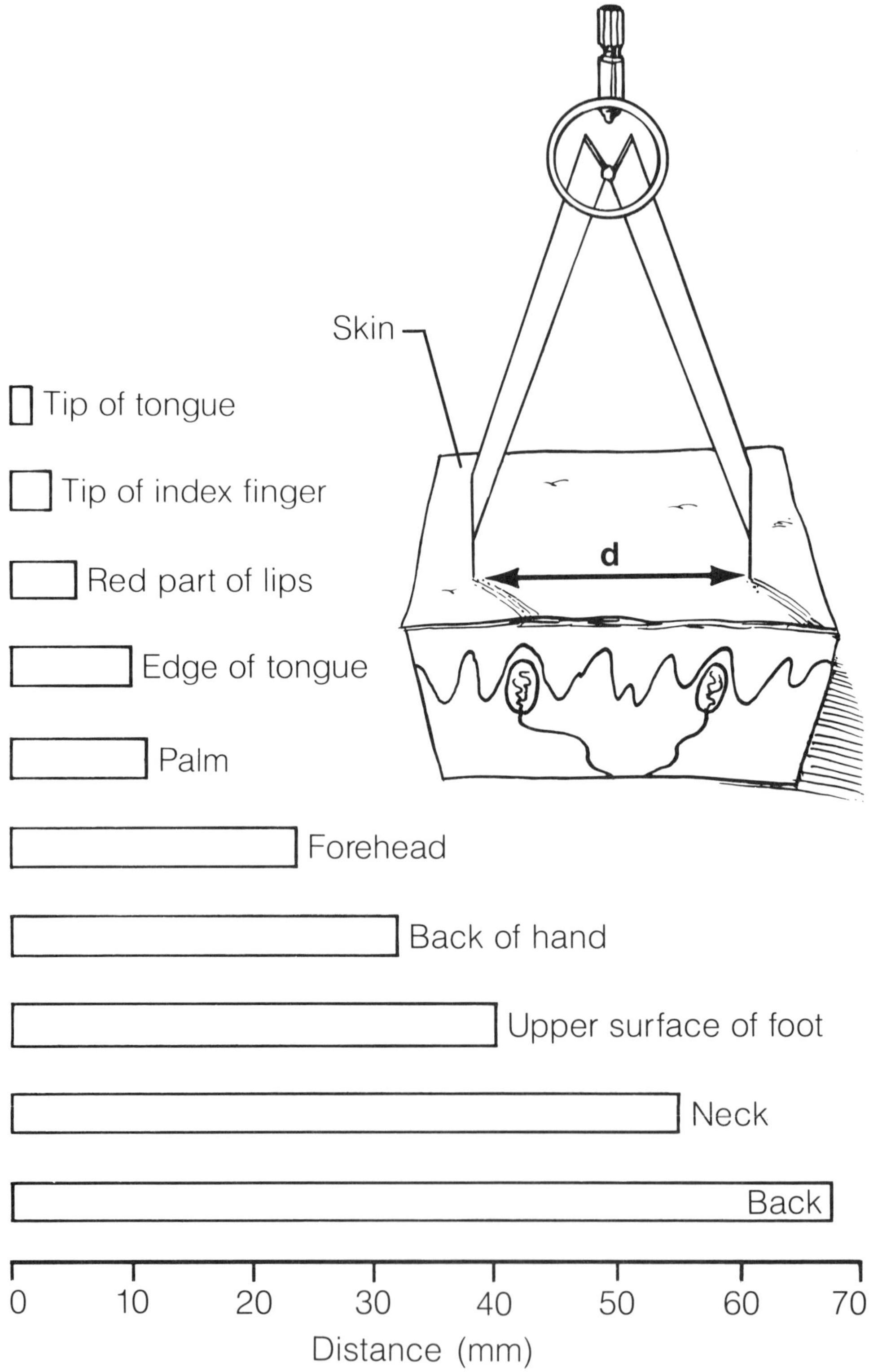

Hierarchy of motor control

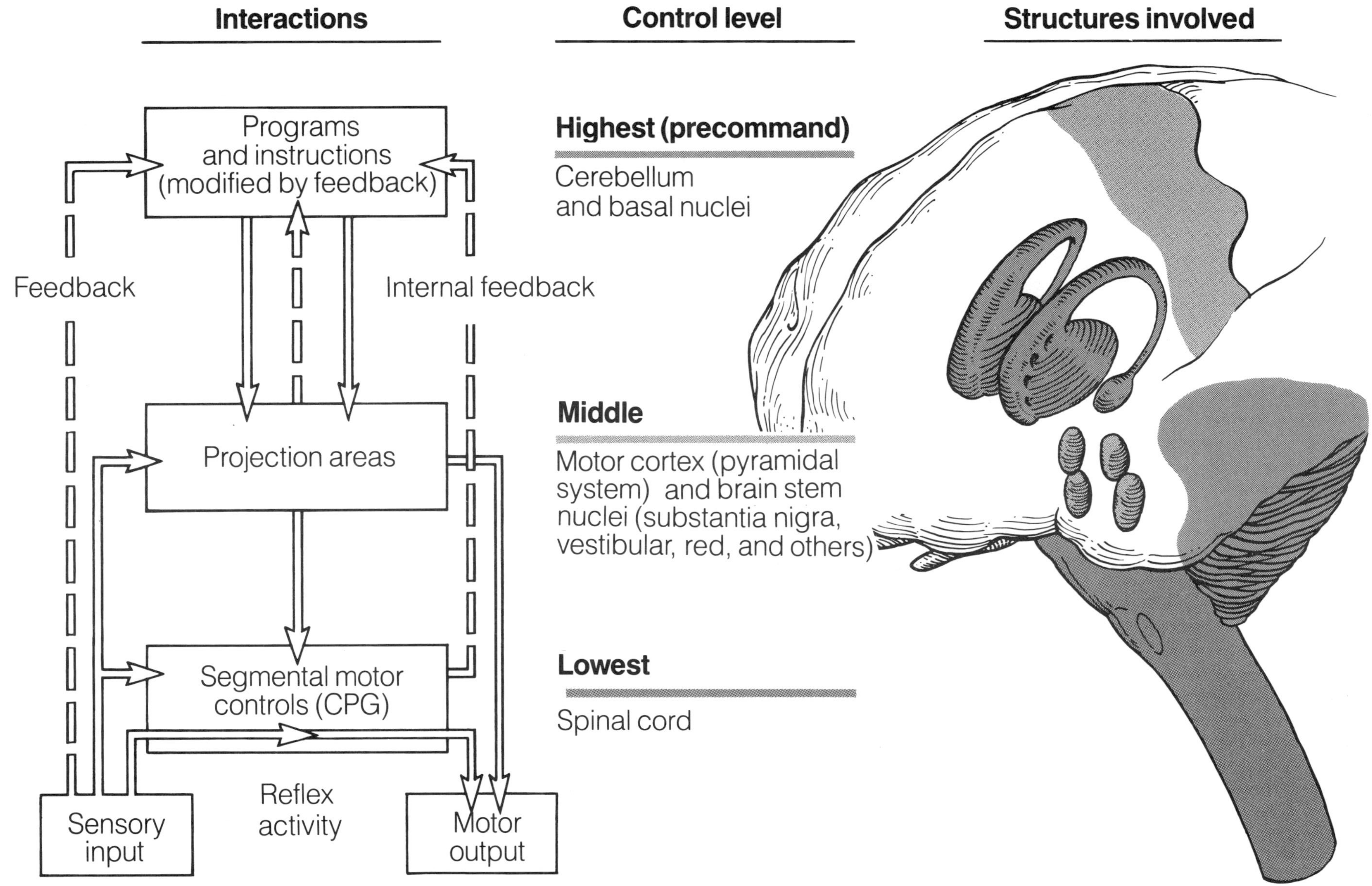

Electroencephalography and brain waves

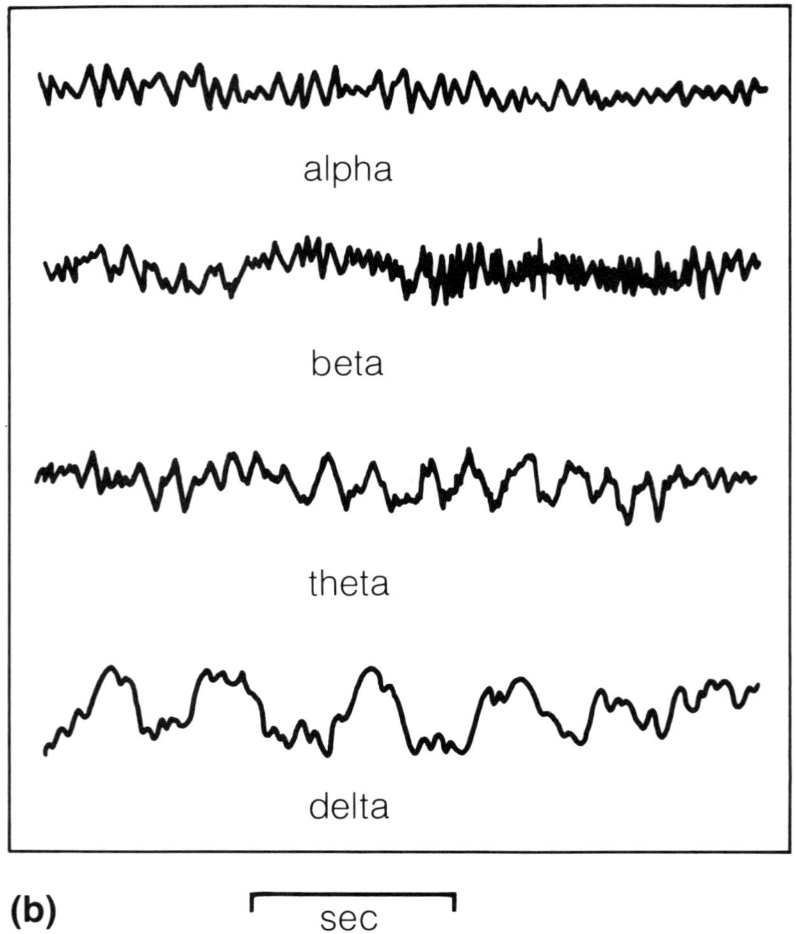

(b)

Memory processing

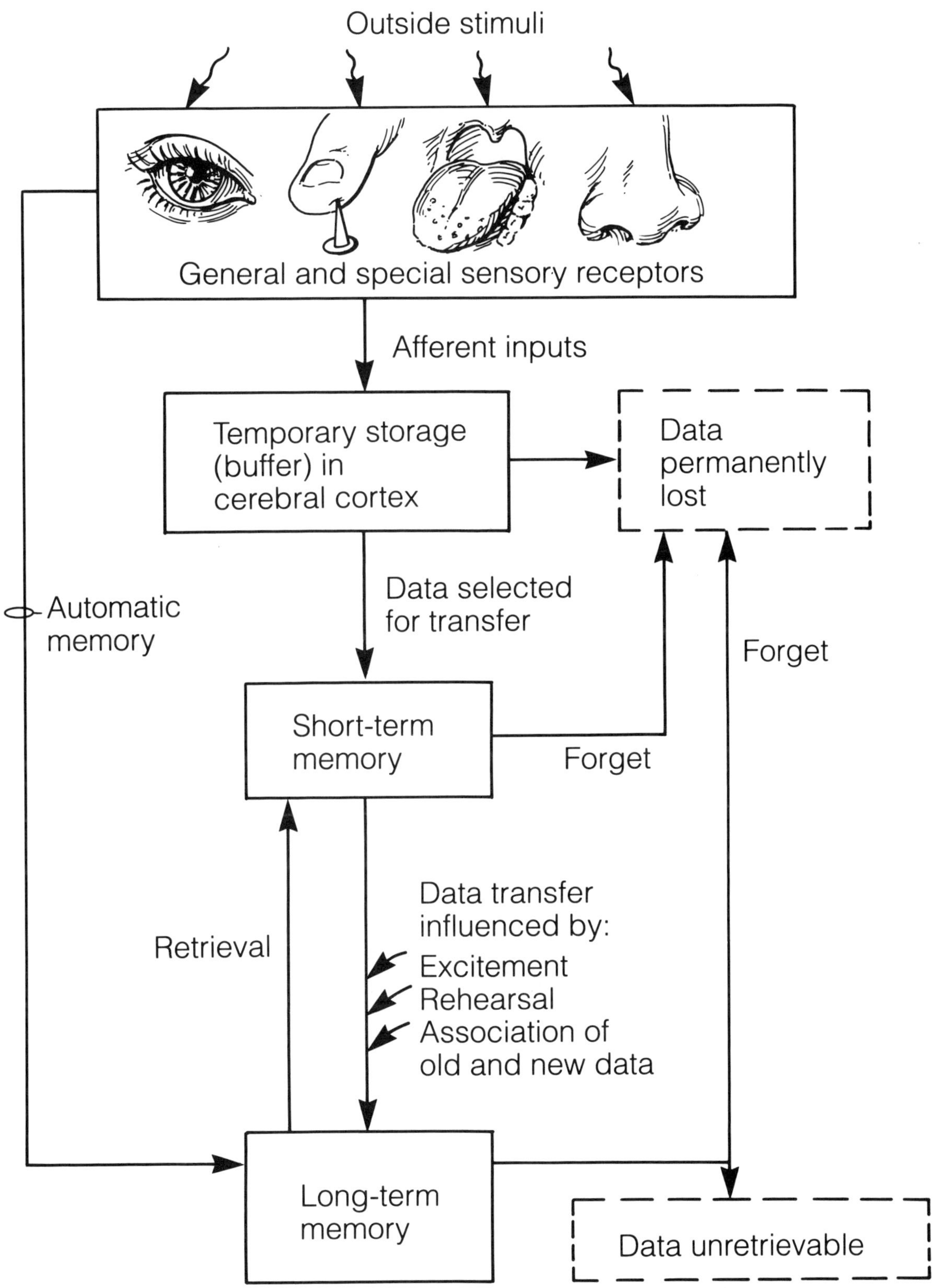

Proposed memory circuits

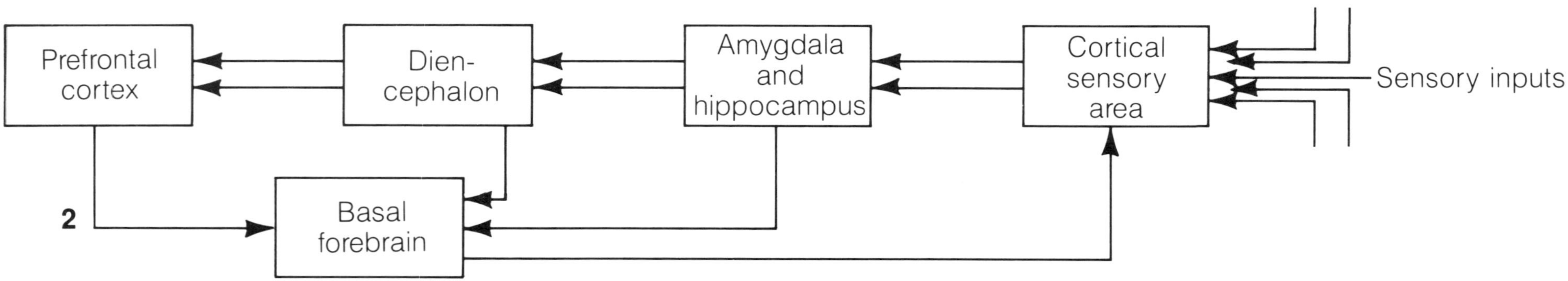

2

Hypothetical language areas

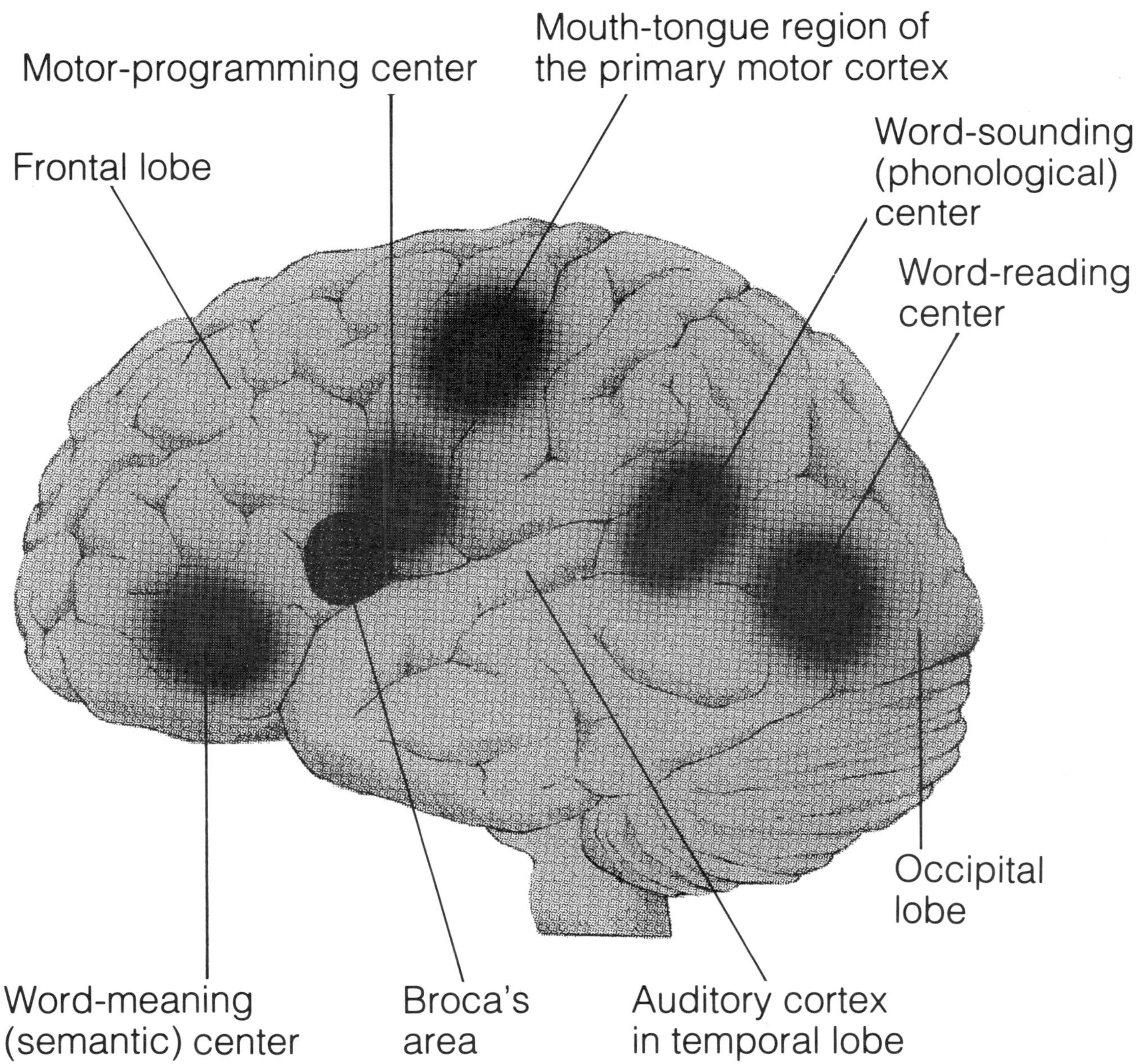

A CLOSER LOOK: Transmission of pain impulses

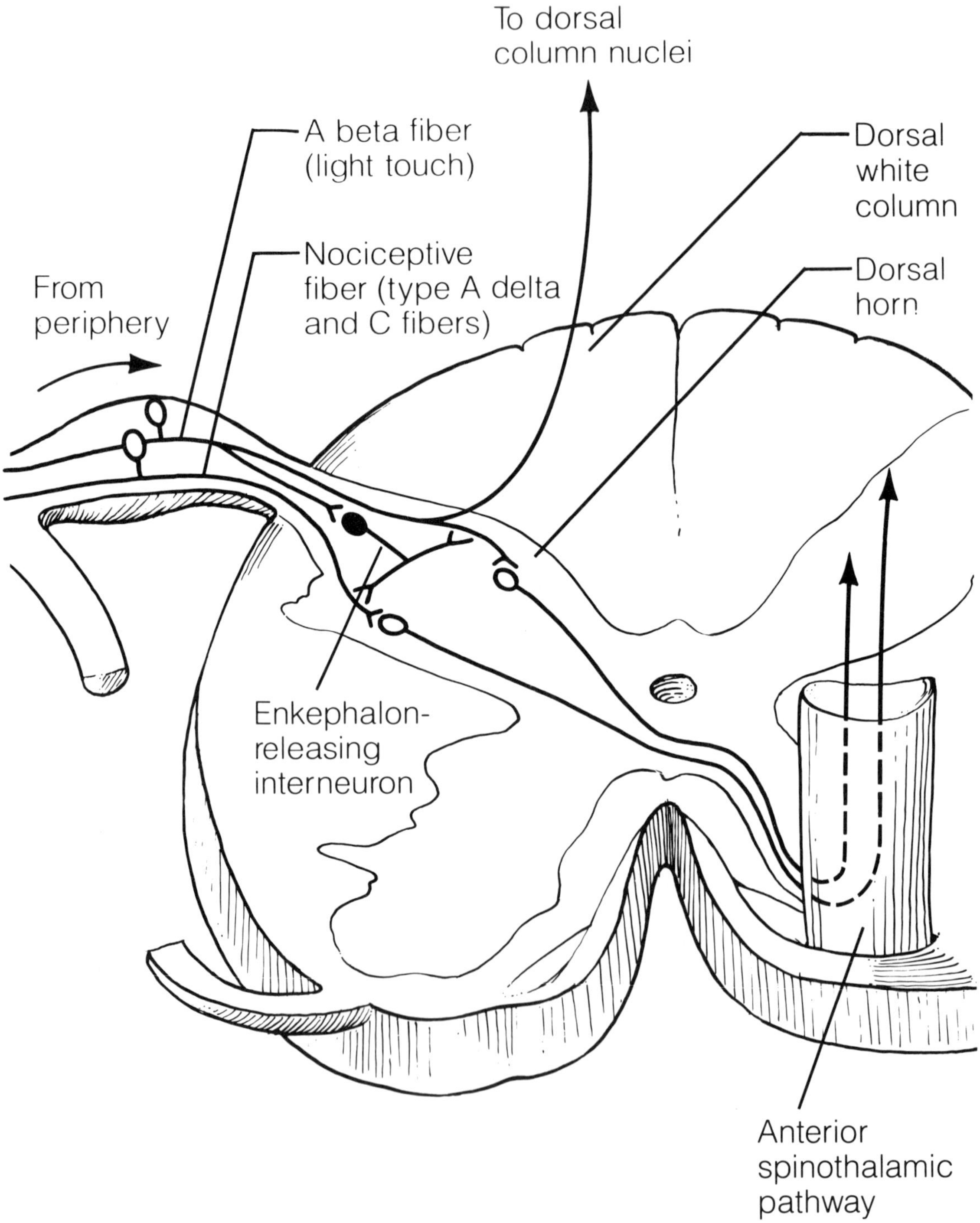

Relative patterns of taste sensitivity on the tongue dorsum

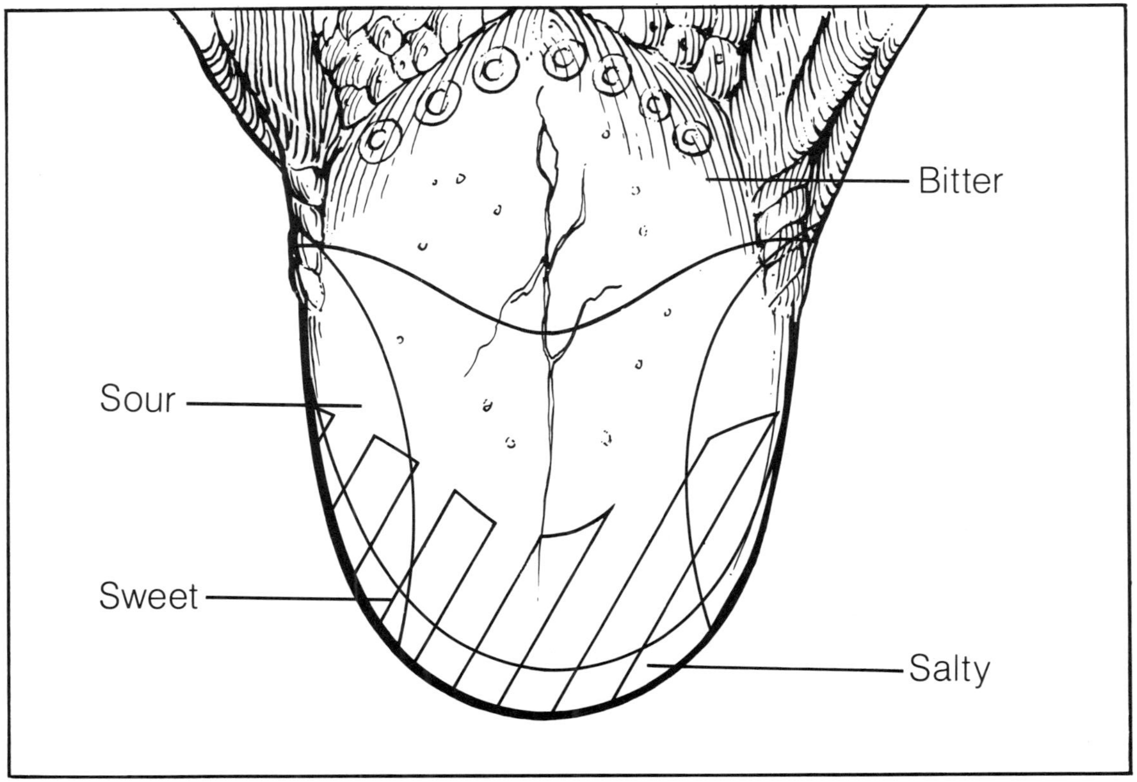

Olfactory receptors

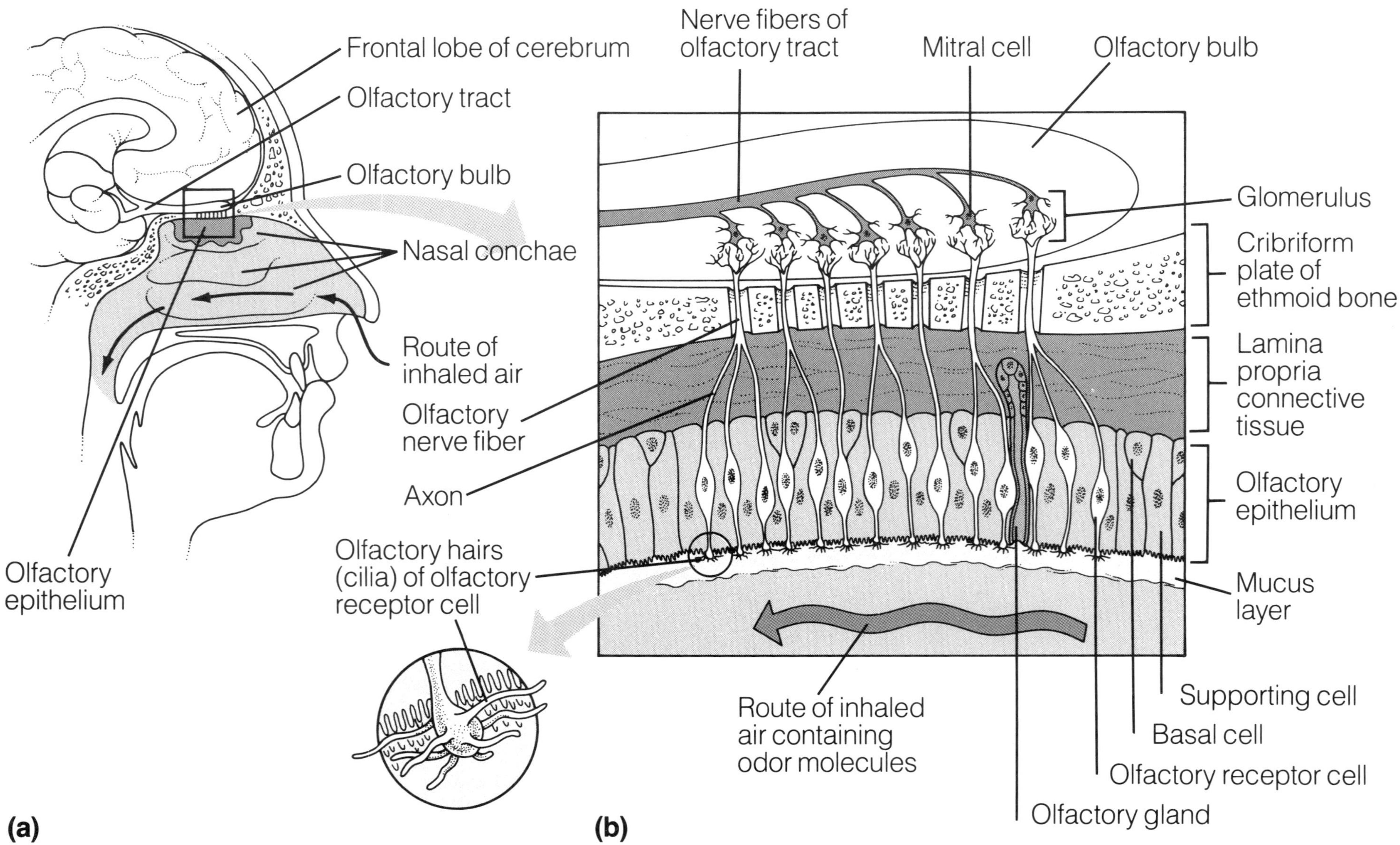

Sensory layer of the retina

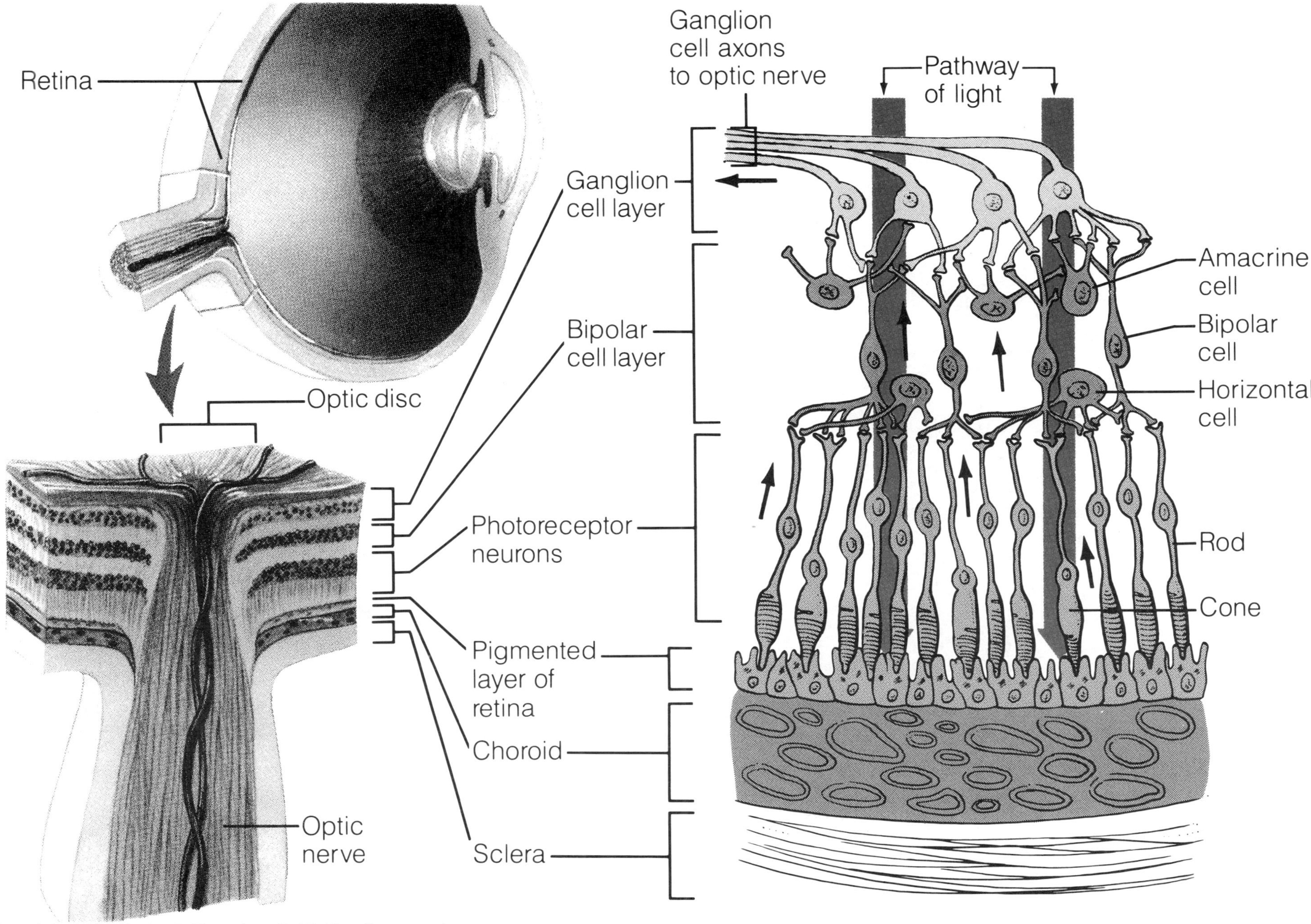

Structures seen in a transverse section through the anterior portion of the eye

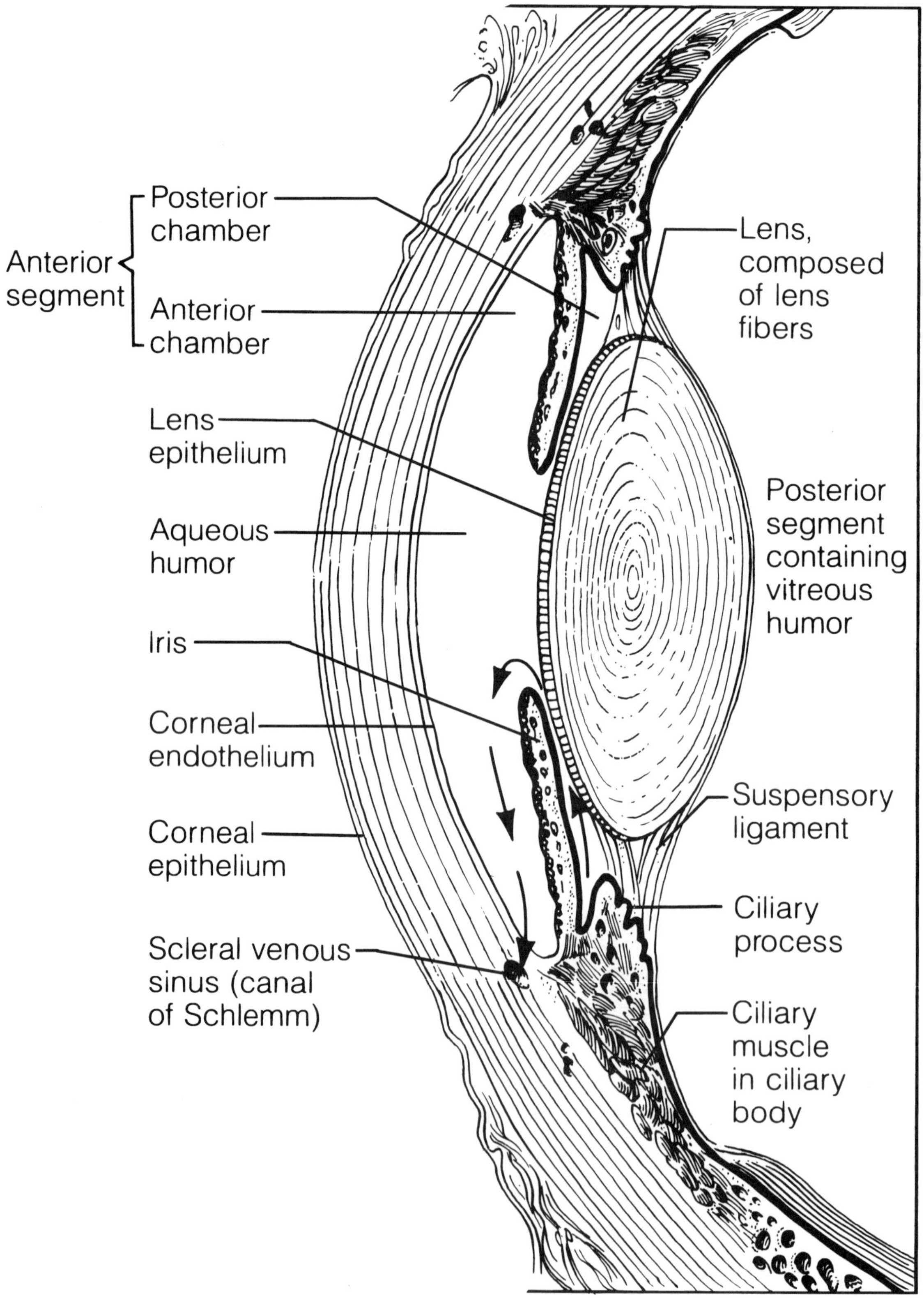

Focusing for distant and close vision

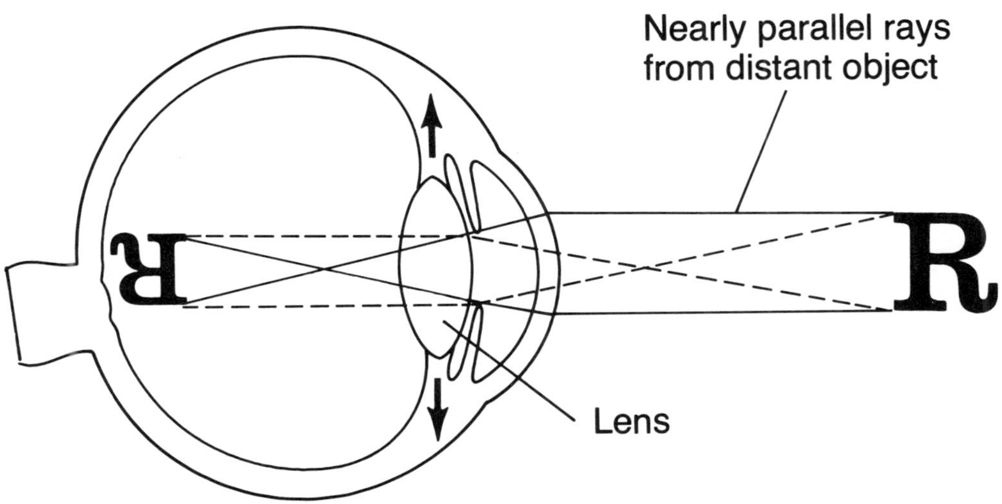

(a) Lens is flattened for distant vision

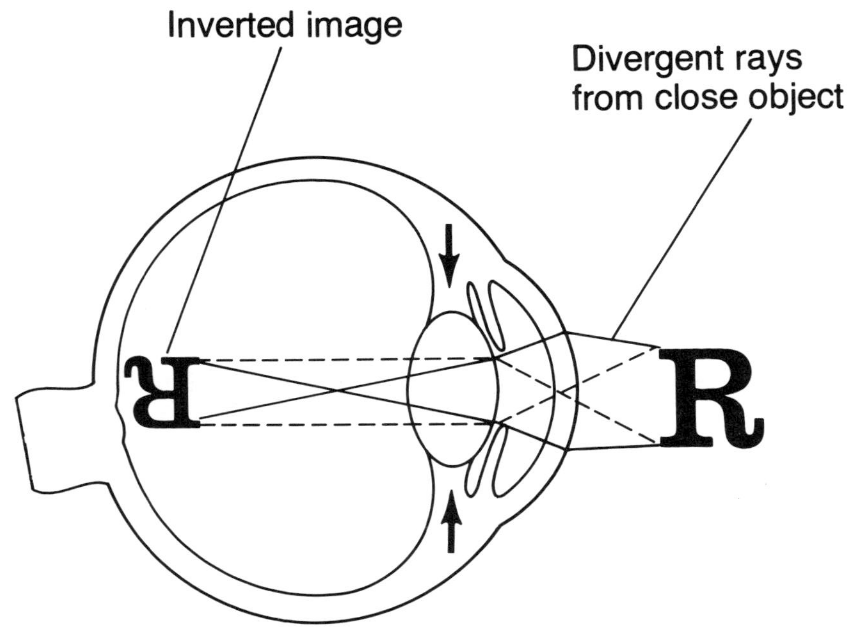

(b) Lens bulges for close vision

Problems of refraction

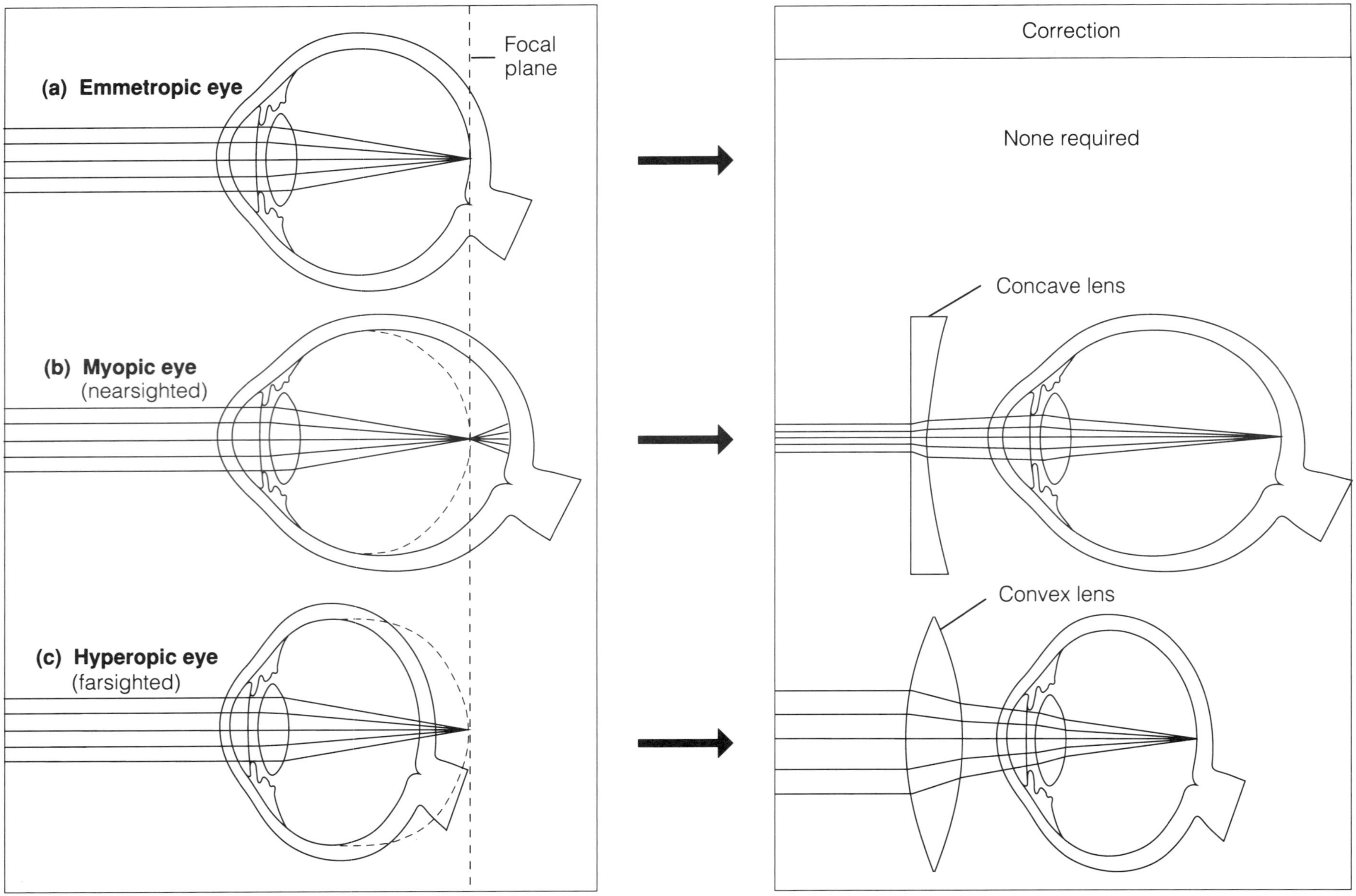

Sequence of chemical events in the formation and breakdown of rhodopsin

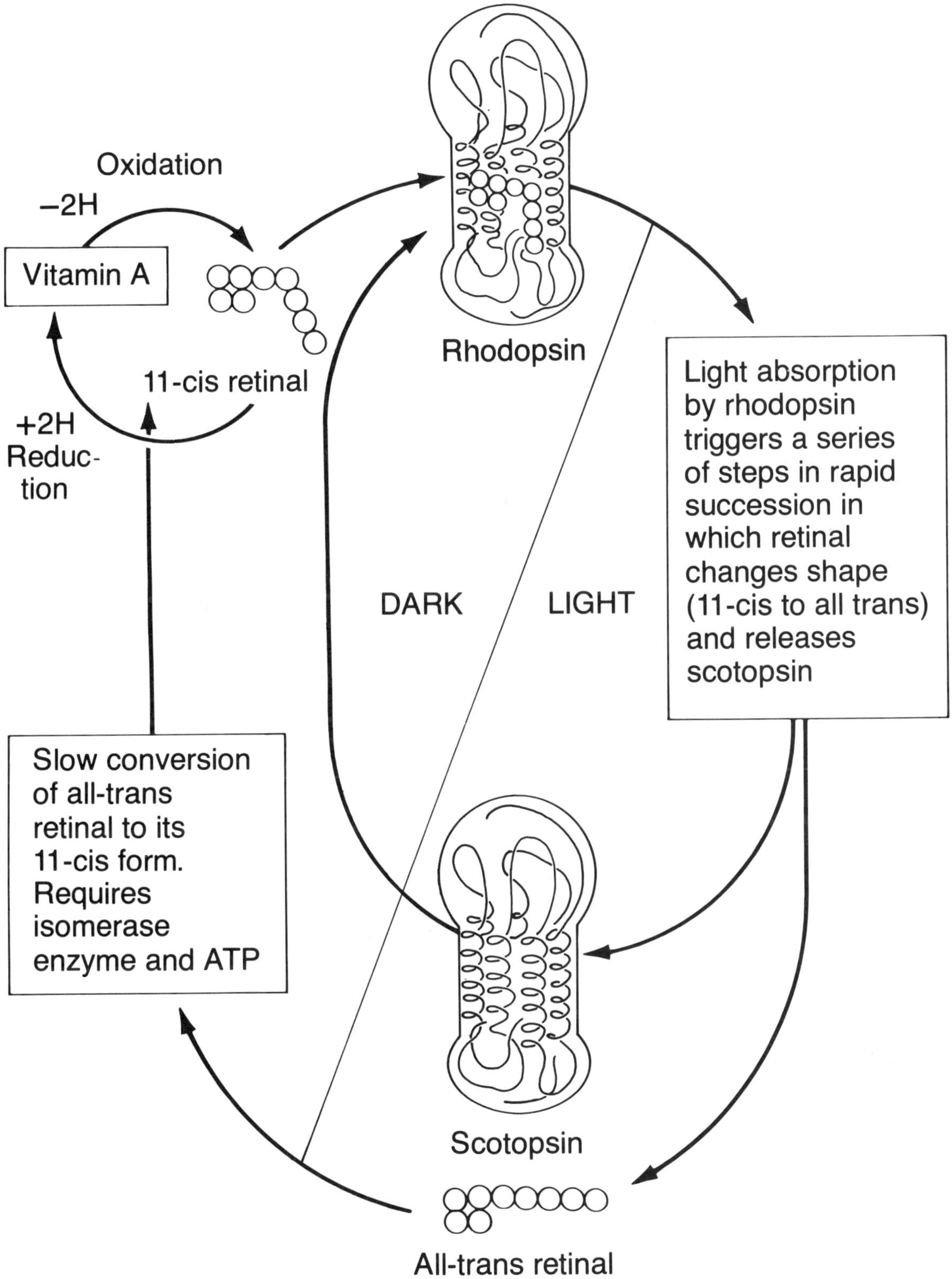

Sensitivities of the three cone types to the different wavelengths of the visible spectrum

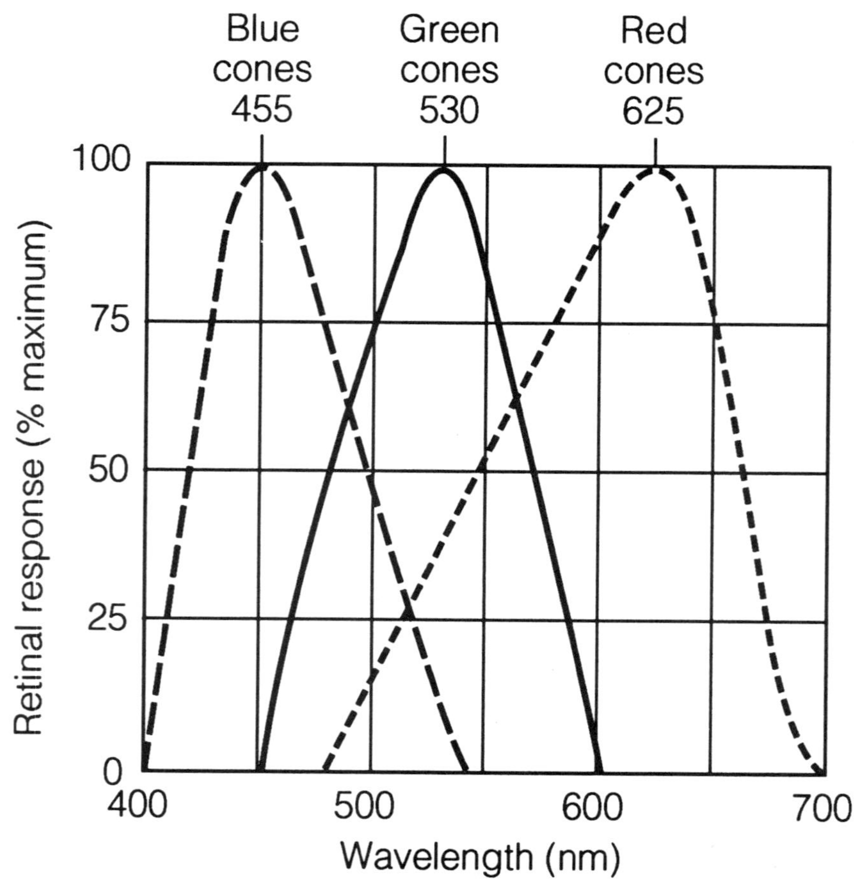

Visual fields of the eyes and visual pathway to the brain, inferior view

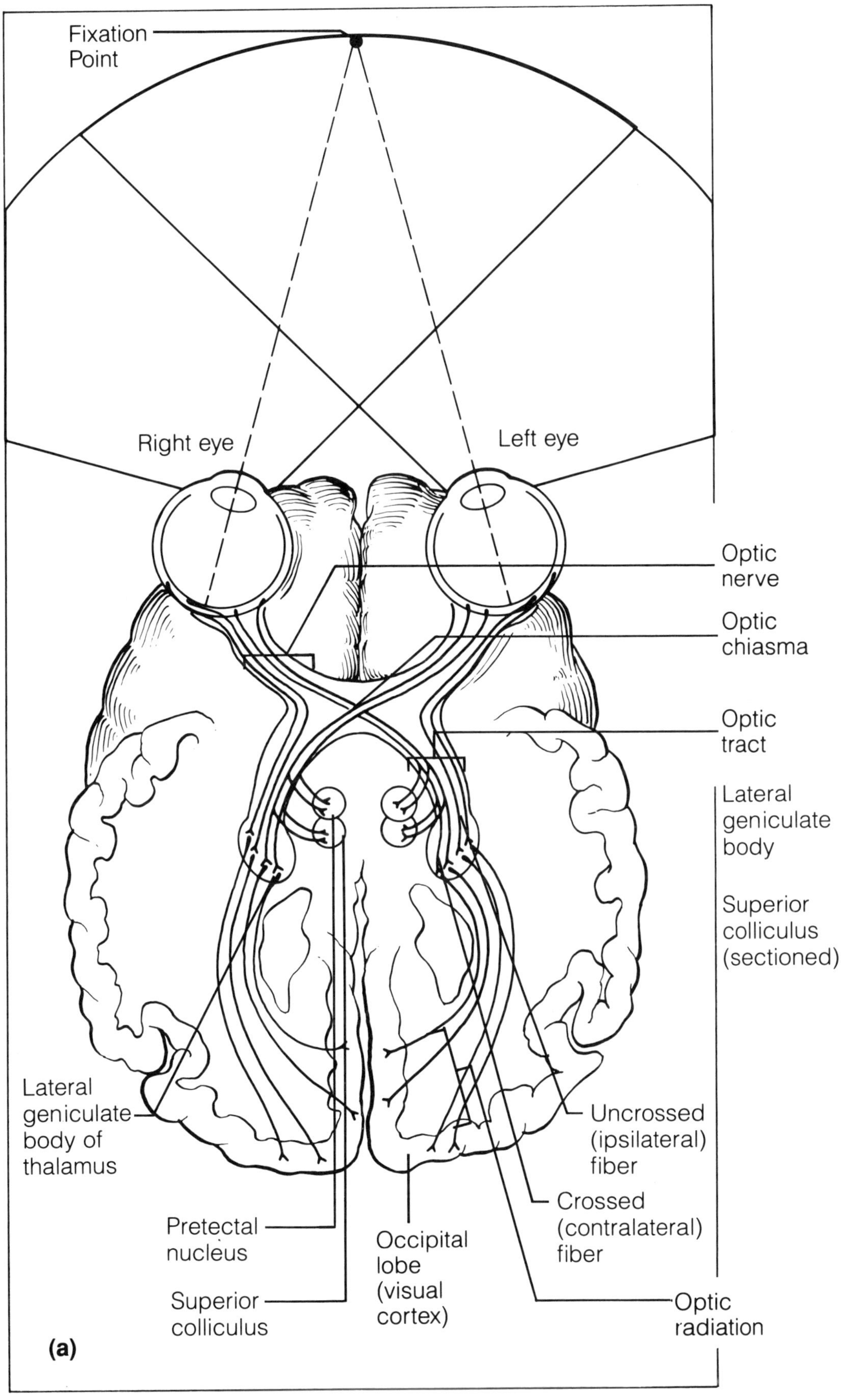

Sound: sources and propagation

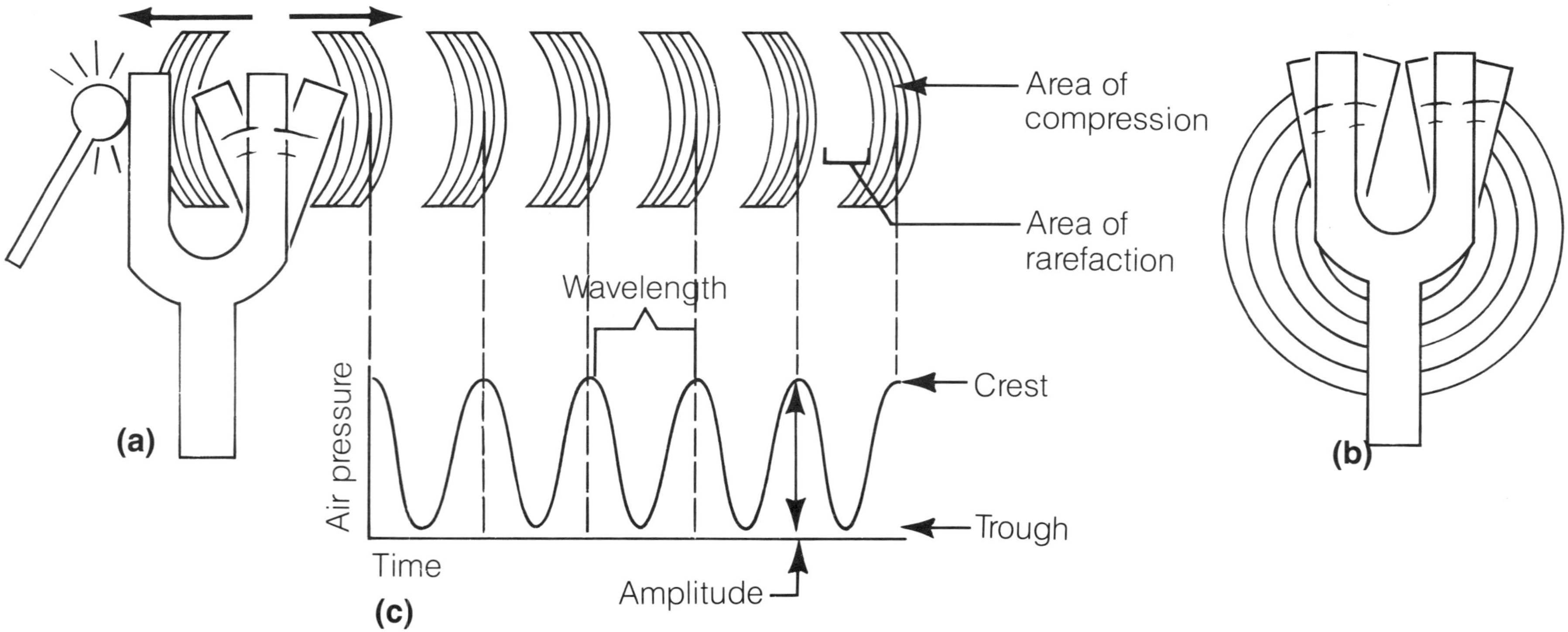

Frequency and amplitude of sound waves

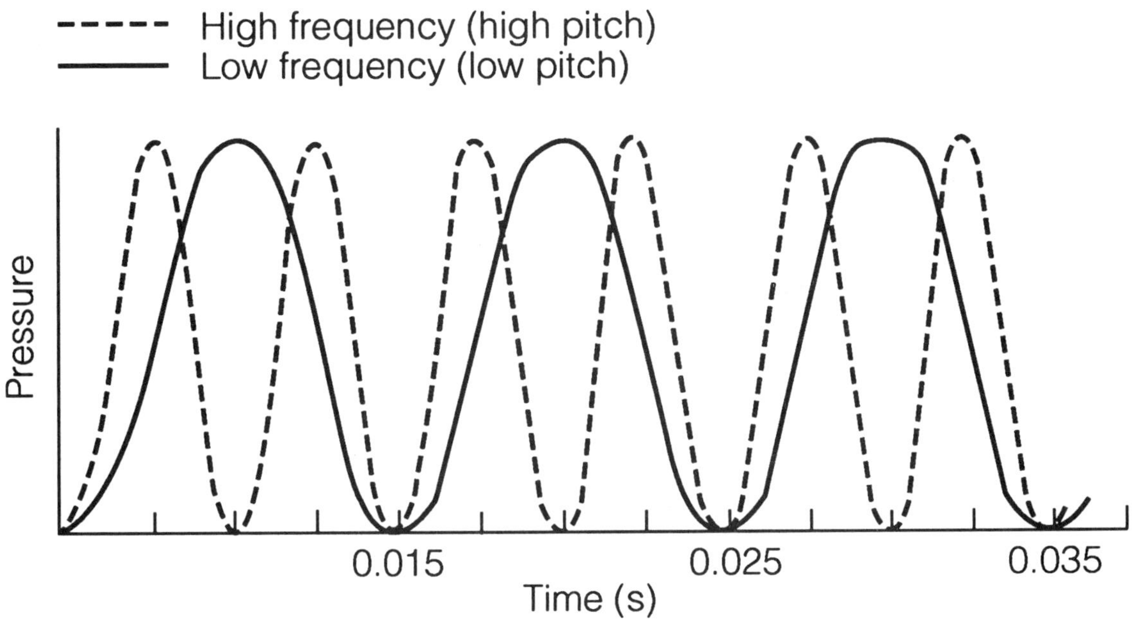

(a)

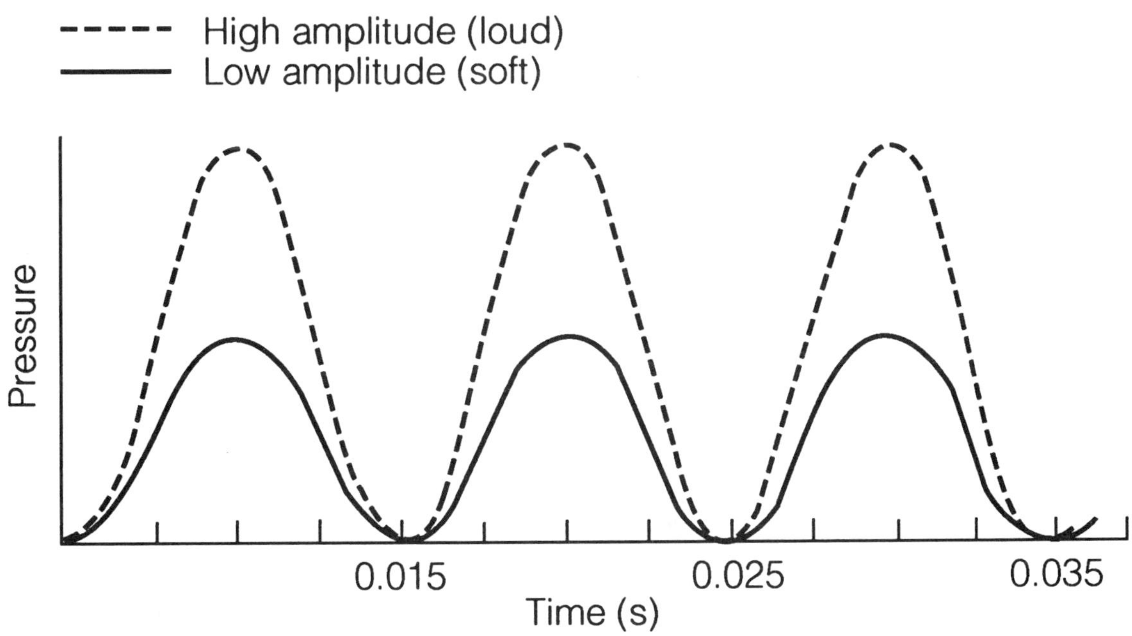

(b)

Resonance of the basilar membrane and activation of the cochlear hair cells

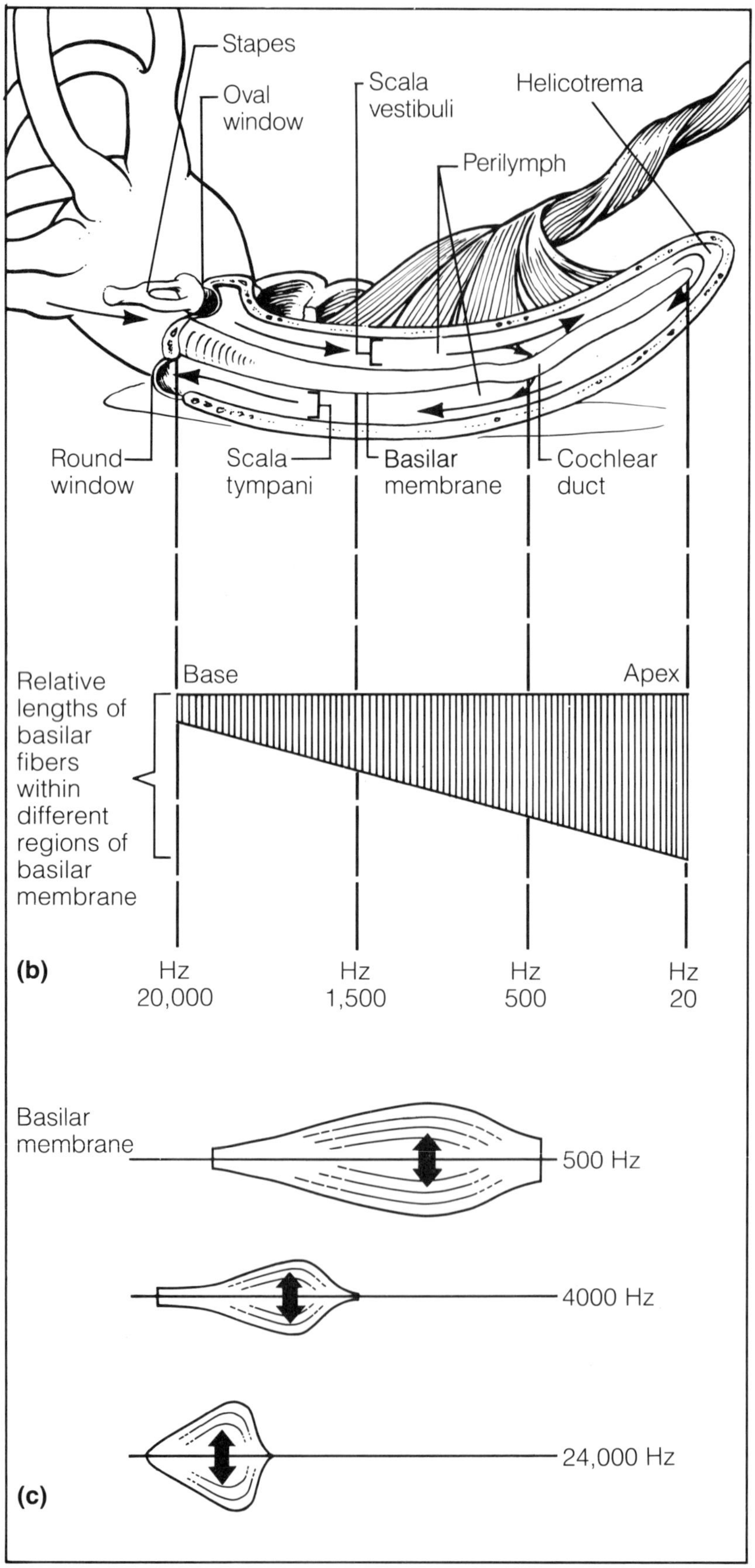

Simplified diagram of the auditory pathway from the organ of Corti to the auditory cortex in the temporal lobe of the brain

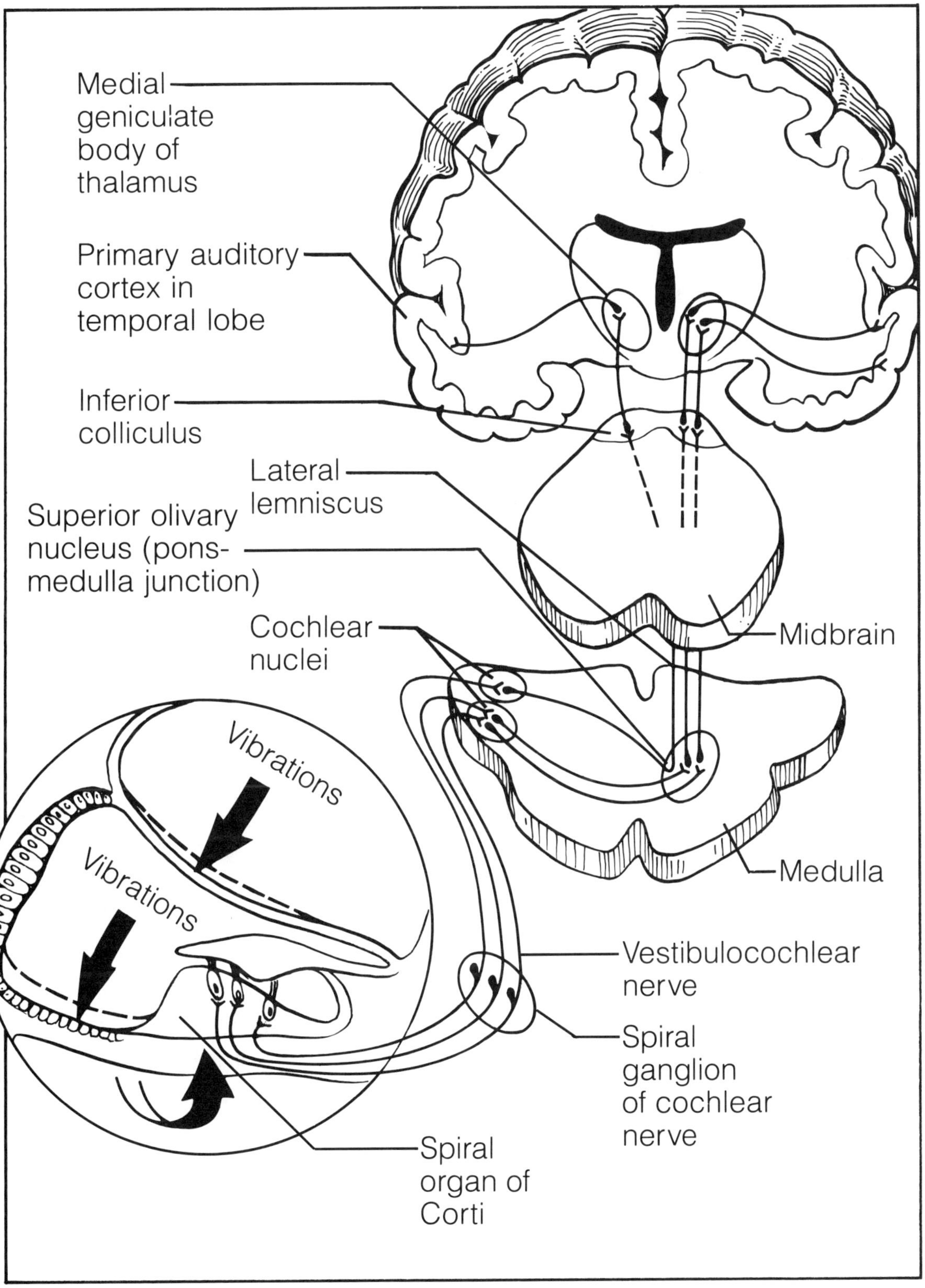

Pathways of the balance and orientation system

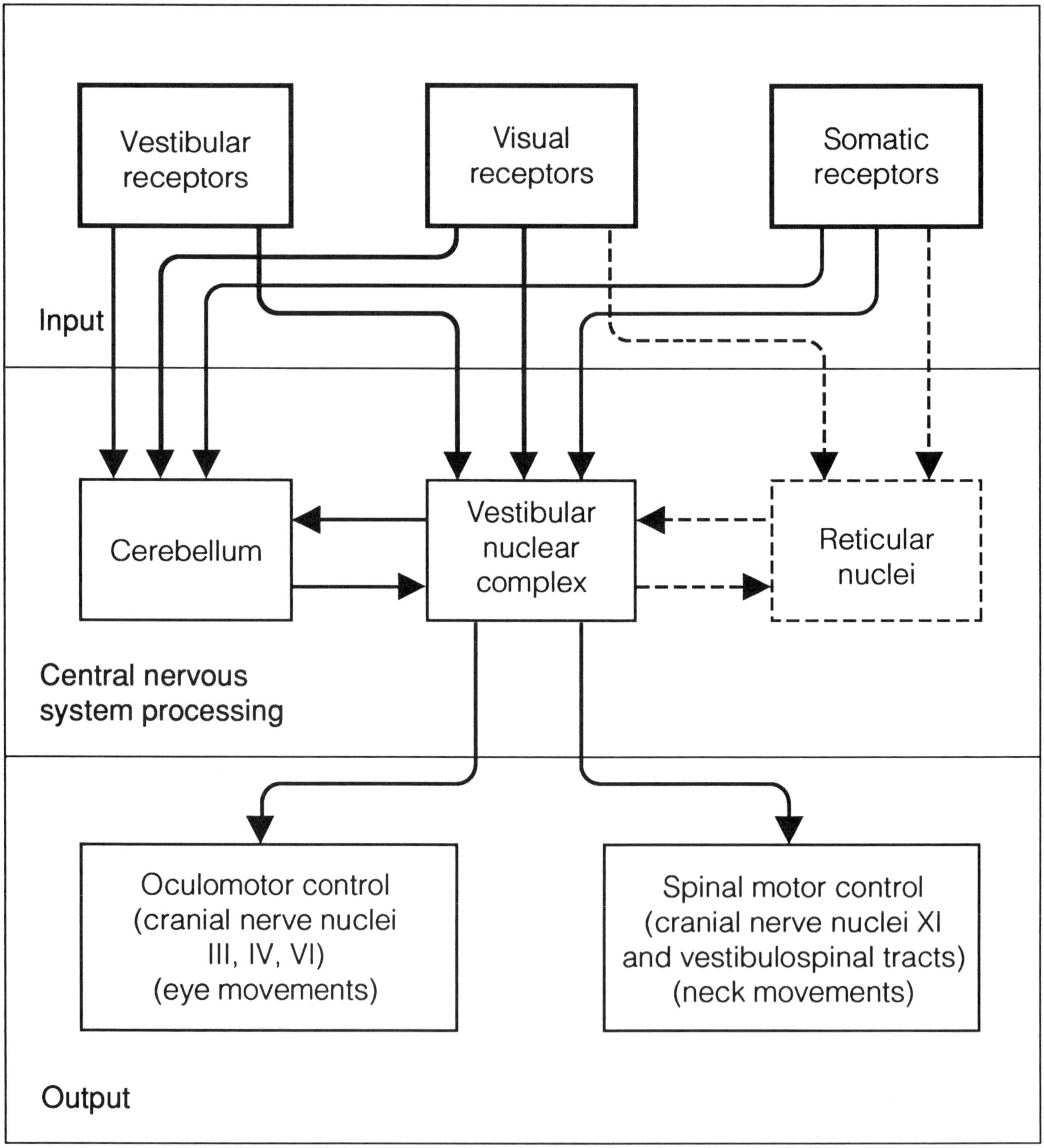

Location of the major endocrine organs of the body

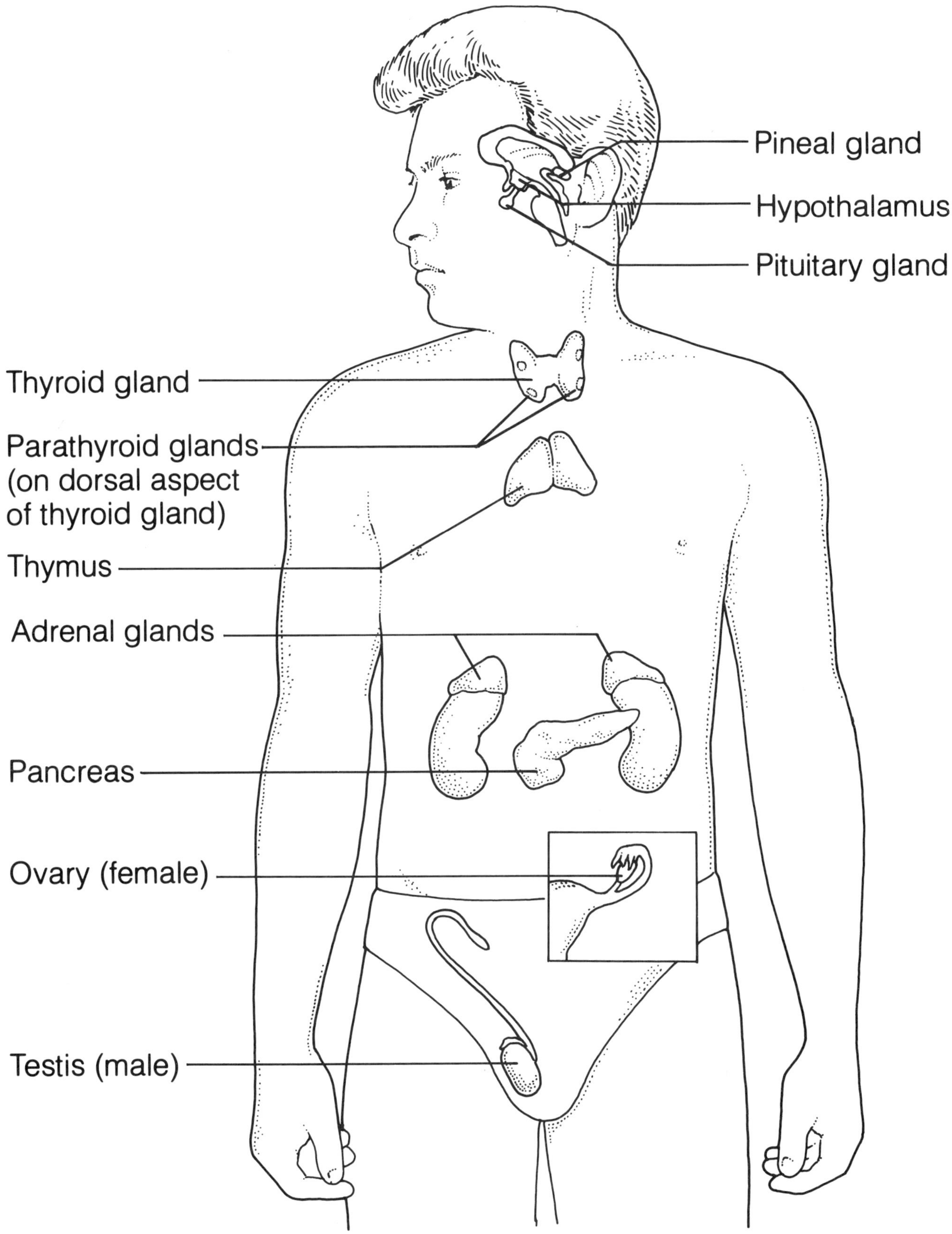

Second messenger mechanisms of protein-peptide hormones

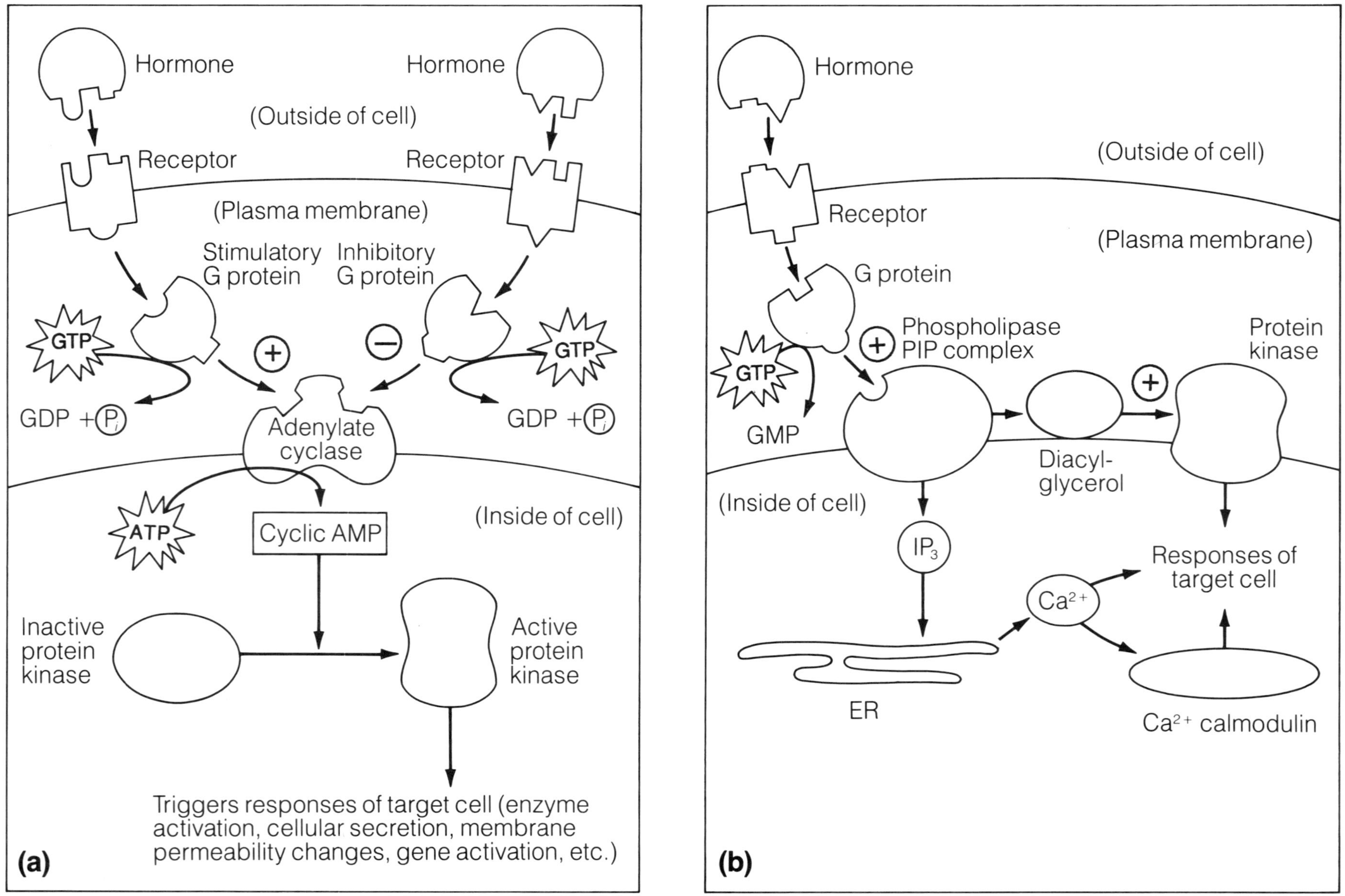

Direct gene activation mechanism of steroid hormones

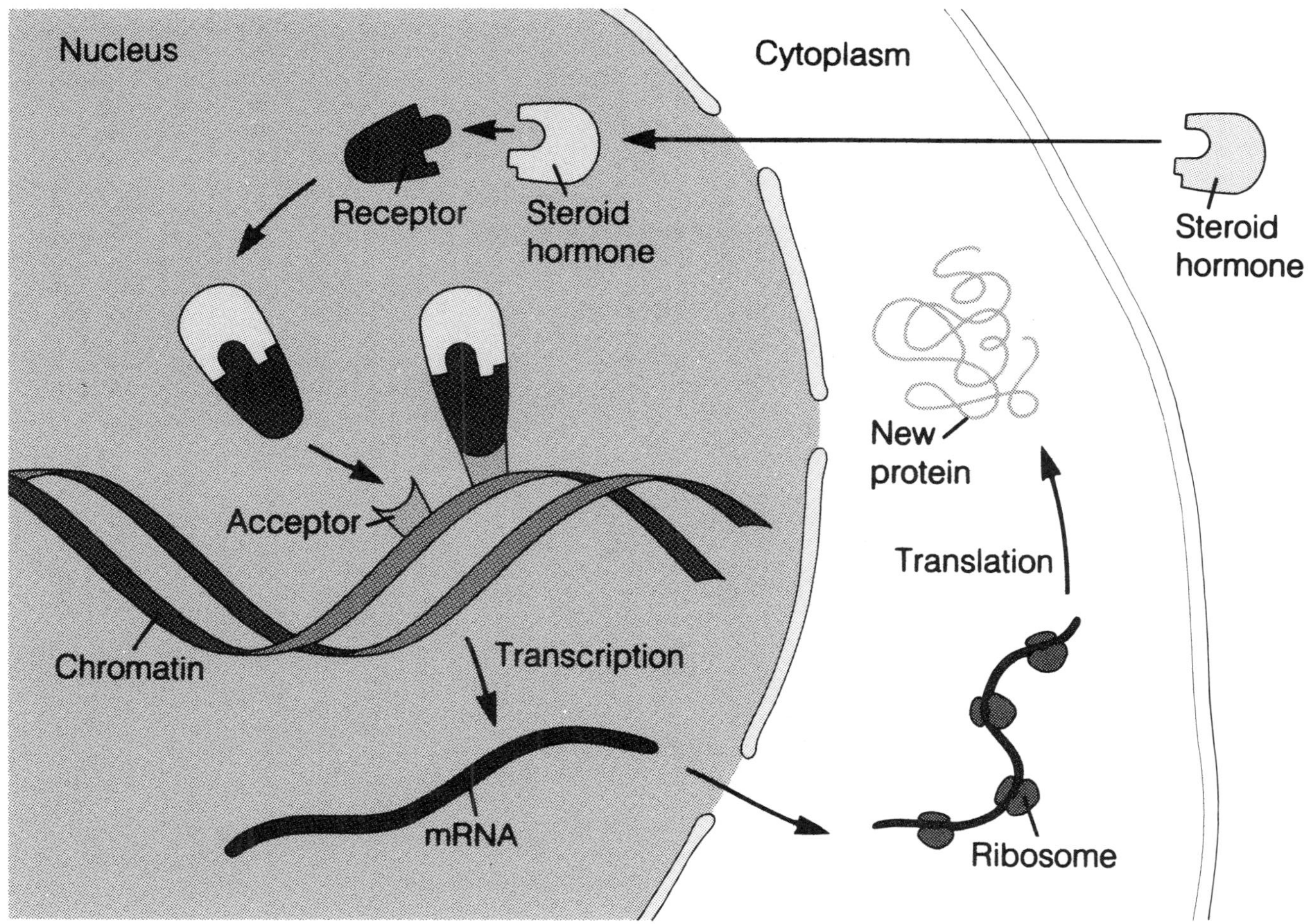

Endocrine gland stimuli

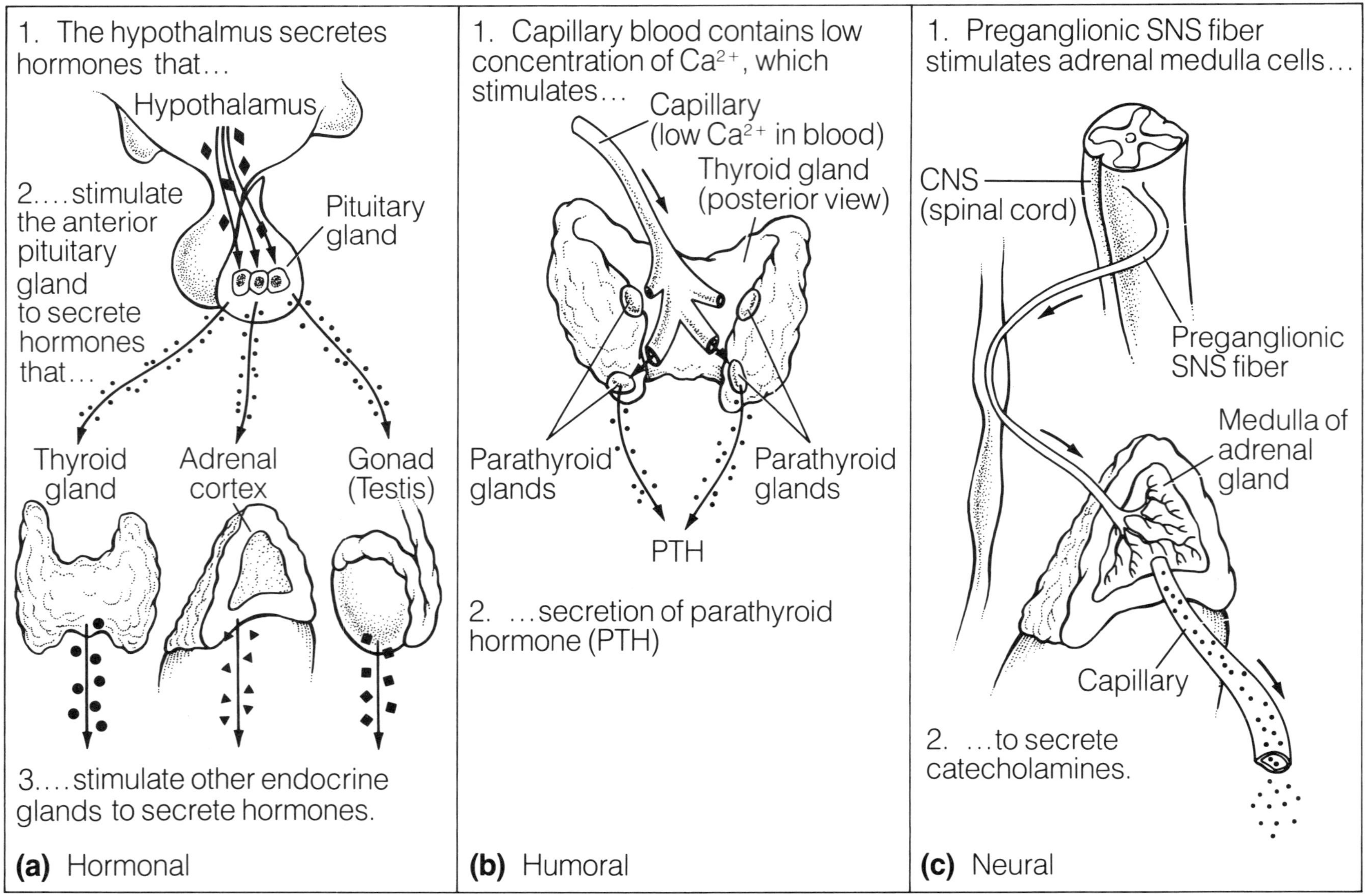

Structural and functional relationships of the pituitary and the hypothalamus

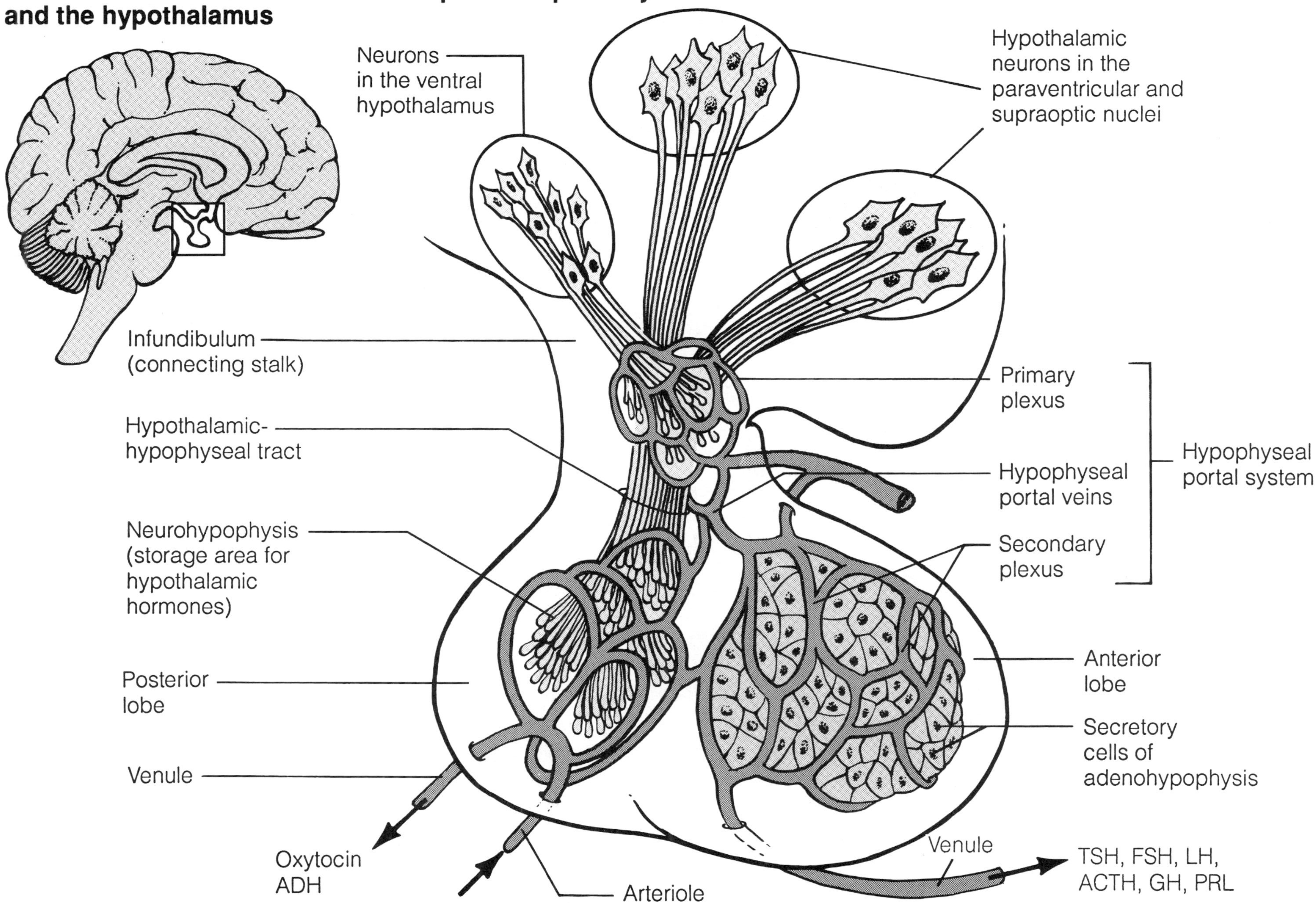

Proposed scheme for classifying the metabolic actions of growth hormone (GH)

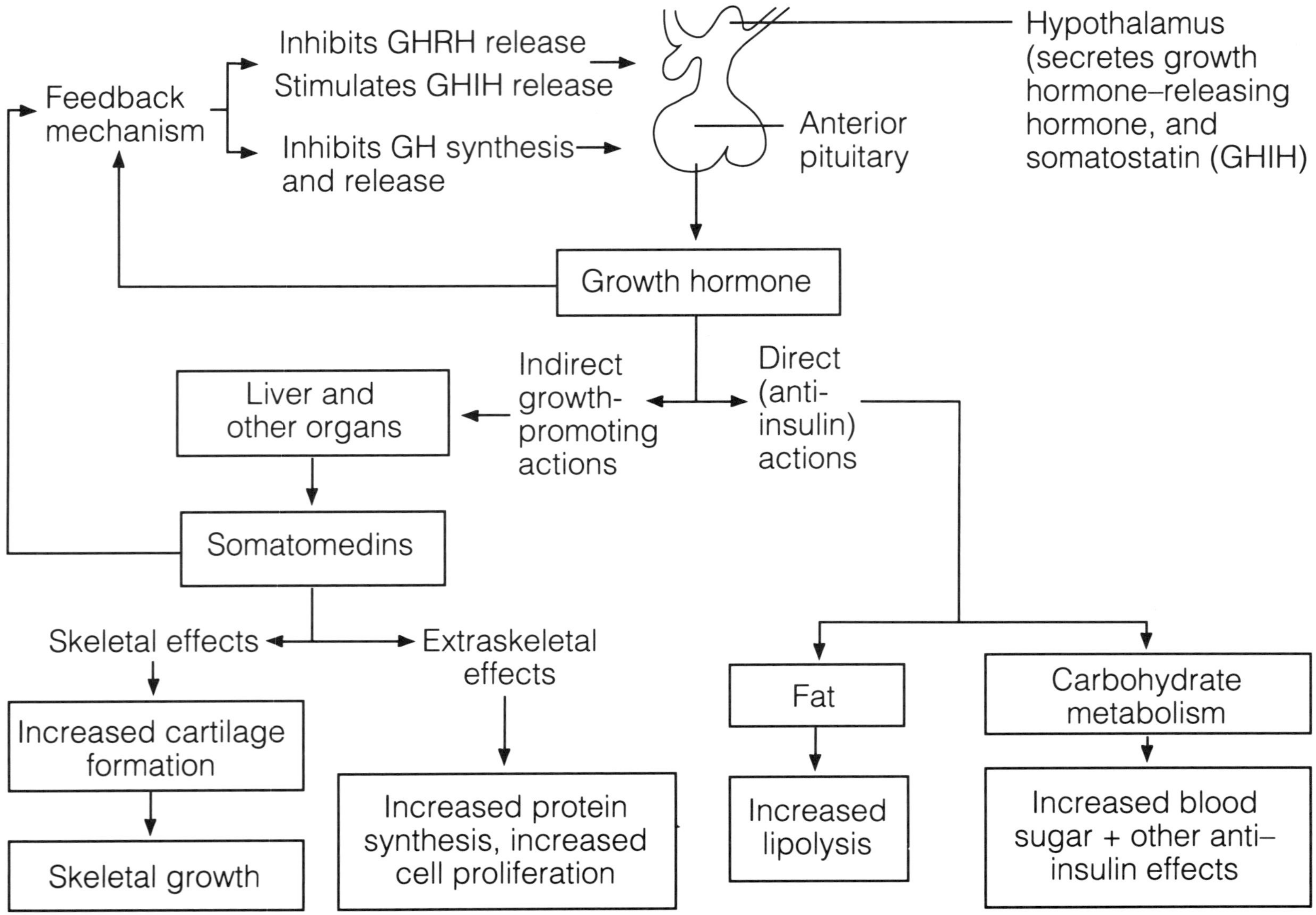

Summary of parathyroid hormone's effects on bone, the kidneys, and the intestine

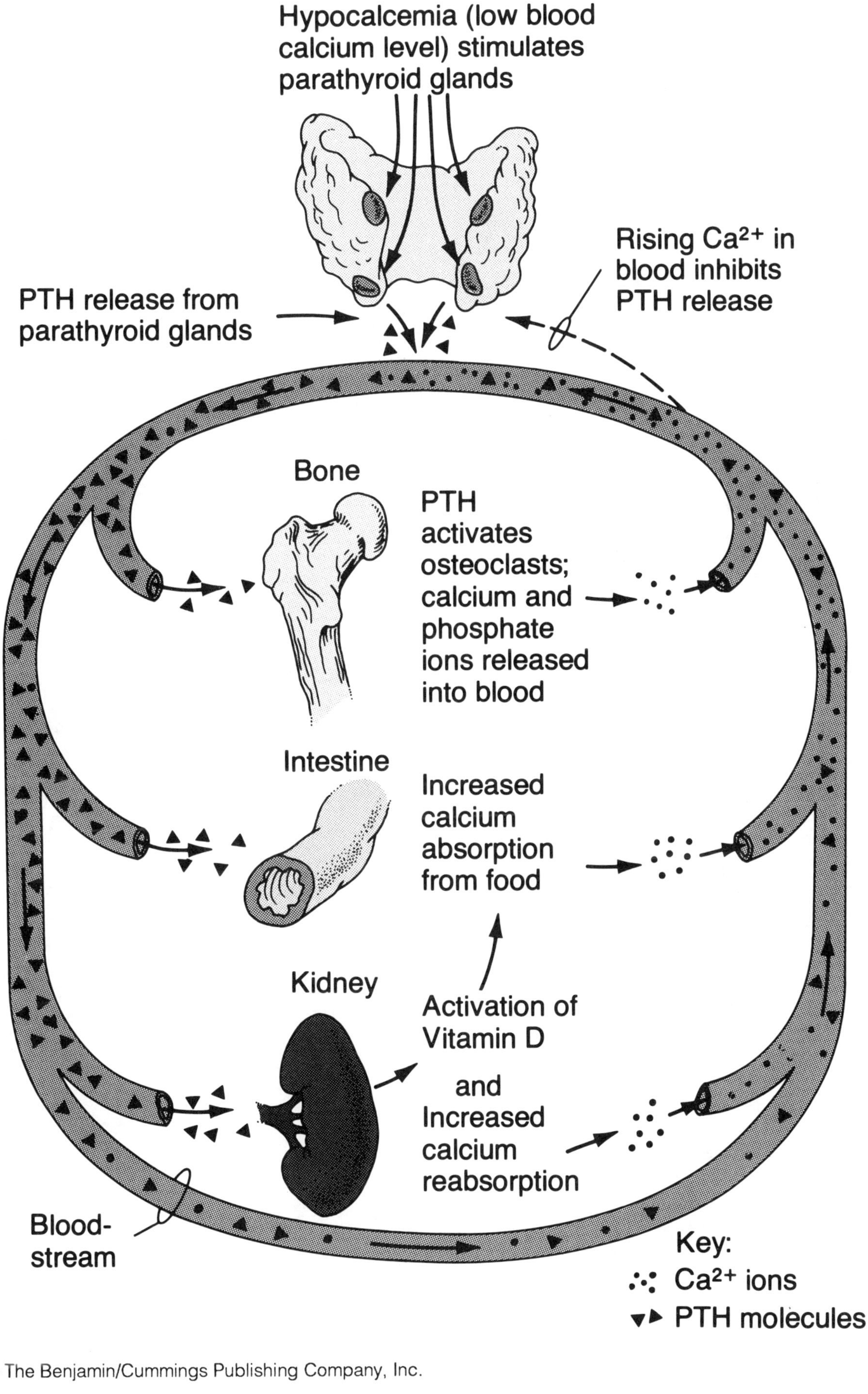

Control of aldosterone release from the adrenal cortex

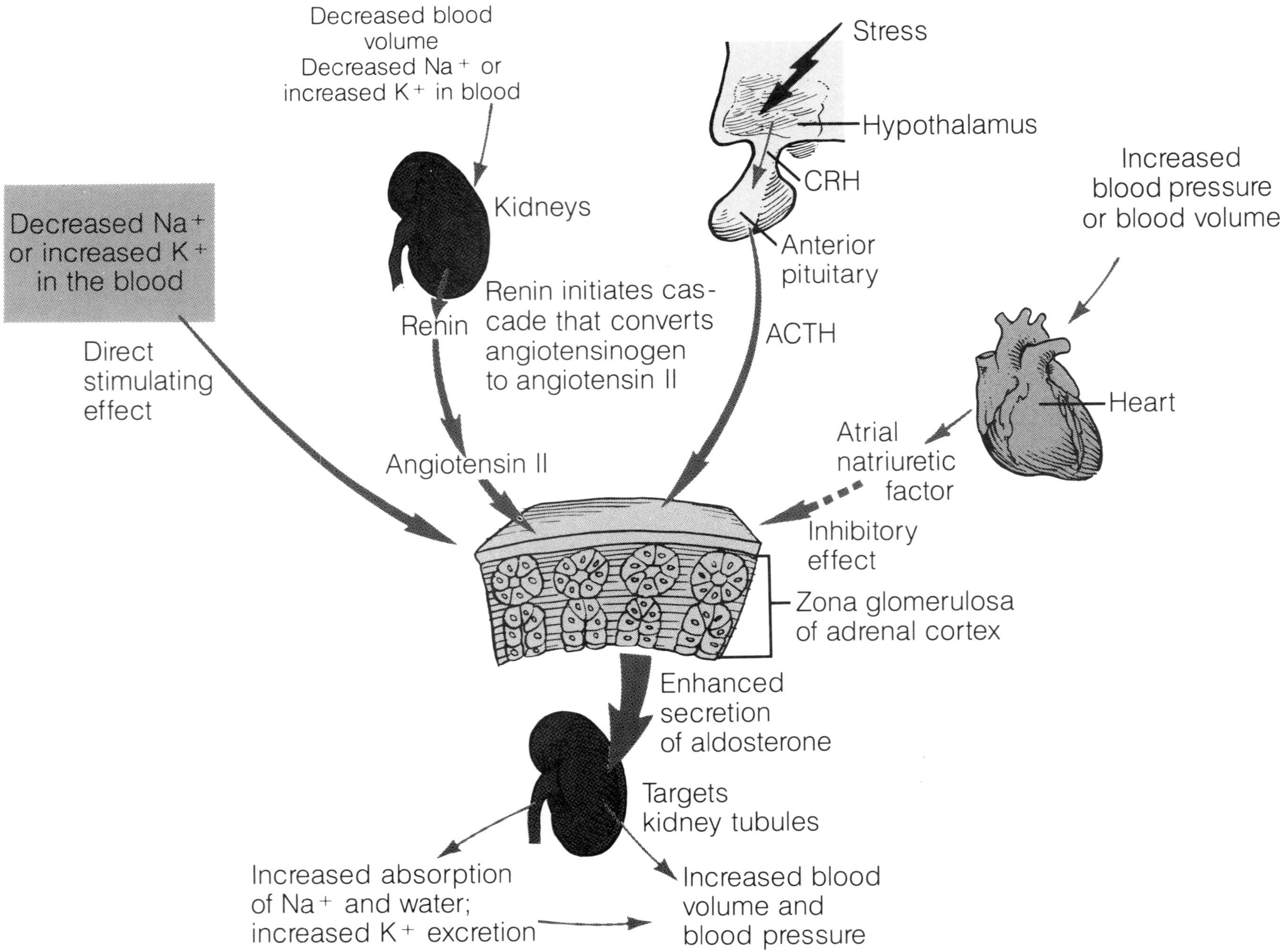

The stress response

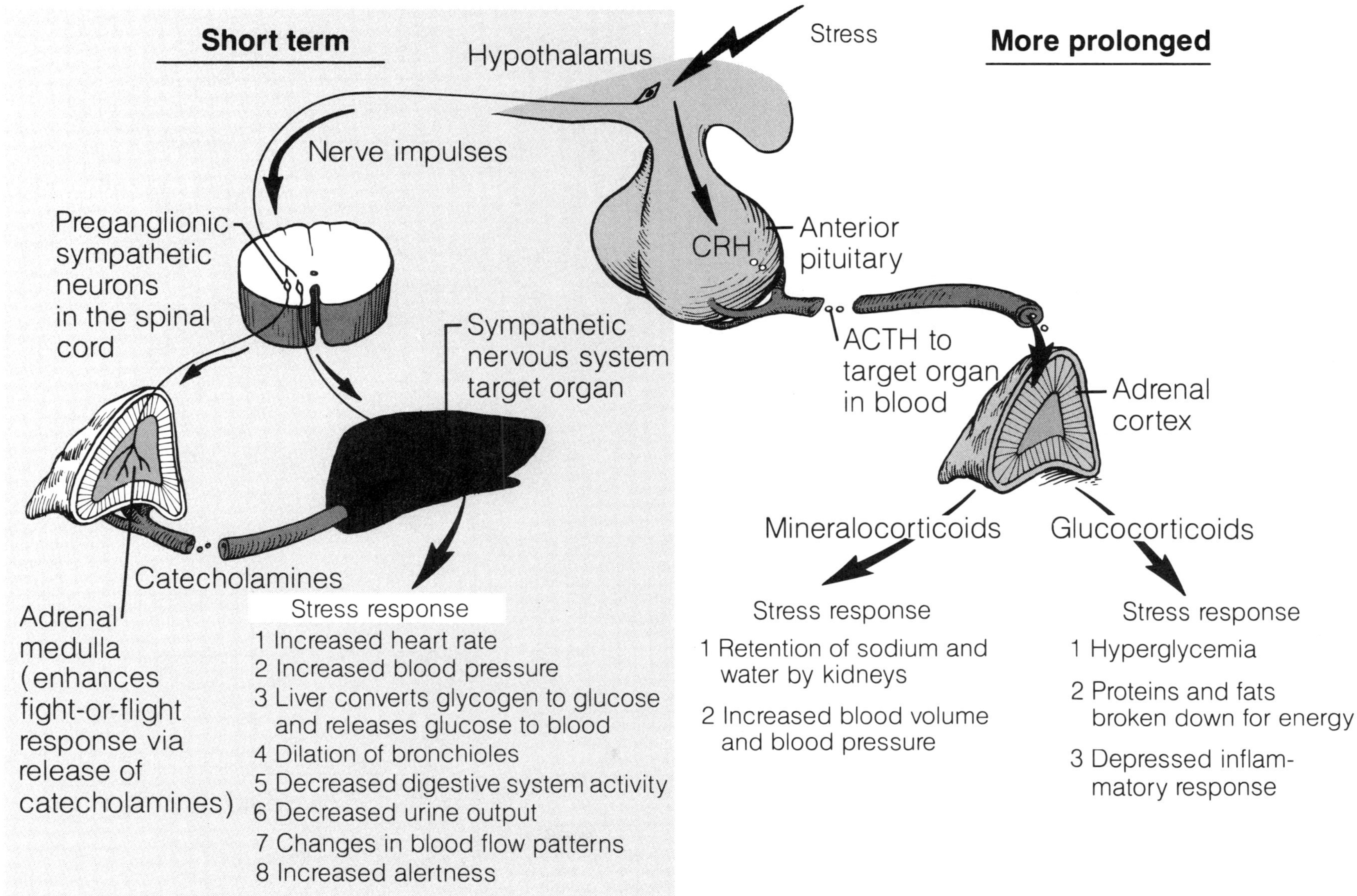

Regulation of blood sugar levels by insulin and glucagon

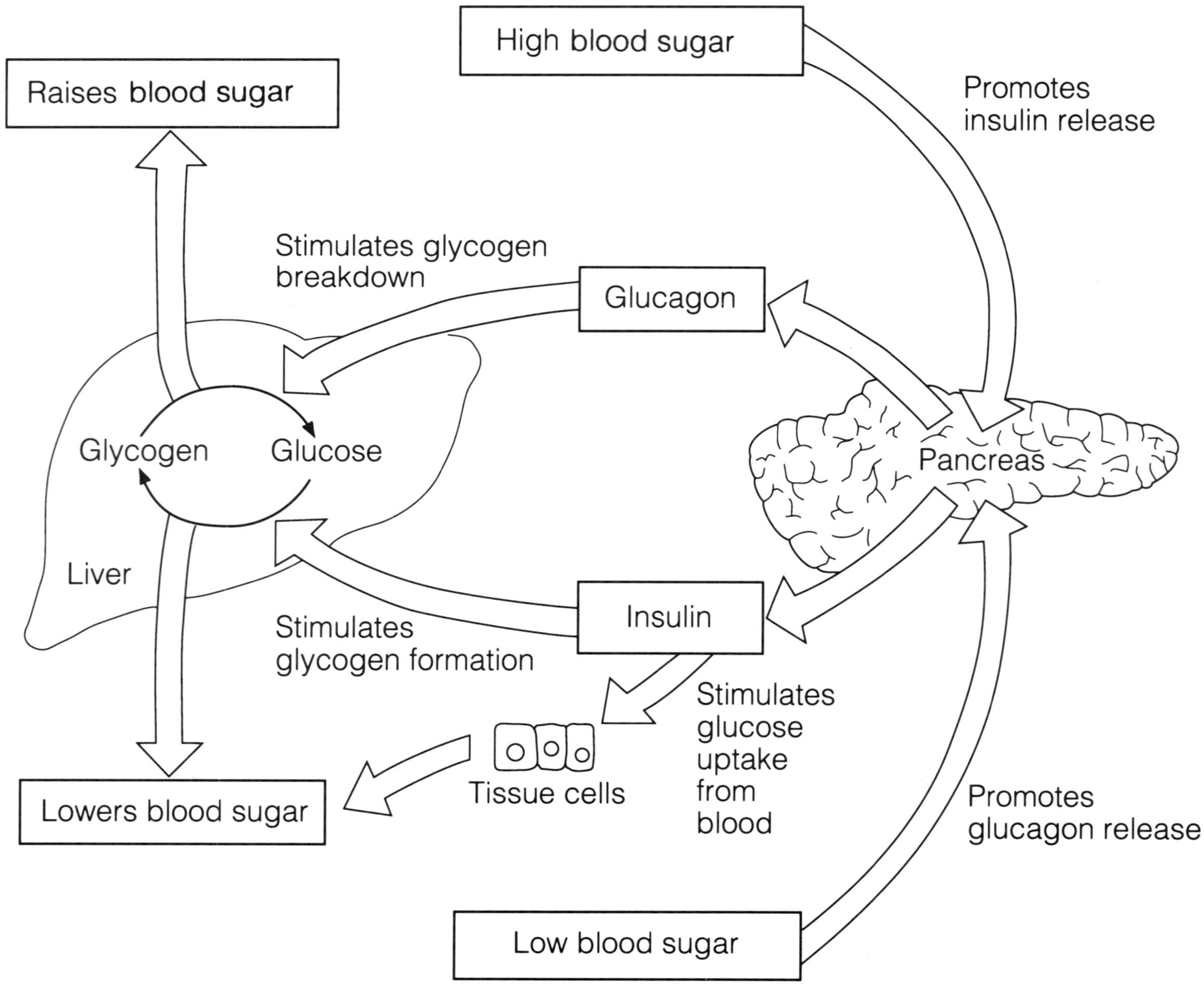

Symptomatic results of insulin deficit (diabetes mellitus)

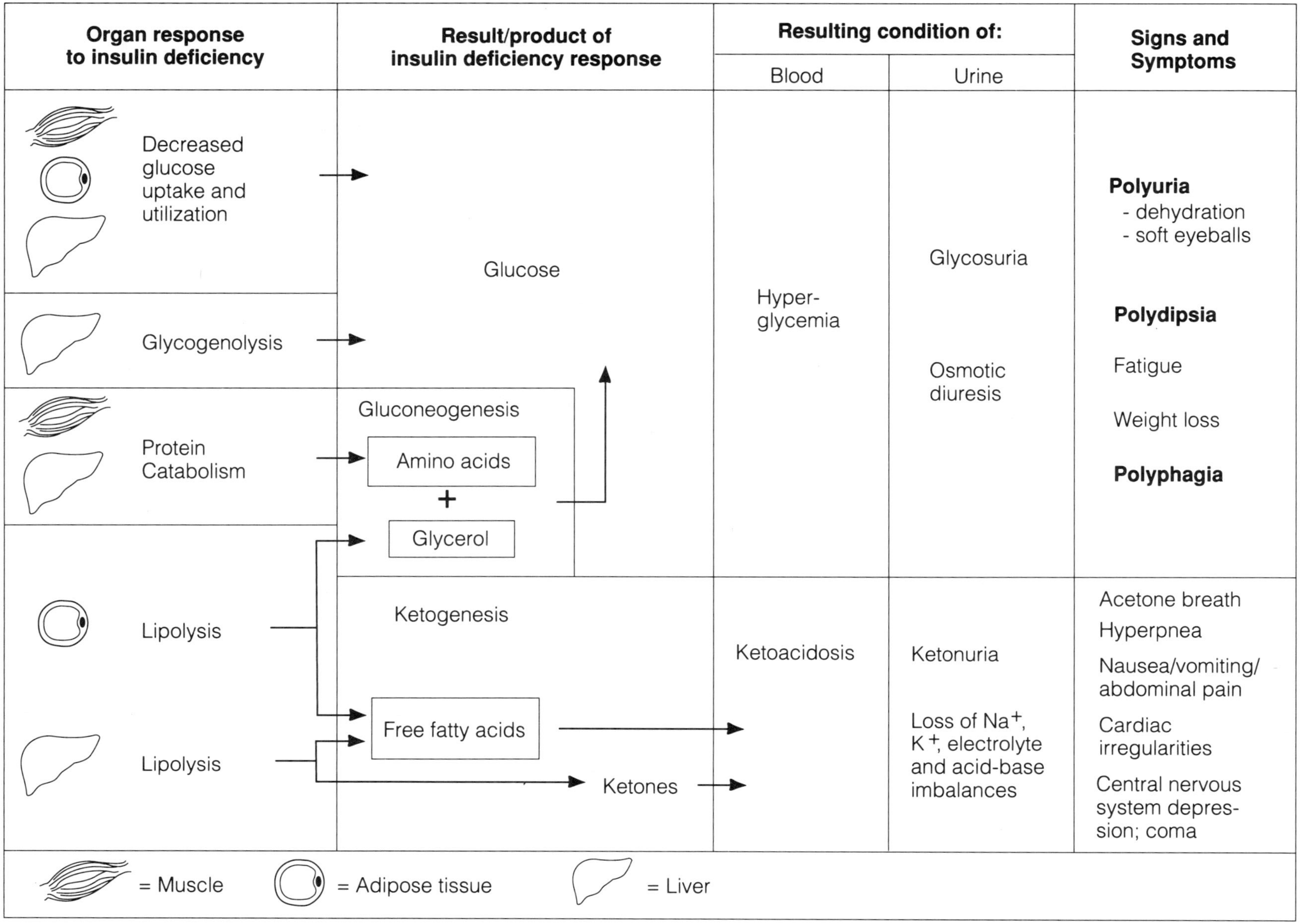

Structure of hemoglobin

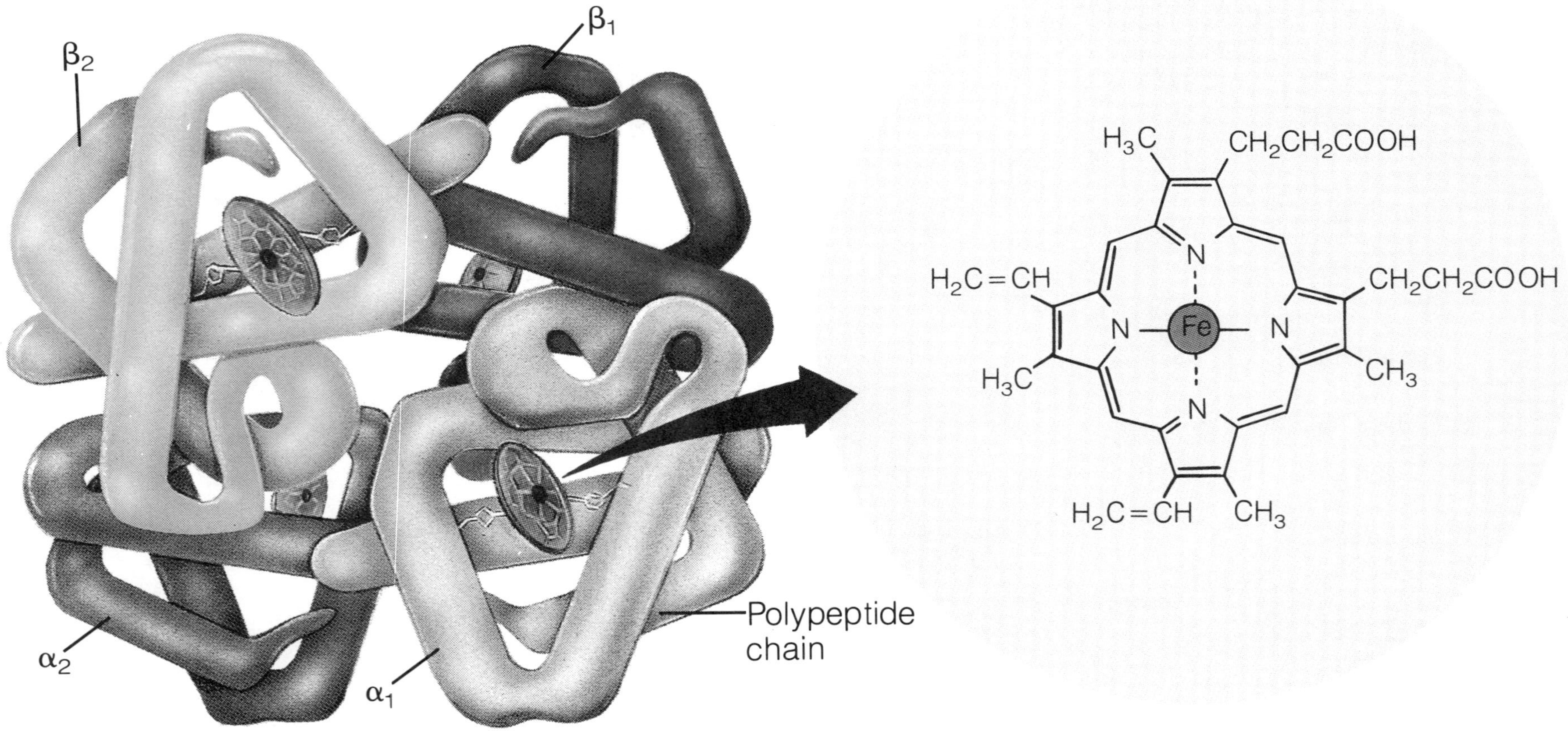

(a) Hemoglobin **(b) Iron-containing heme group**

Erythropoietin mechanism for regulating the rate of erythropoiesis

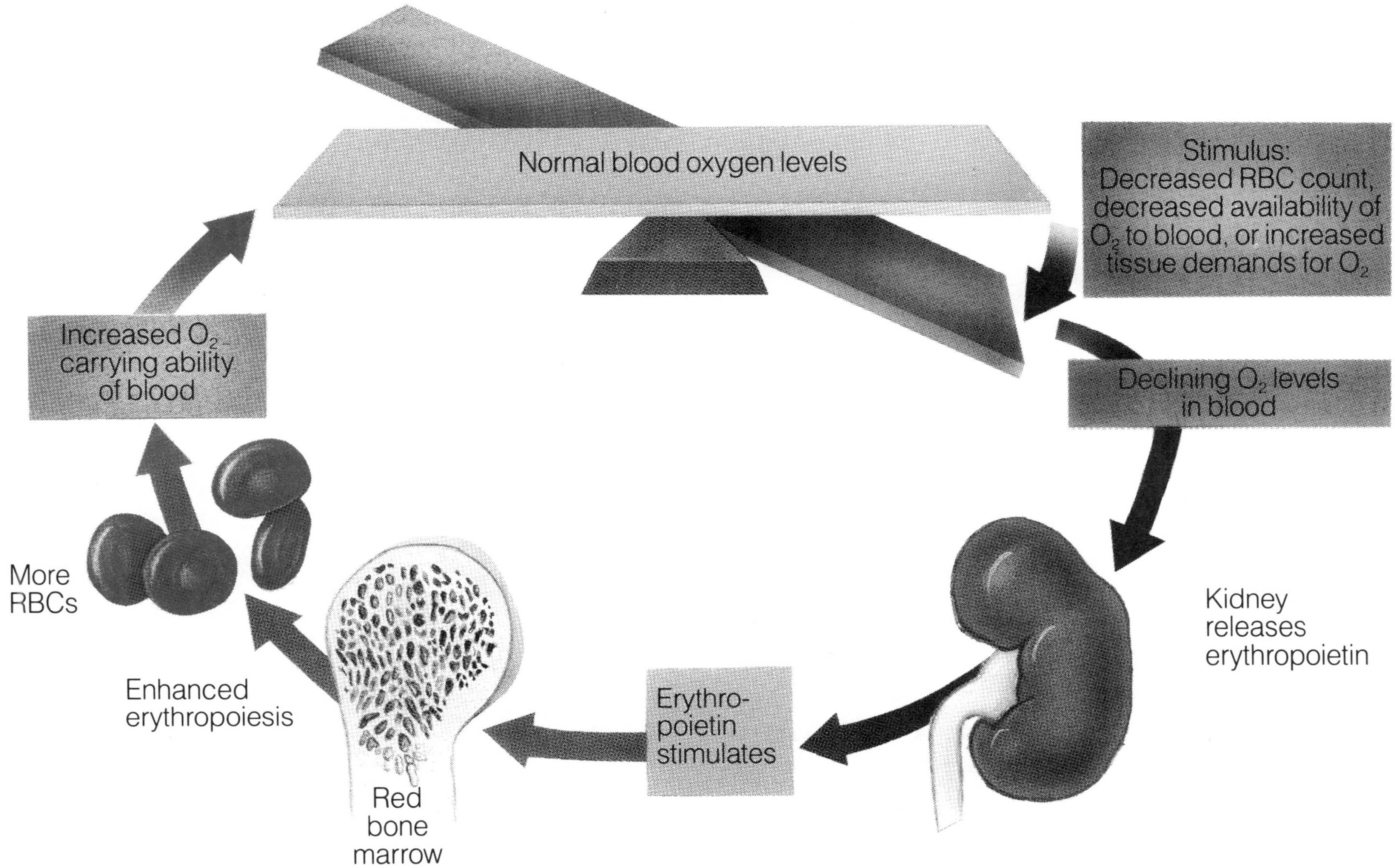

Life cycle of red blood cells

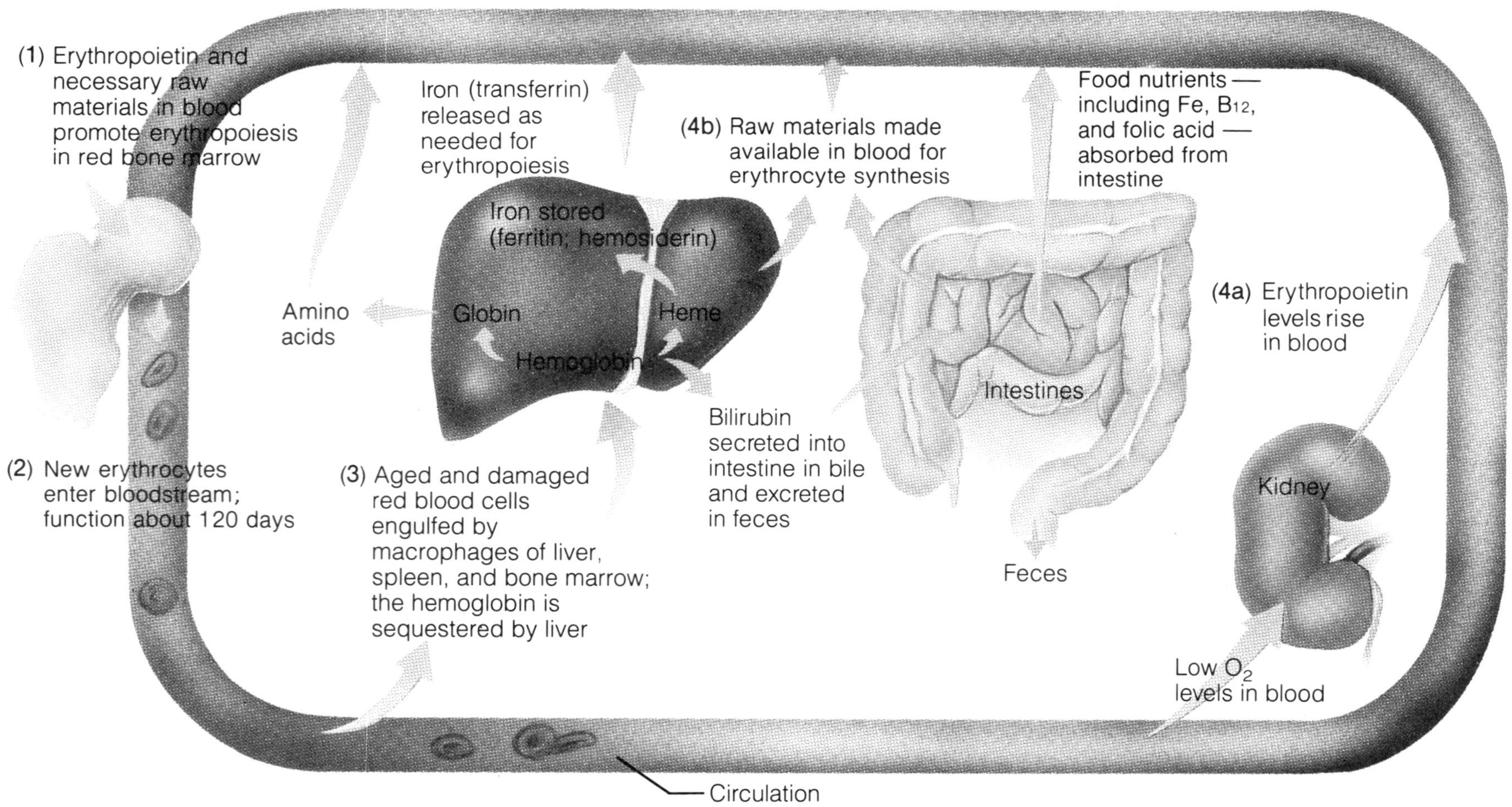

Types of leukocytes

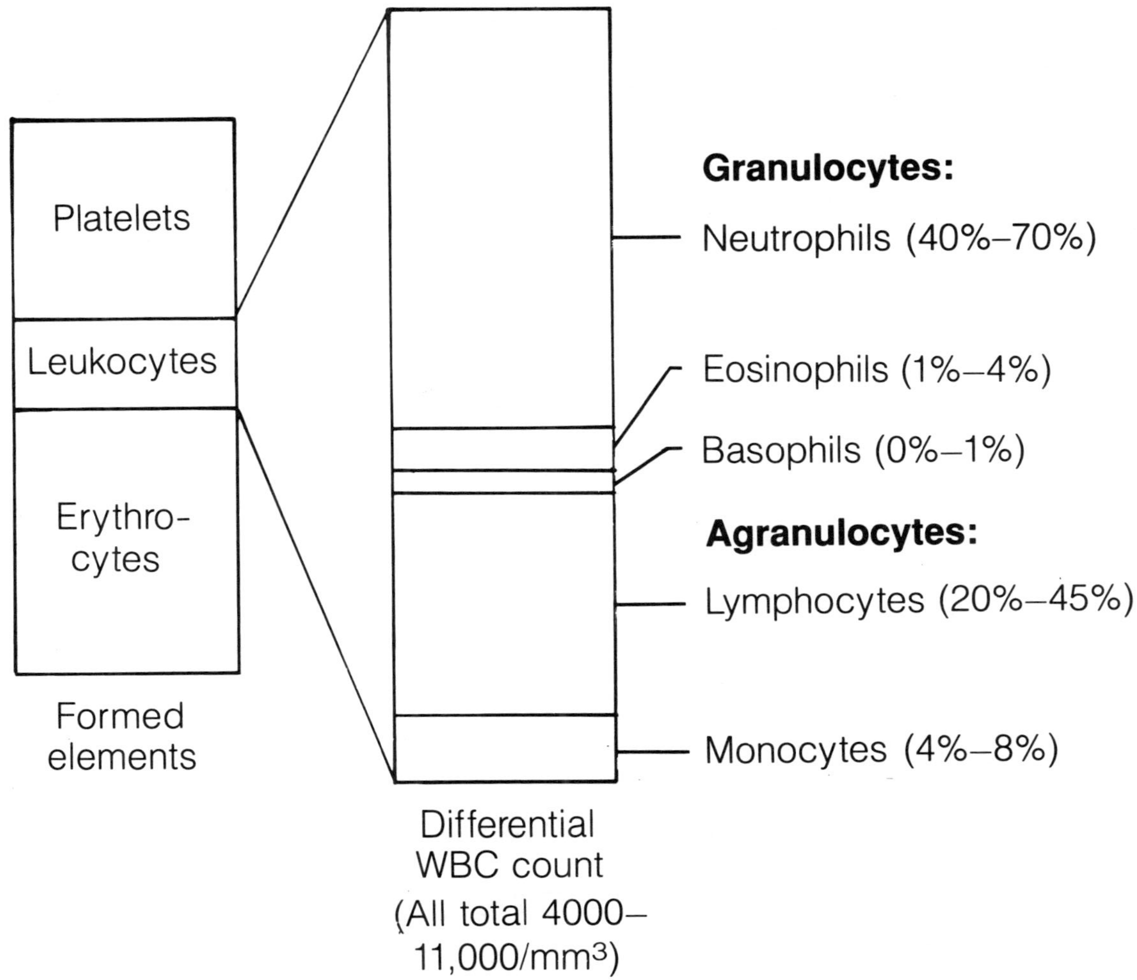

Events of platelet plug formation and blood clotting

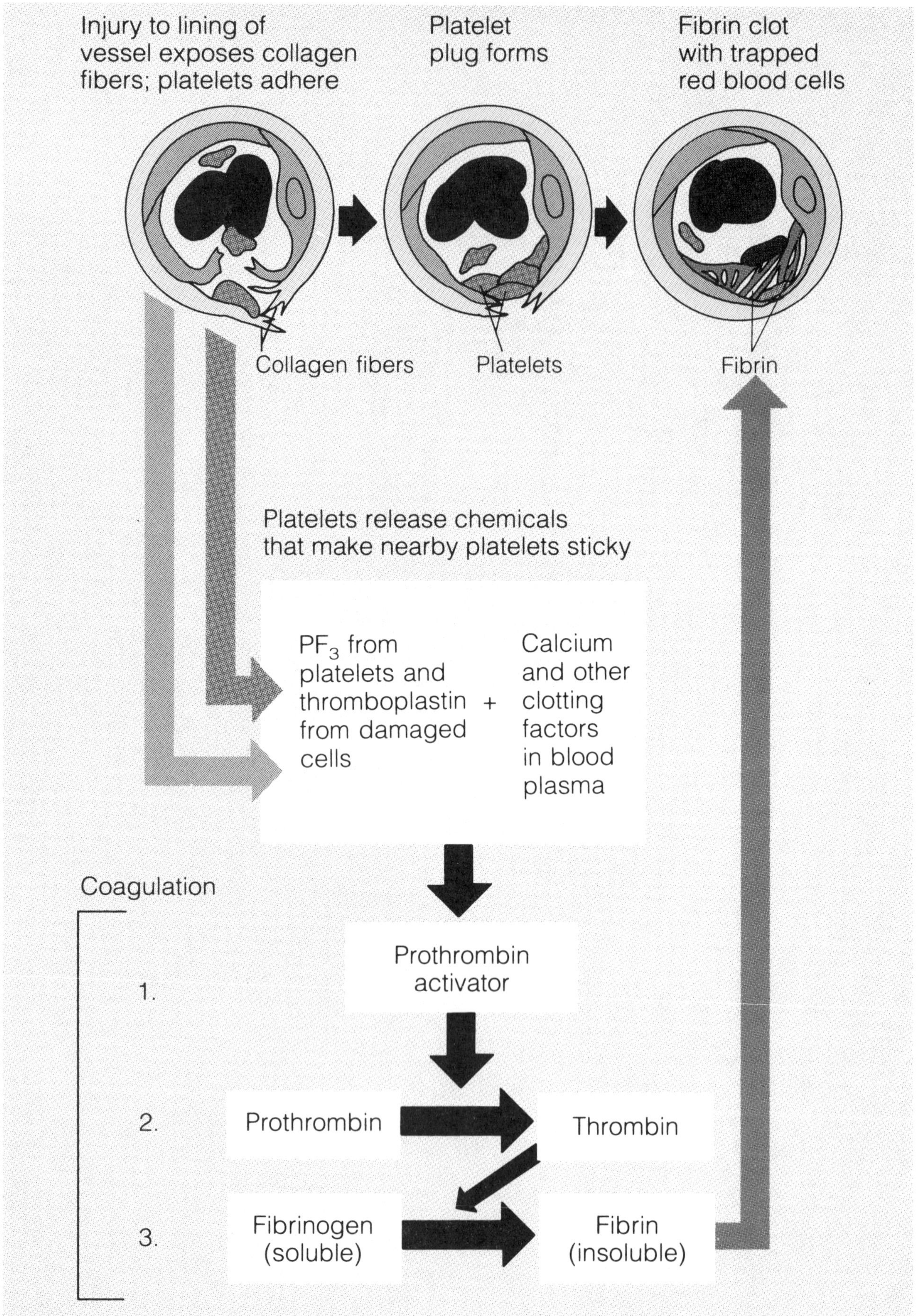

ABO blood groups

Blood group	Frequency (% U.S. population) White Black Asian			RBC antigens (agglutinogens)	Illustration	Plasma antibodies (agglutinins)	Blood that can be received
AB	3	4	5	A B	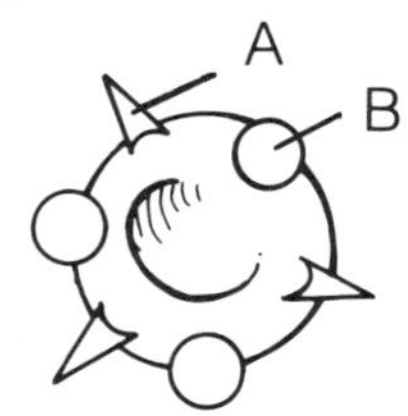 	None	A, B, AB, O Universal recipient
B	9	20	27	B	 	Anti-A	B, O
A	41	27	28	A	 	Anti-B	A, O
O	47	49	40	None	 	Anti-A Anti-B	O Universal donor

Microscopic anatomy of cardiac muscle

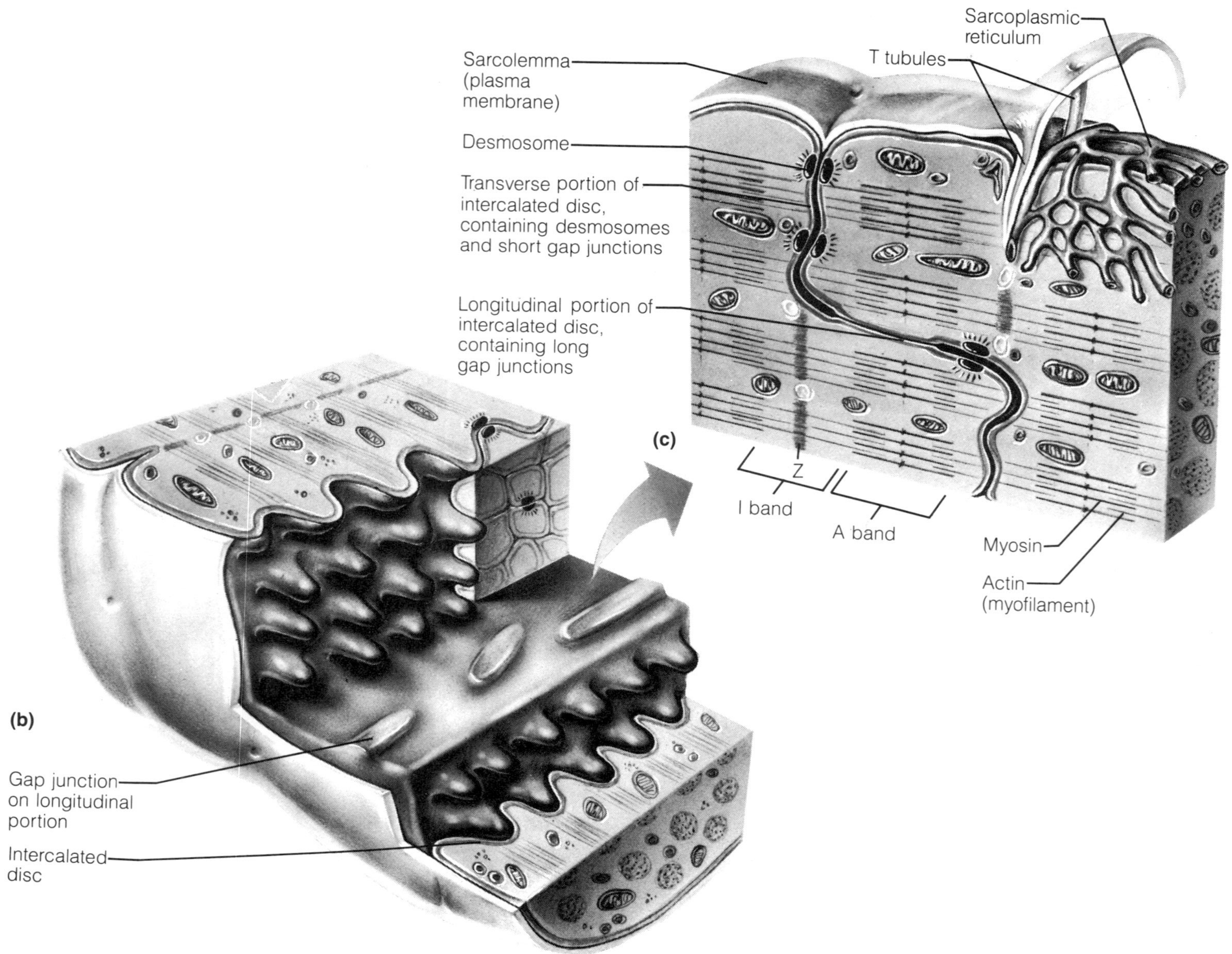

Changes in membrane potential and membrane permeability during action potentials of contractile cardiac muscle cells (Figure 19.12)

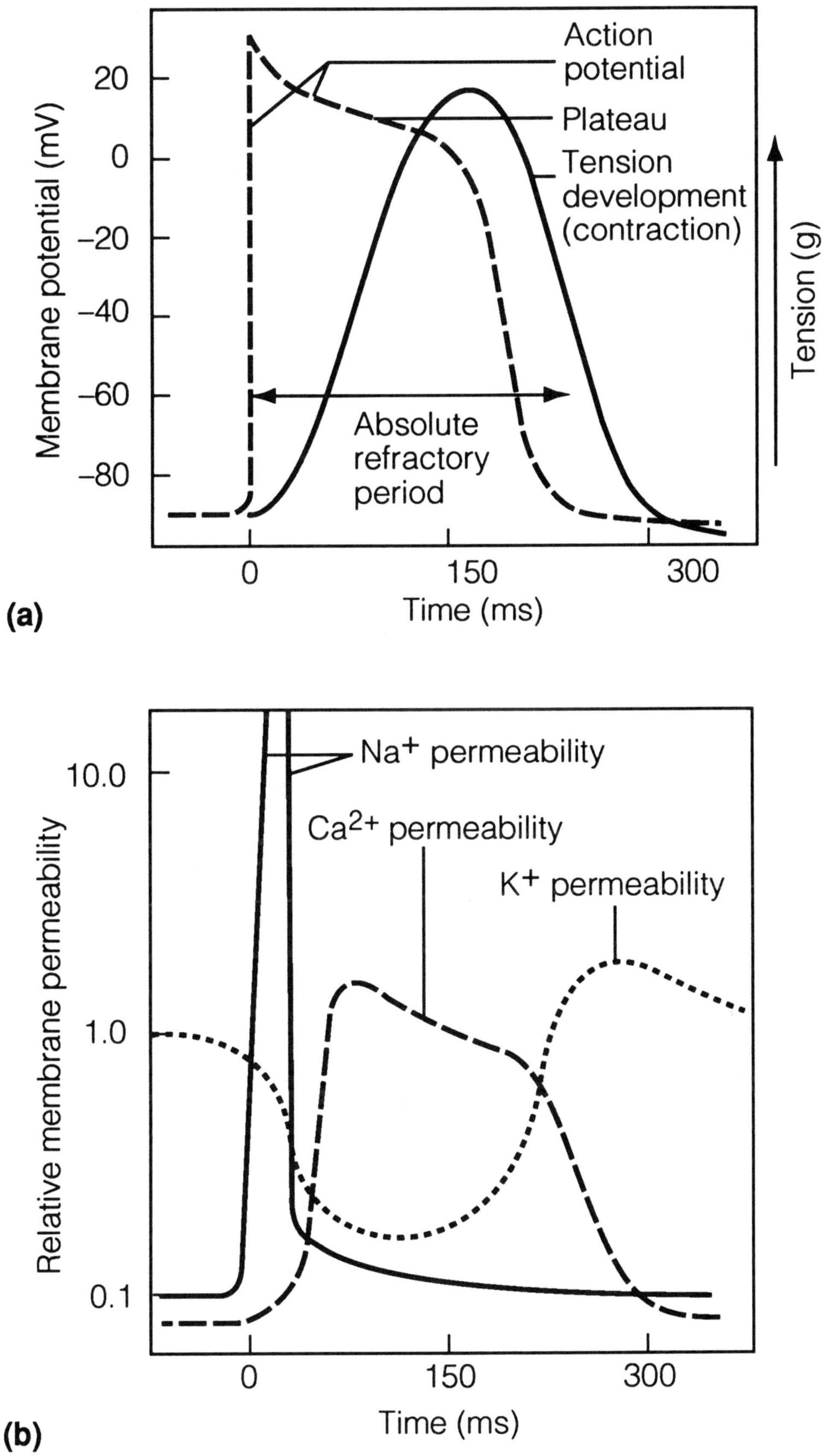

Pacemaker and action potentials of autorhythmic cells of the heart

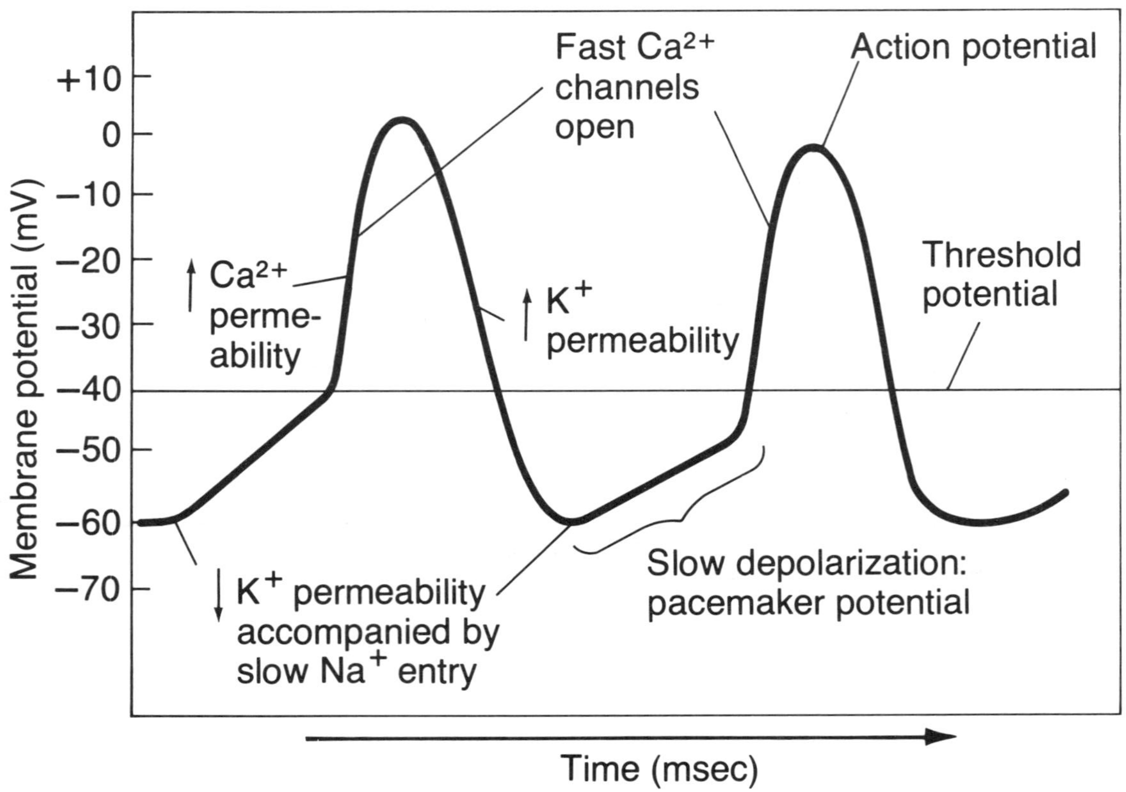

Intrinsic conduction system of the heart

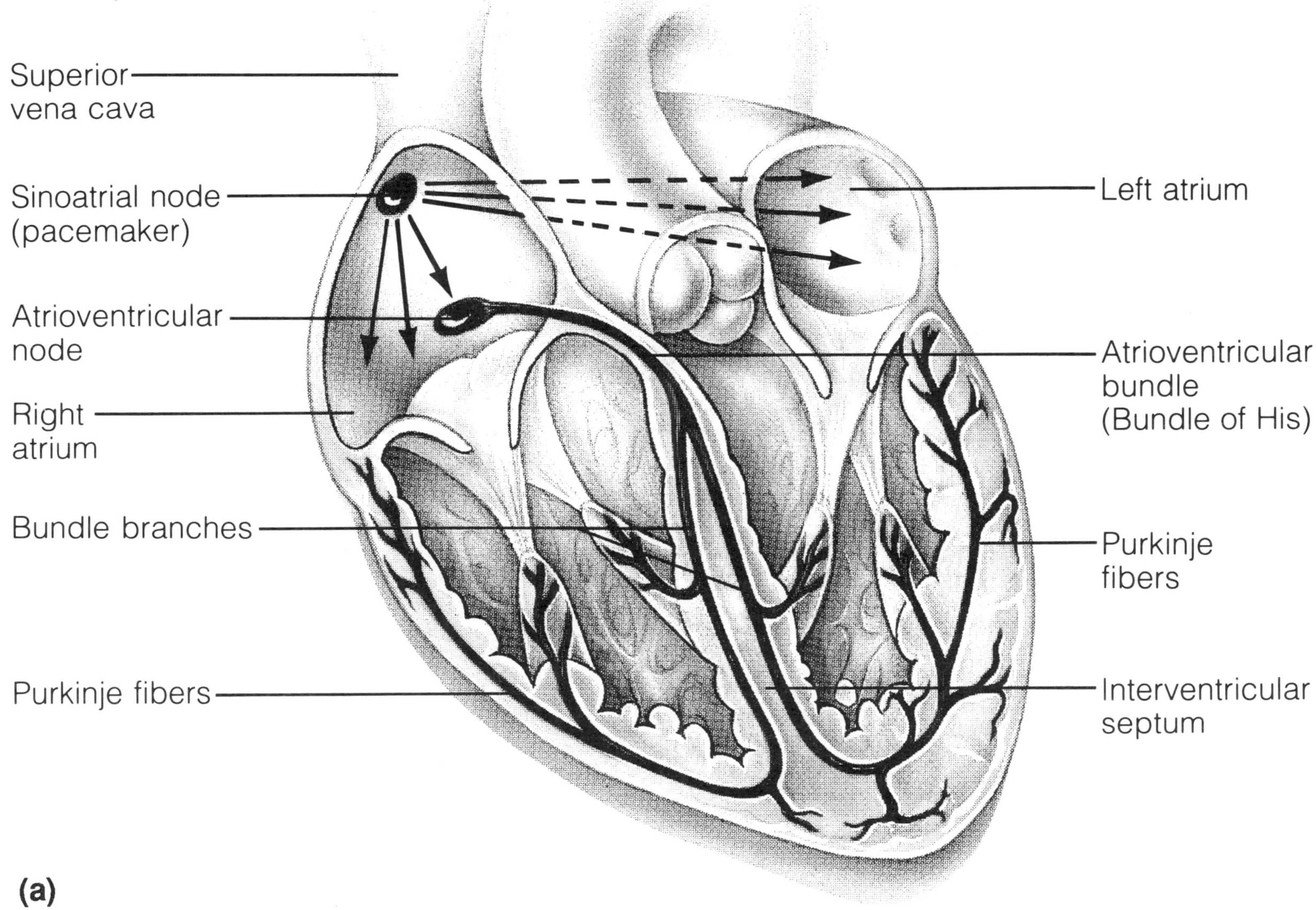

(a)

Intrinsic conduction system of the heart: Sequence of potentials generated across the heart

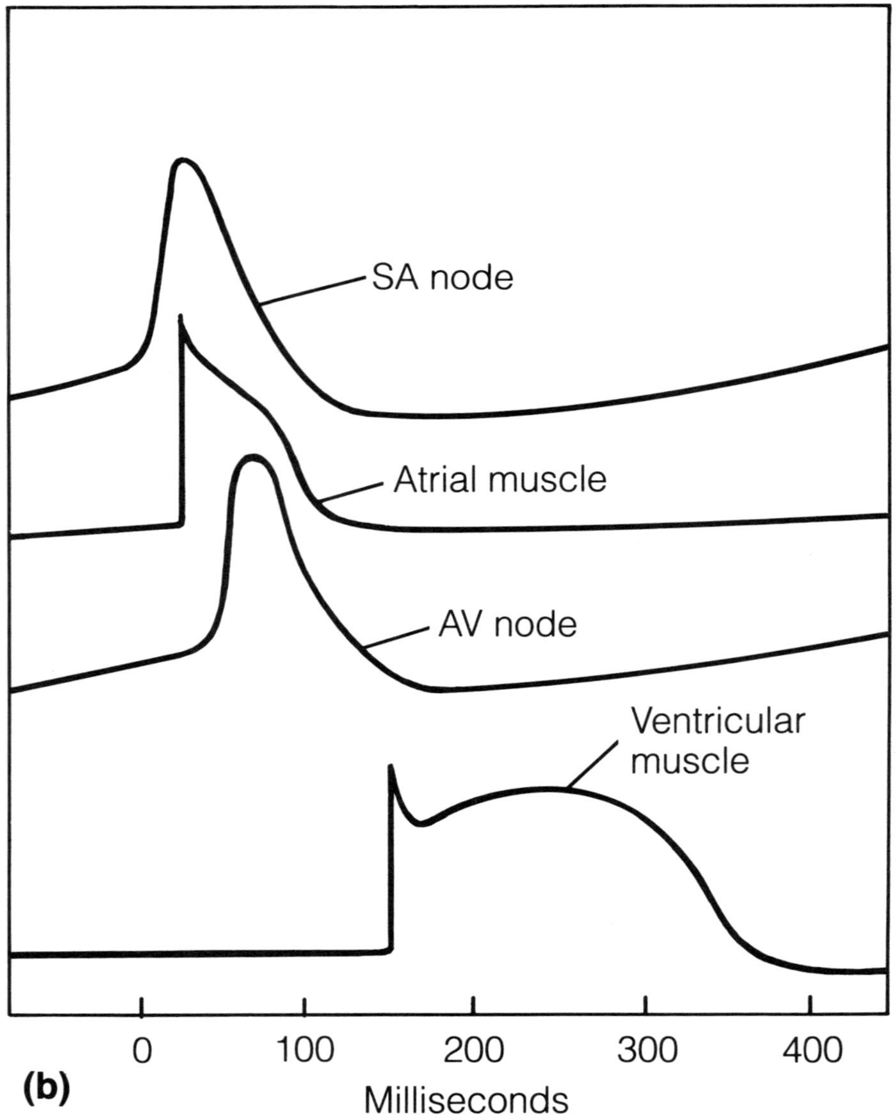

(b)

An electrocardiogram tracing illustrating the three normally recognizable deflection waves and the important intervals

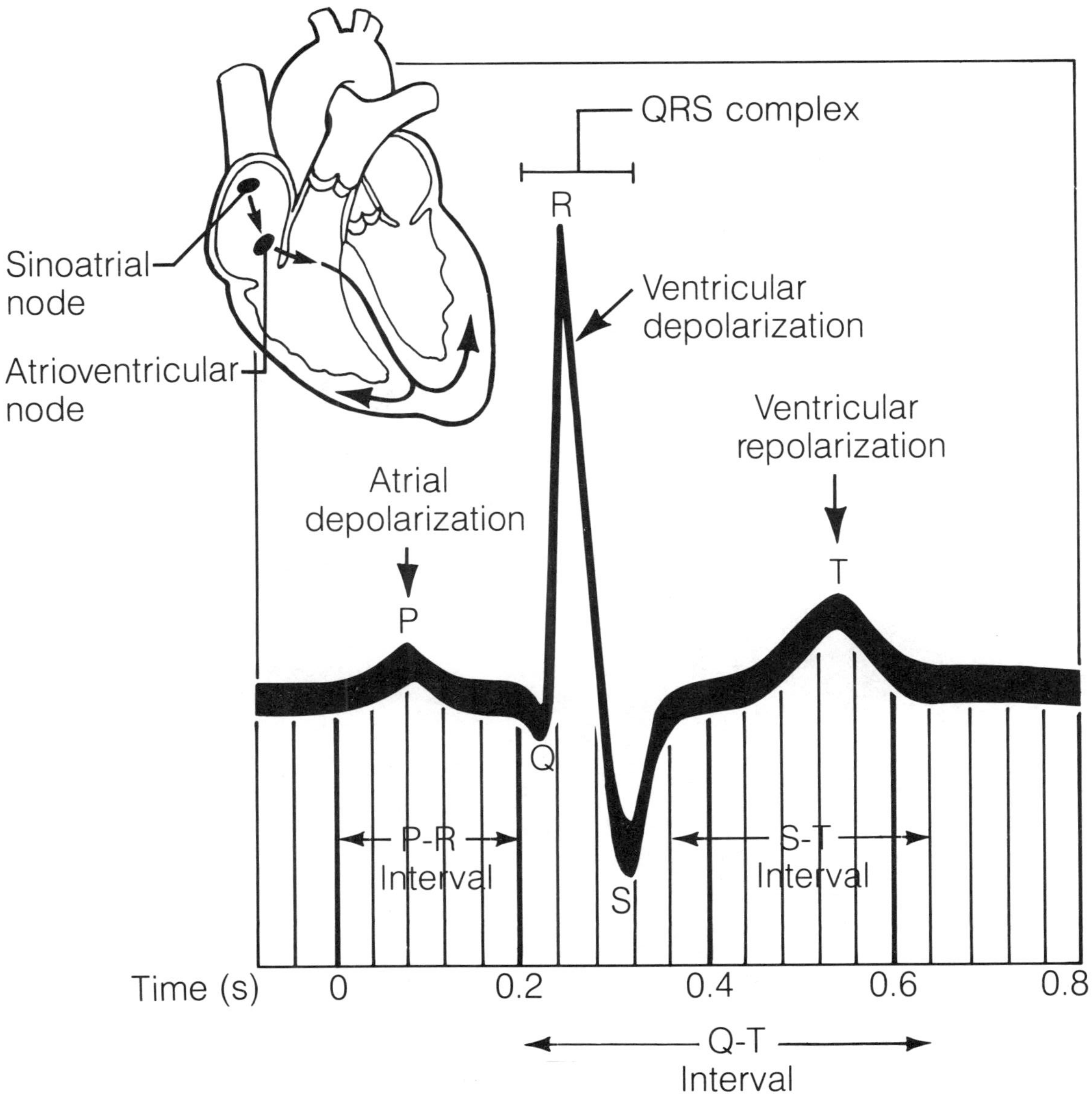

The sequence of excitation of the heart is related to the deflection waves of an ECG tracing

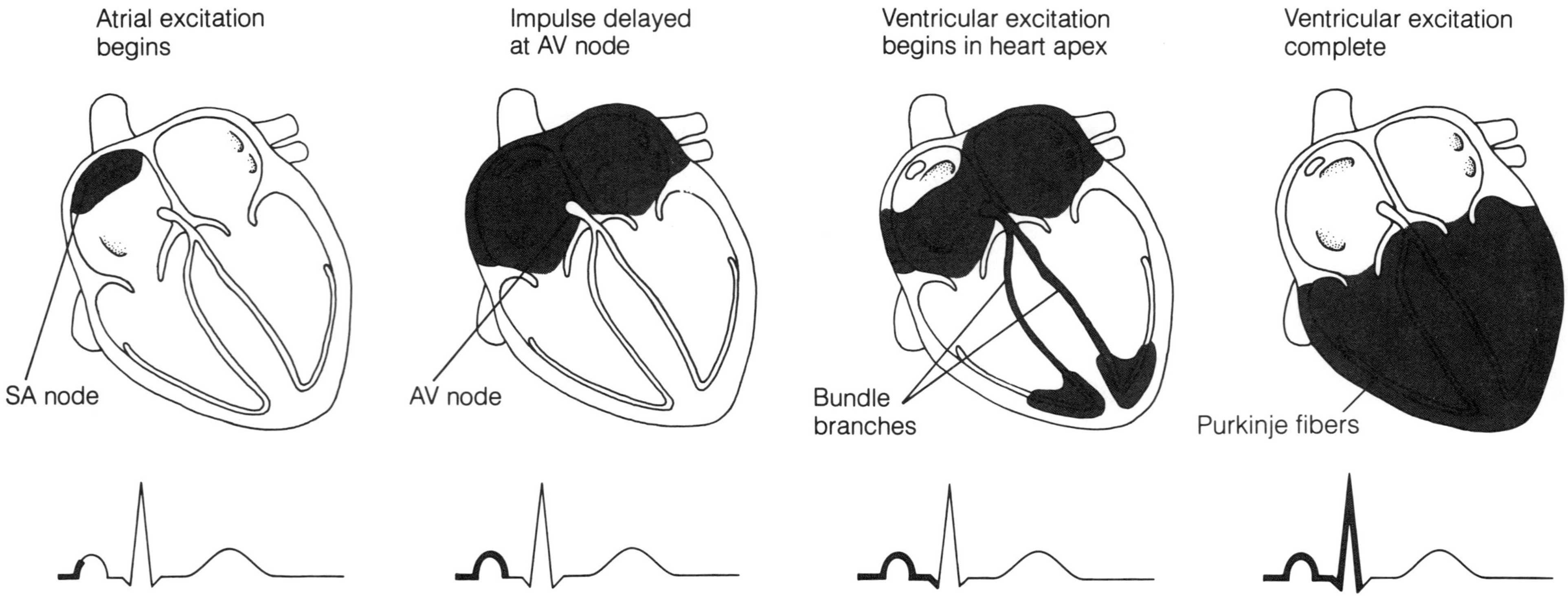

Normal and abnormal ECG tracings

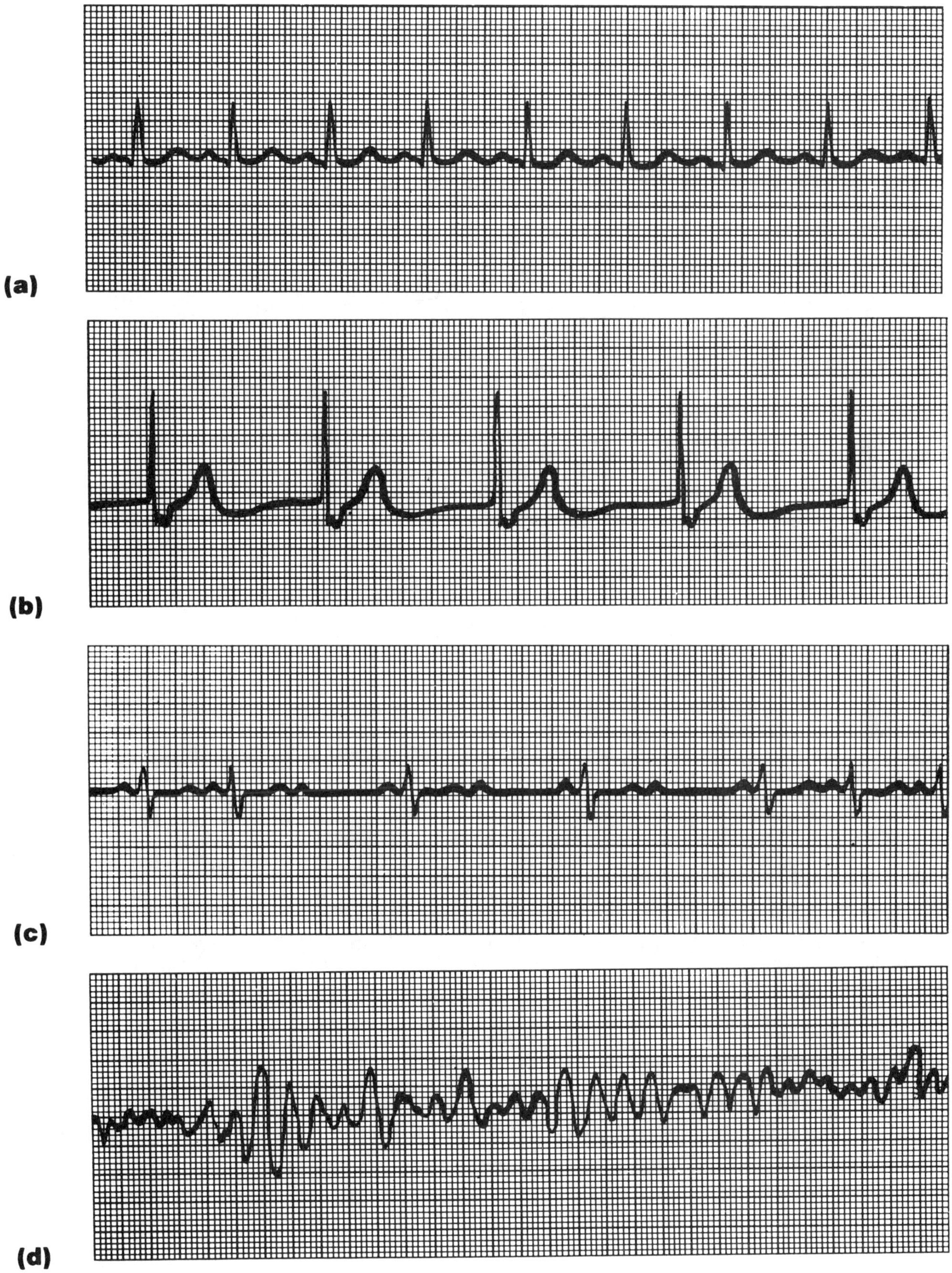

Summary of events occurring in the heart during the cardiac cycle

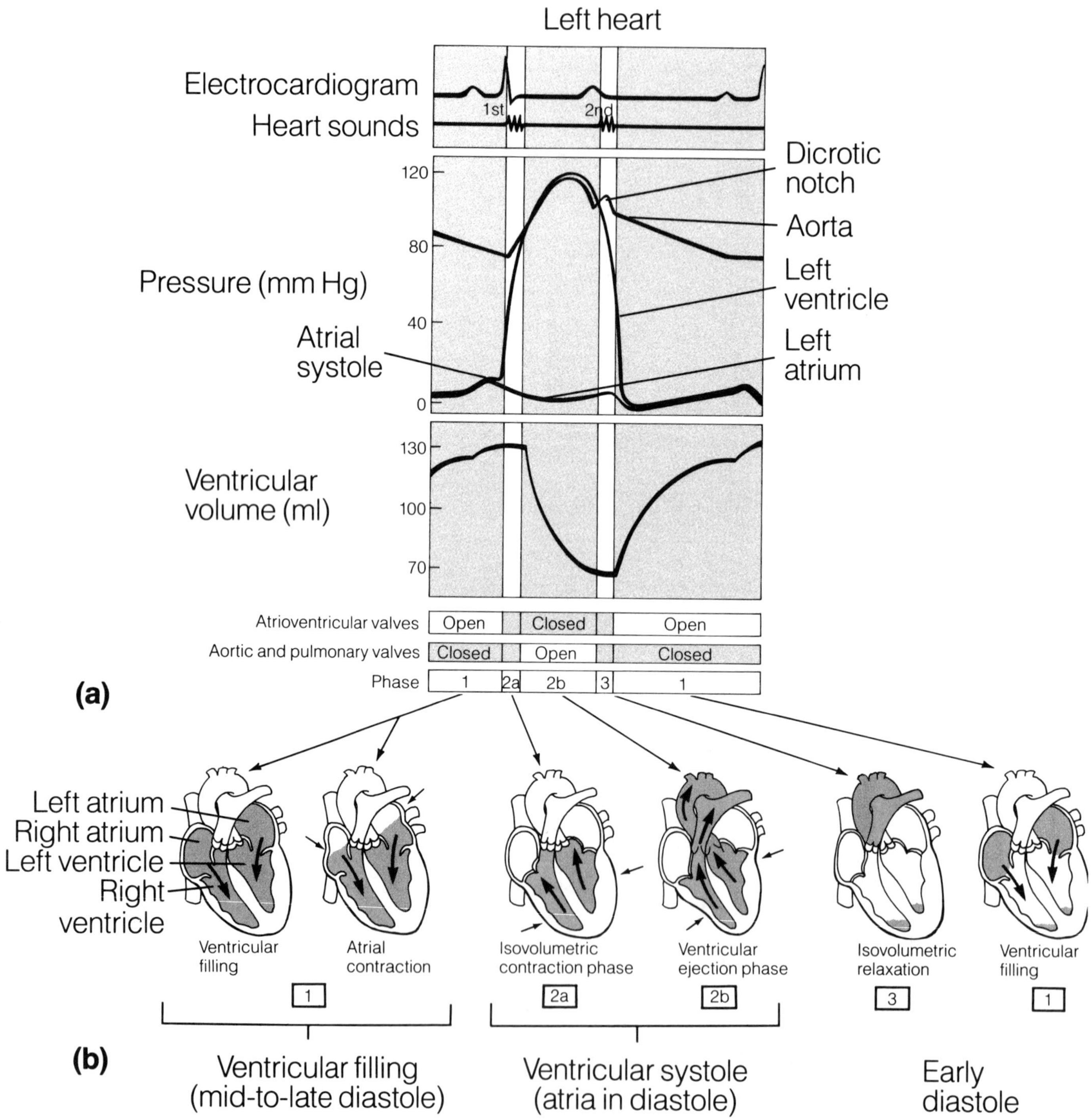

Areas of the thoracic surface where the heart sounds can be best detected

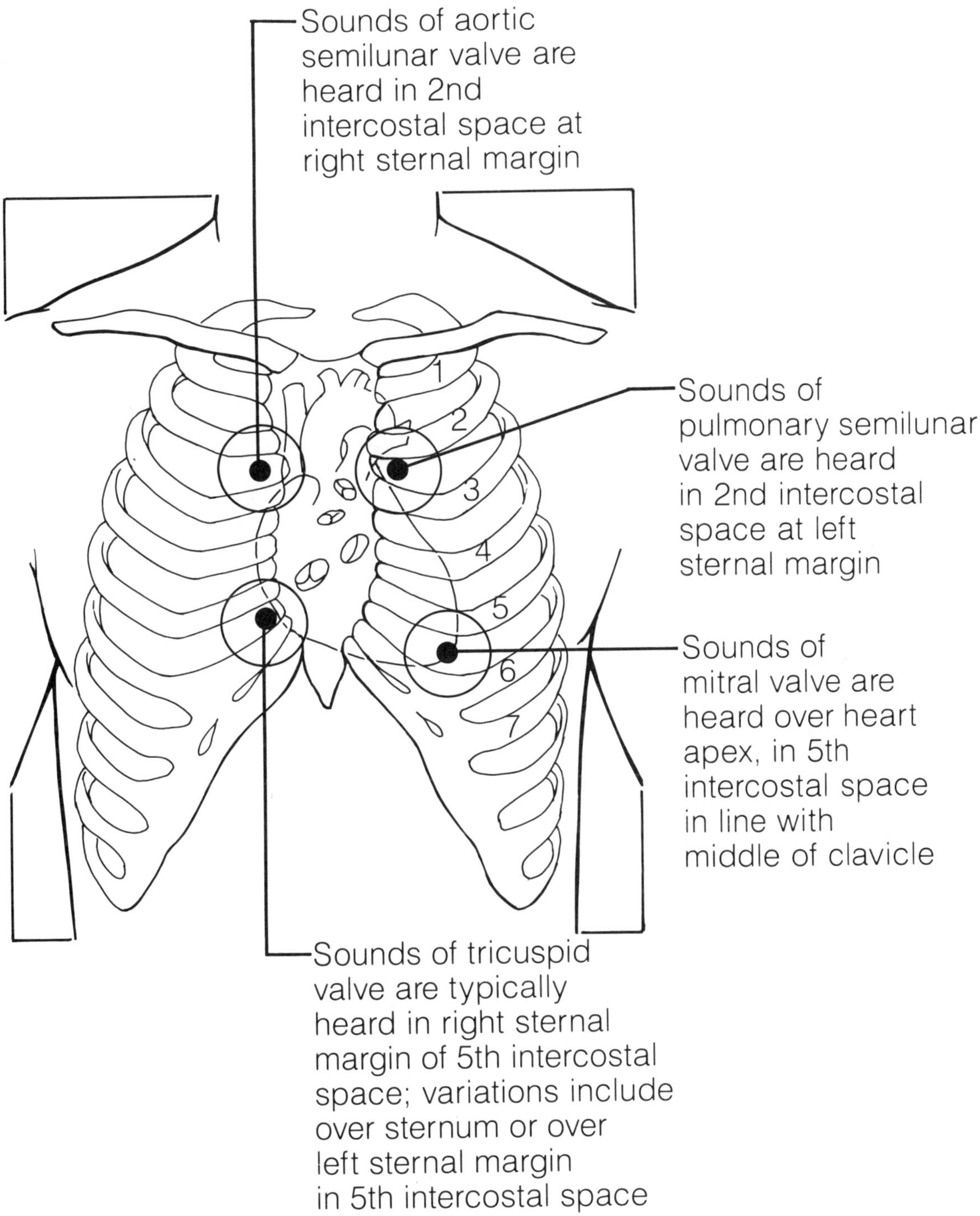

Influence of the autonomic nervous system on cardiac output

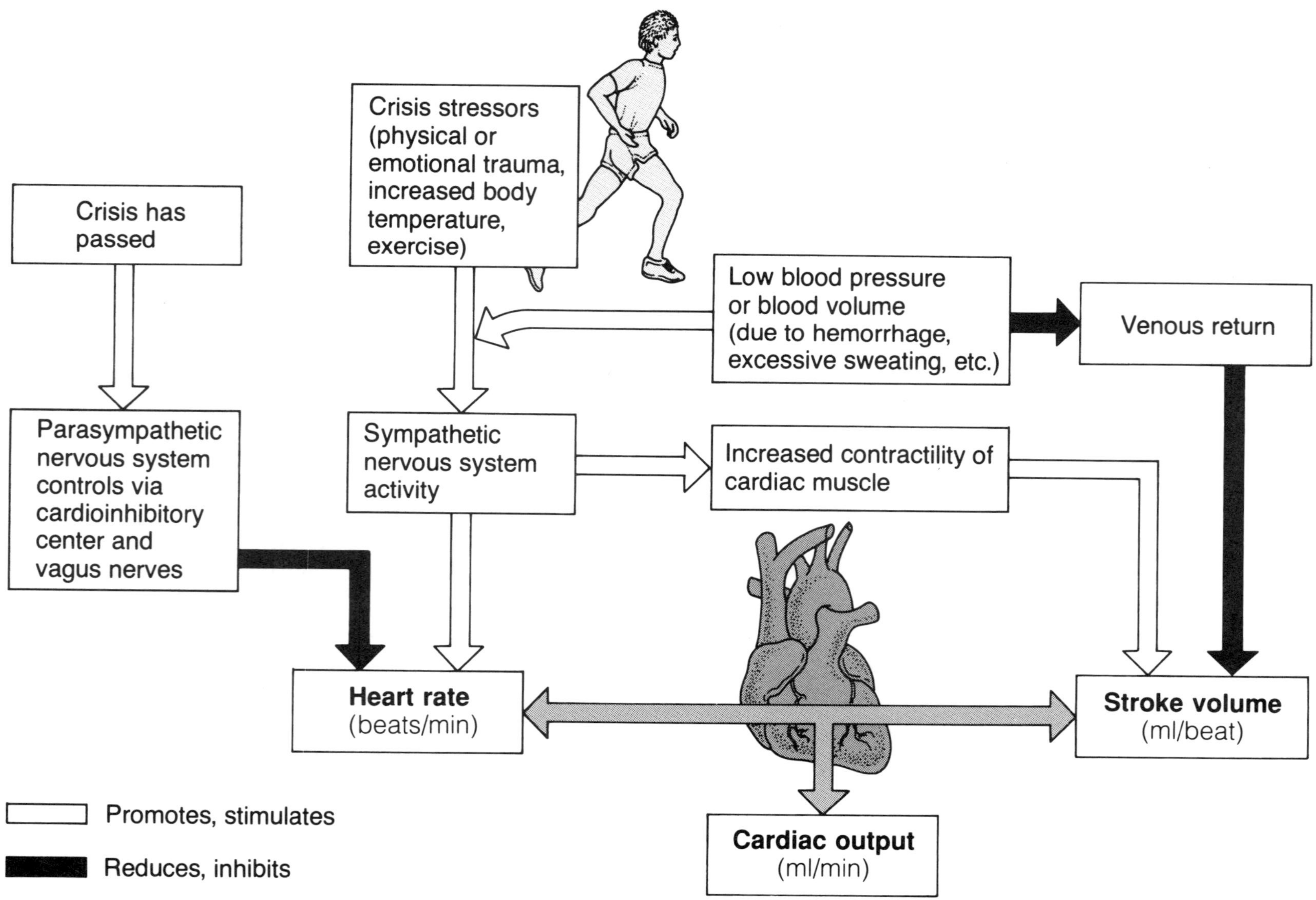

Capillary structure and transport mechanisms

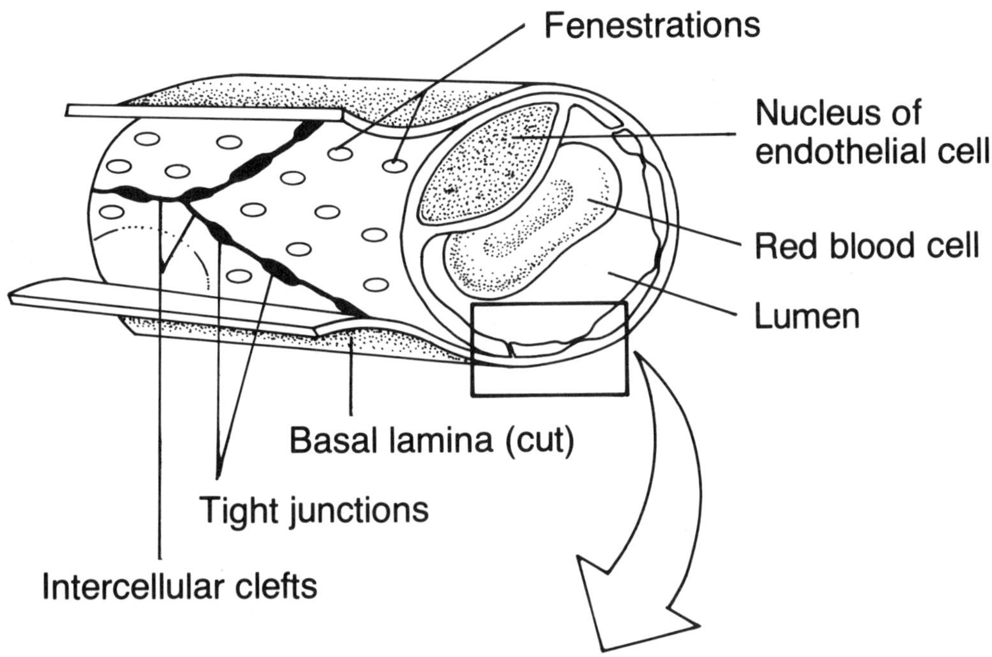

(a)

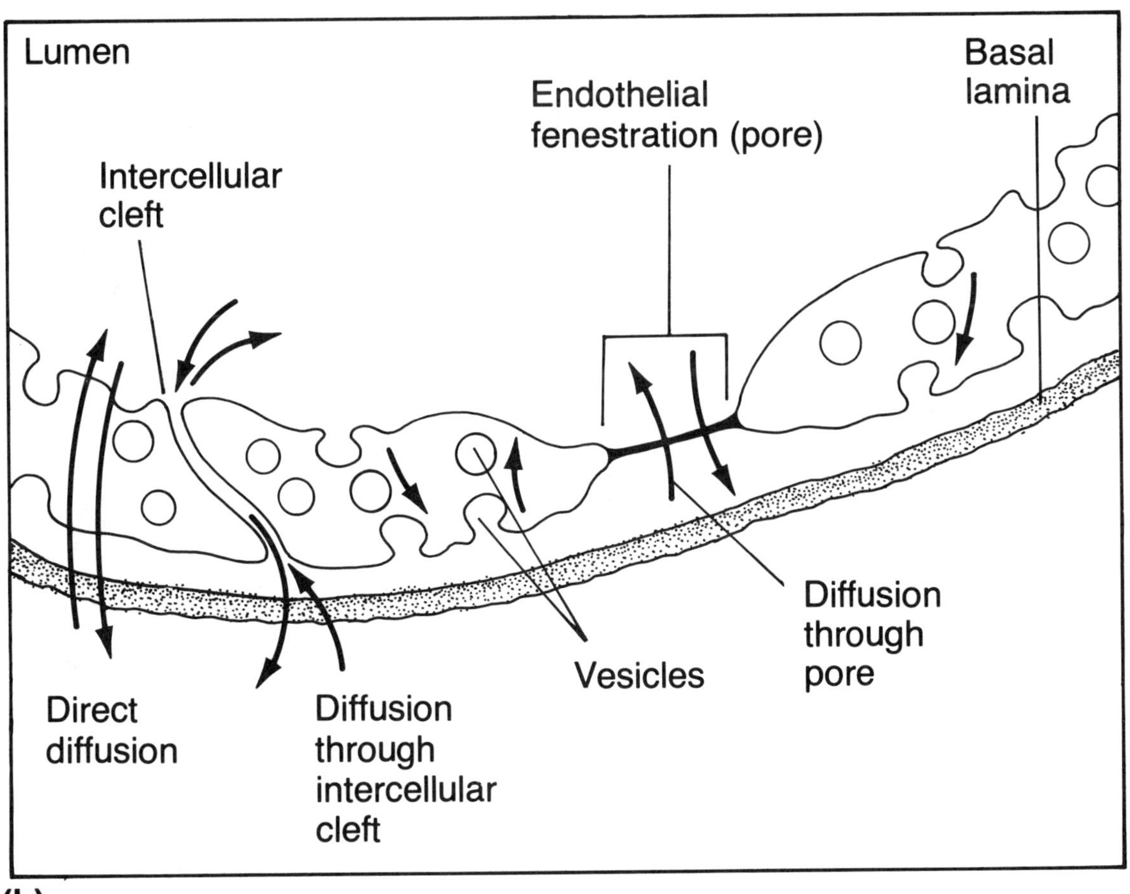

(b)

Relationship between blood pressure, blood flow velocity, and total cross-sectional area of the systemic circulation

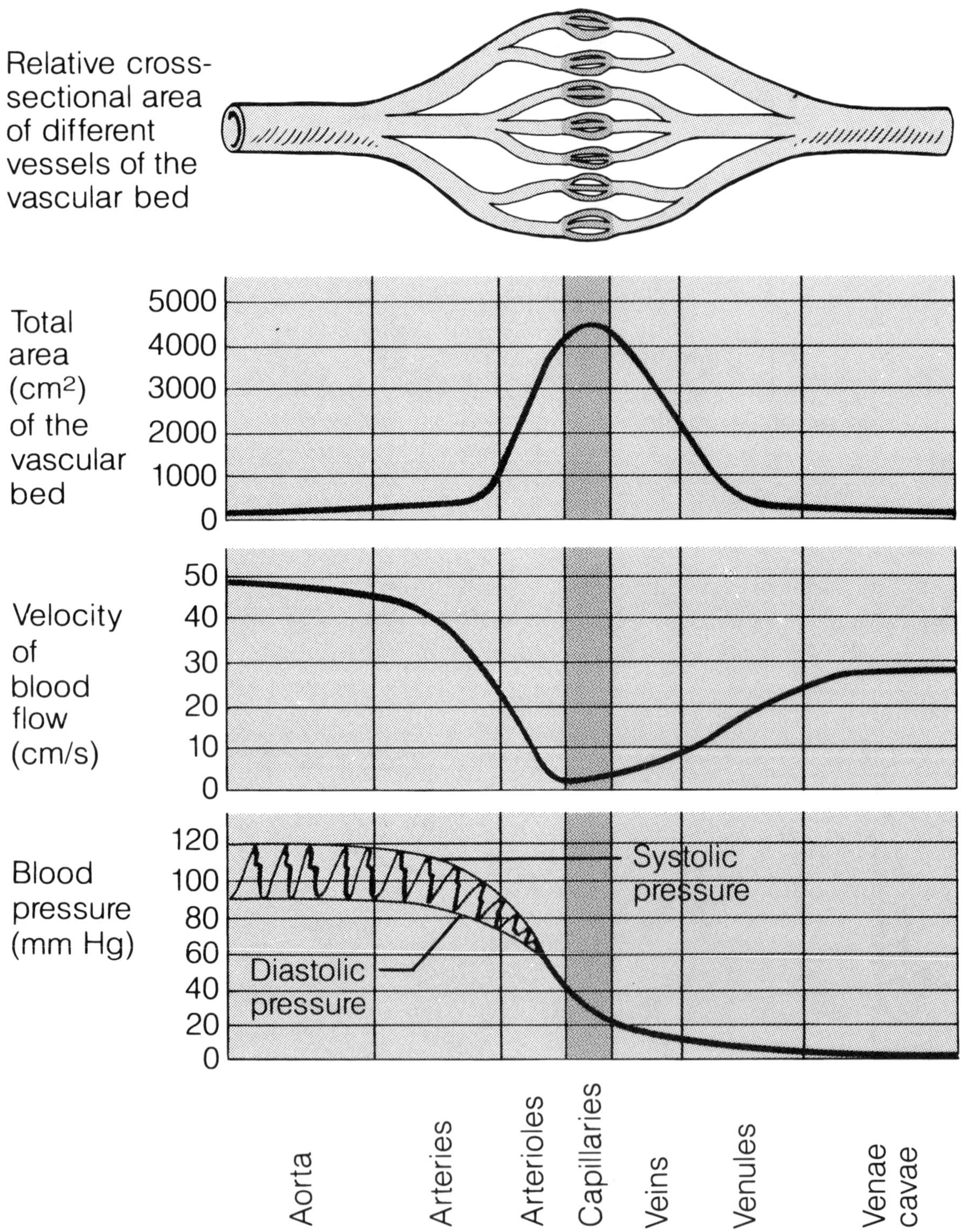

Operation of the muscular pump

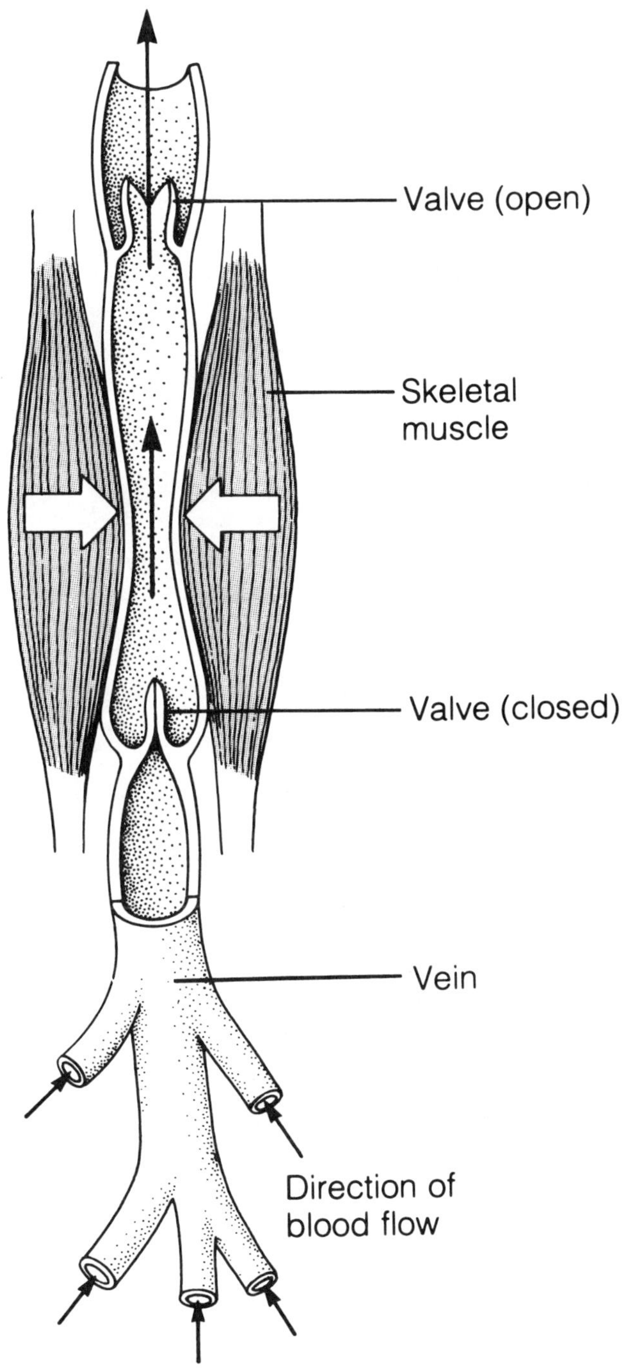

Baroreceptor reflexes that help to maintain blood pressure homeostasis

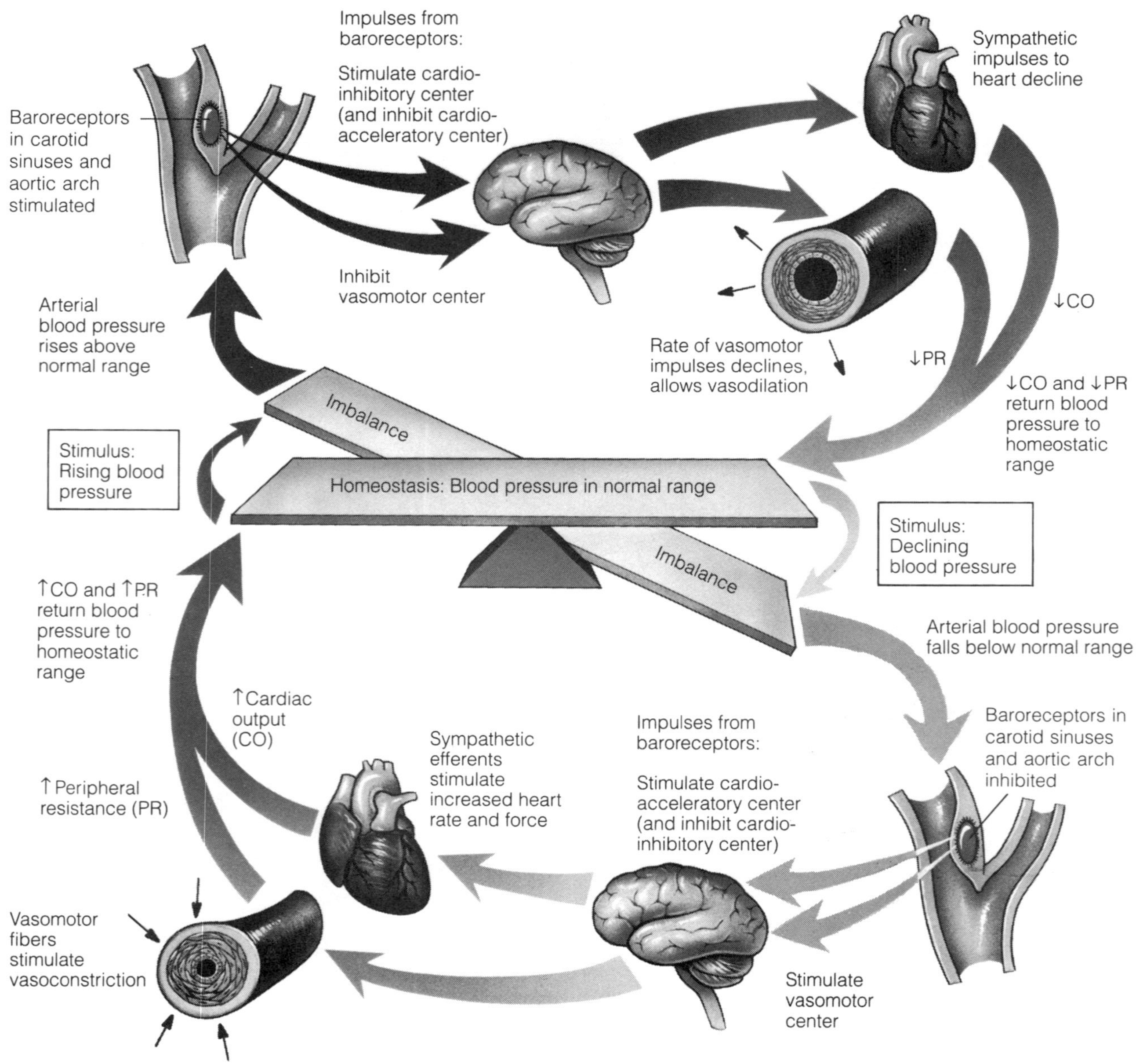

Summary of factors causing an increase in mean systemic arterial blood pressure

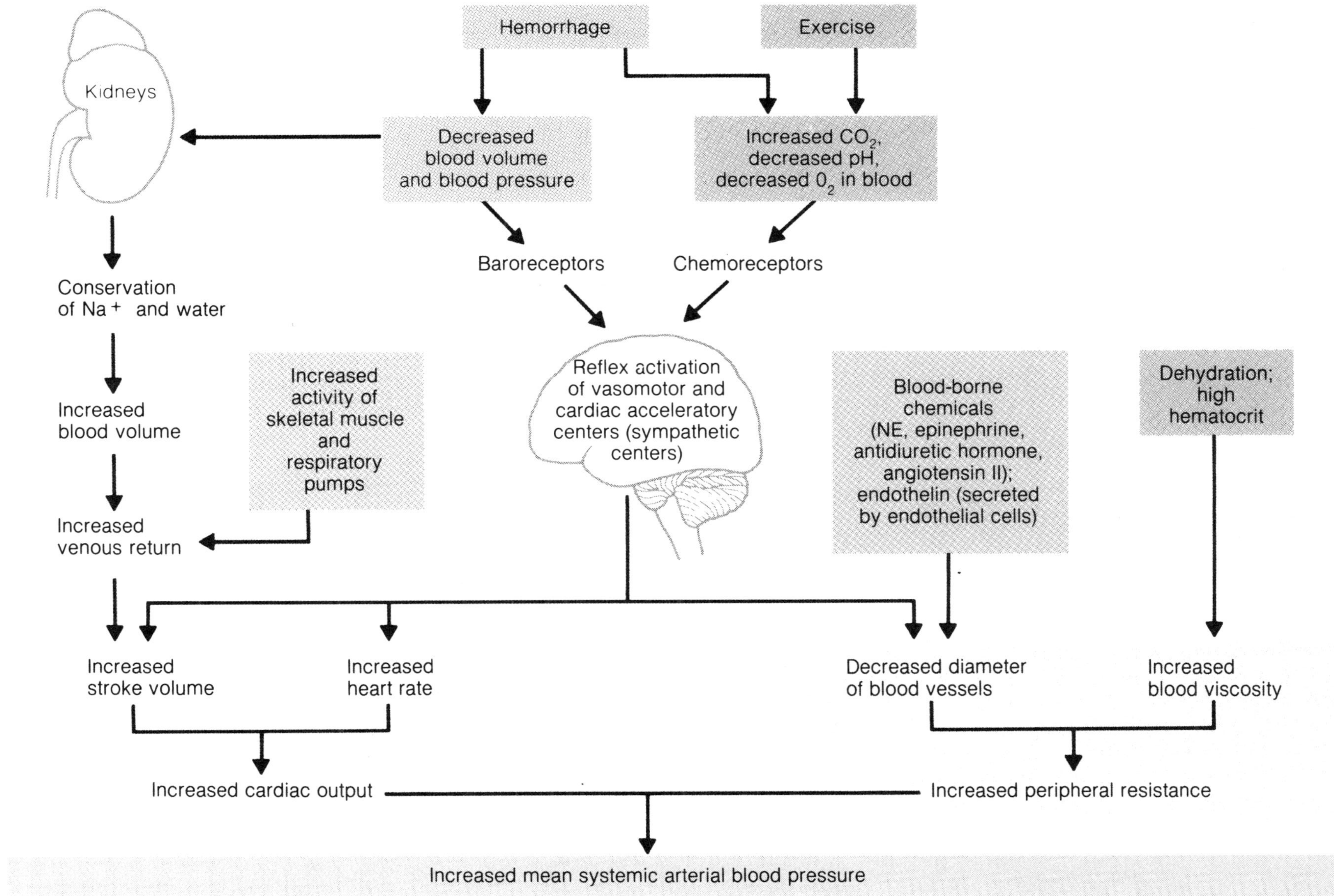

Distribution of blood flow to selected body organs at rest and during strenuous excercise

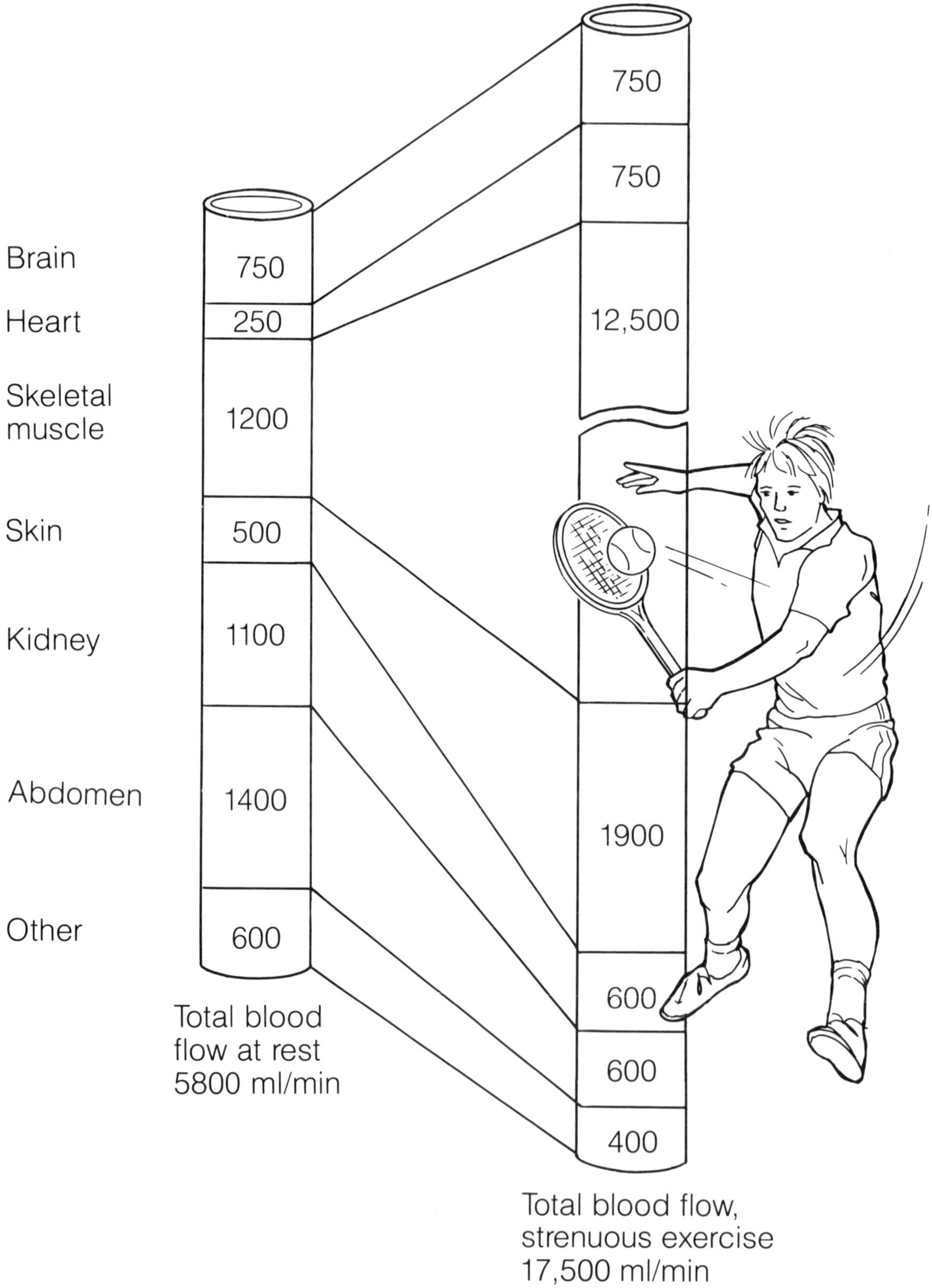

Forces acting at the capillary bed determine the direction of fluid flows

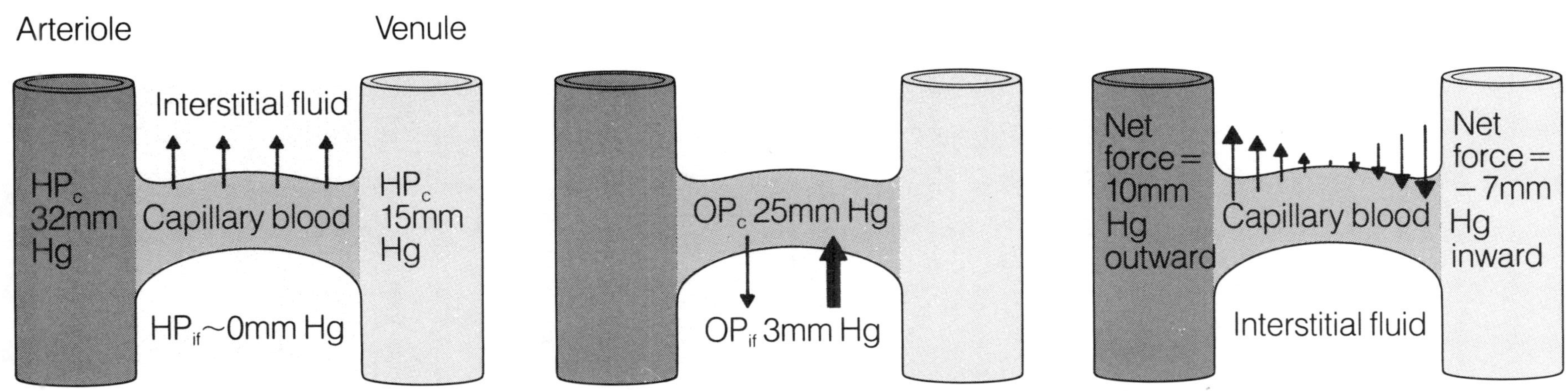

Net arteriolar HP = 32mm Hg (Net HP = HP_c (32) − HP_{if} (0))

Net venule HP = 15mm Hg (Net HP = HP_c (15) − HP_{if} (0))

(a) Hydrostatic pressures

Net OP = 22mm Hg (Net OP = OP_c (25) − OP_{if} (3))

(b) Osmotic pressure

(Net force = net HP (32) − net OP (22) = 10mm Hg outward)

(Net force = net HP (15) − net OP (22) = −7mm Hg inward)

(c) Direction of fluid flows

Events and signs of hypovolemic shock

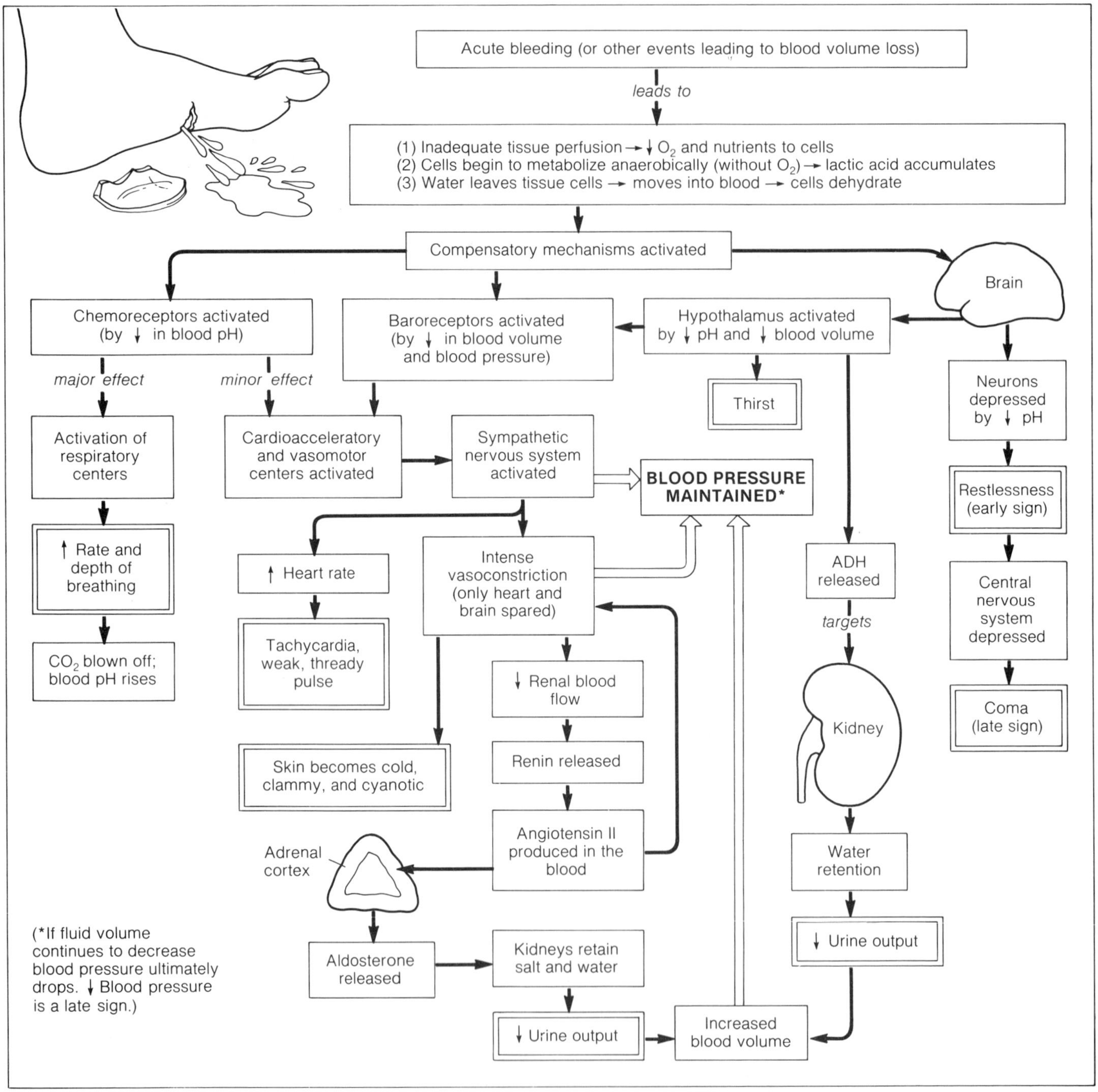

Distribution and special structural features of lymphatic capillaries

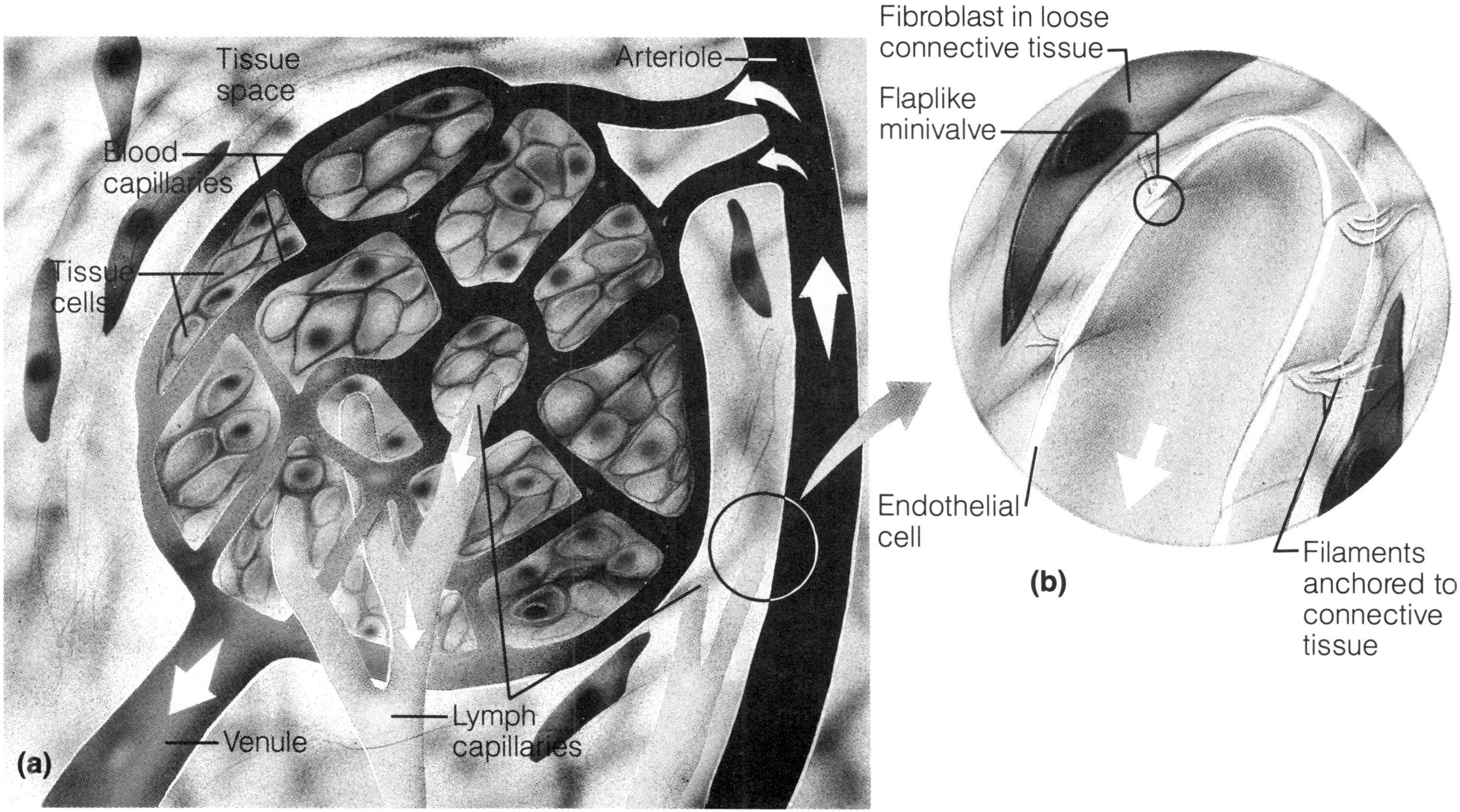

The lymphatic system

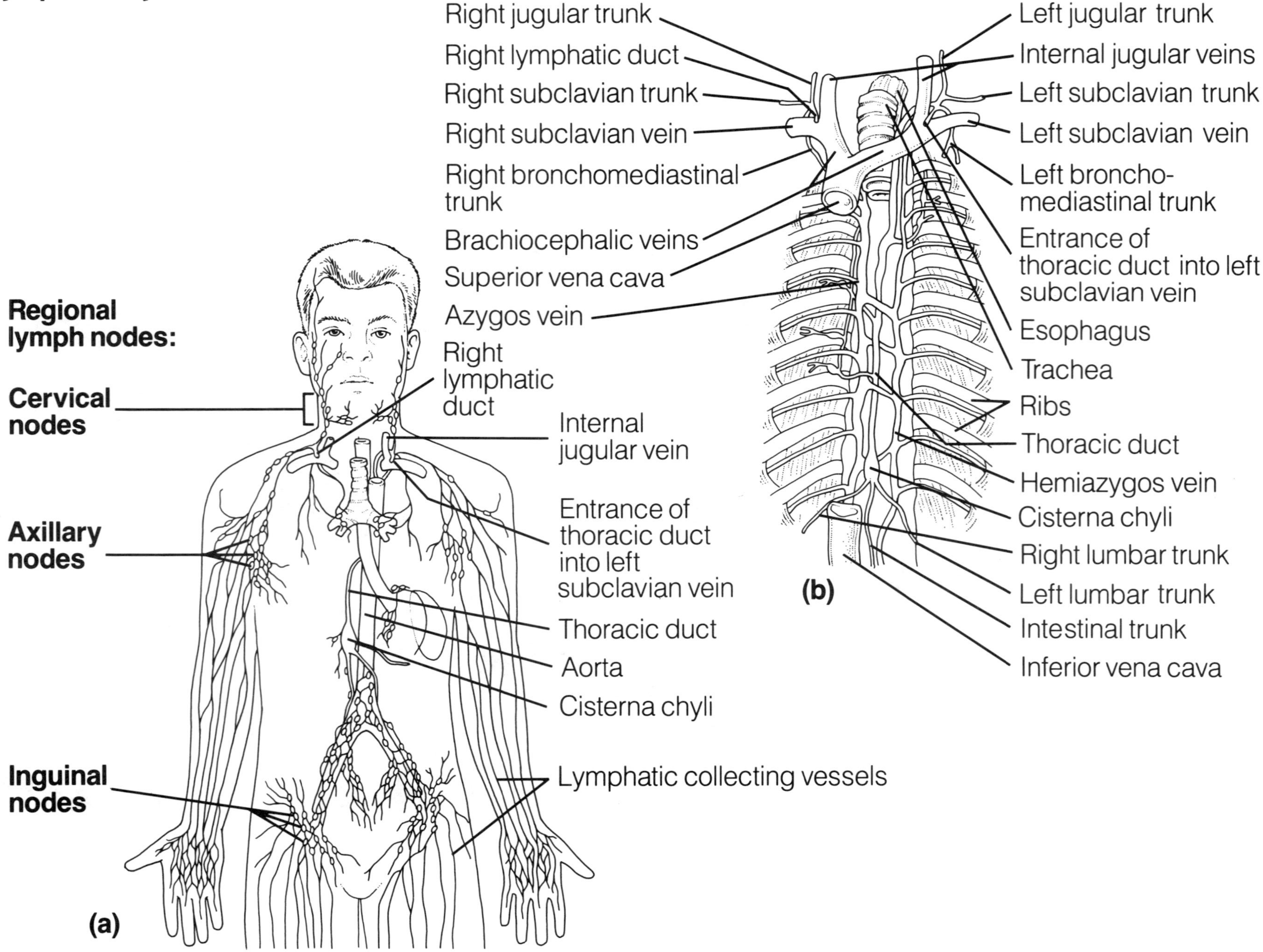

Lymphoid organs

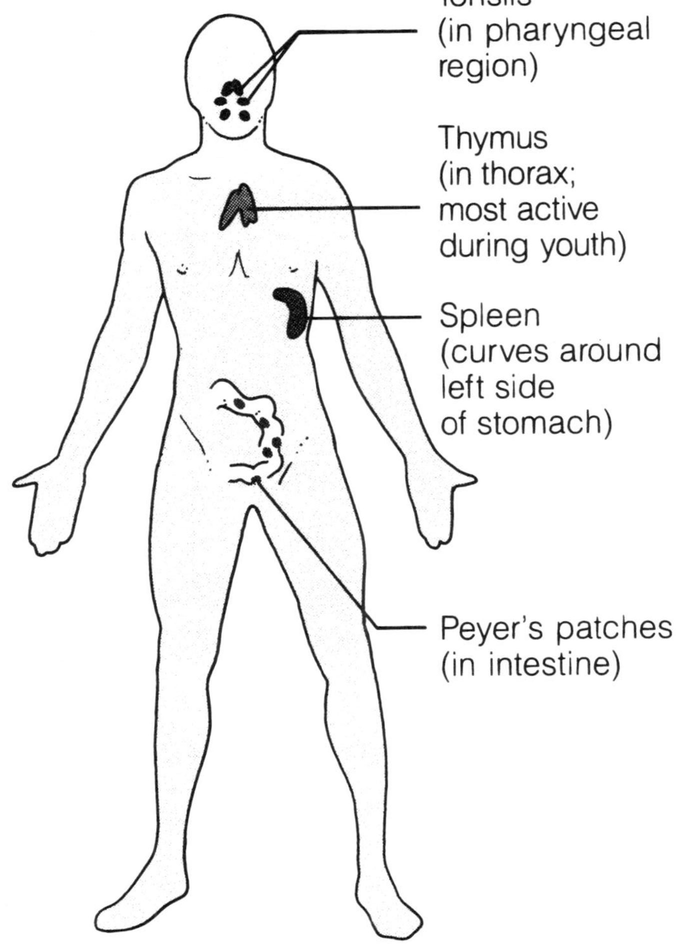

Events in inflammation

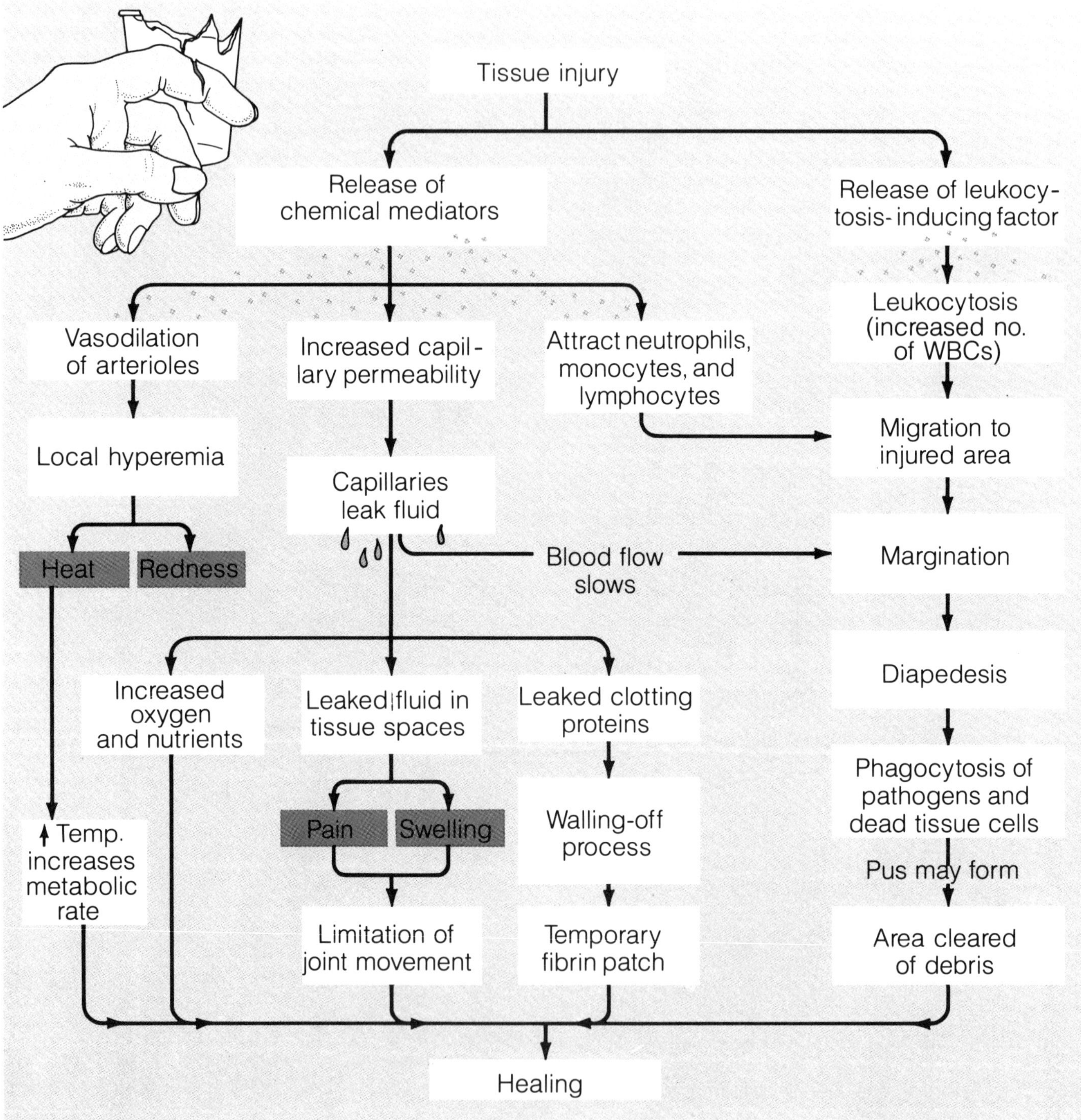

Lymphocyte traffic

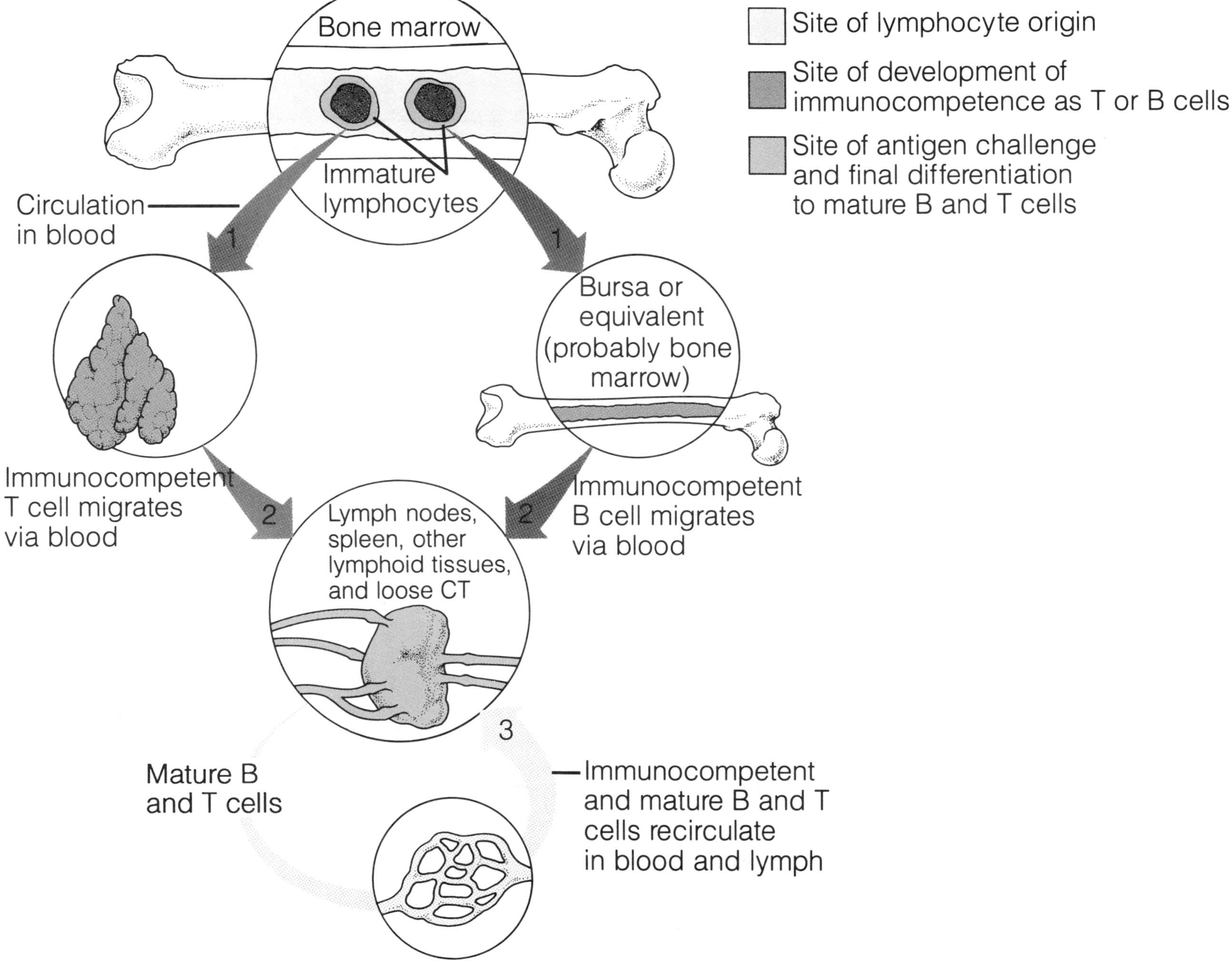

Primary and secondary humoral responses

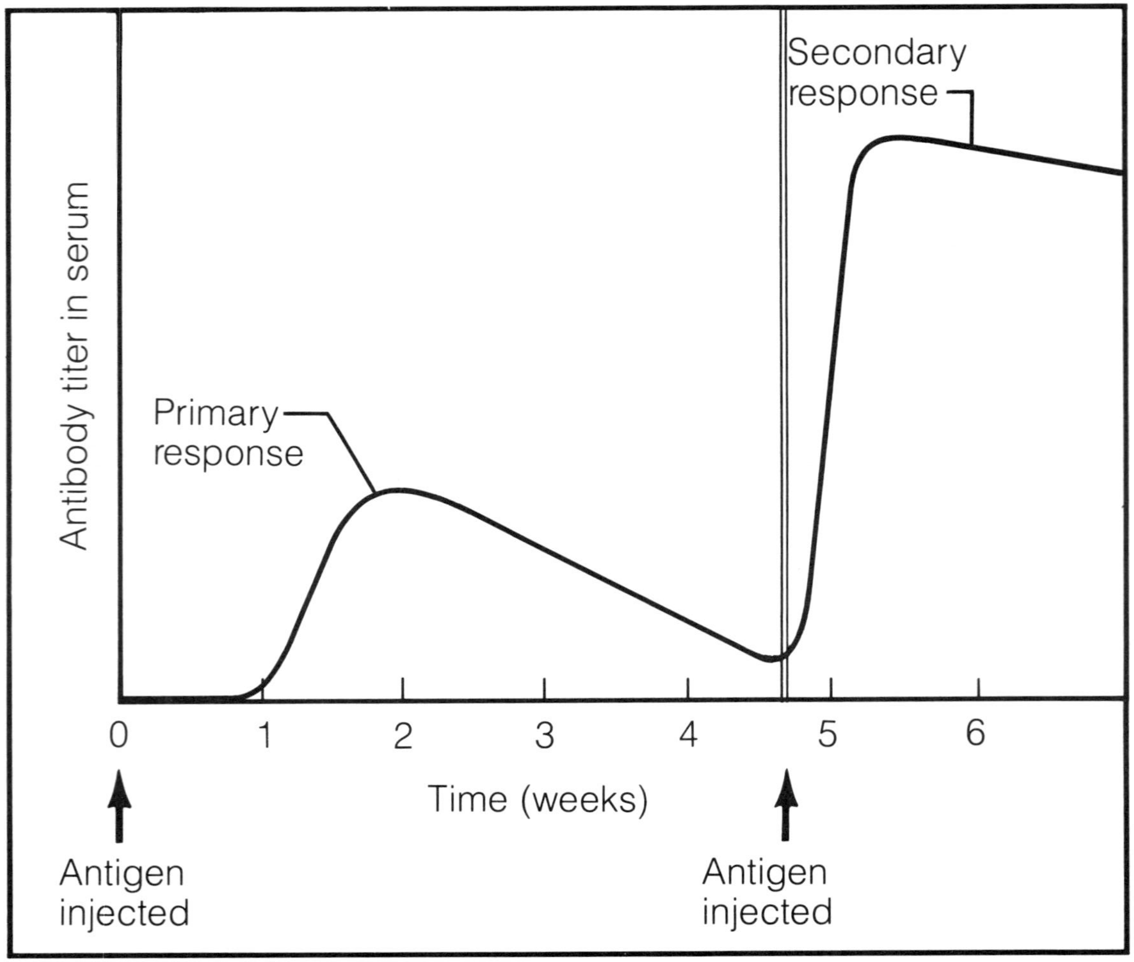

Types of acquired immunity

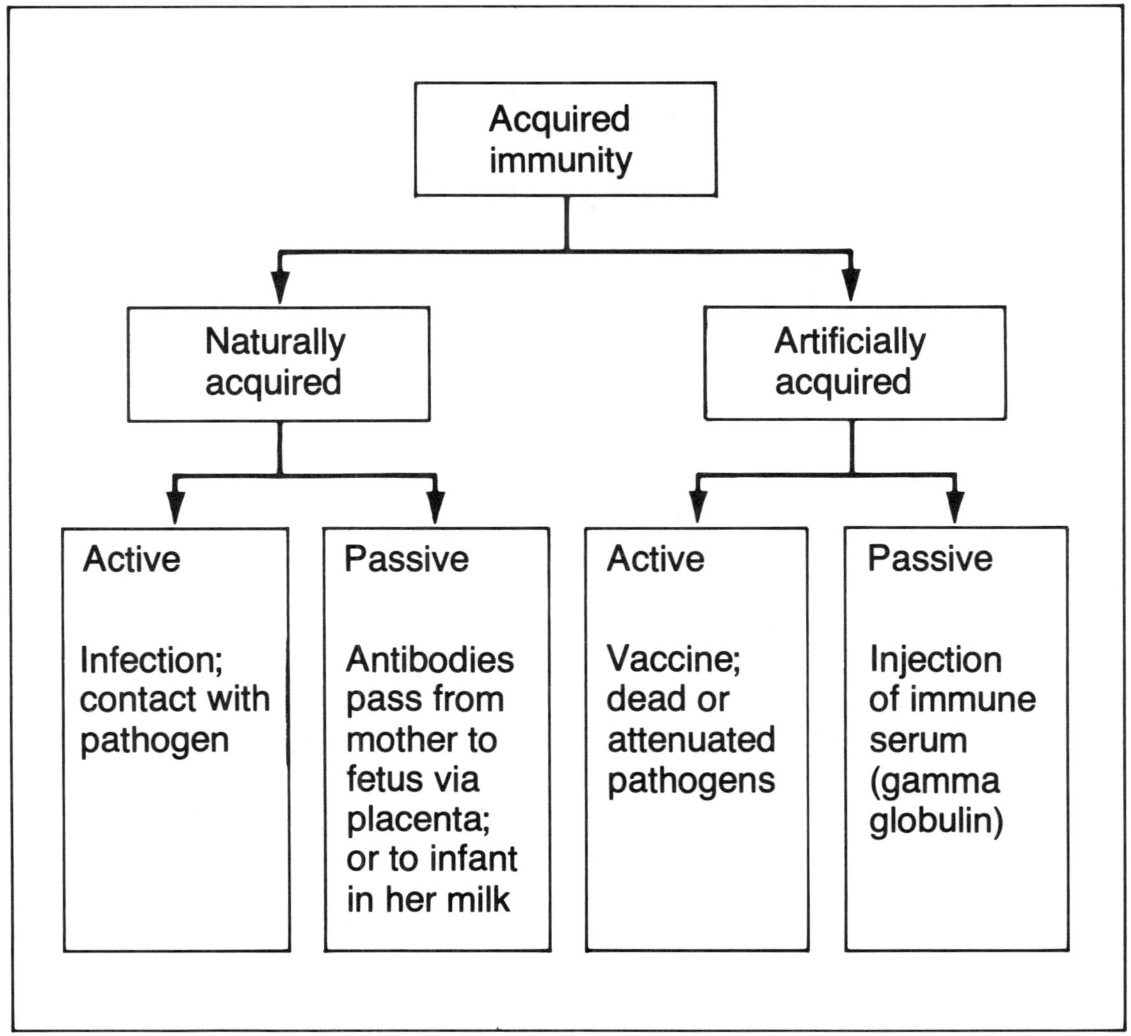

Basic antibody structure

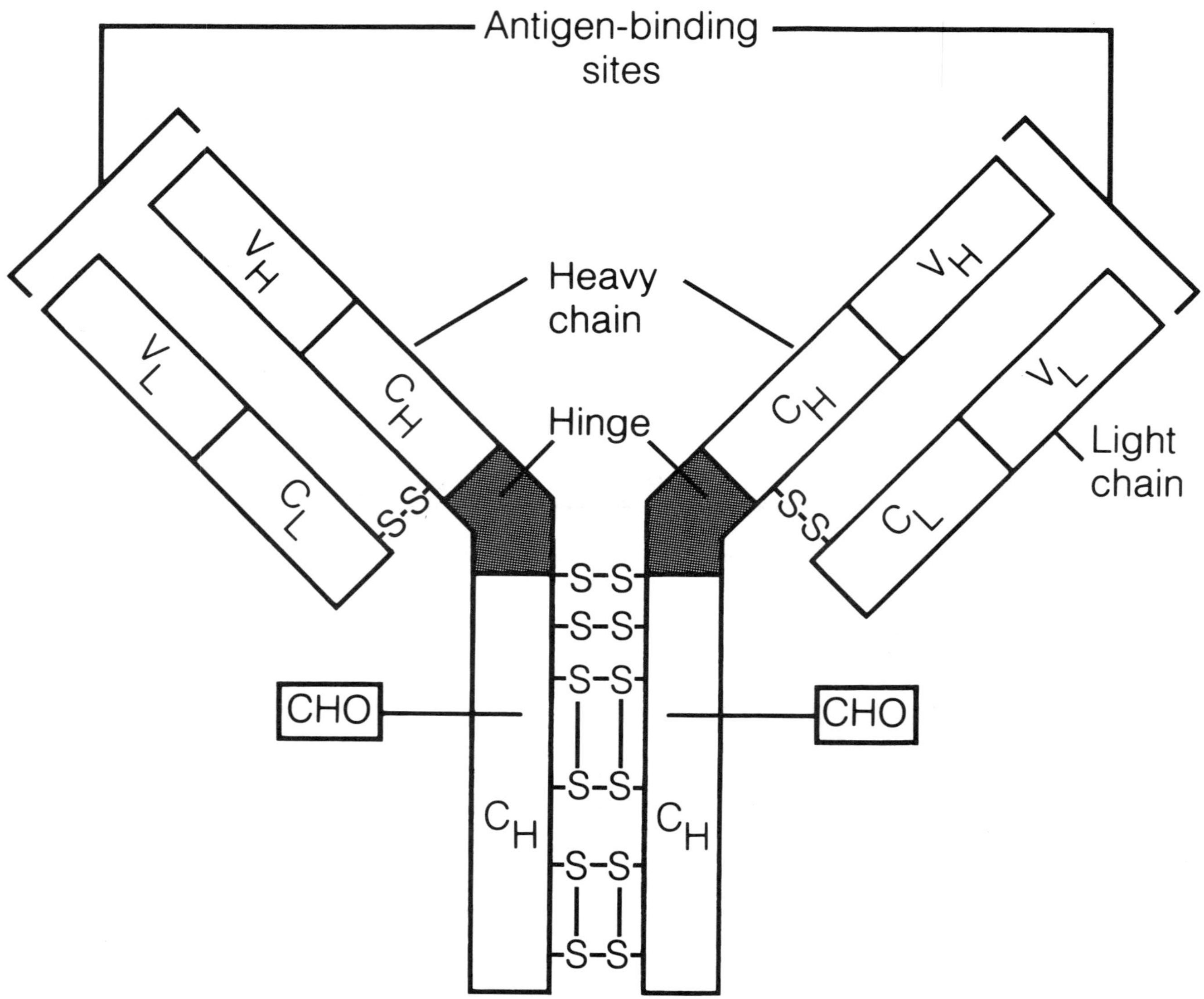

Mechanisms of antibody action

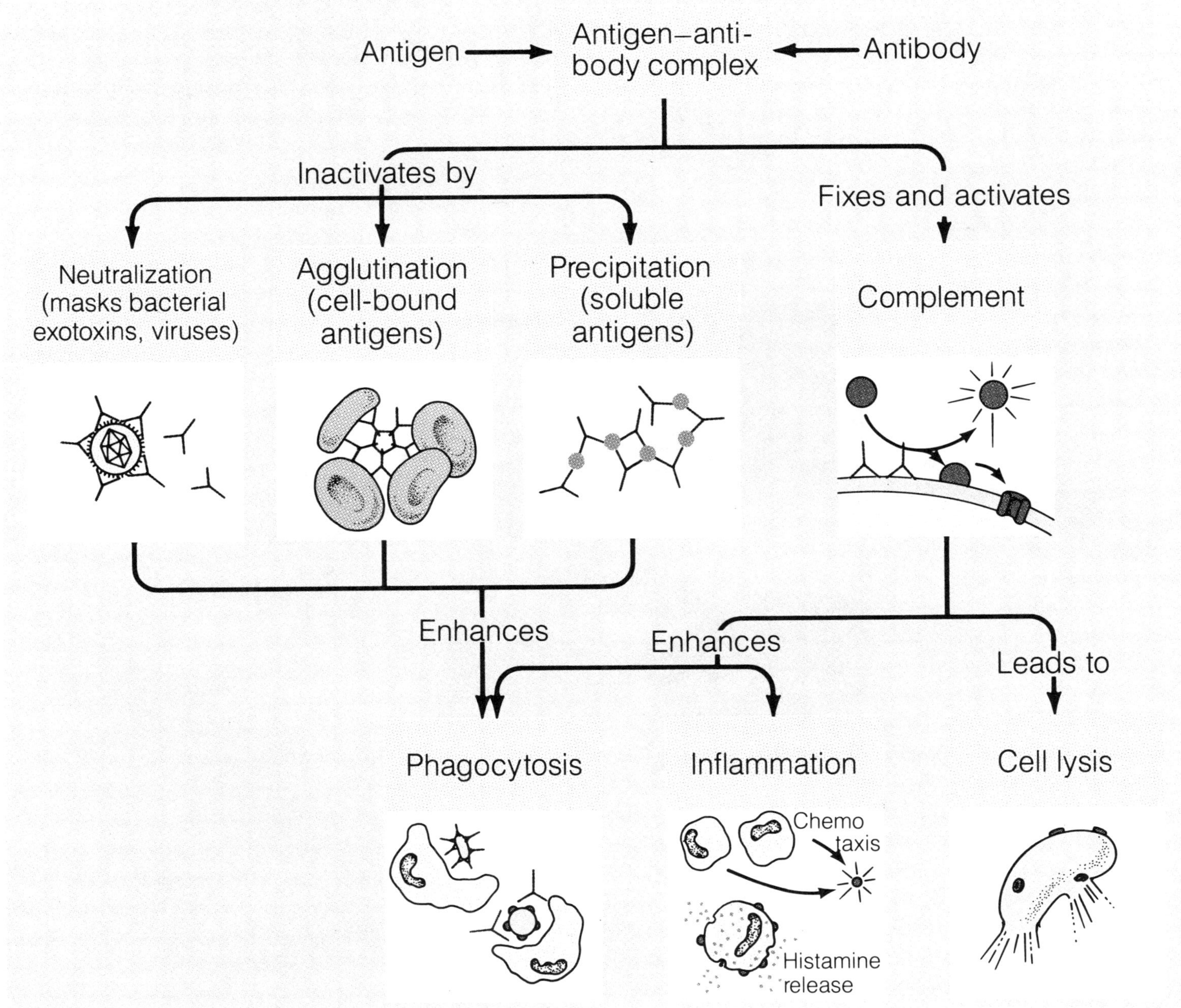

The central role of helper T cells

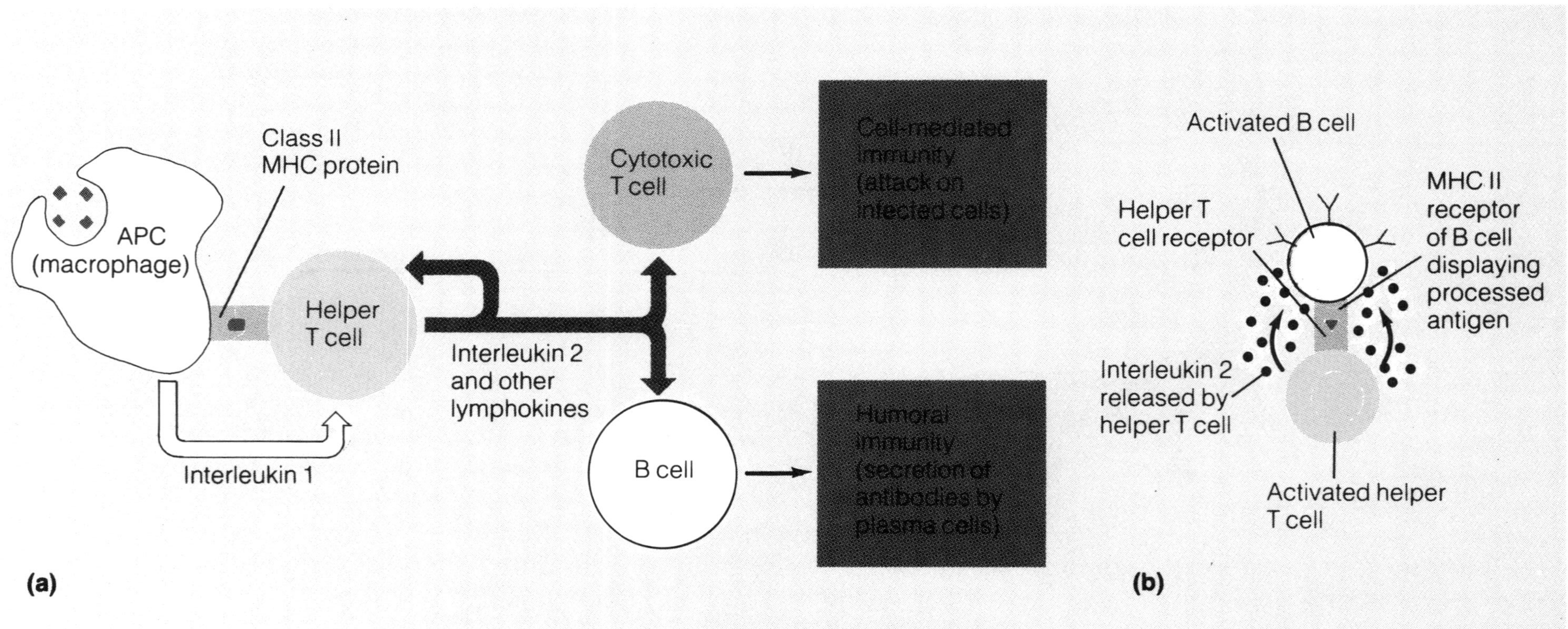

Anatomical relationships of organs in the thoracic cavity

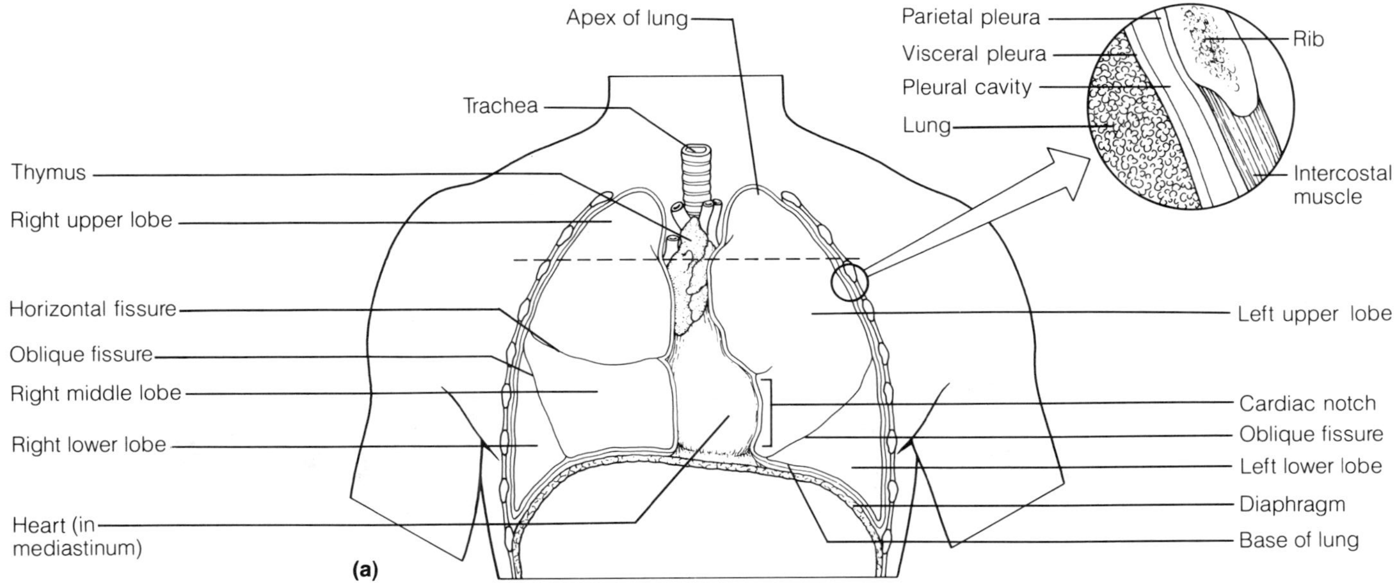

(a)

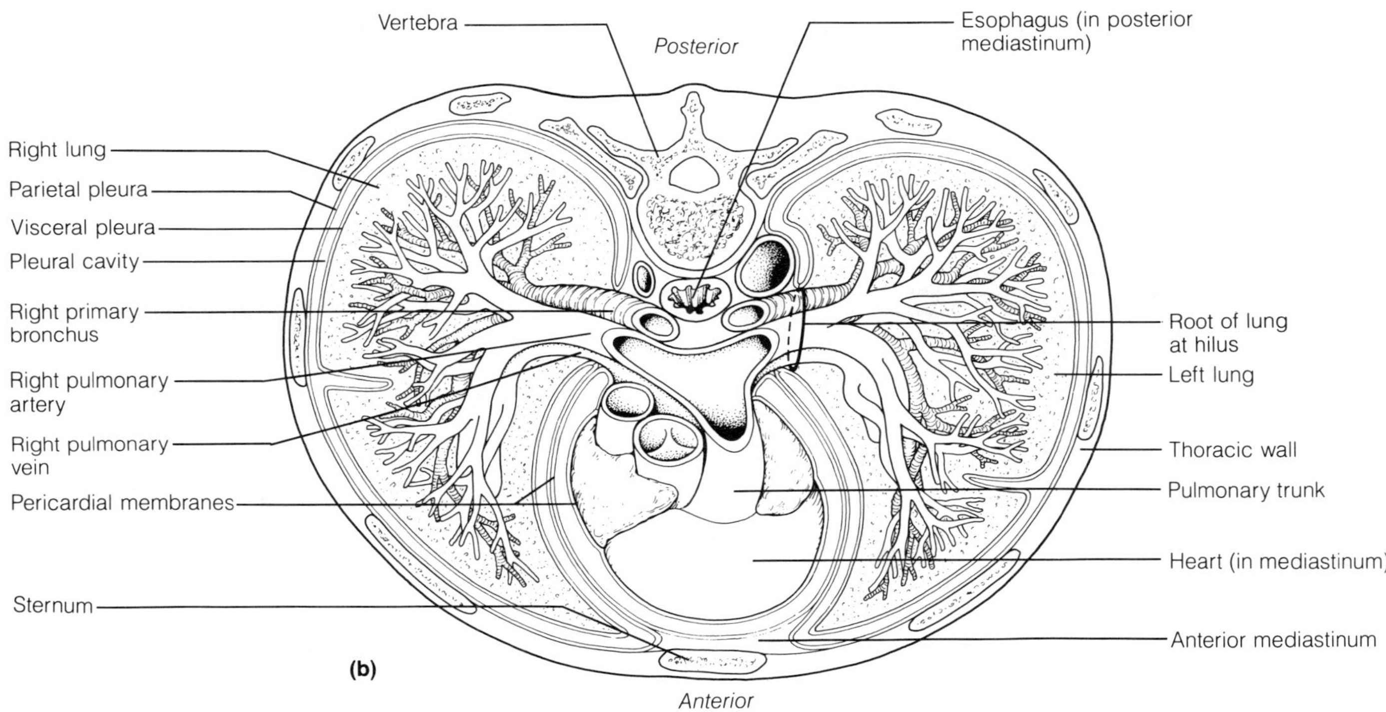

(b)

Intrapulmonary and intrapleural pressure relationships

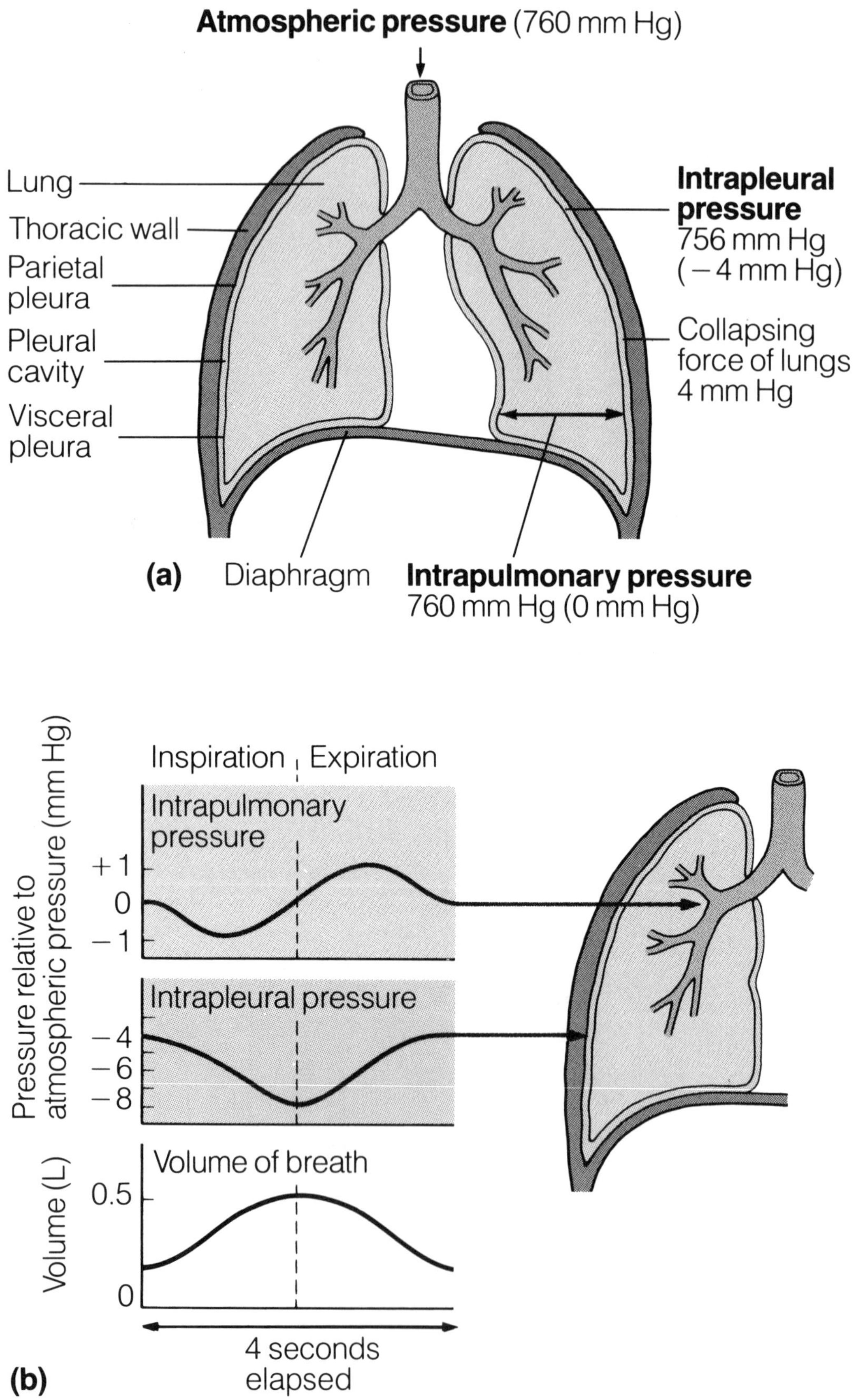

Changes in thoracic volume that lead to the flow of air (gases) during breathing

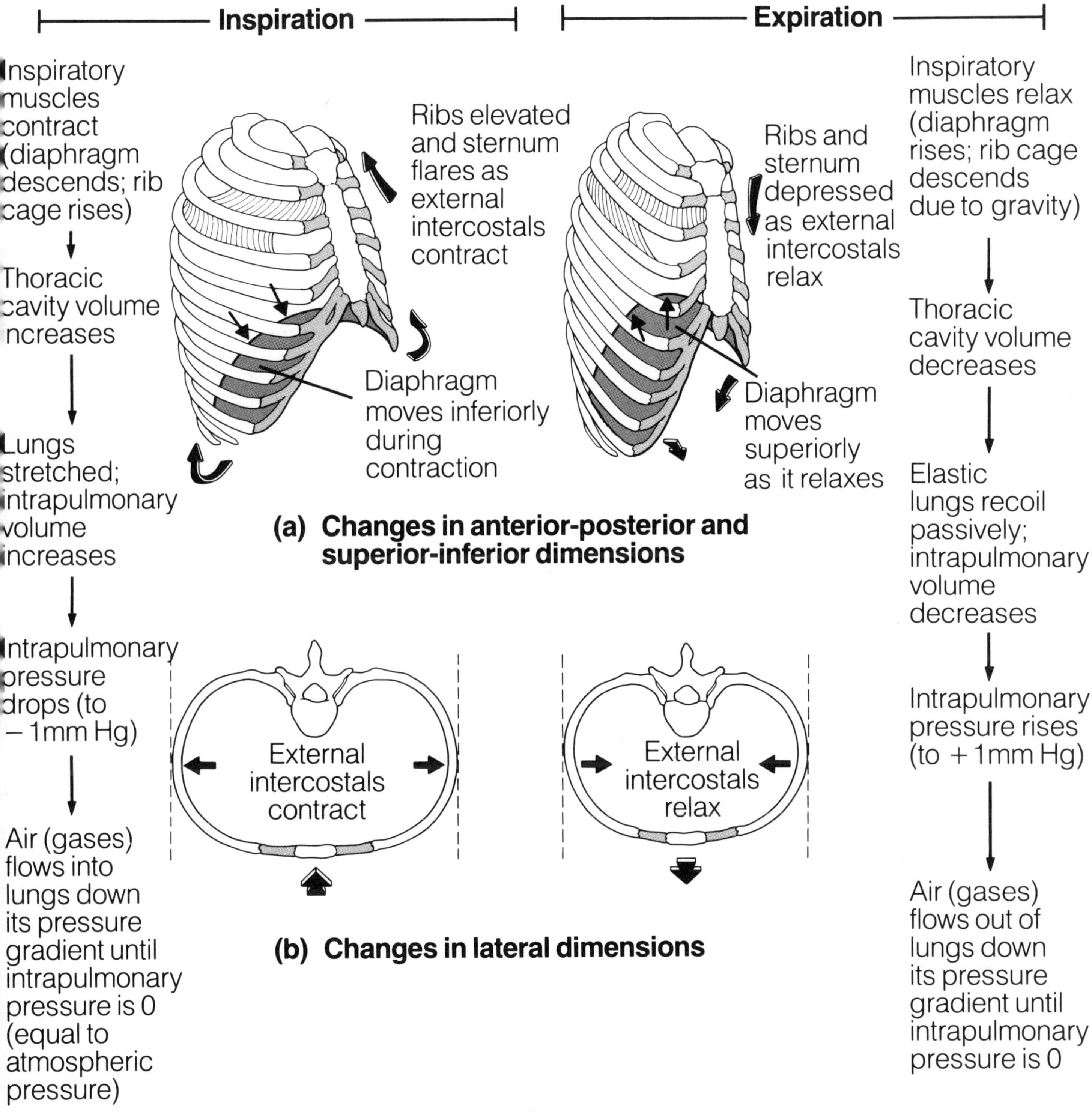

(a) Changes in anterior-posterior and superior-inferior dimensions

(b) Changes in lateral dimensions

Resistance of the various segments of the respiratory passageways

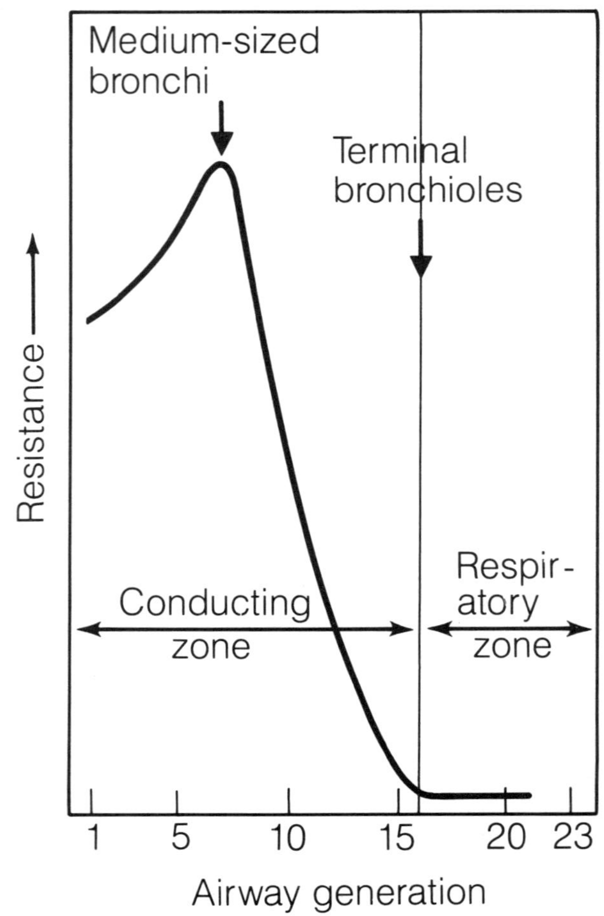

Respiratory volumes and capacities

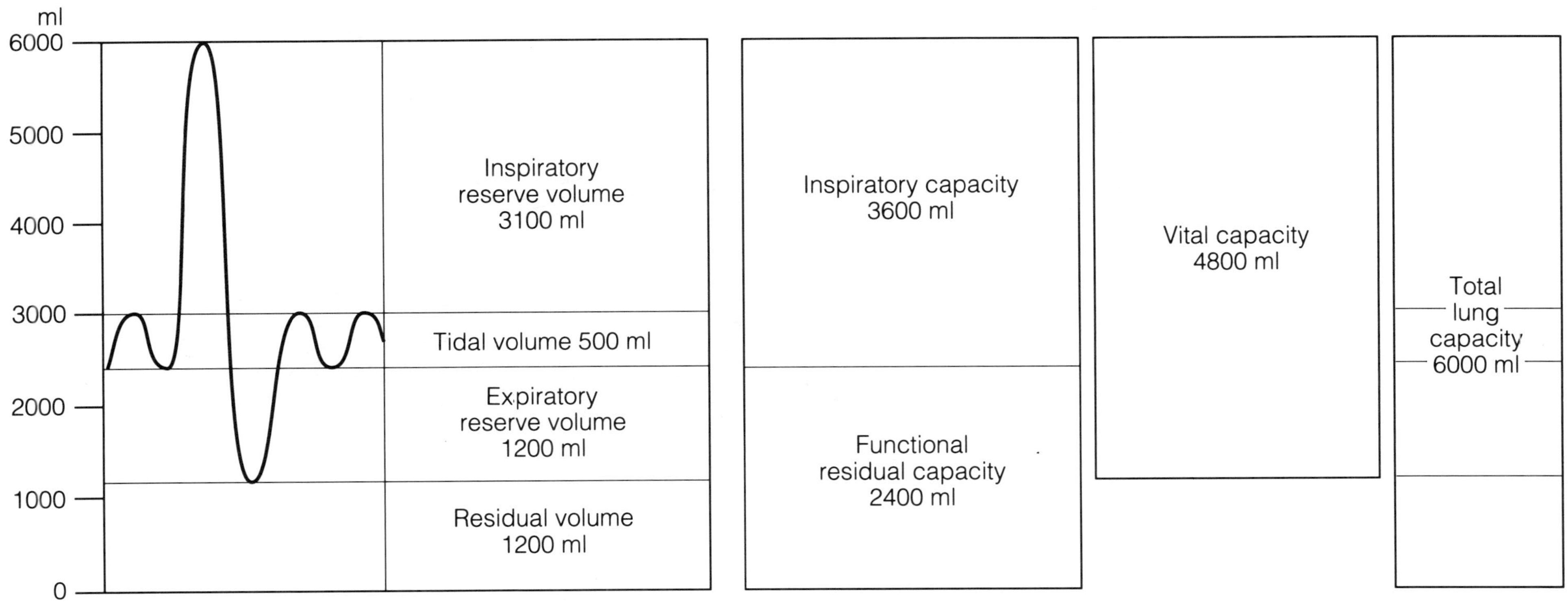

(a) Spirographic record

Partial pressure gradients promoting gas movements in the body

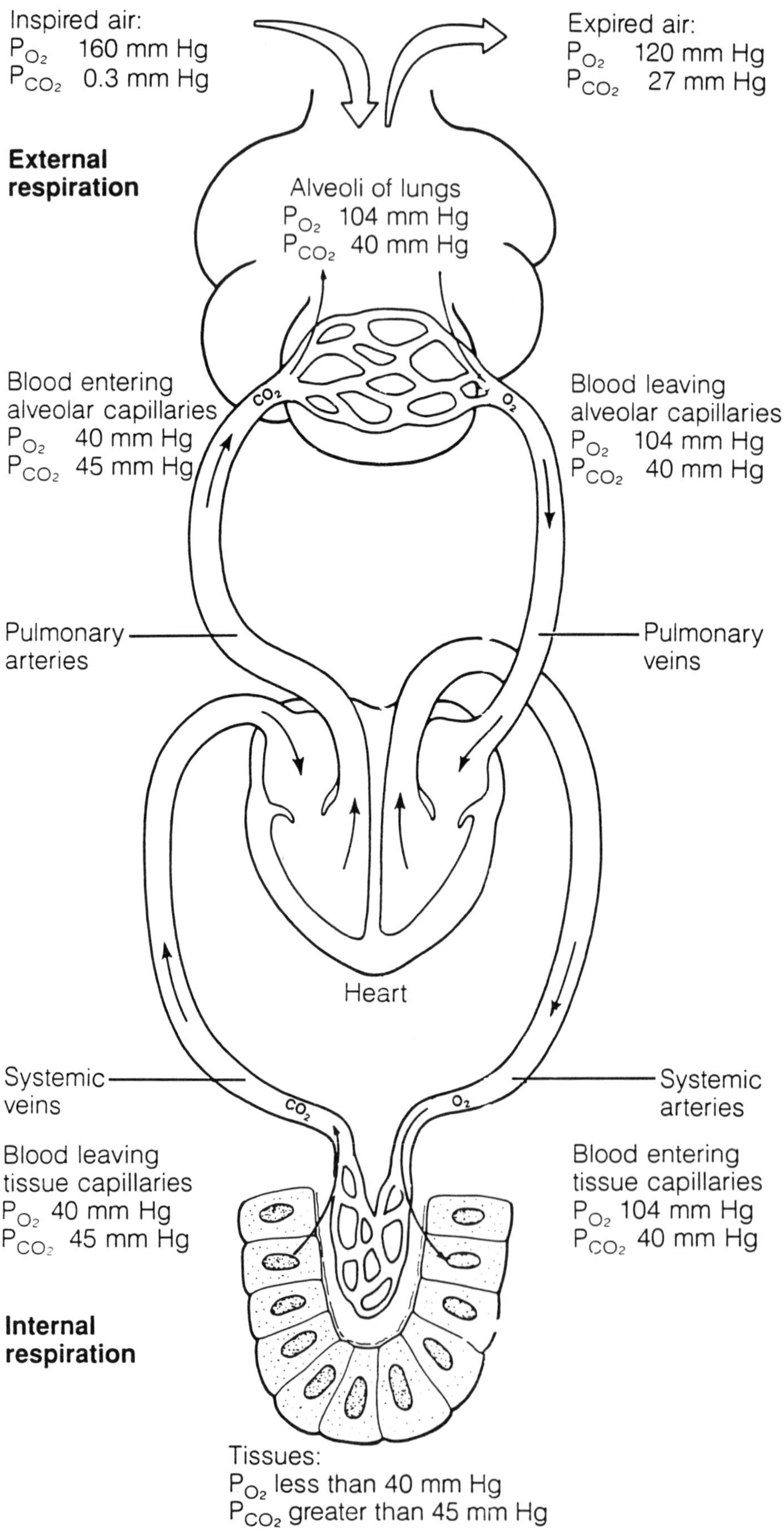

Oxygenation of blood in the pulmonary capillaries

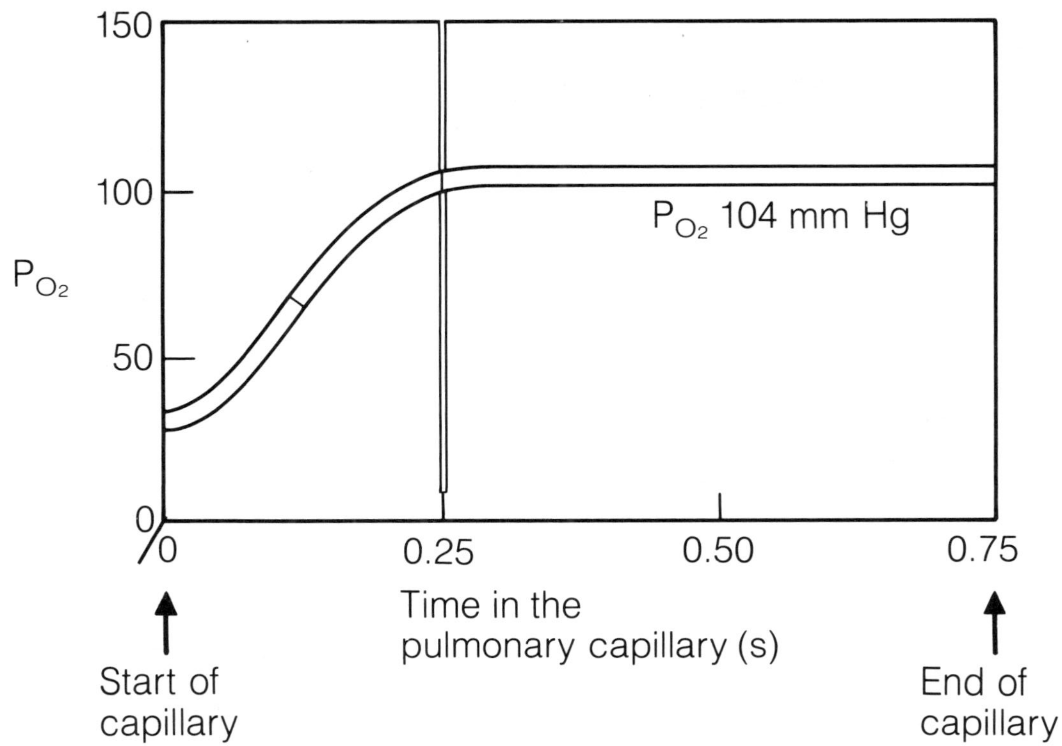

Ventilation-Perfusion coupling

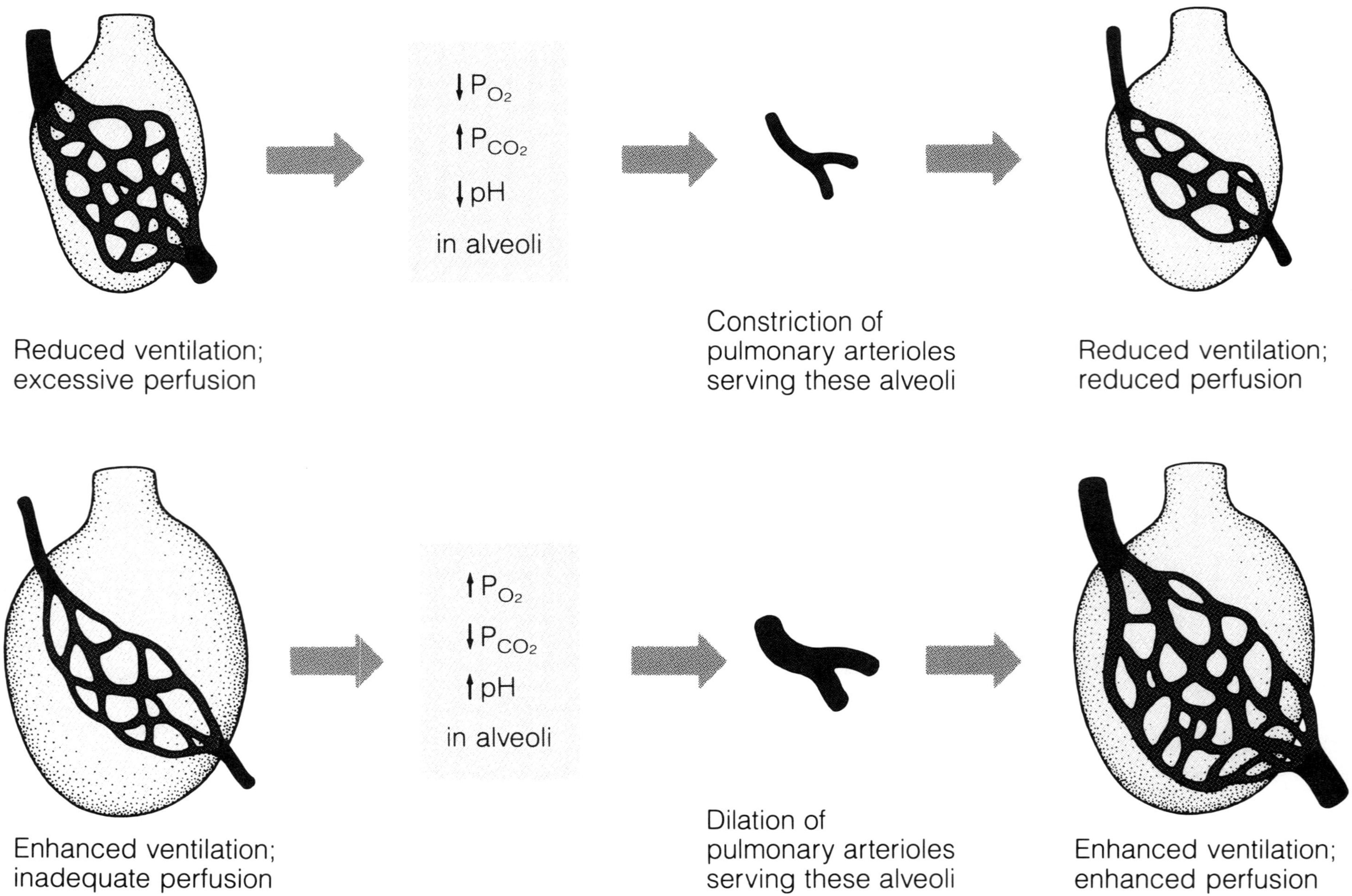

Oxygen-hemoglobin dissociation curve

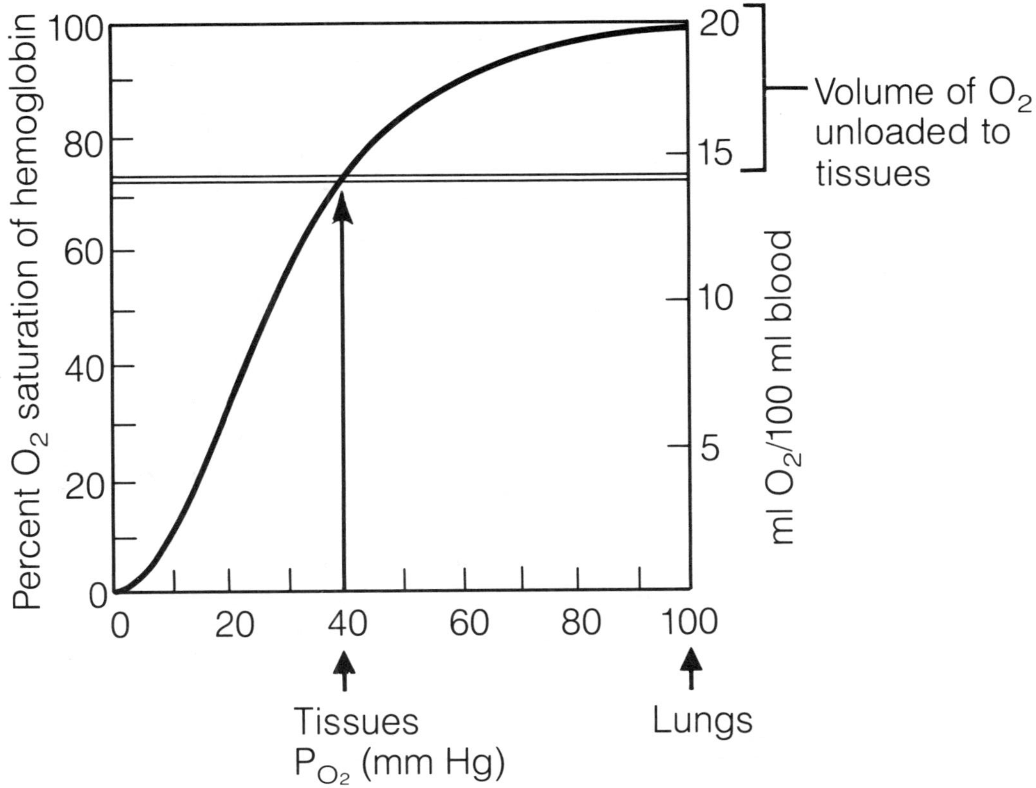

Effects of temperature, P_{CO_2}, and blood pH on the oxygen–hemoglobin dissociation curve

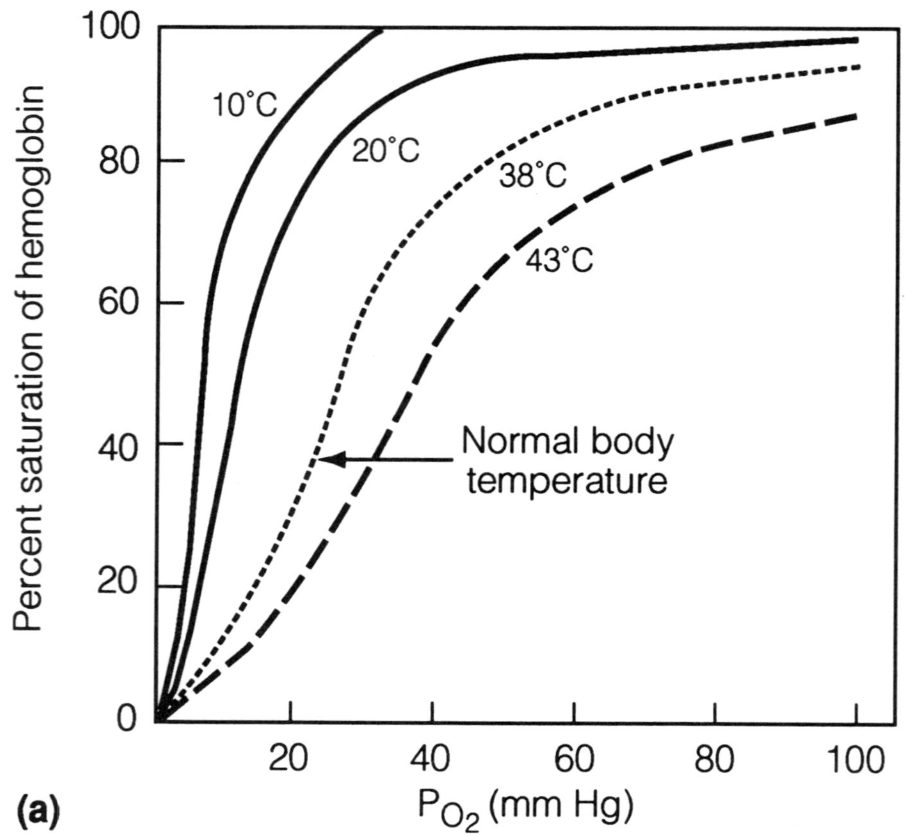

(a)

Effects of temperature, P_{CO_2}, and blood pH on the oxygen–hemoglobin dissociation curve

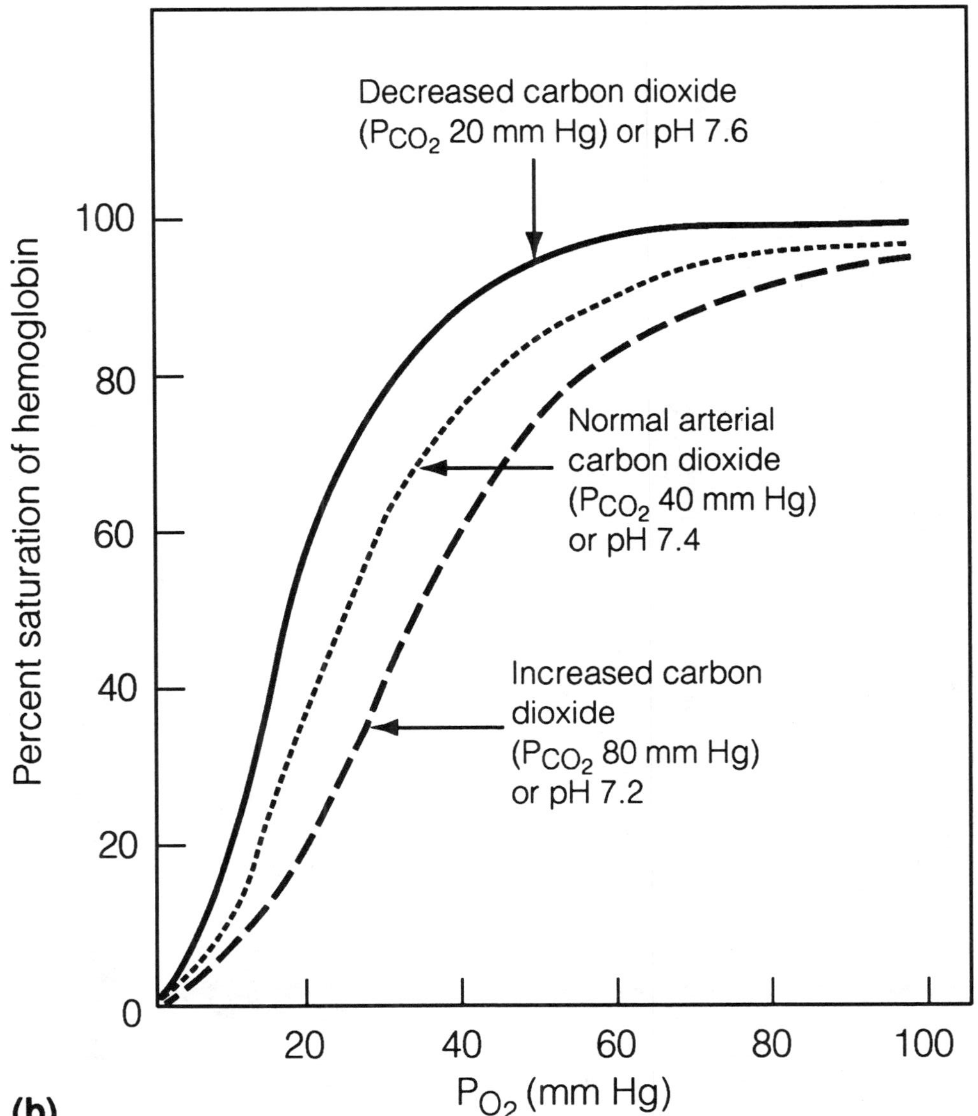

(b)

Transport and exchange of carbon dioxide and oxygen

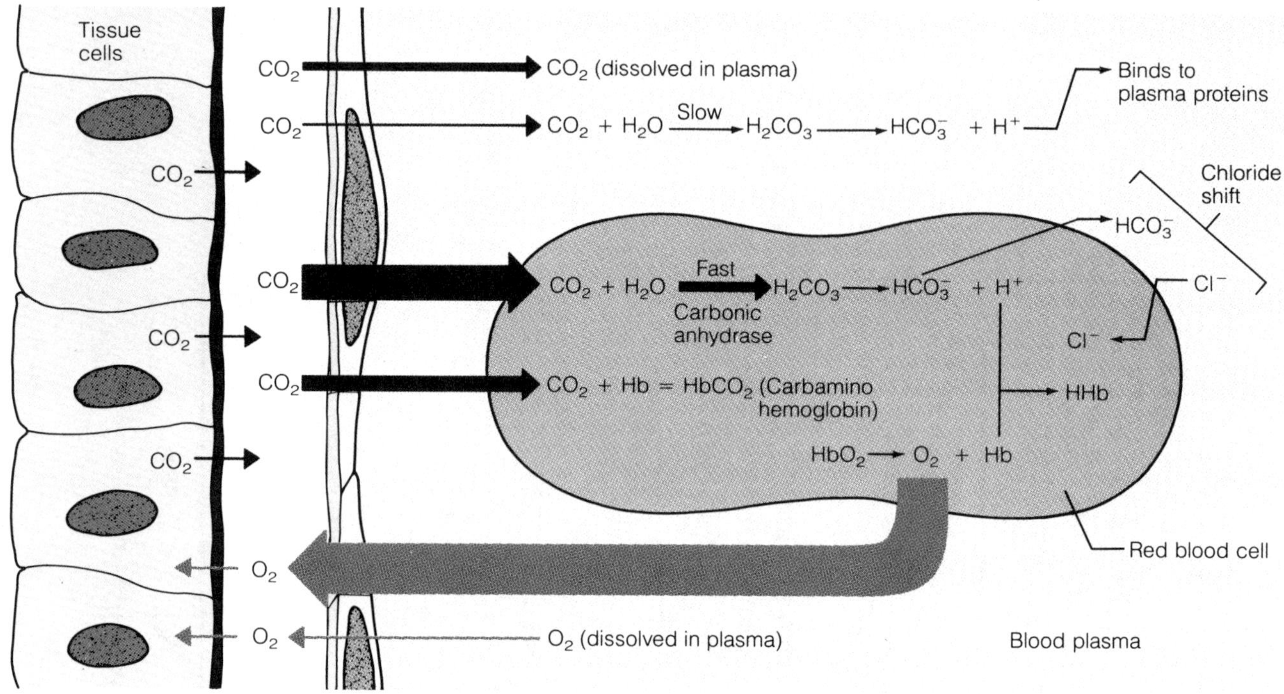

(a)

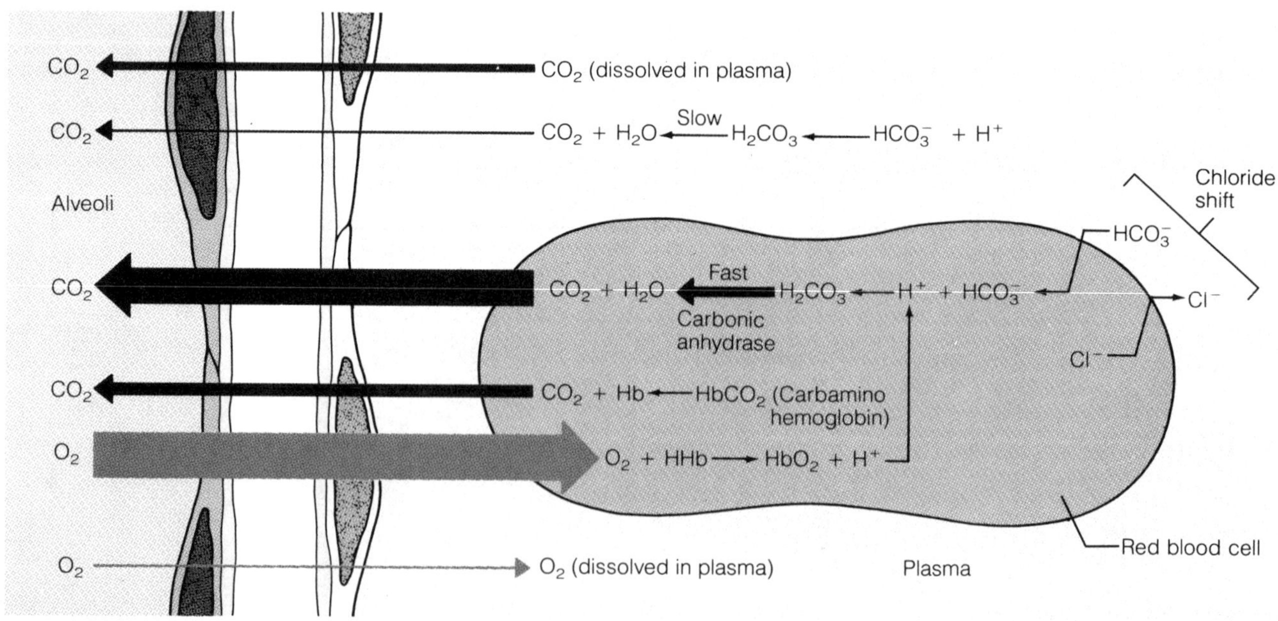

(b)

Neural pathways involved in the control of respiratory rhythm

Key:
(+) = Positive effect (stimulation)
(−) = Negative effect (inhibition)

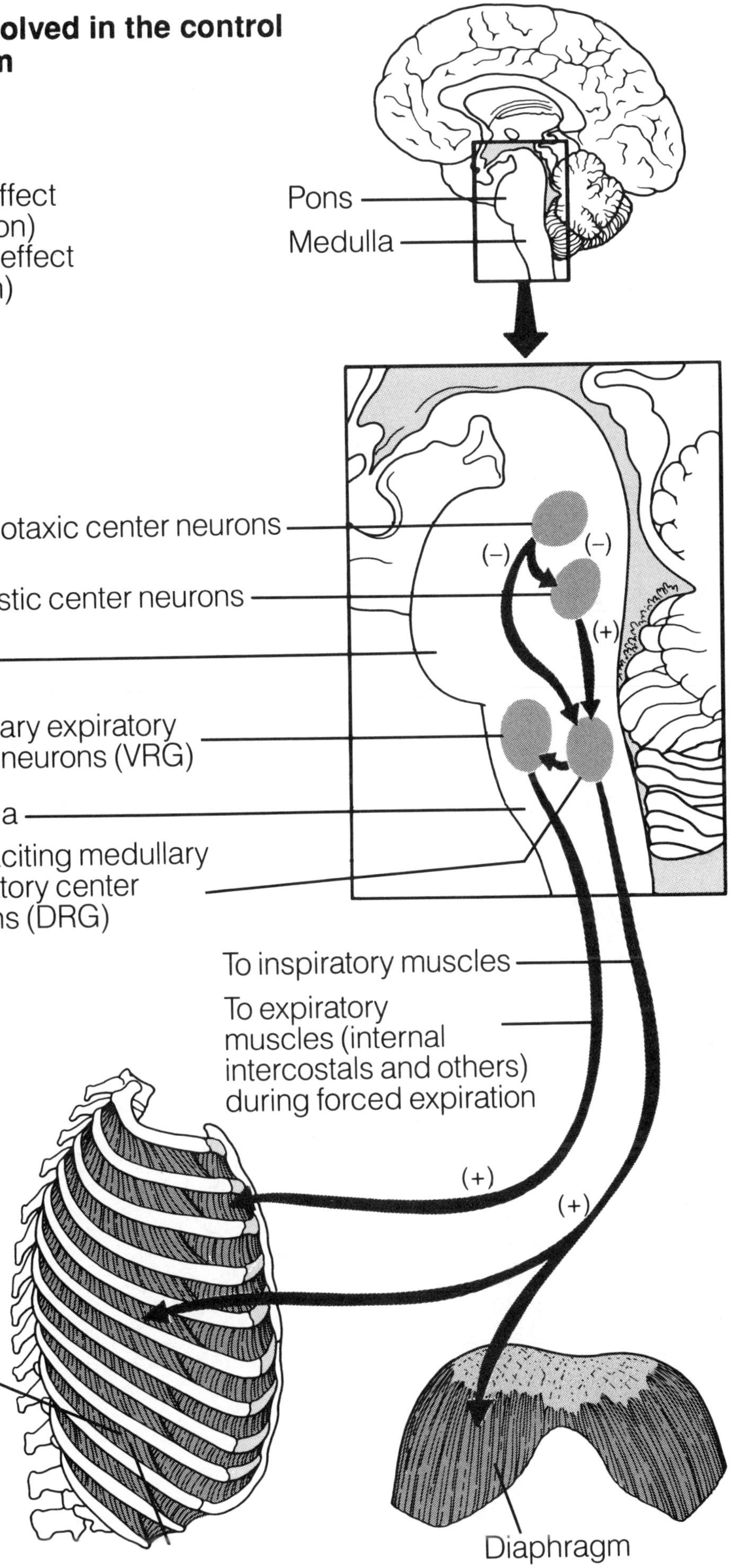

Neural and chemical influences on the respiratory centers in the medulla

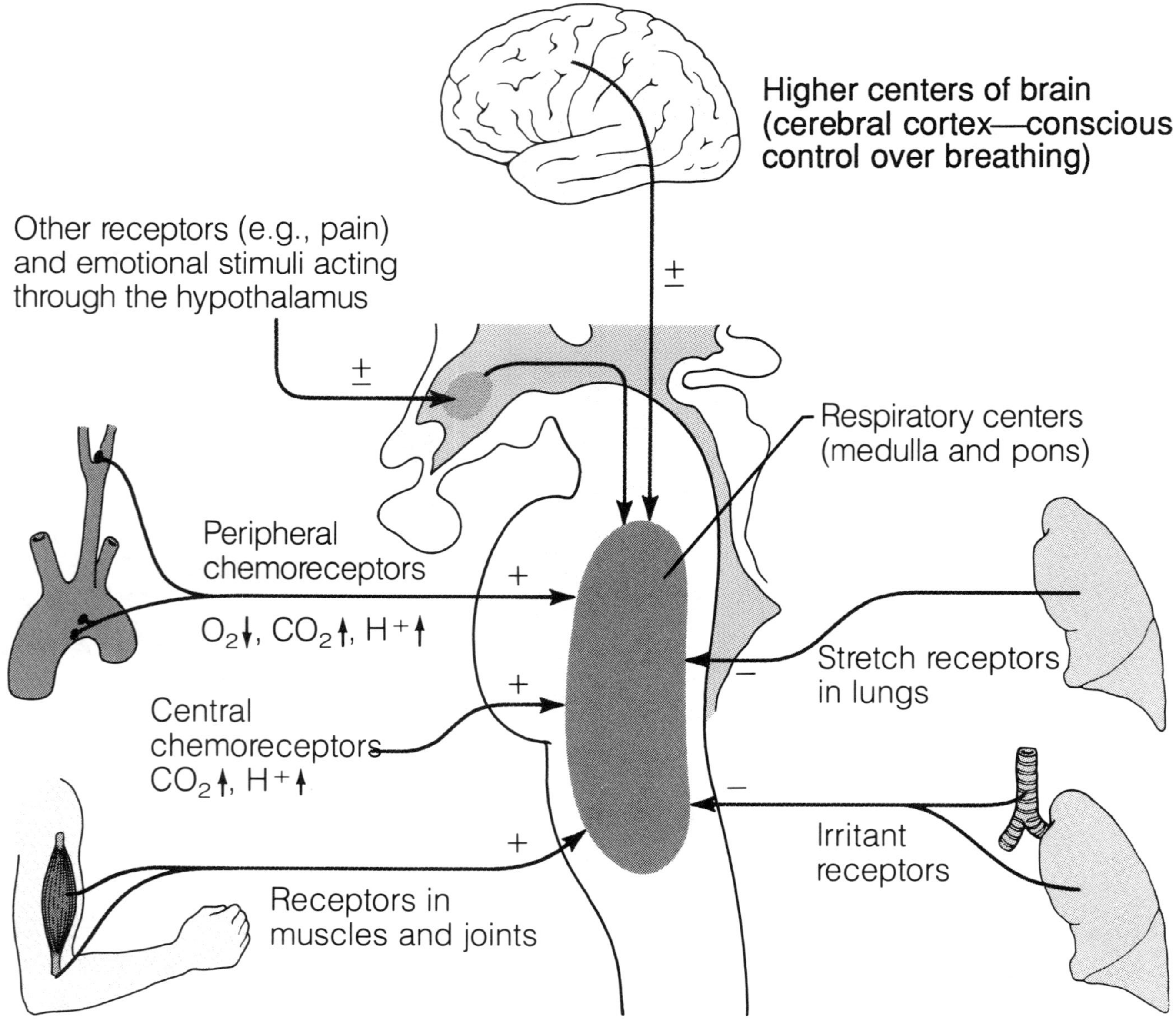

Mechanism for controlling the pH of cerebrospinal fluid by regulation of ventilation

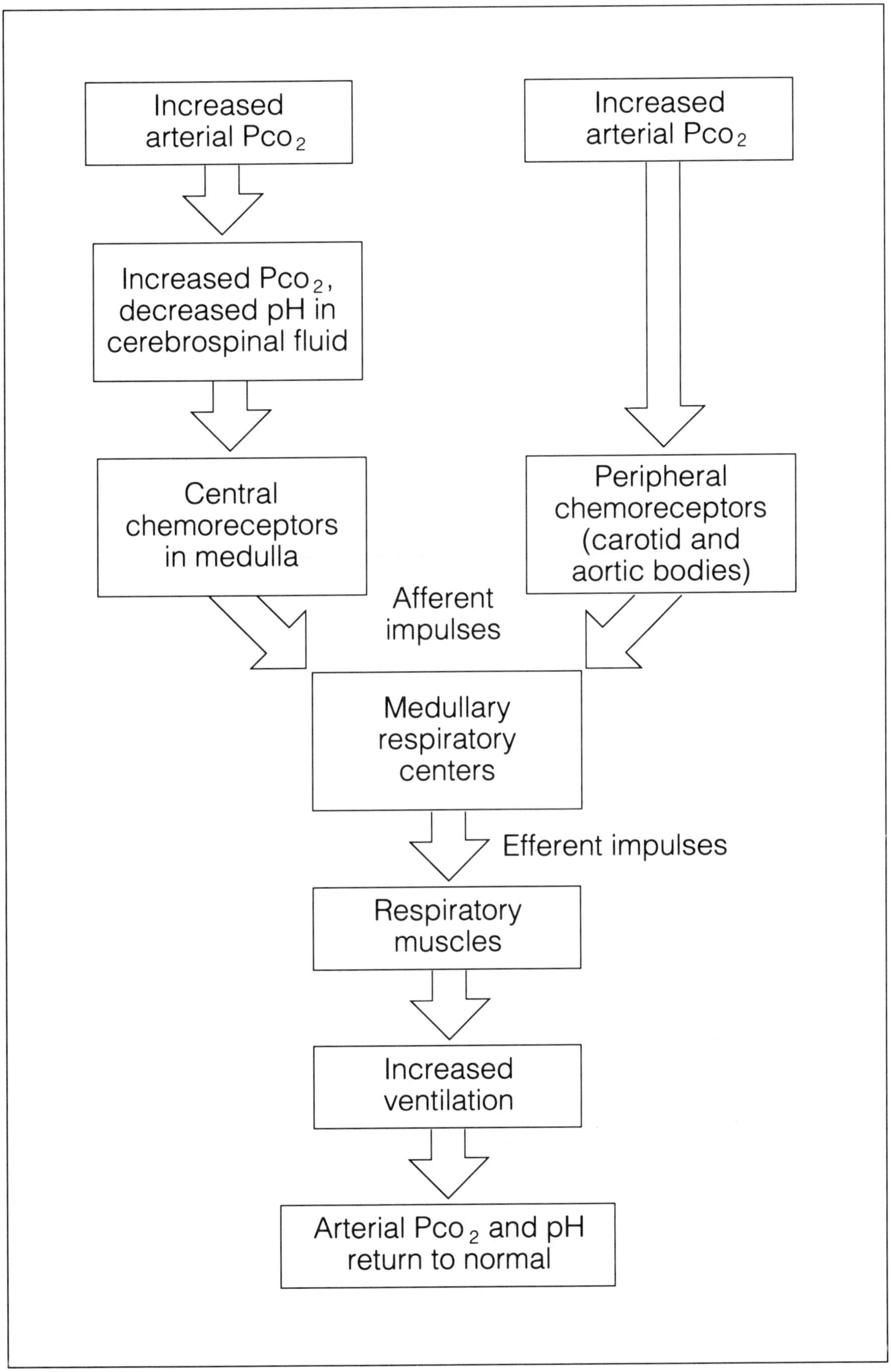

Summary of gastrointestinal tract activities

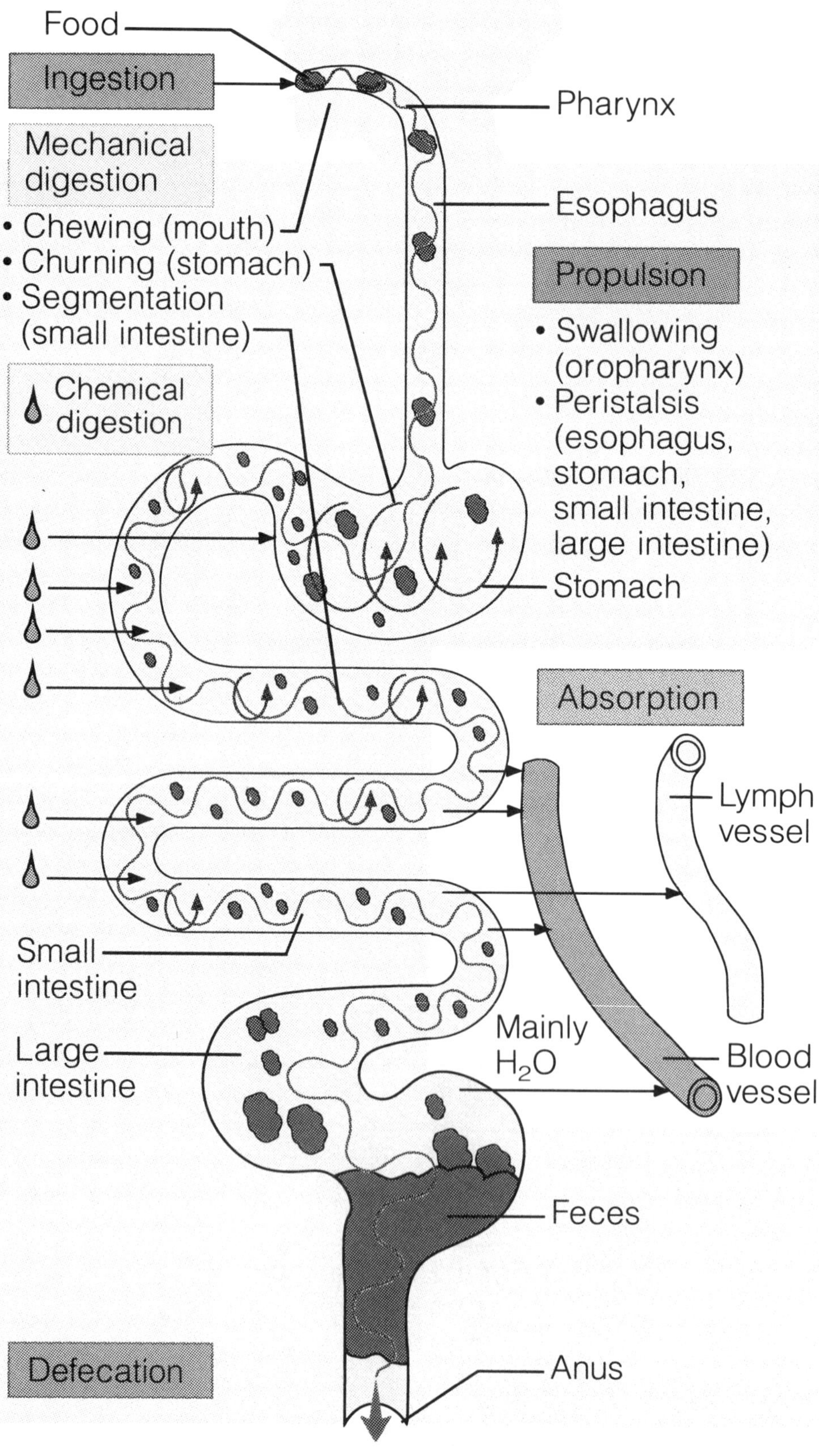

Peristalsis and segmentation

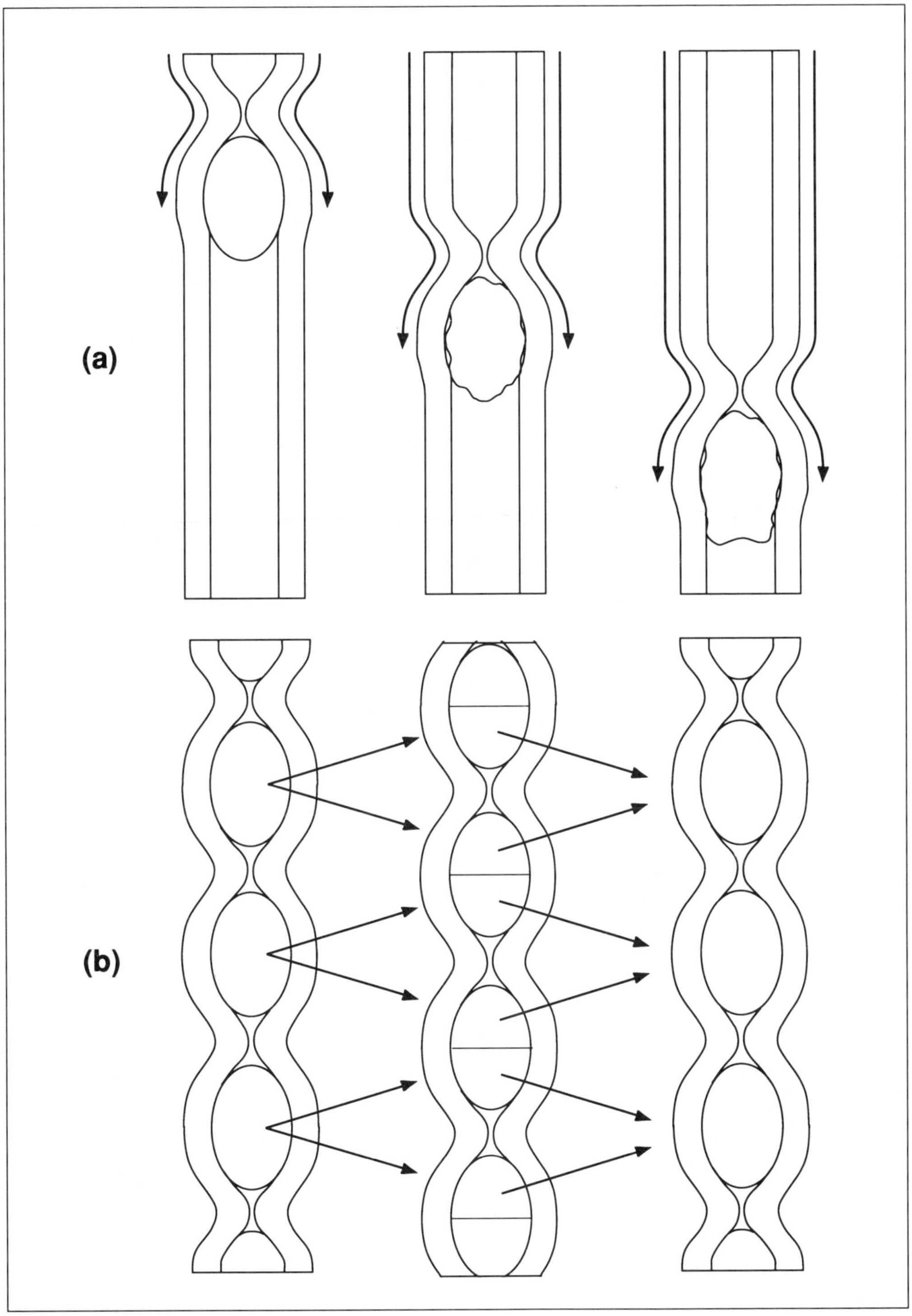

Schematic of the long and short neural reflex pathways initiated by stimuli in the gastrointestinal tract

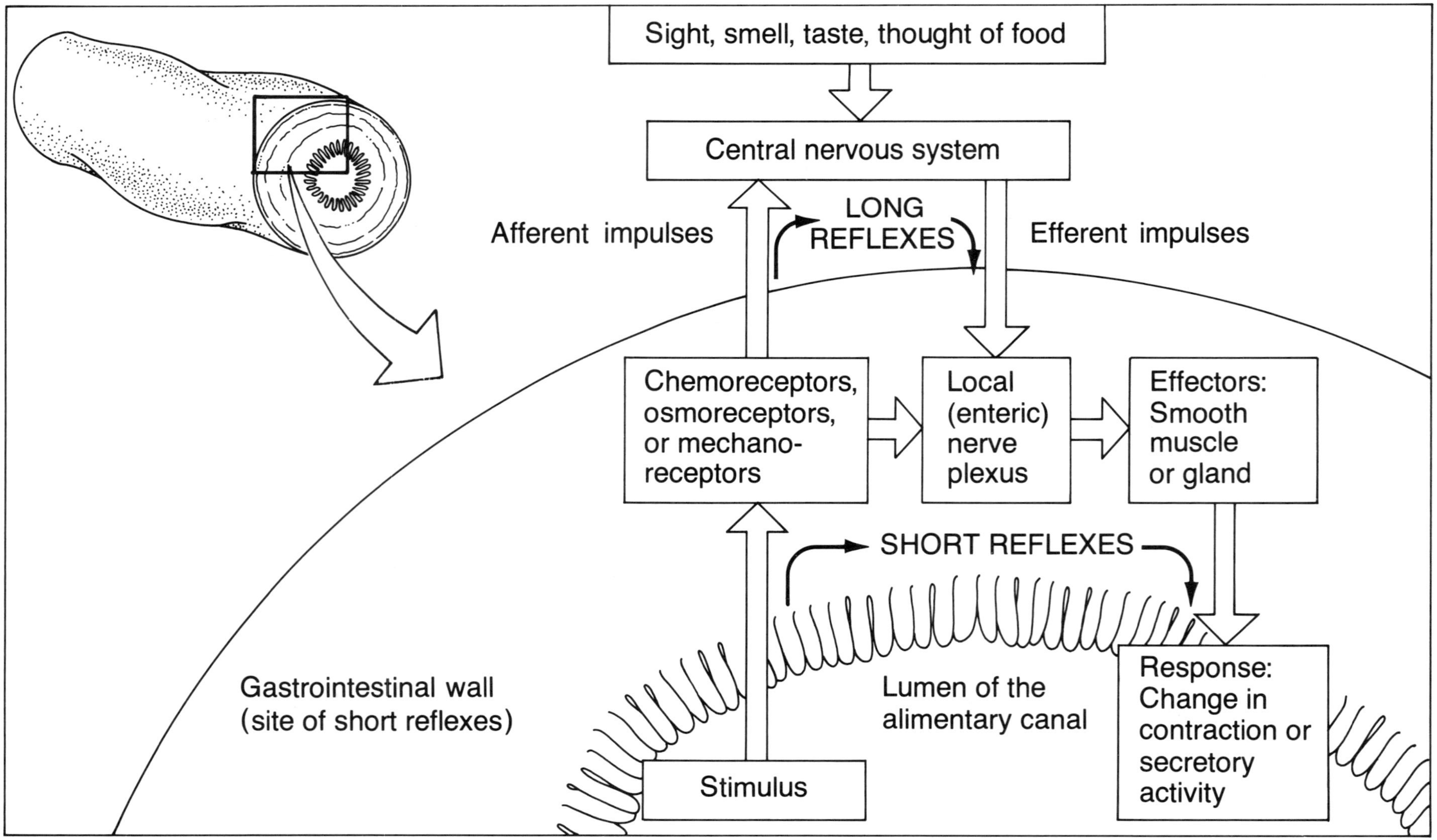

The basic structure of the wall of the alimentary canal

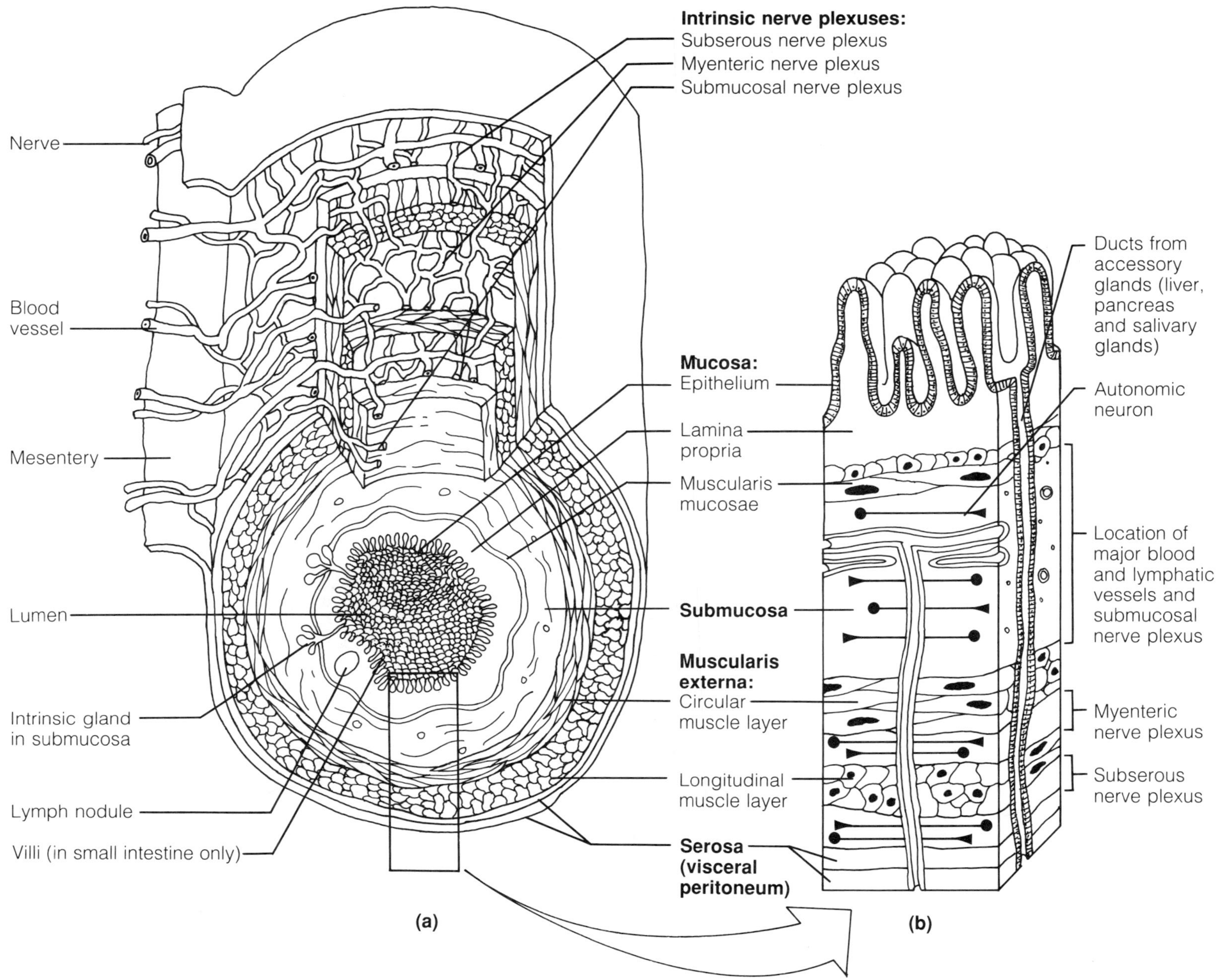

Deglutition (swallowing)

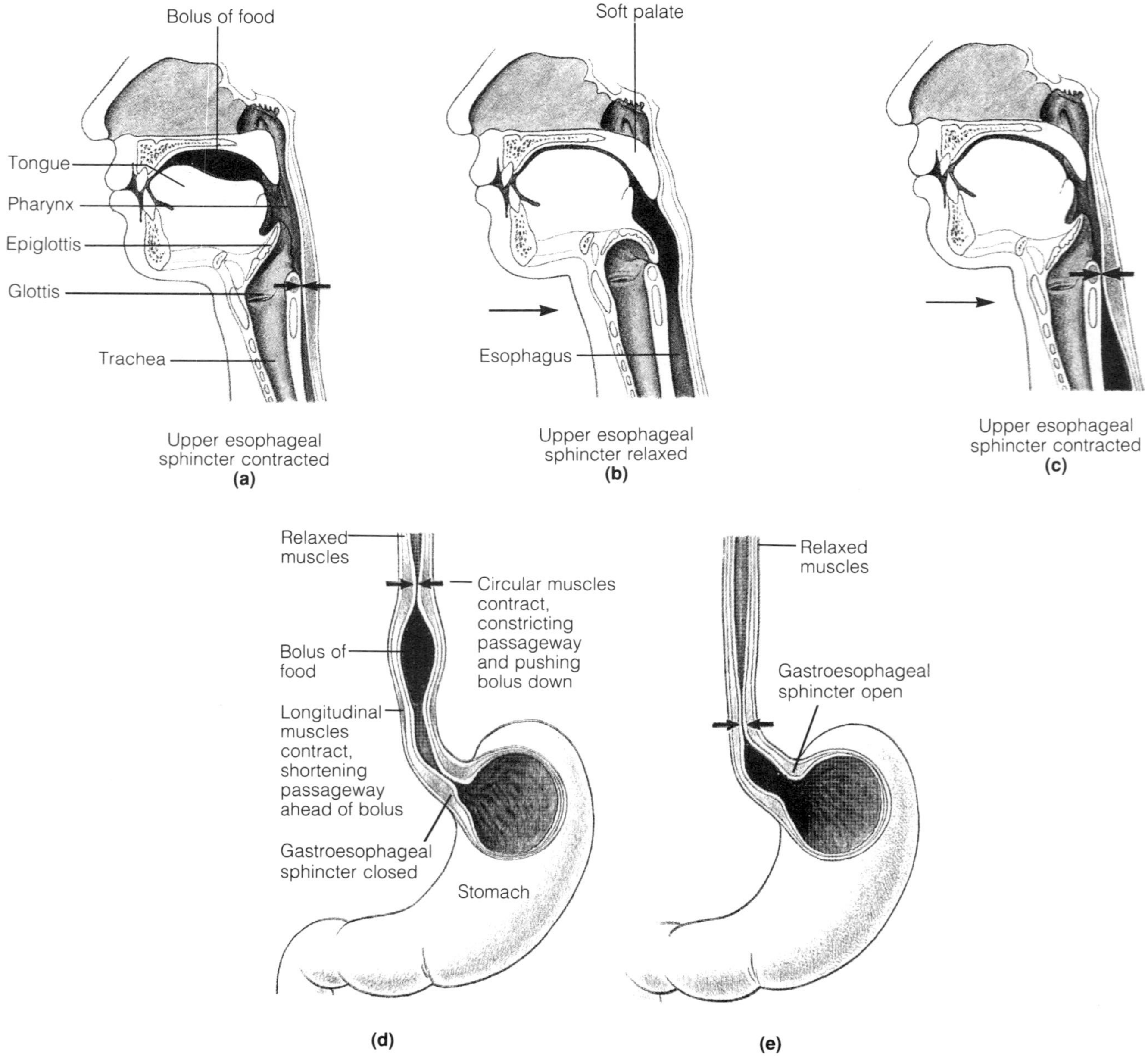

Neural and hormonal mechanisms that regulate the release of gastric juice

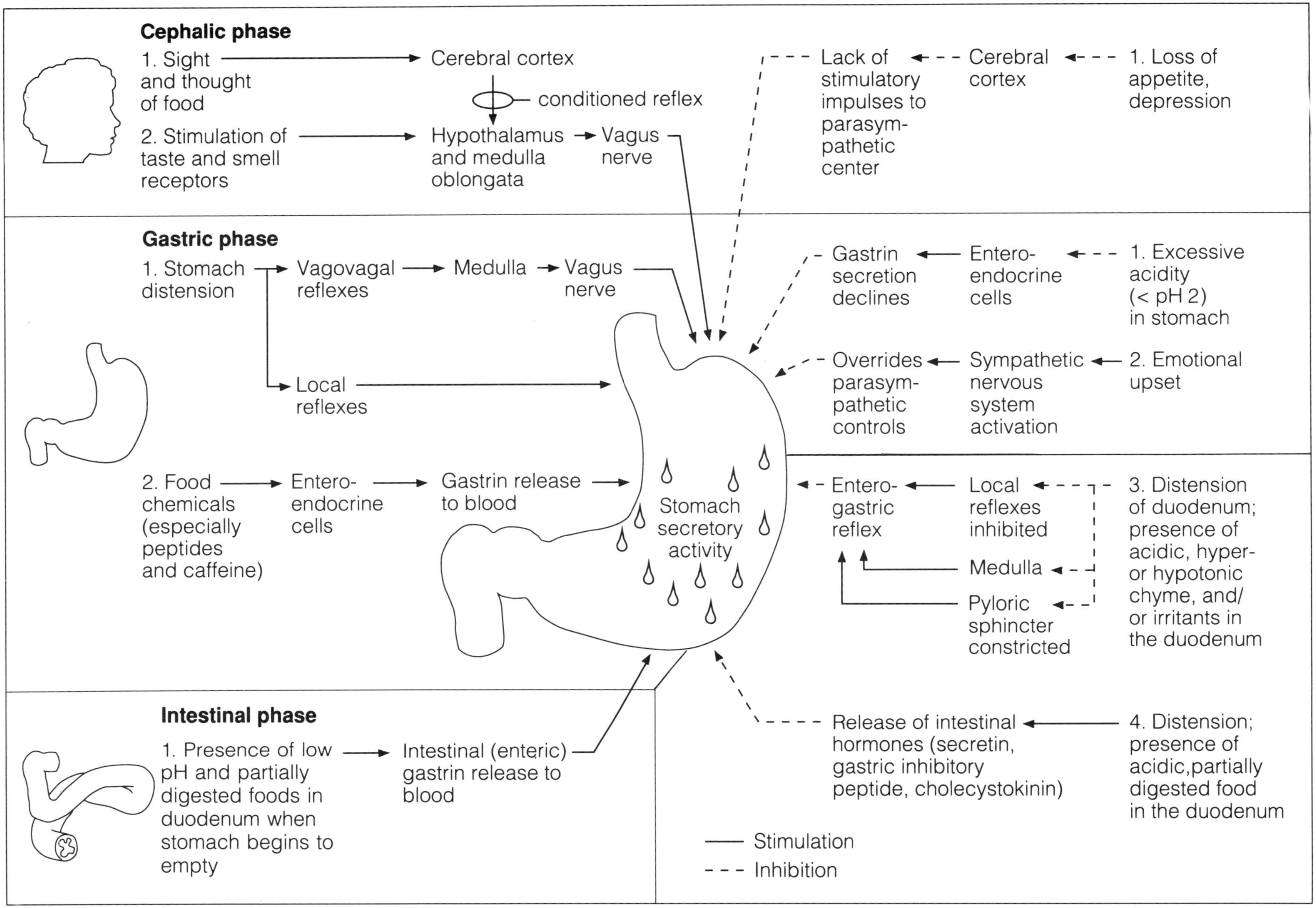

Peristaltic waves act primarily in the inferior portion of the stomach to mix and move chyme through the pyloric valve

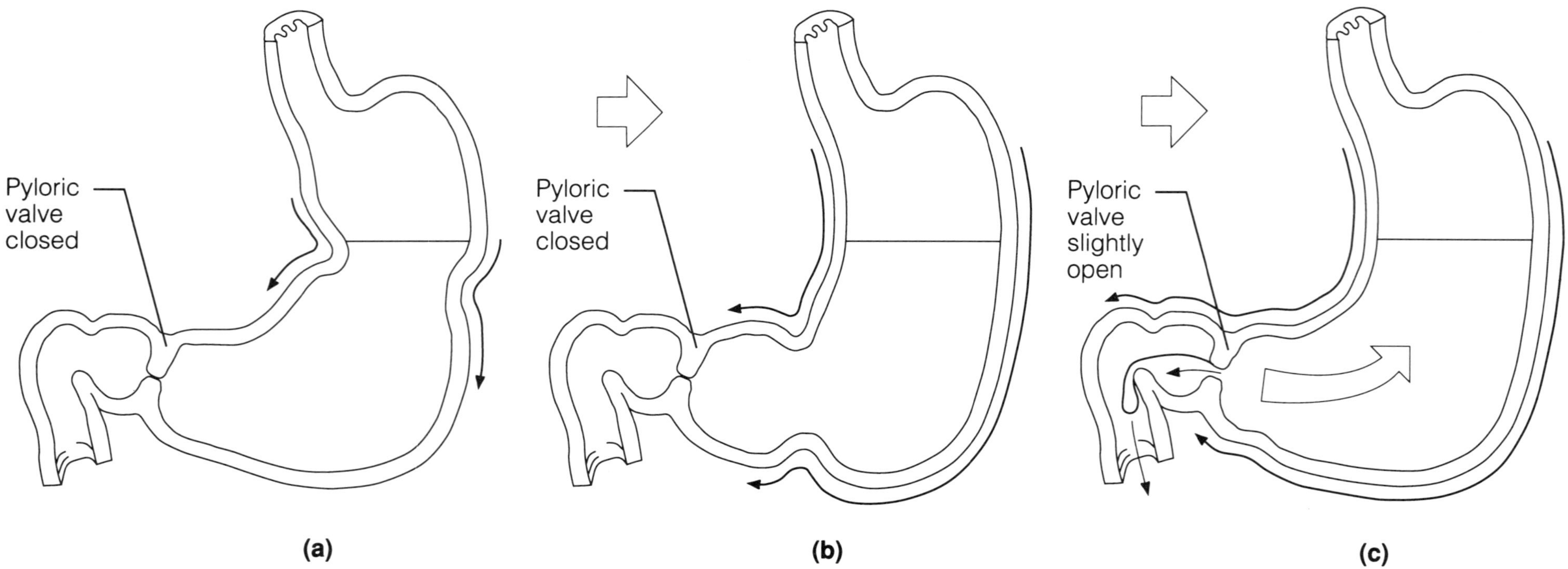

Neural and hormonal factors inhibiting gastric emptying

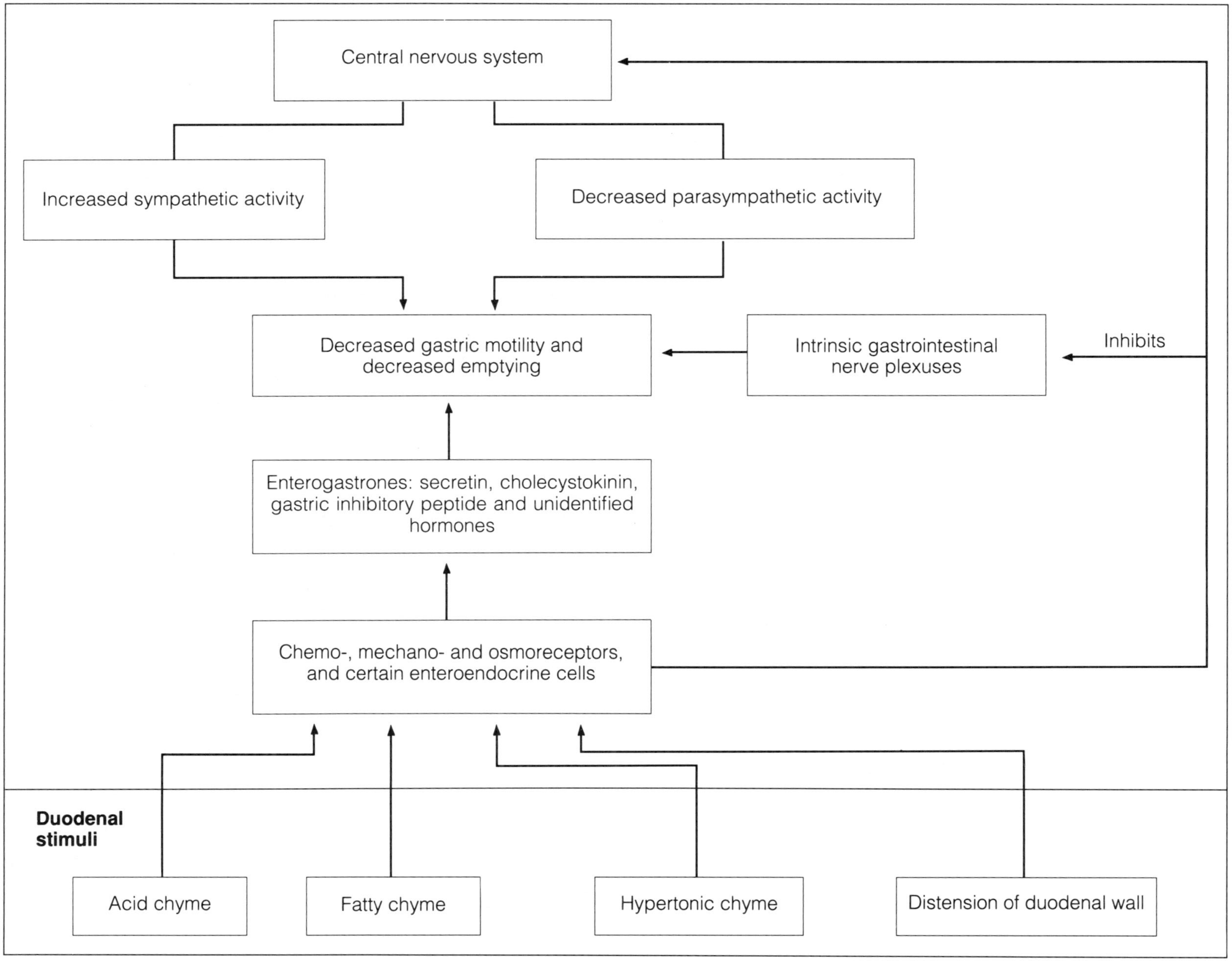

Structural modifications of the small intestine that increase its surface area for digestion and absorption

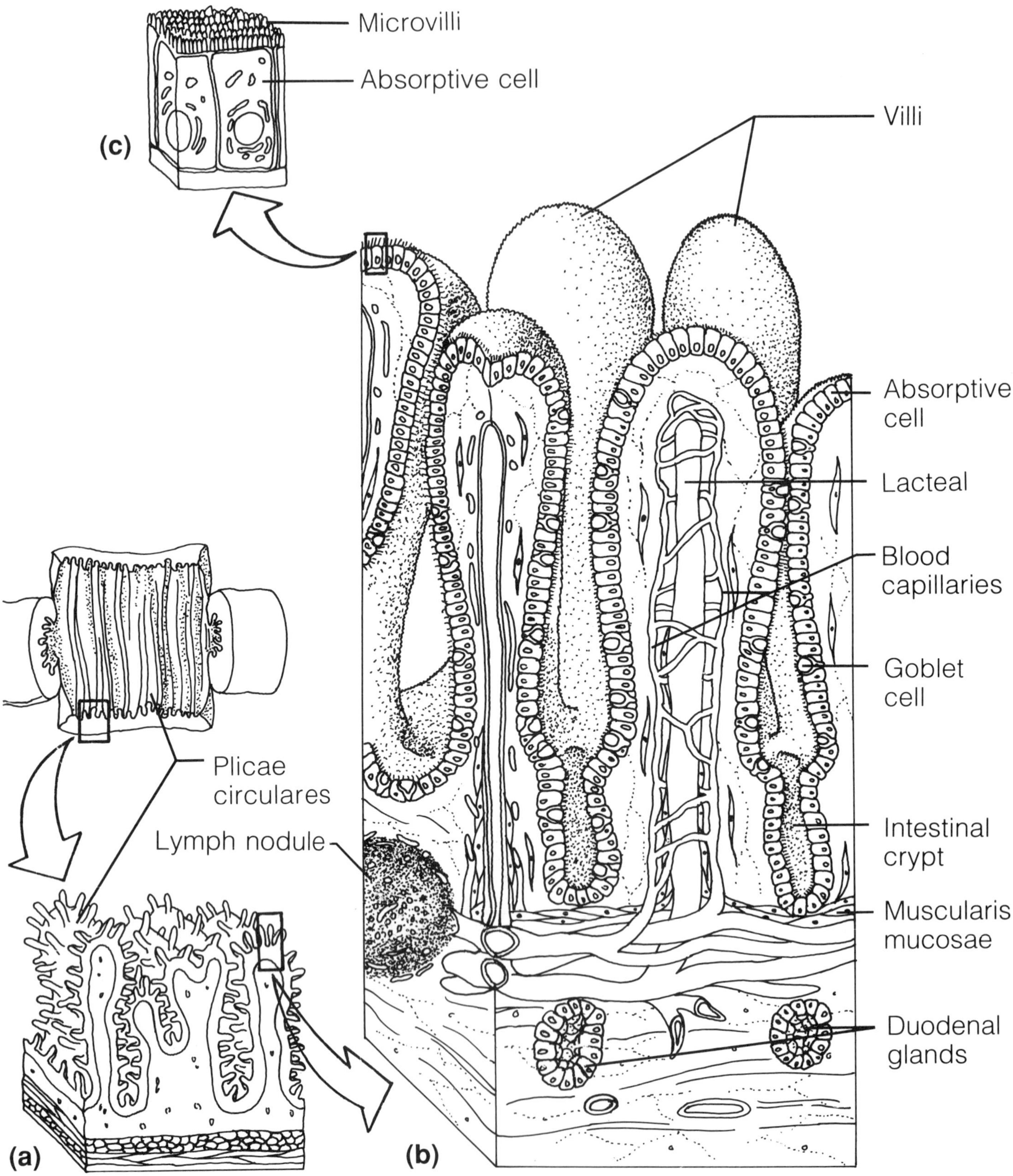

Microscopic anatomy of the liver

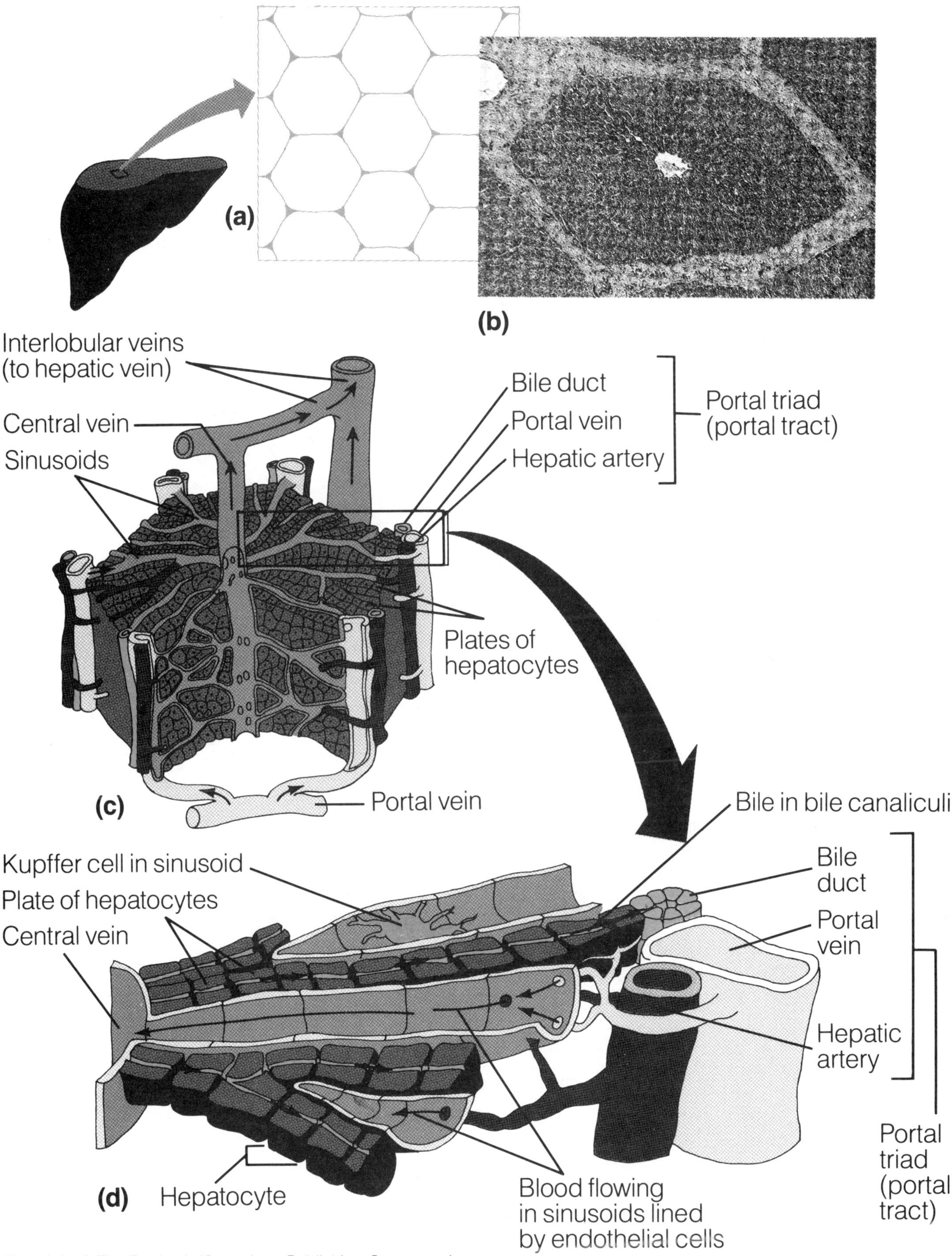

Mechanisms promoting the secretion of bile and its entry into the duodenum

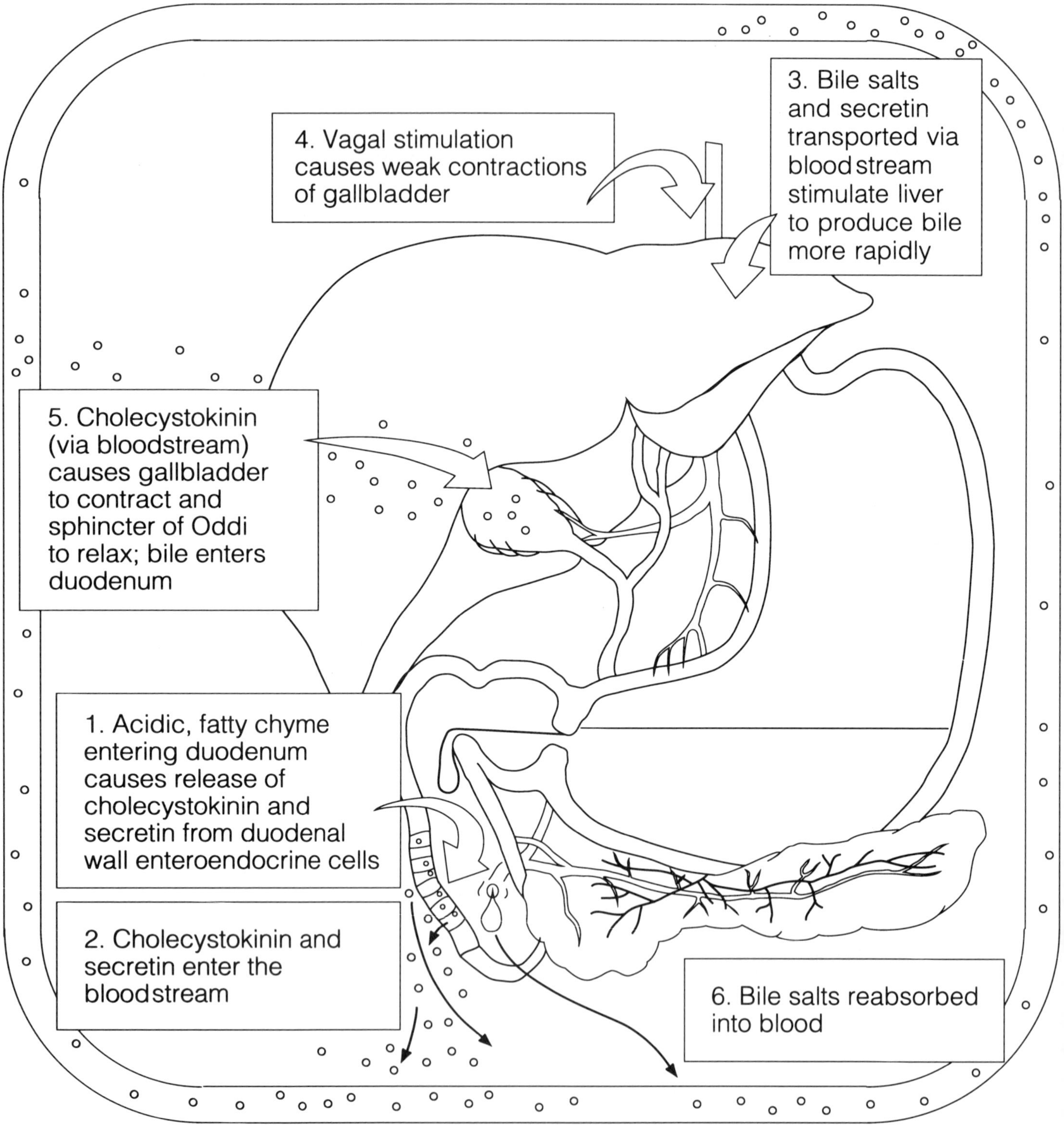

Activation of pancreatic proteases in the small intestine

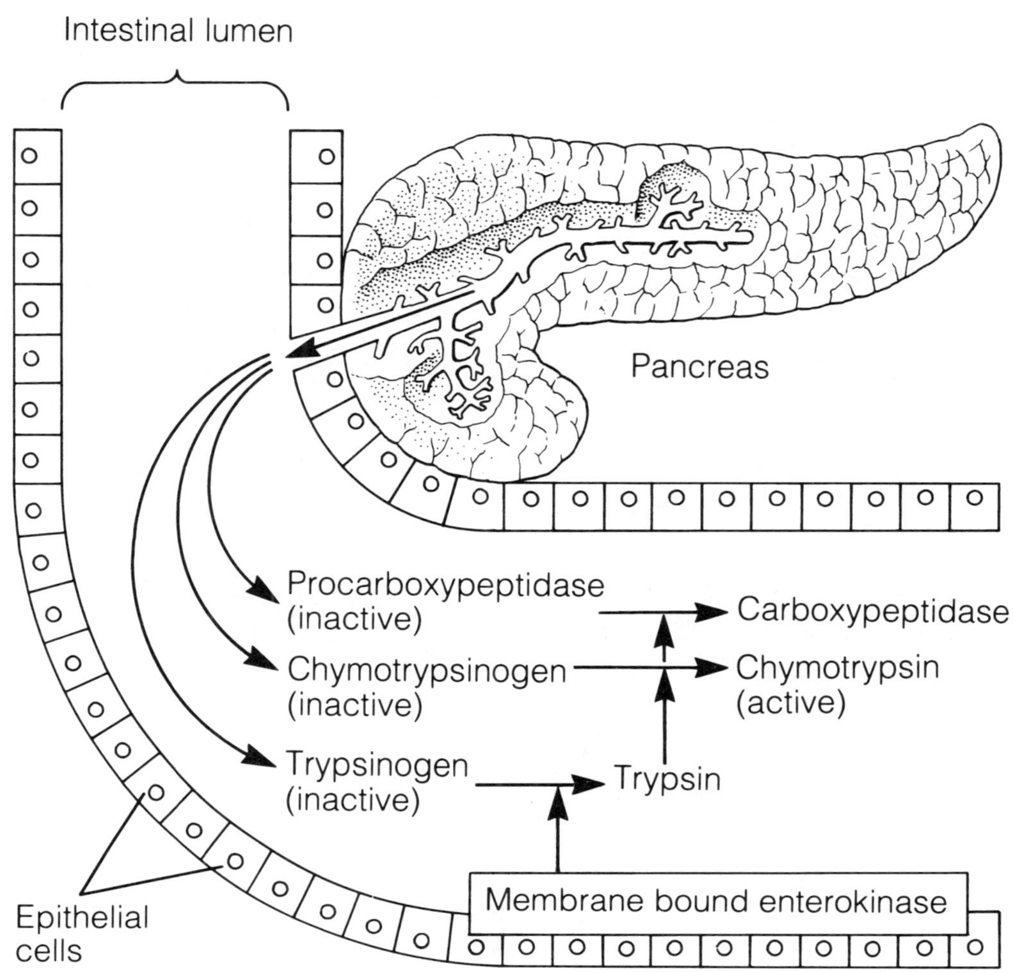

Regulation of pancreatic juice secretion by neural and hormonal factors

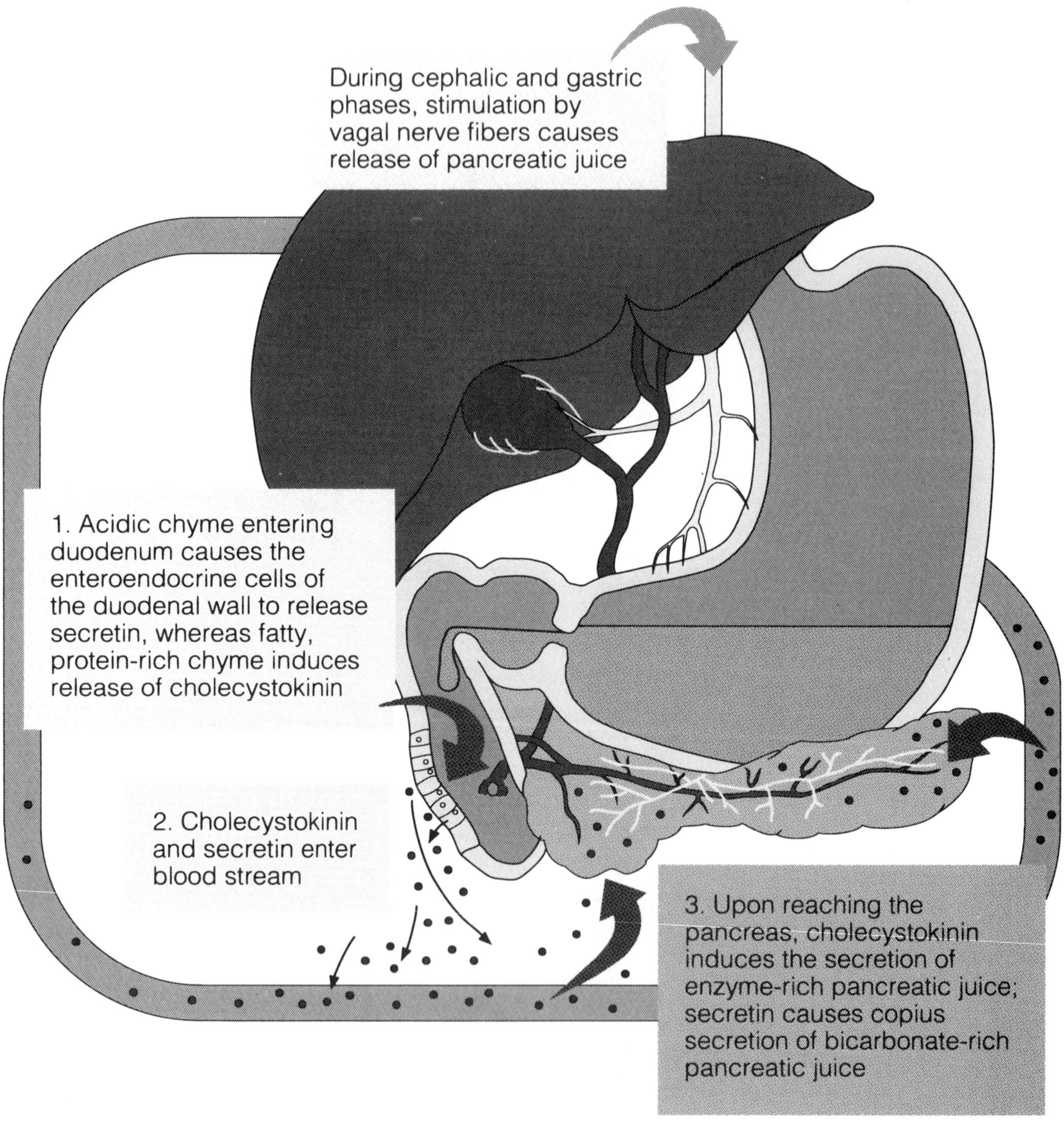

Defecation reflex

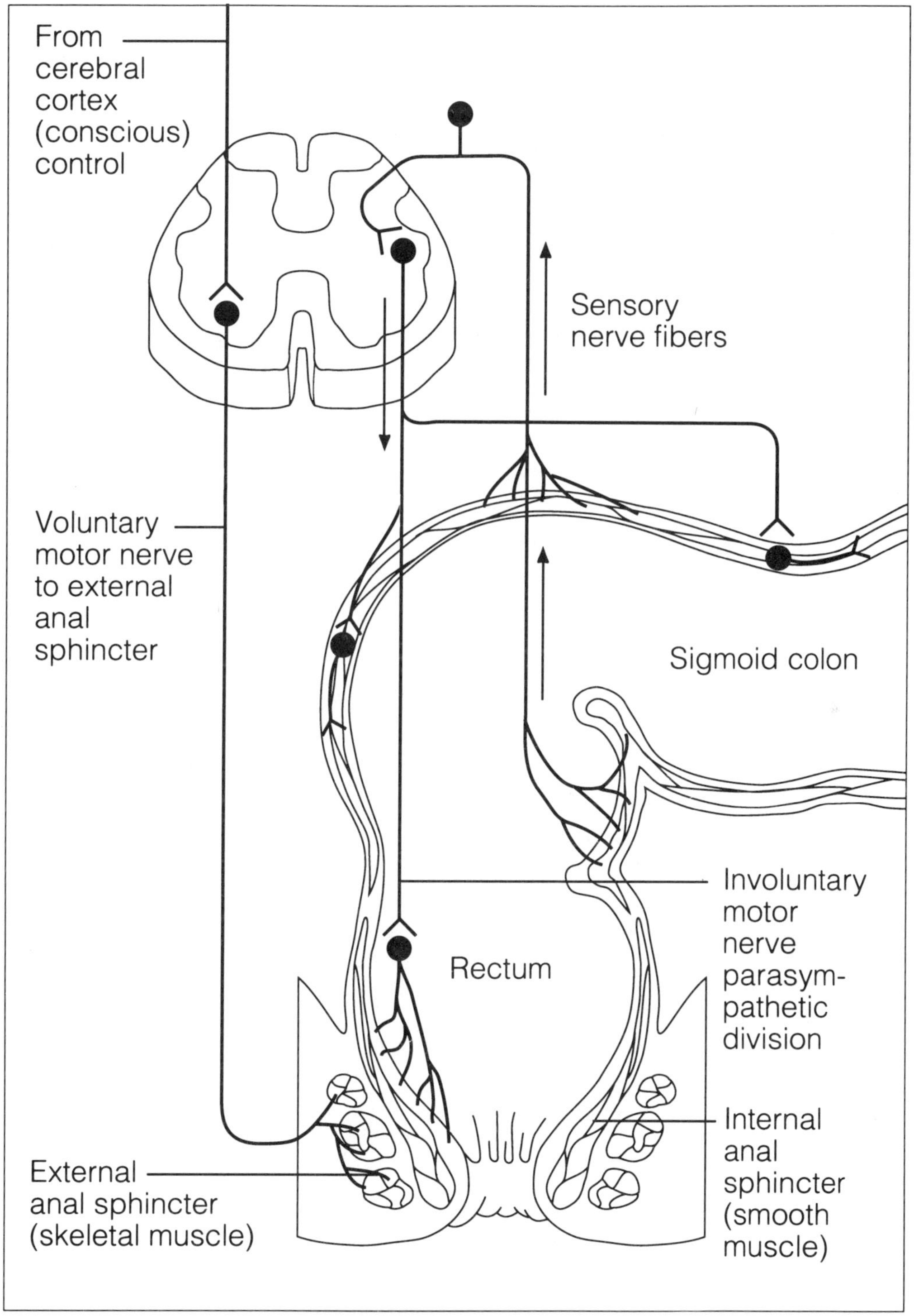

Protein digestion and absorption in the small intestine

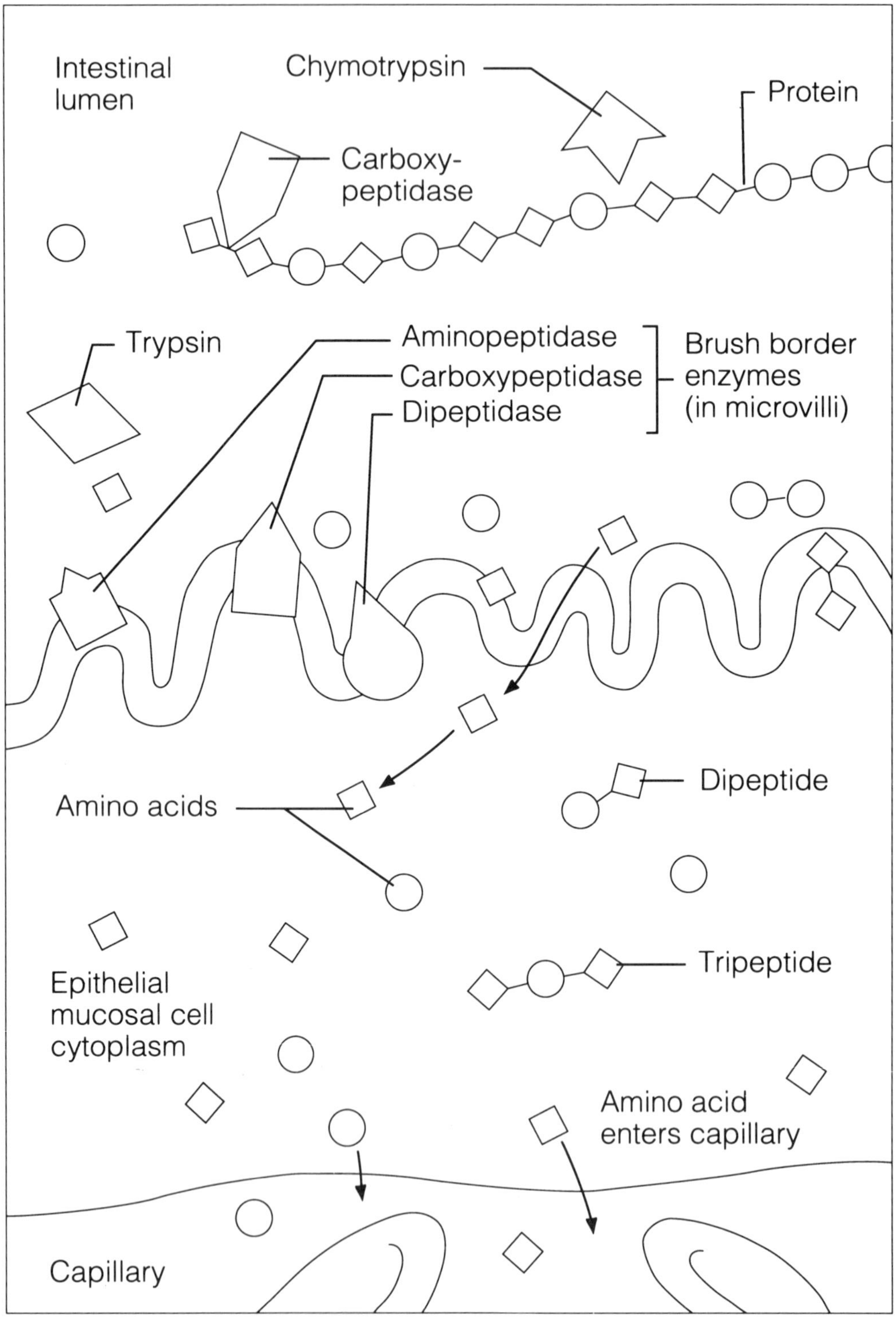

Role of bile salts in fat emulsification

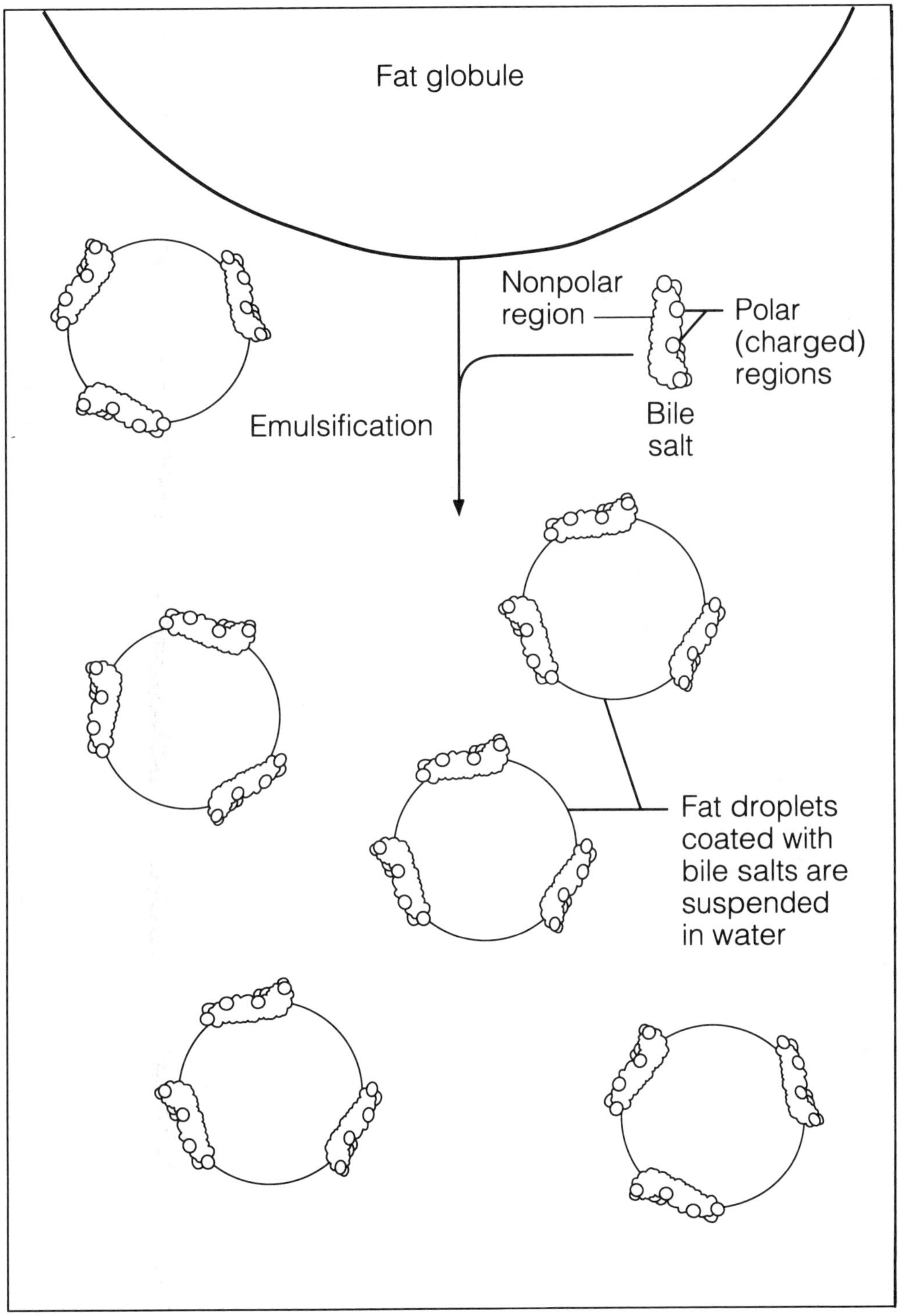

Fatty acid absorption

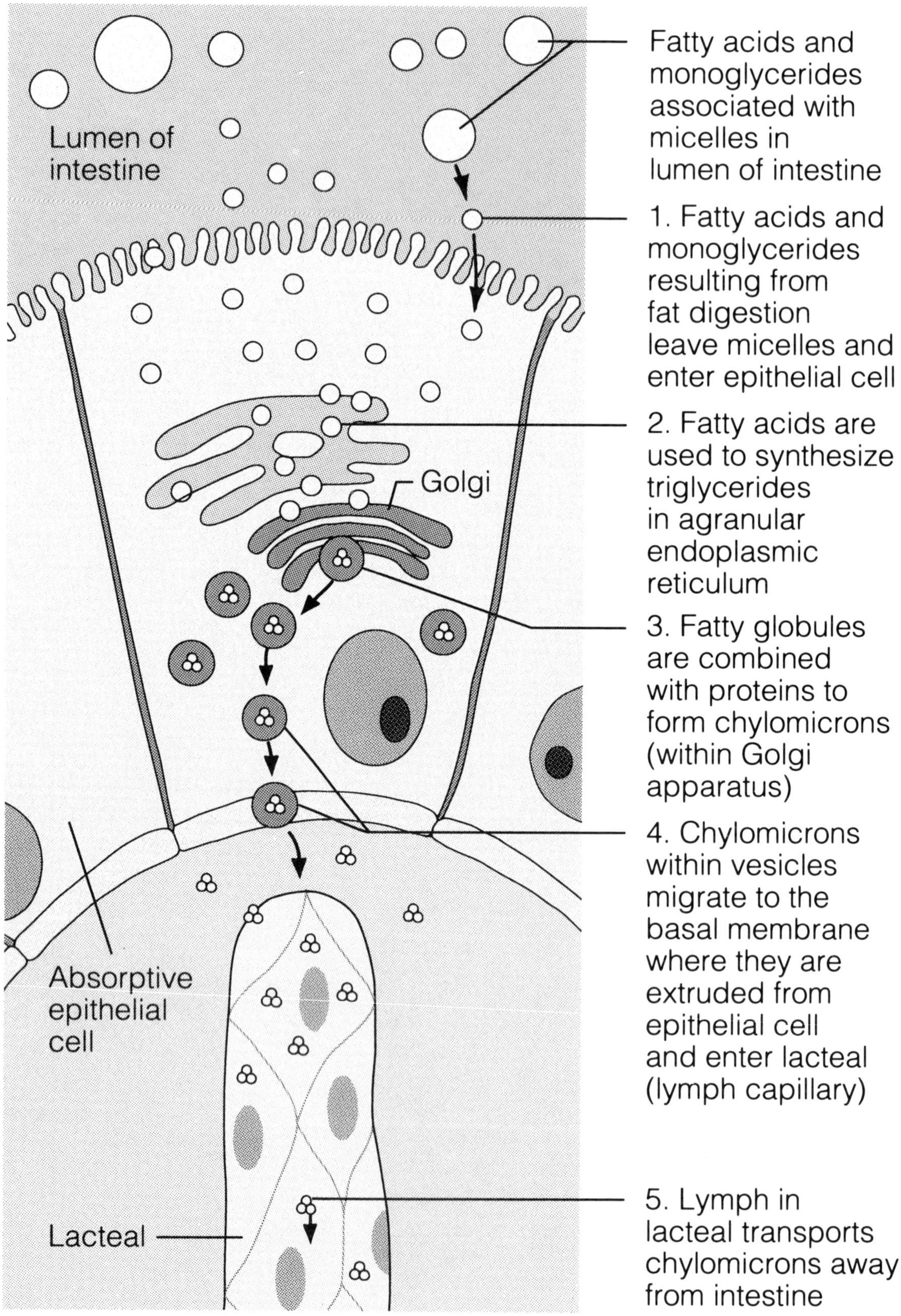

Overview of nutrient use by body cells

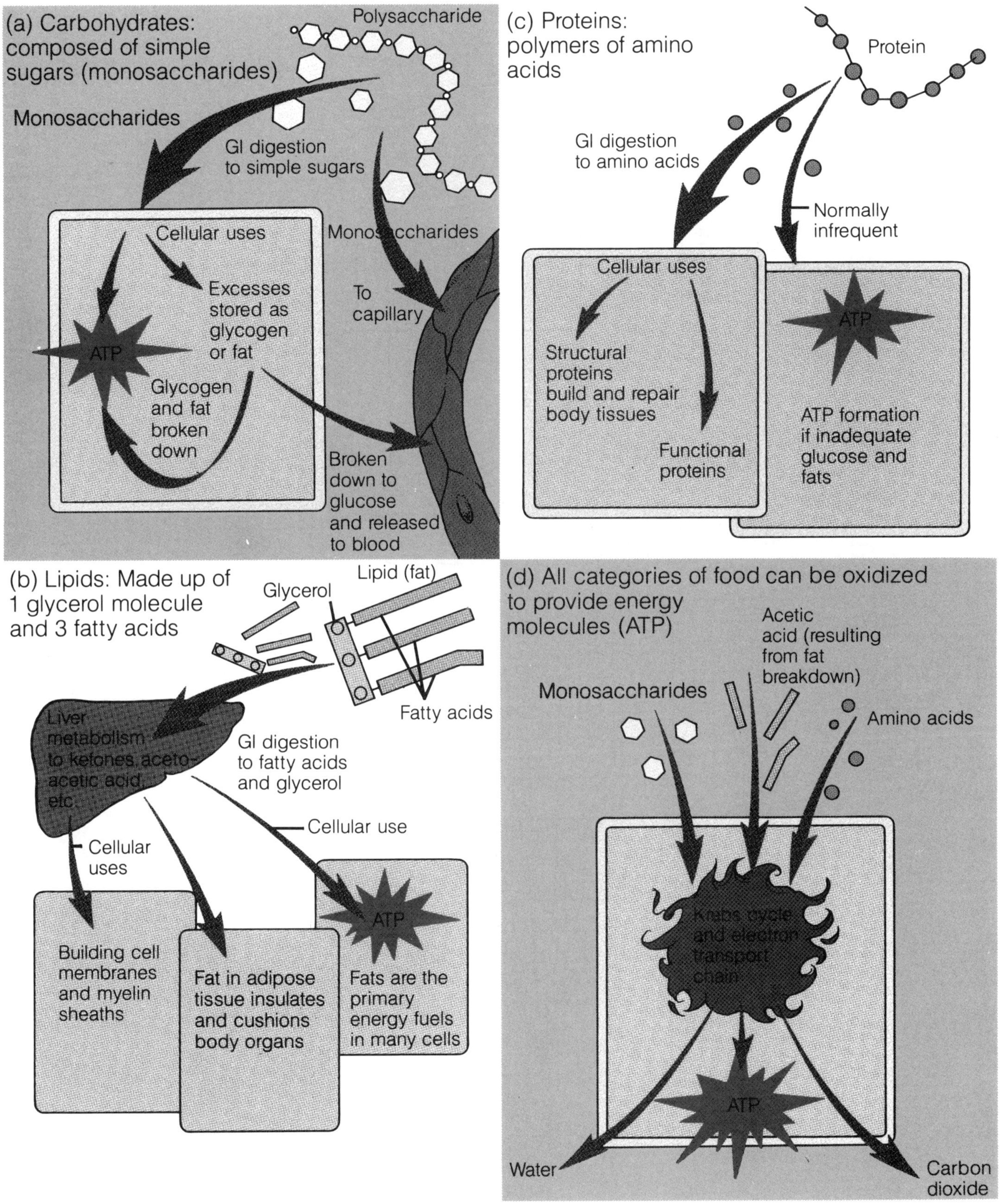

Essential amino acids

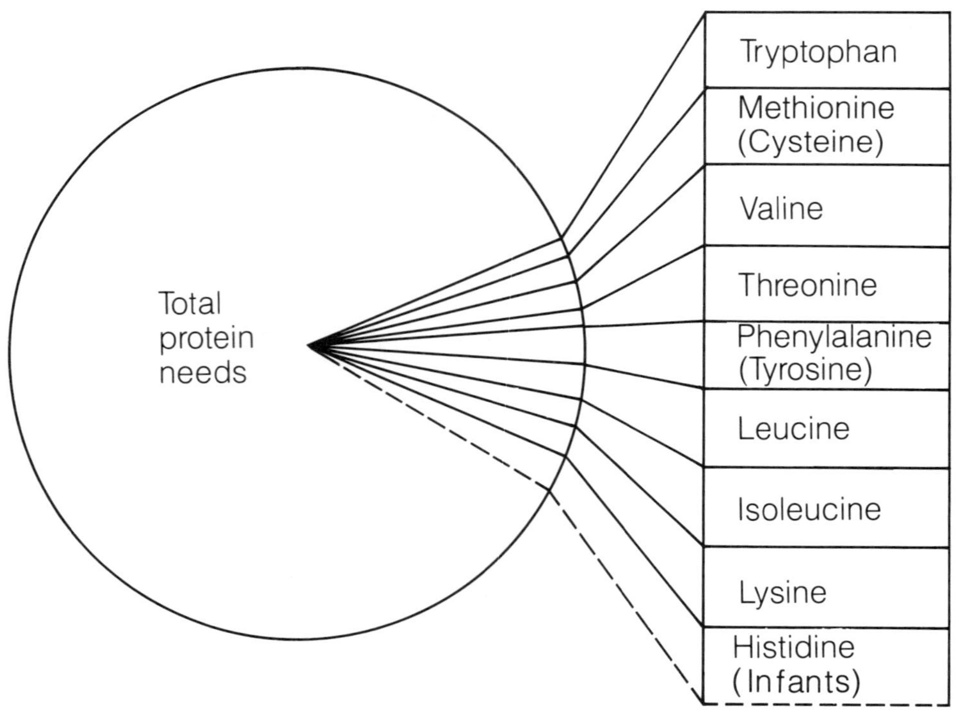

(a) Essential amino acids

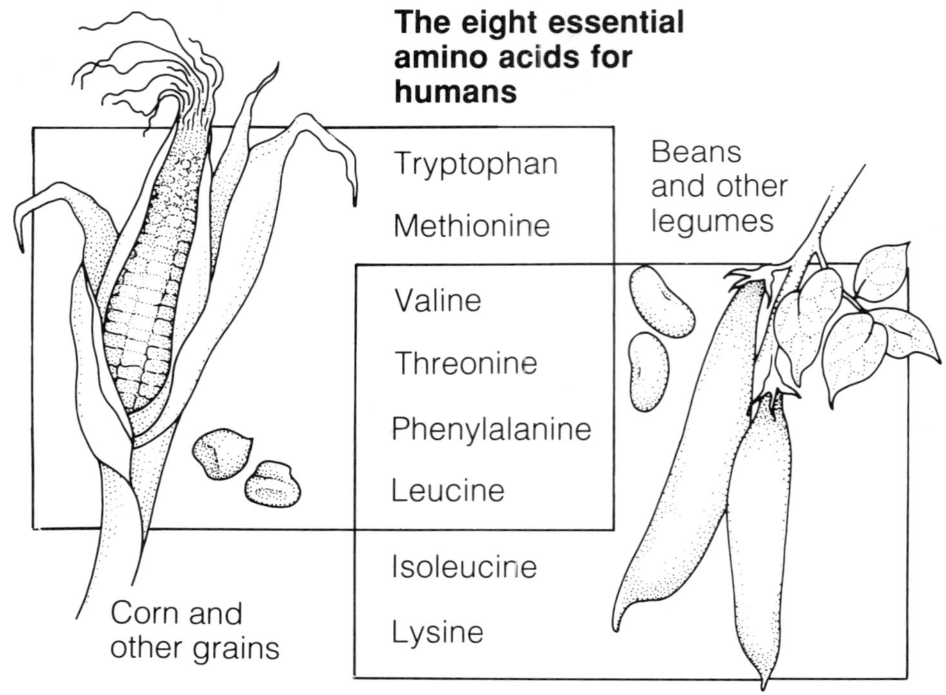

(b) Vegetarian diets providing the eight essential amino acids for humans

The three stages of metabolism of energy-containing nutrients in the body

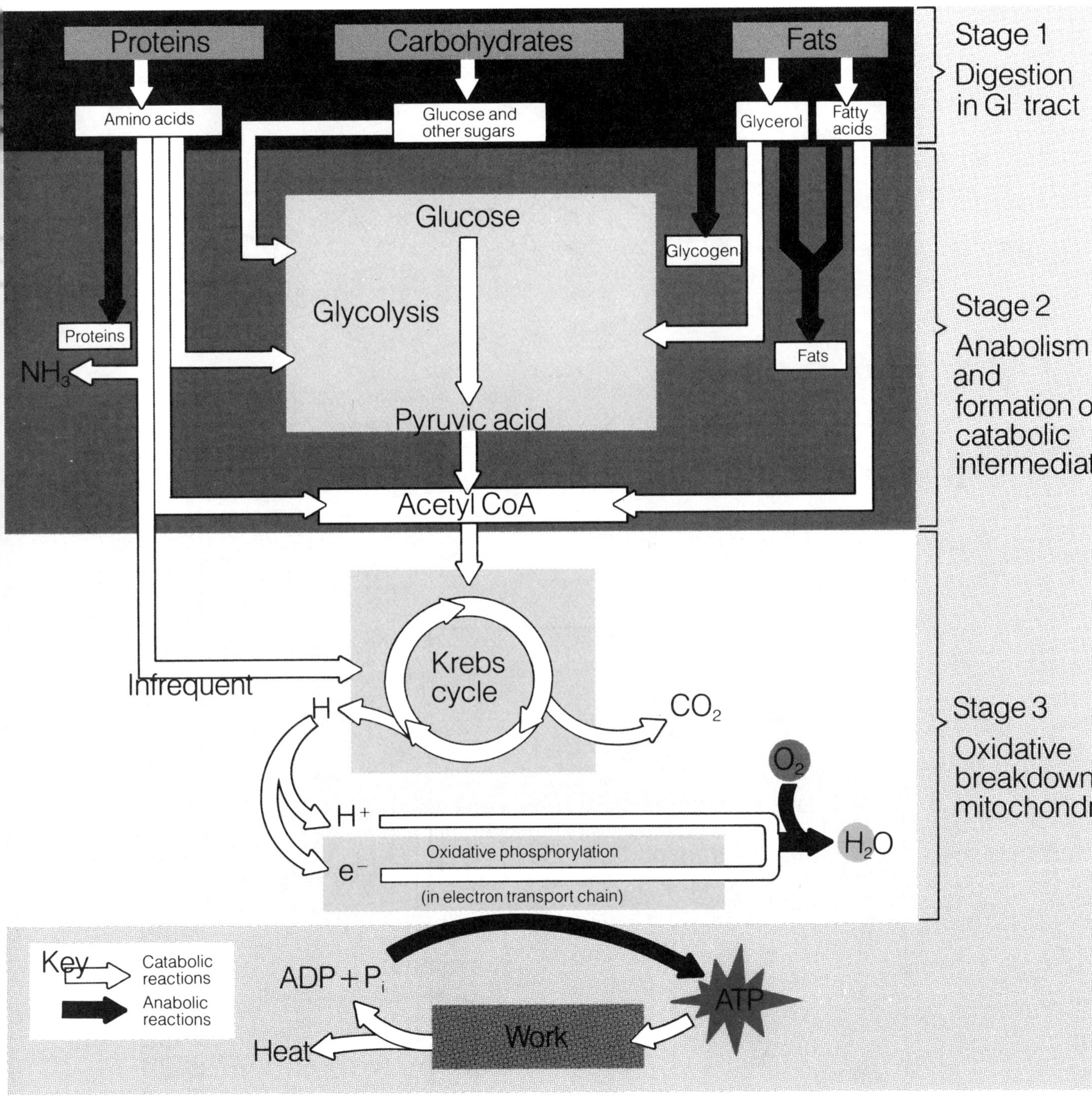

Mechanisms of phosphorylation

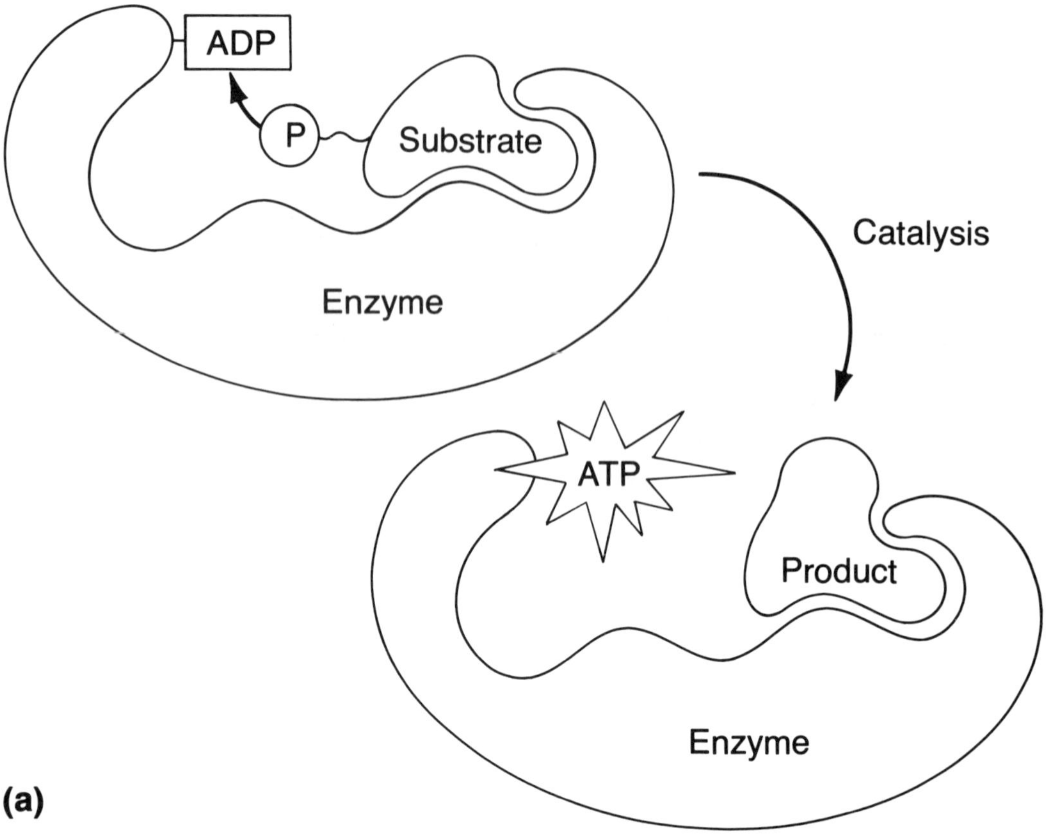

(a)

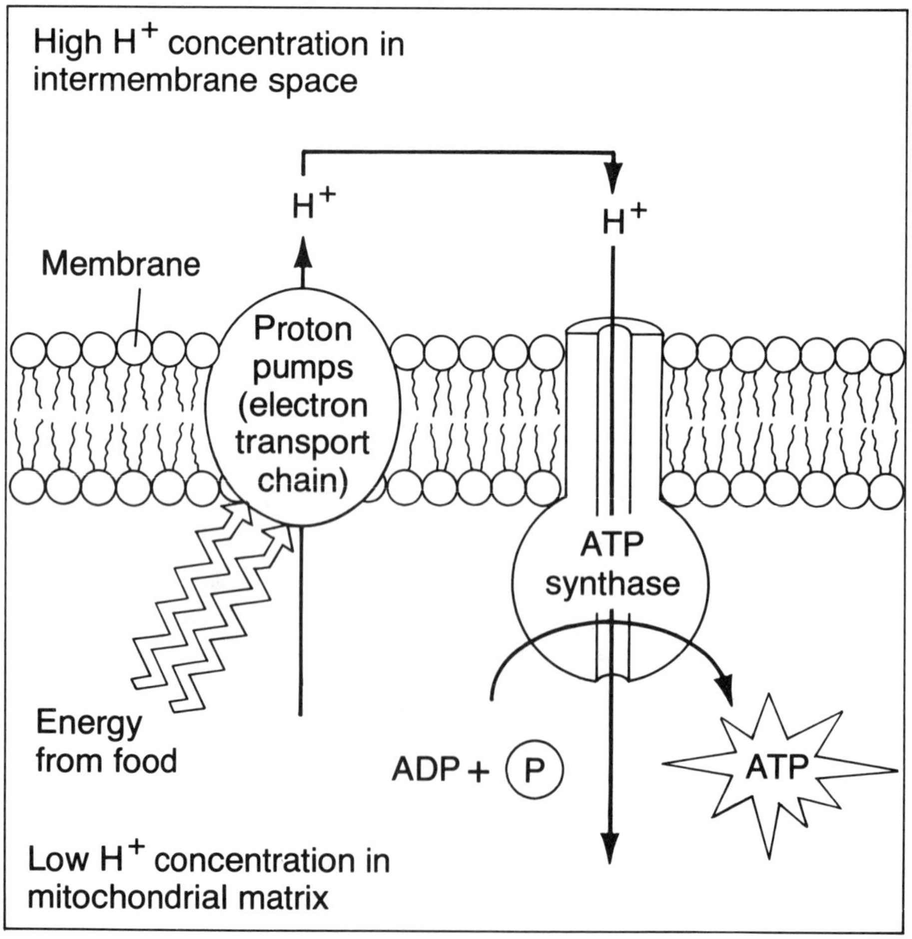

(b)

An overview of sites of ATP formation during cellular respiration

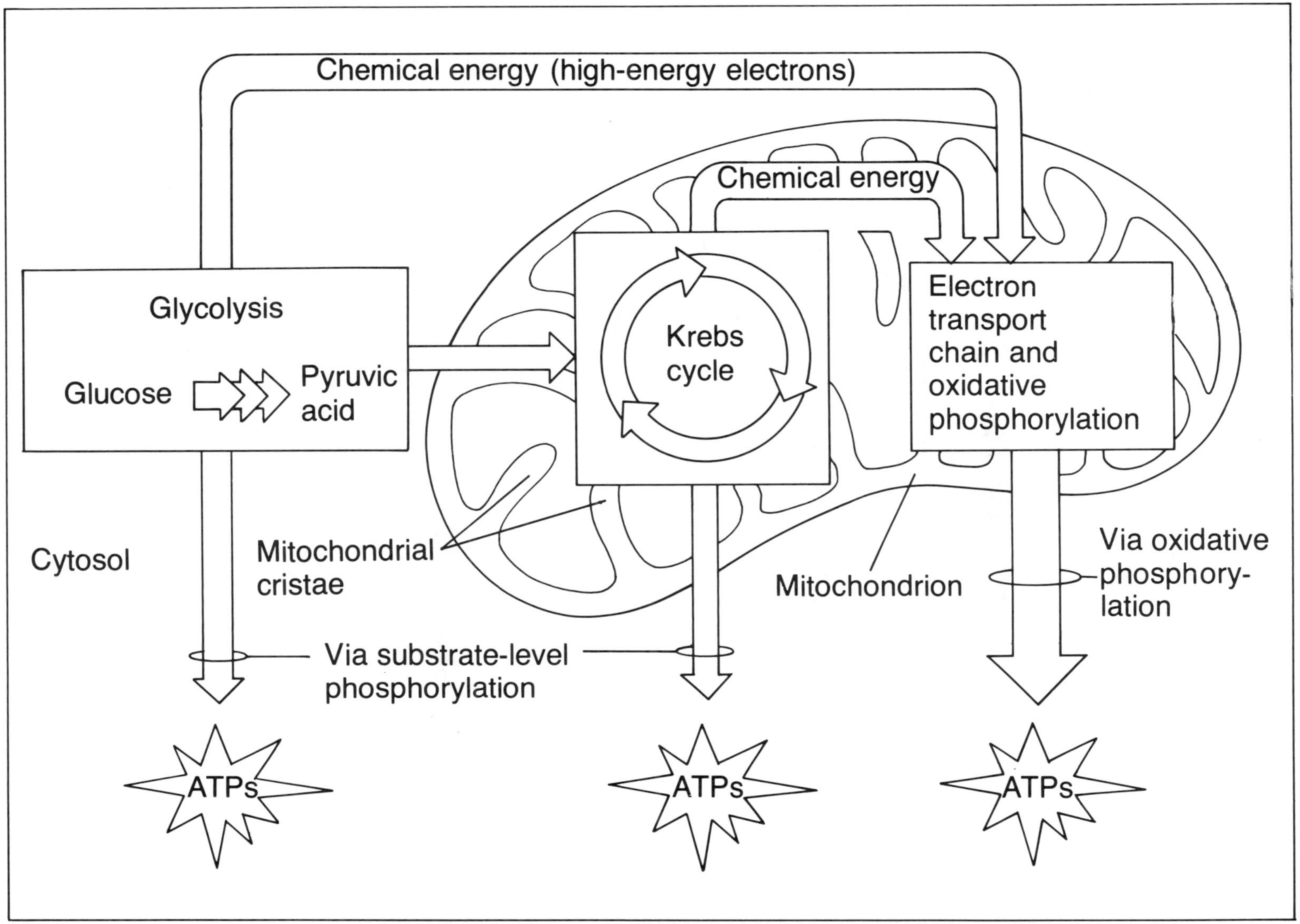

The three major phases of glycolysis

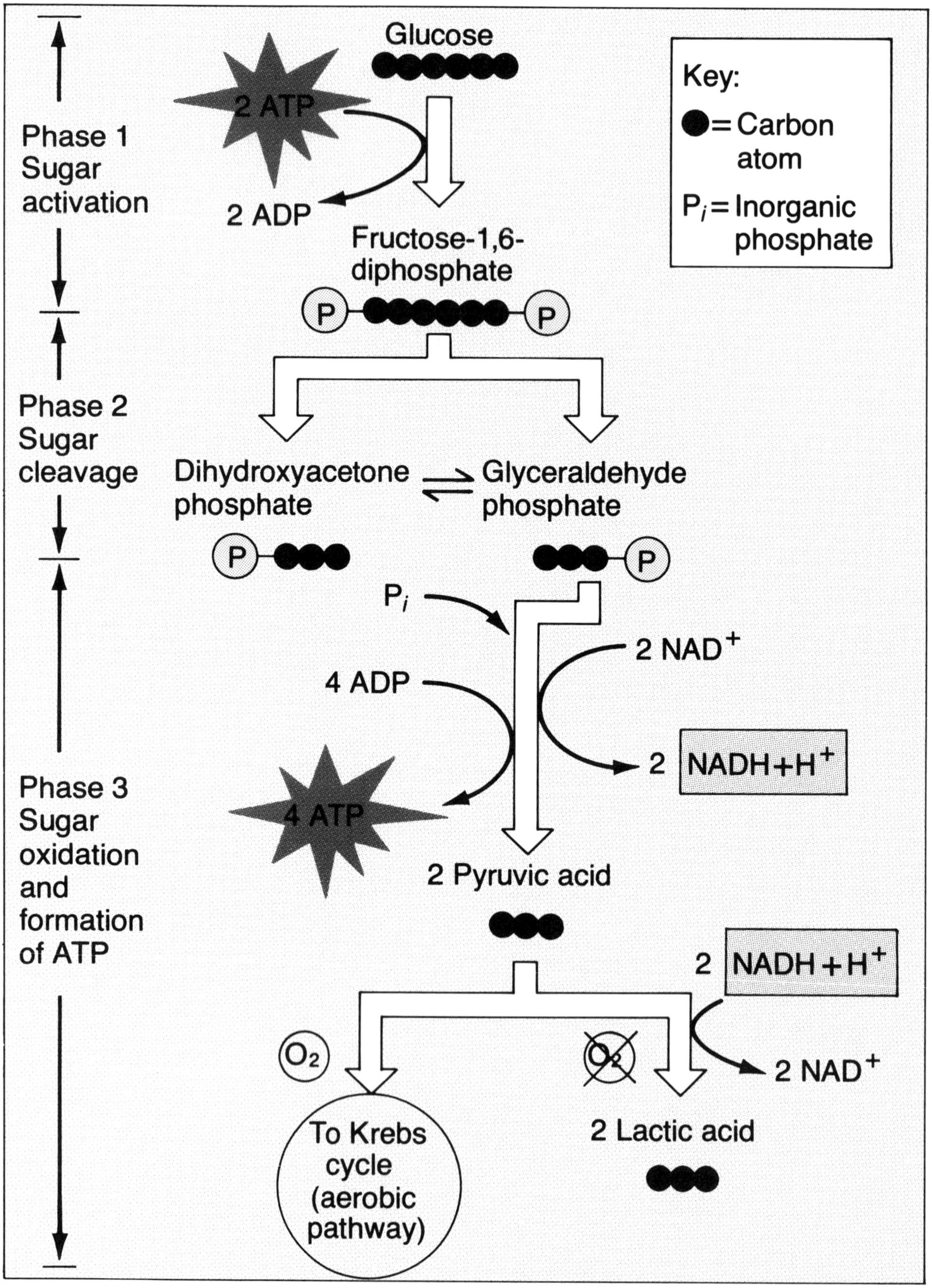

Simplified version of the Krebs cycle

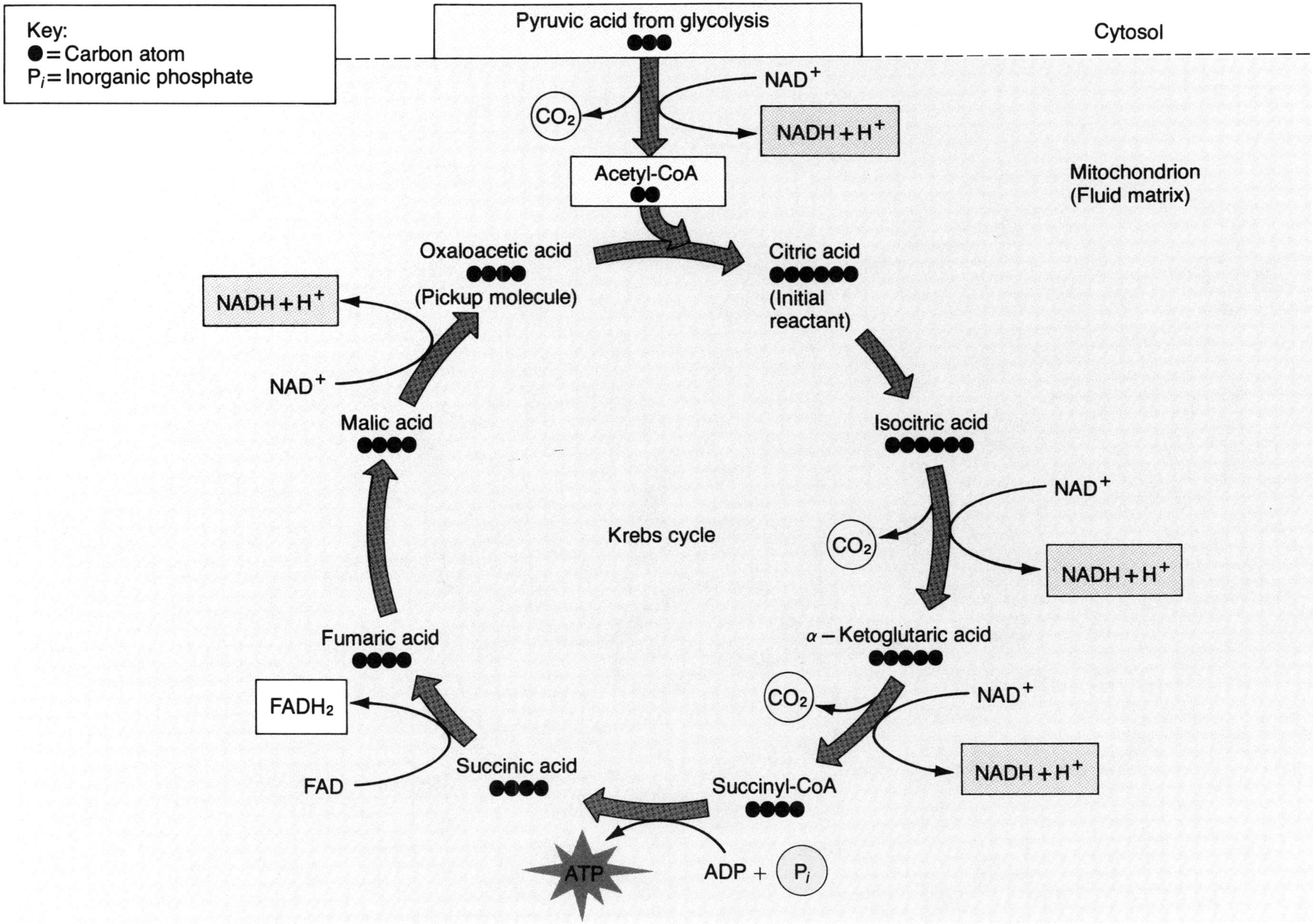

One hypothetical mechanism of oxidative phosphorylation

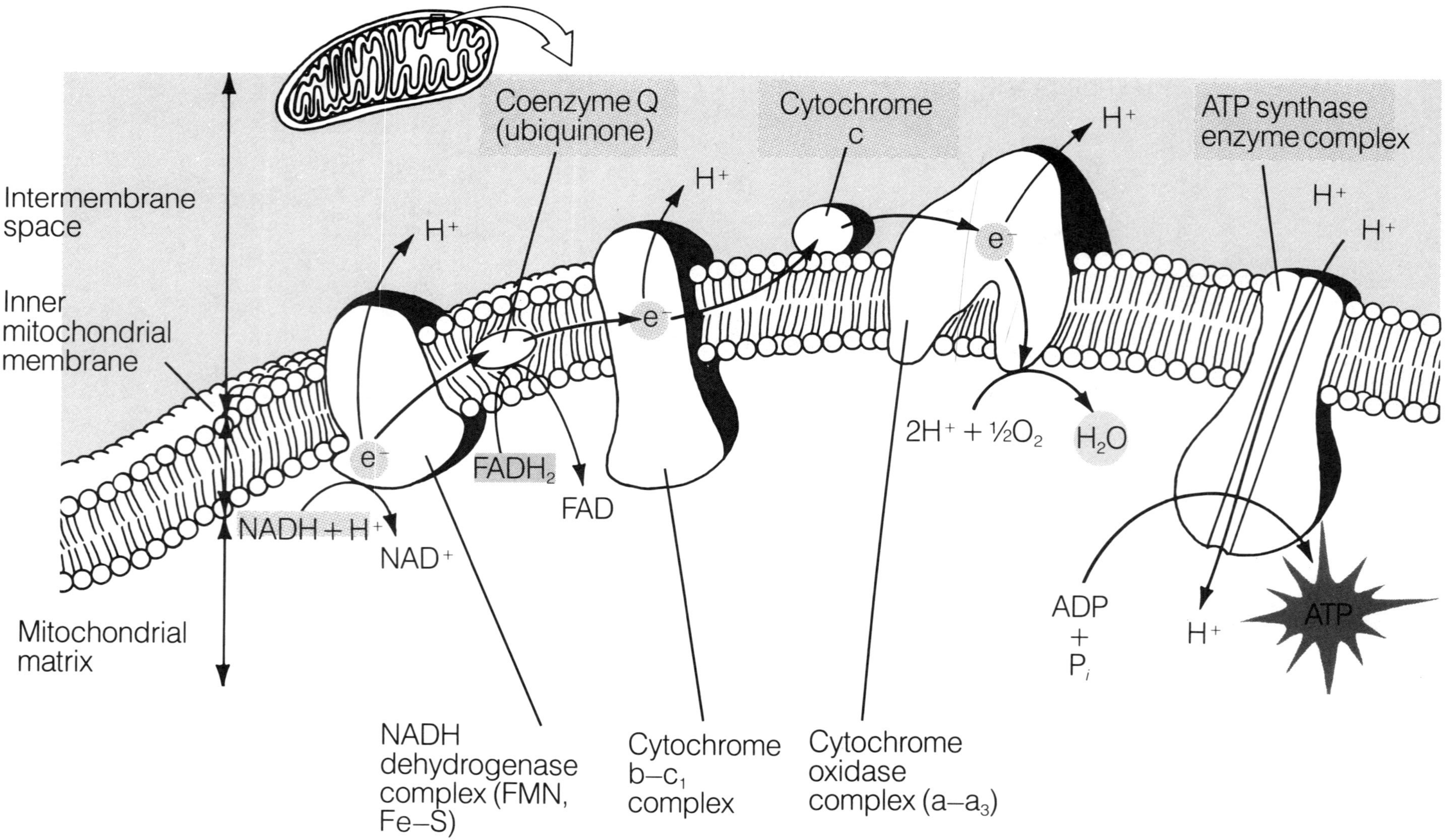

Electronic energy gradient in the electron transport chain

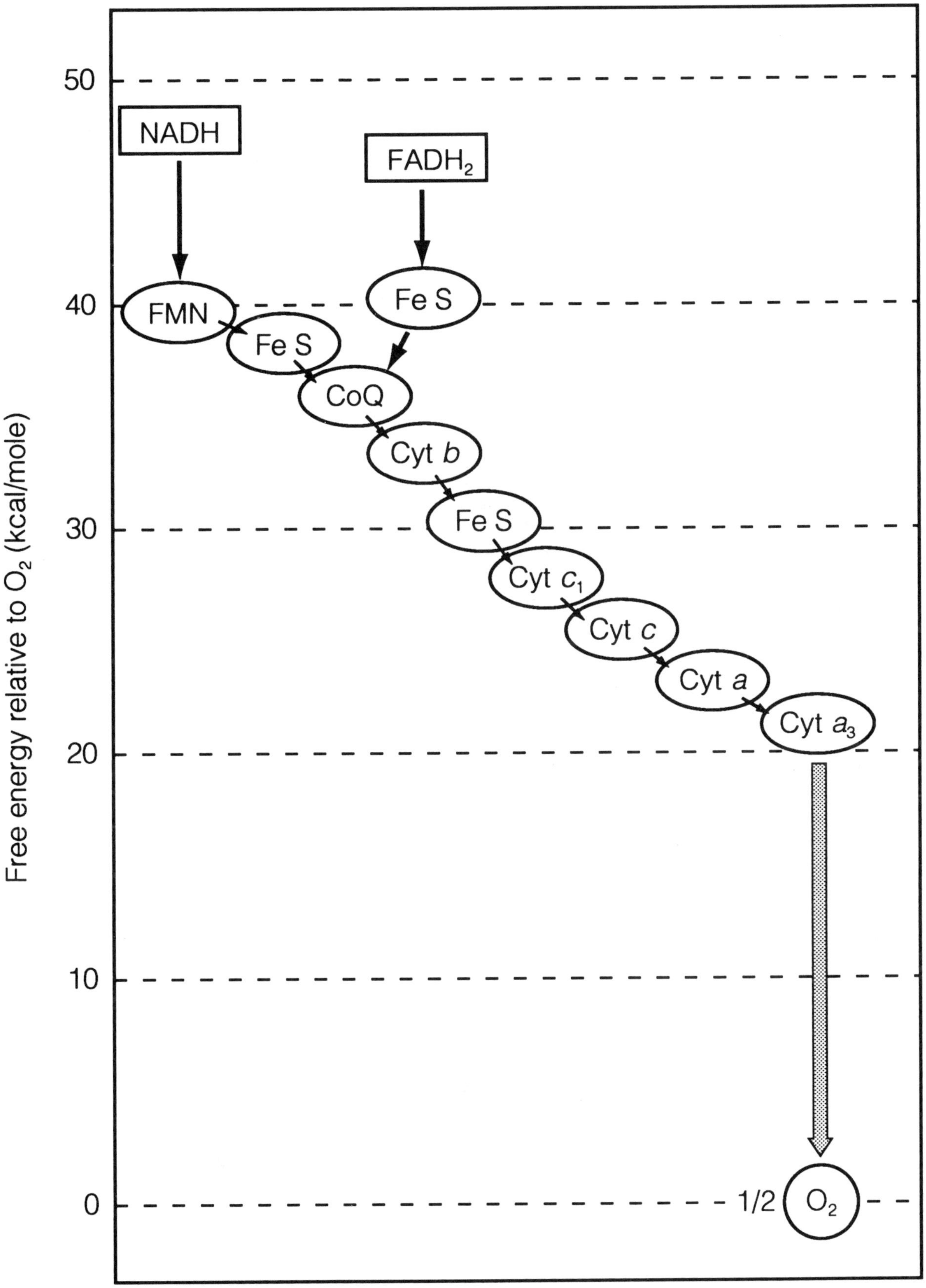

The aerobic oxidation of one molecule of glucose yields a net gain of 36 to 38 ATP molecules

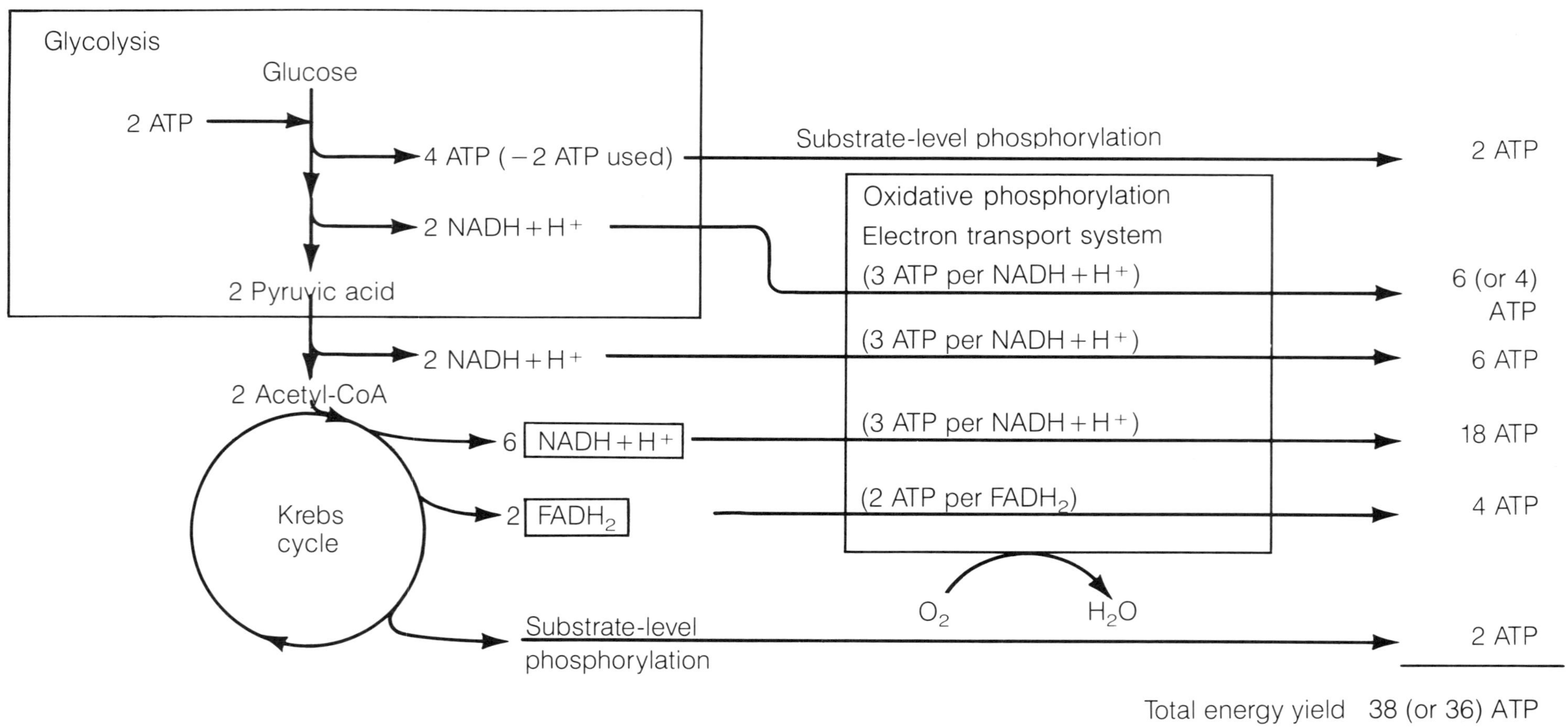

Glycogenesis and glycogenolysis

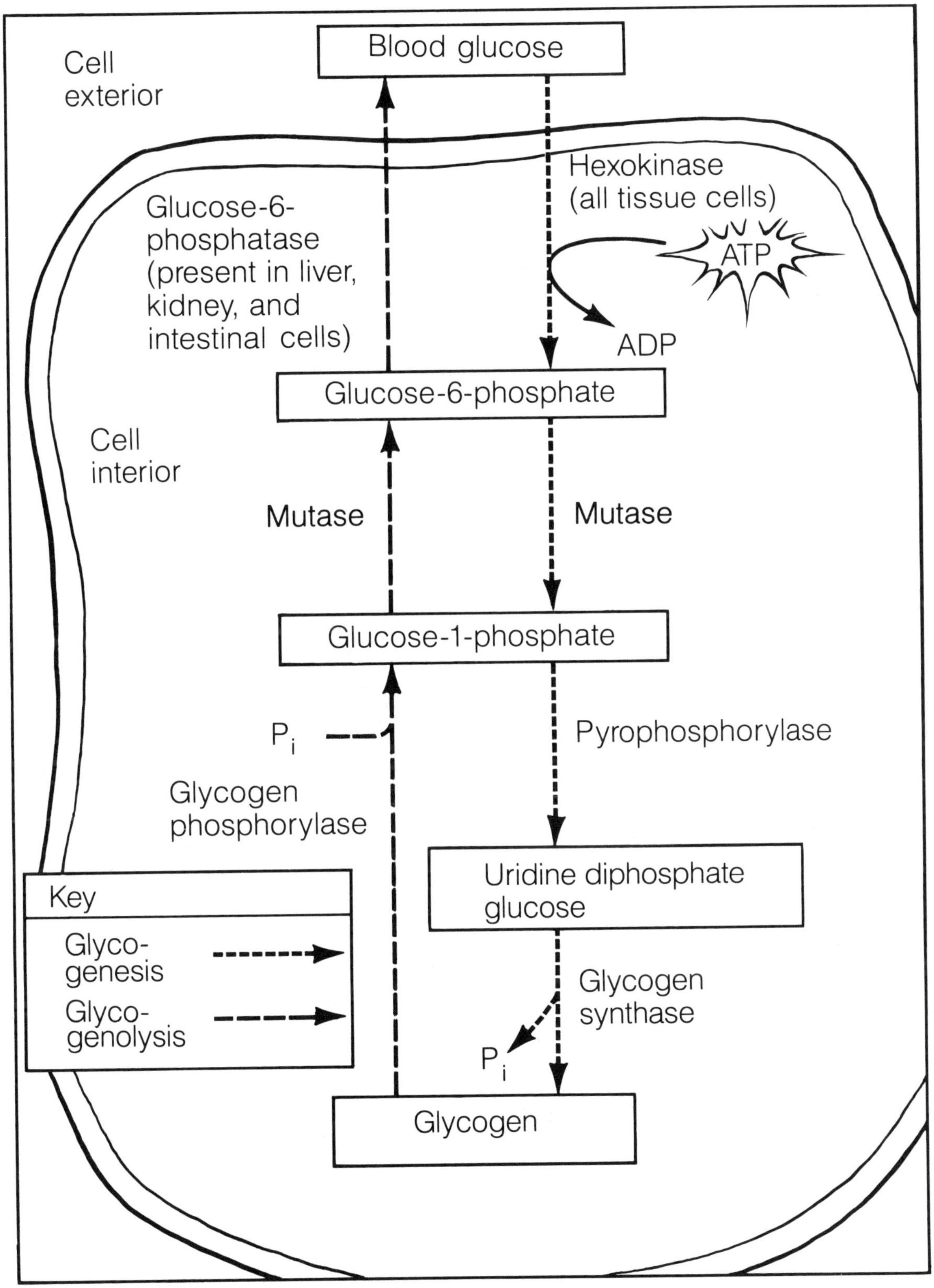

Metabolism of triglycerides

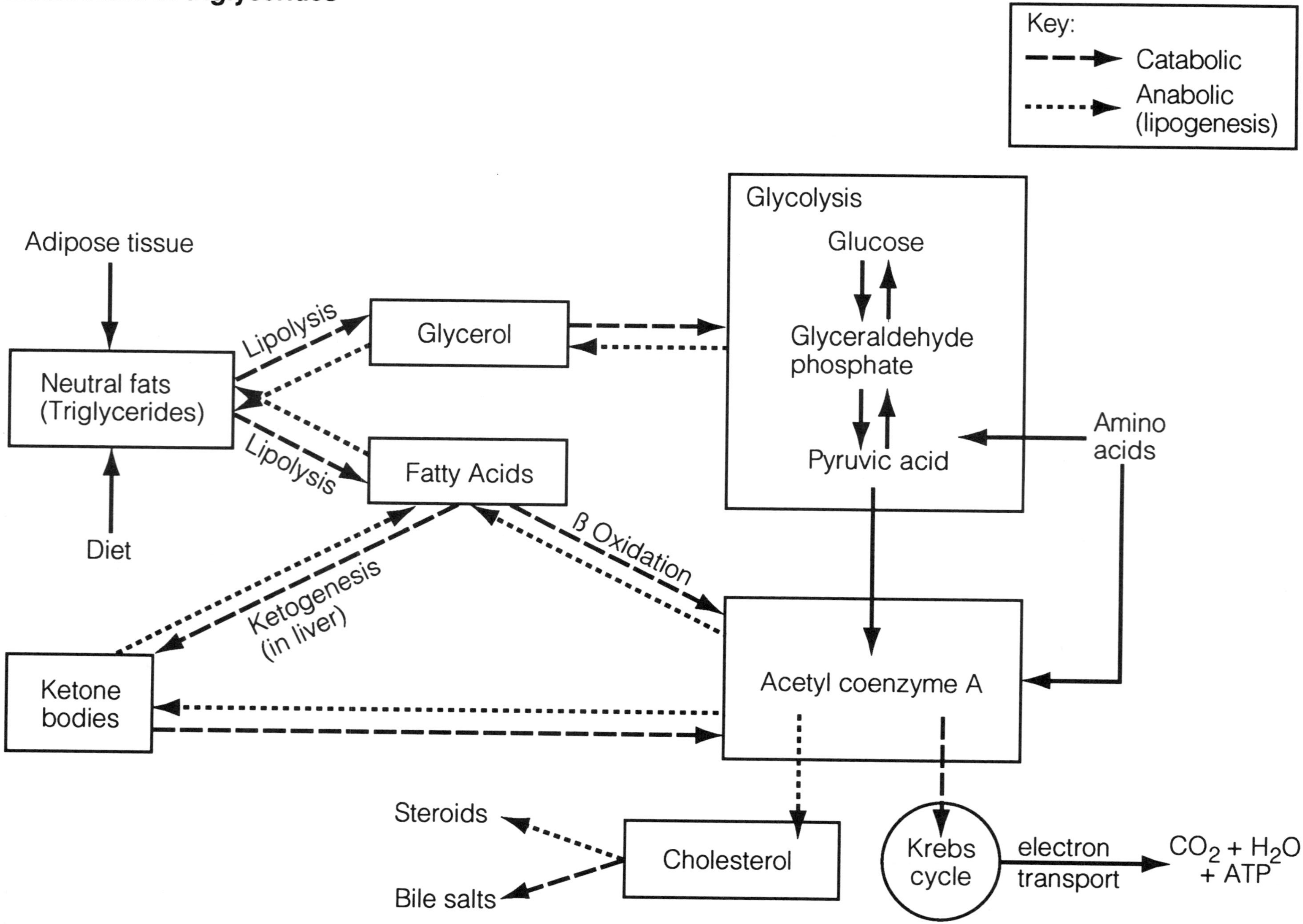

Simplified version of the transamination and oxidative processes that occur when amino acids are oxidized for energy

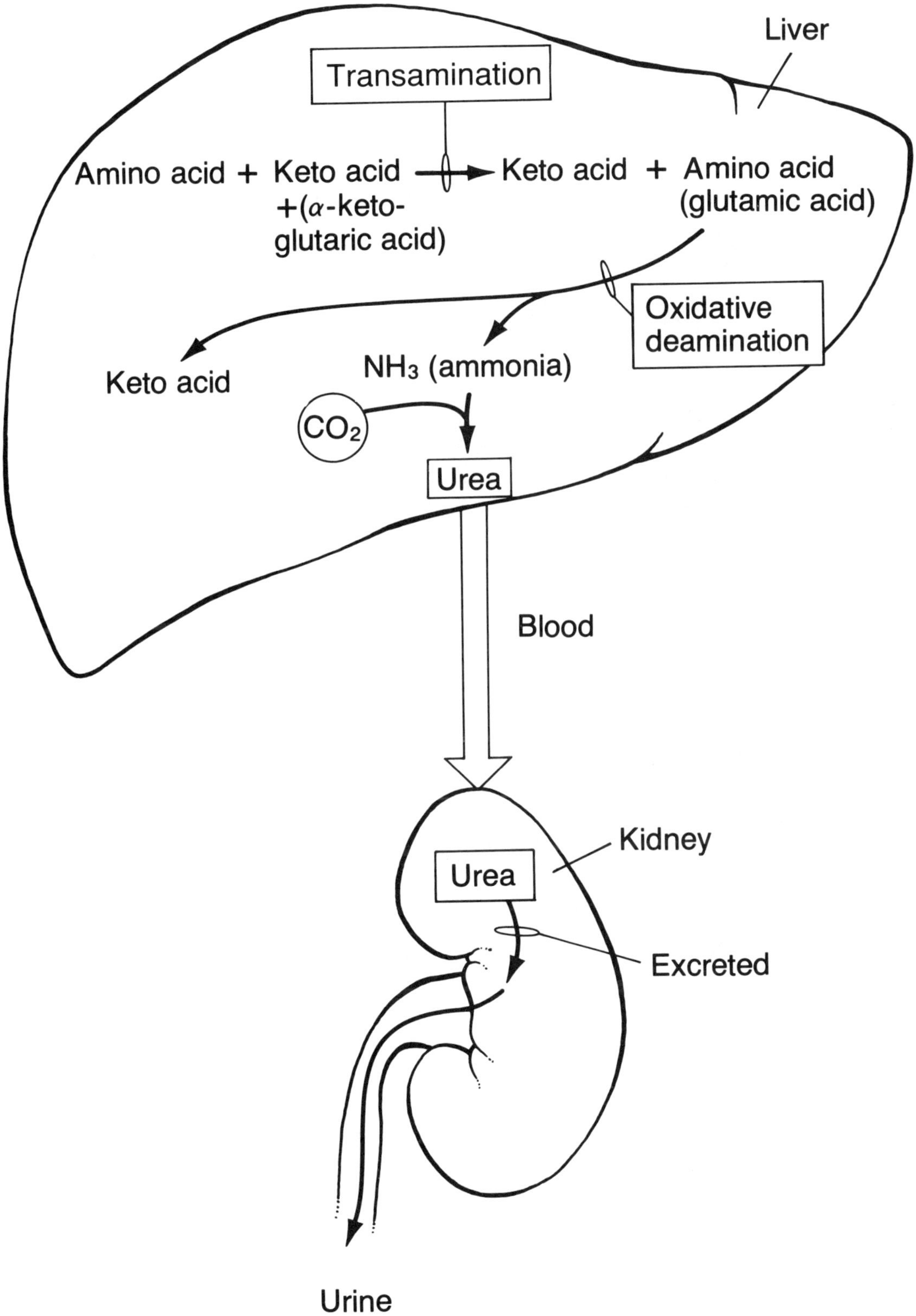

Carbohydrate, fat, and amino acid pools and pathways of interconversion

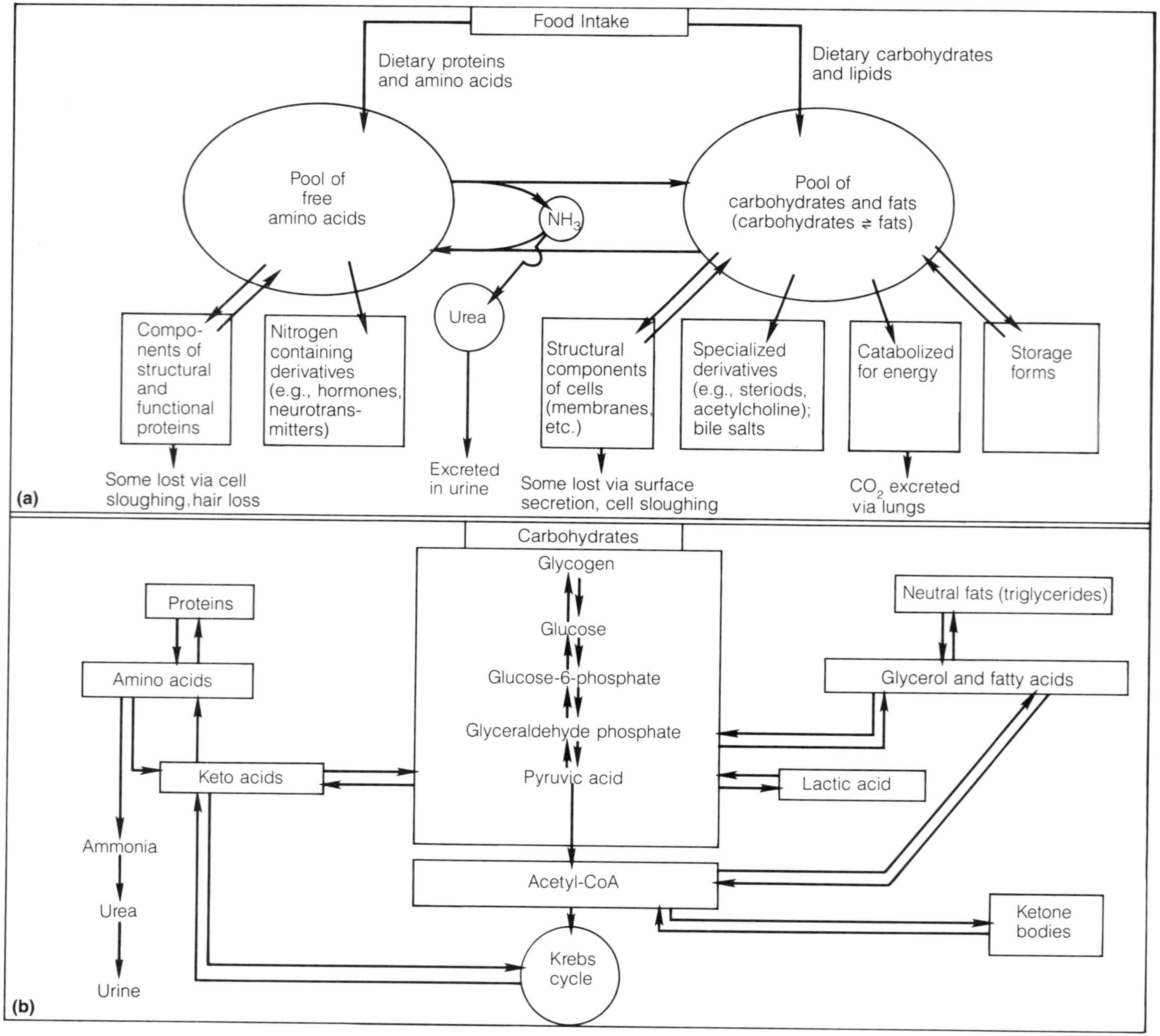

Summary of the major events and principal metabolic pathways of the absorptive state

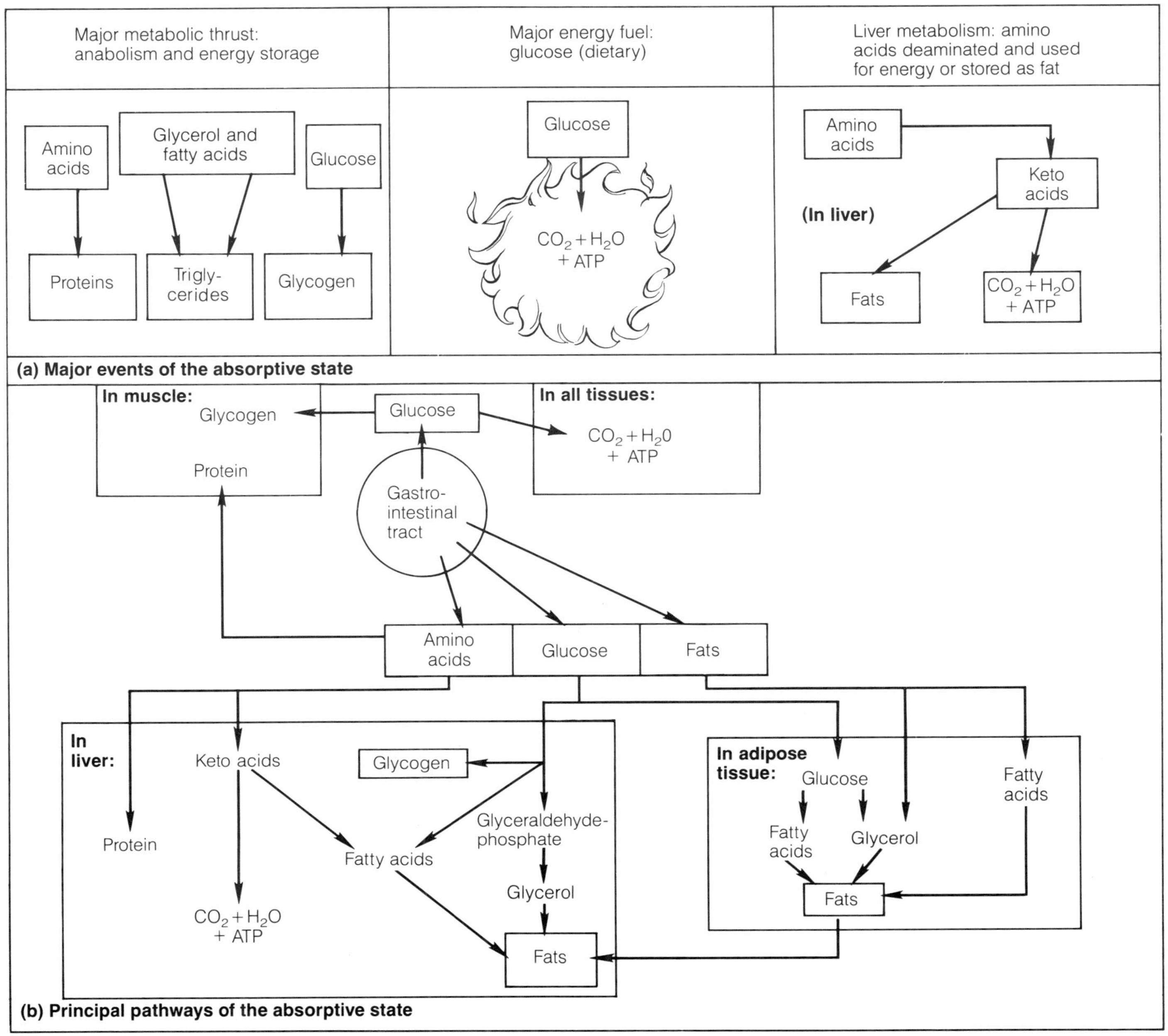

(a) Major events of the absorptive state

(b) Principal pathways of the absorptive state

Summary of the major events and principal metabolic pathways of the postabsorptive state

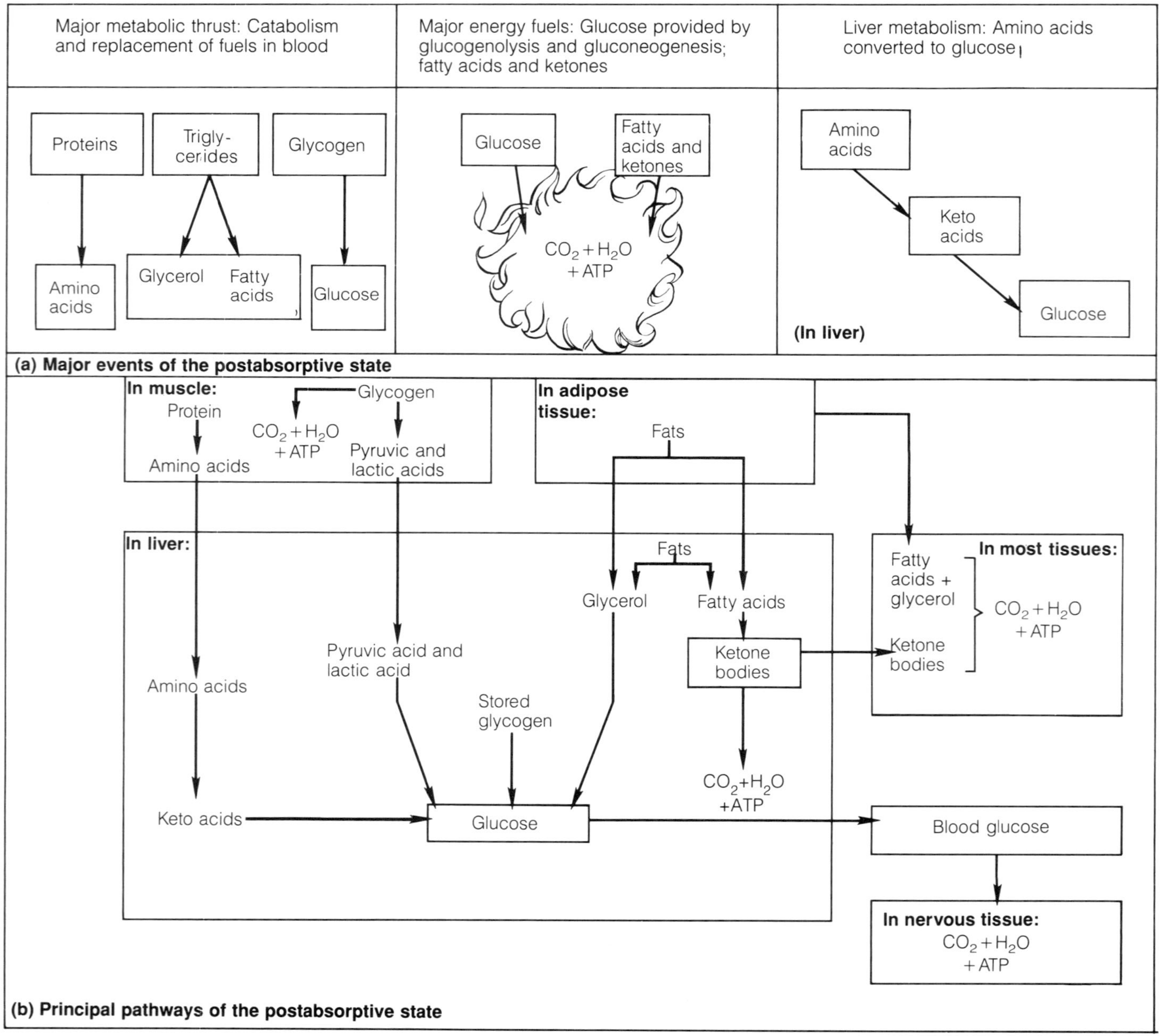

Influence of glucagon on plasma glucose concentrations

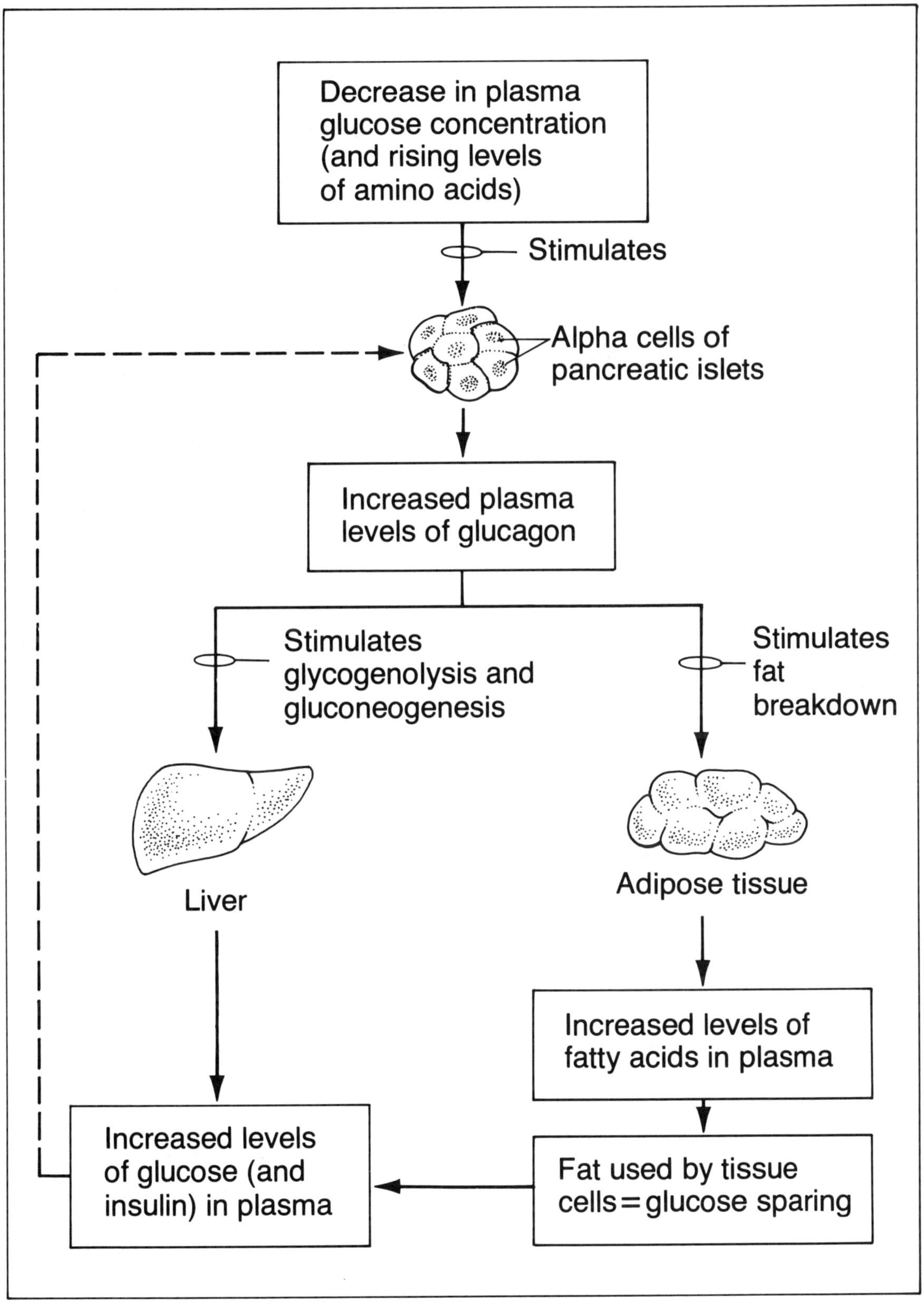

Sympathetic nervous system–epinephrine mechanism for providing glucose to the blood during the postabsorptive state

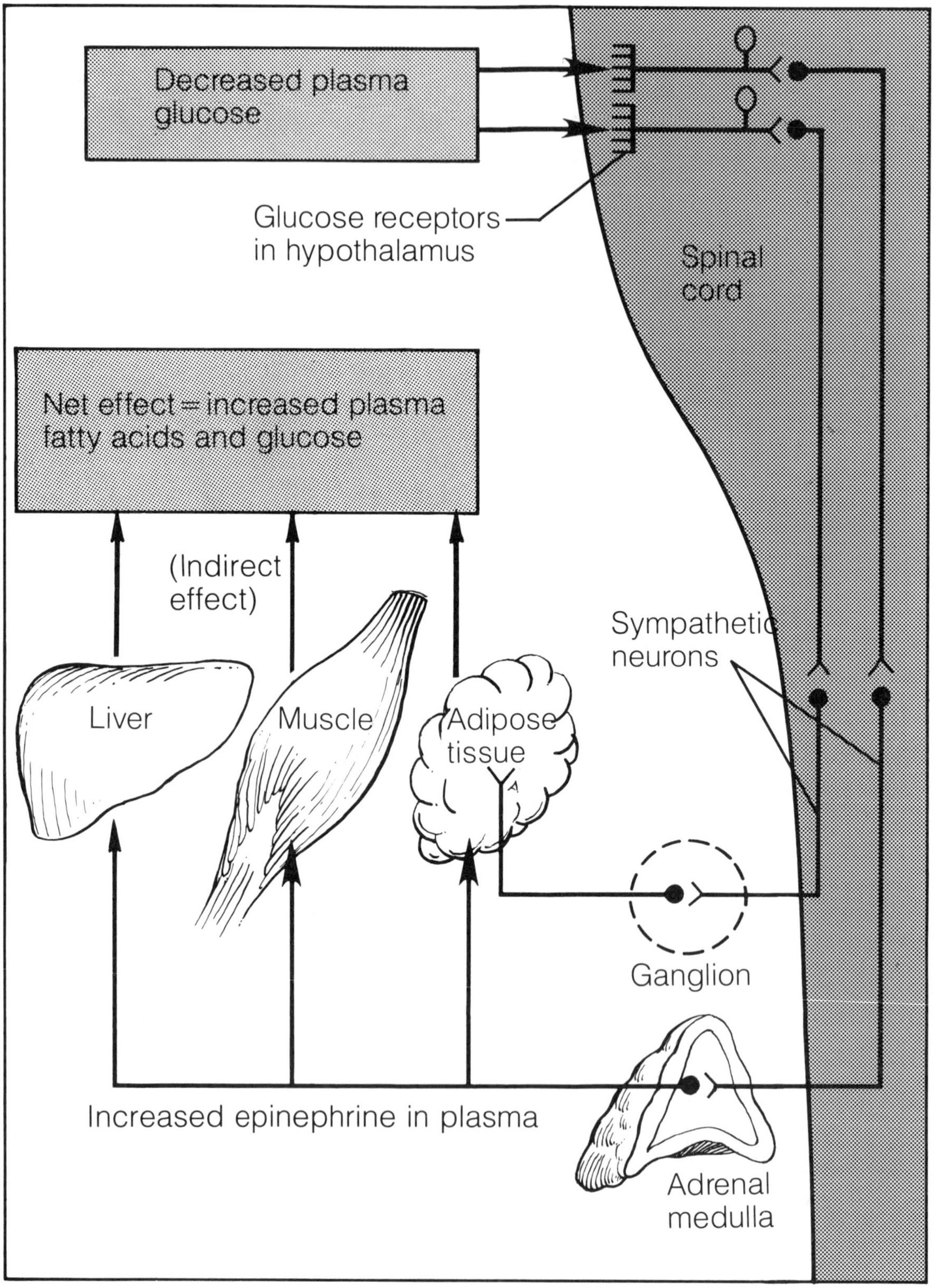

Composition of lipoproteins that transport lipids in body fluids

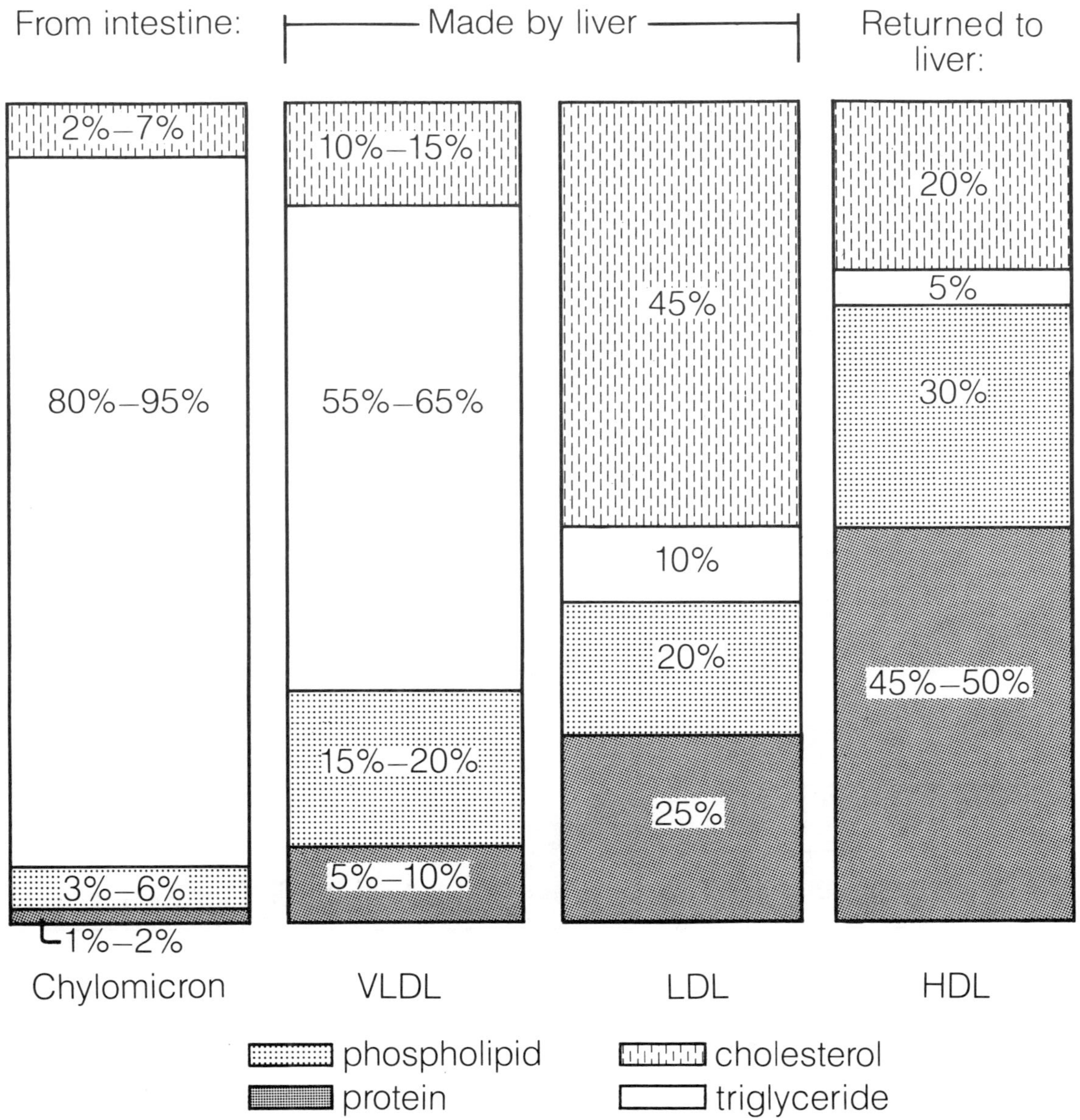

As long as heat production and heat loss are properly balanced, body temperature remains constant

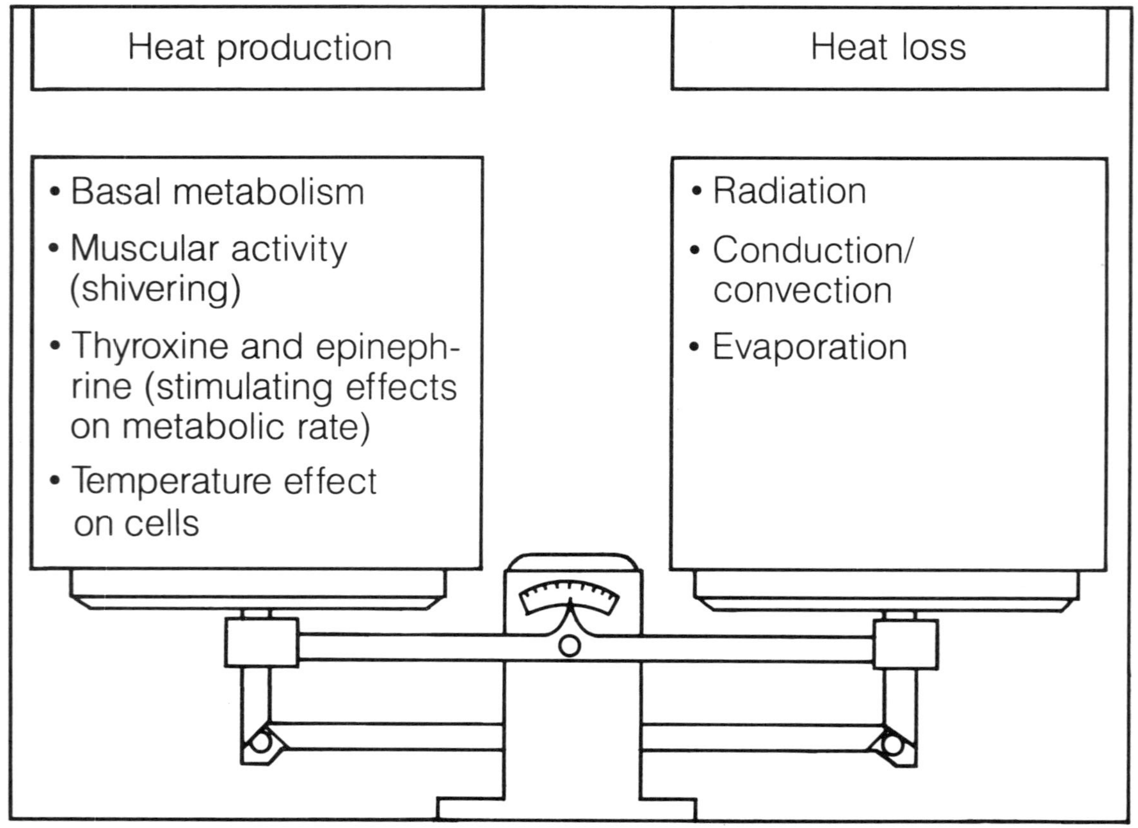

Mechanisms of heat exchange between the body and the external environment

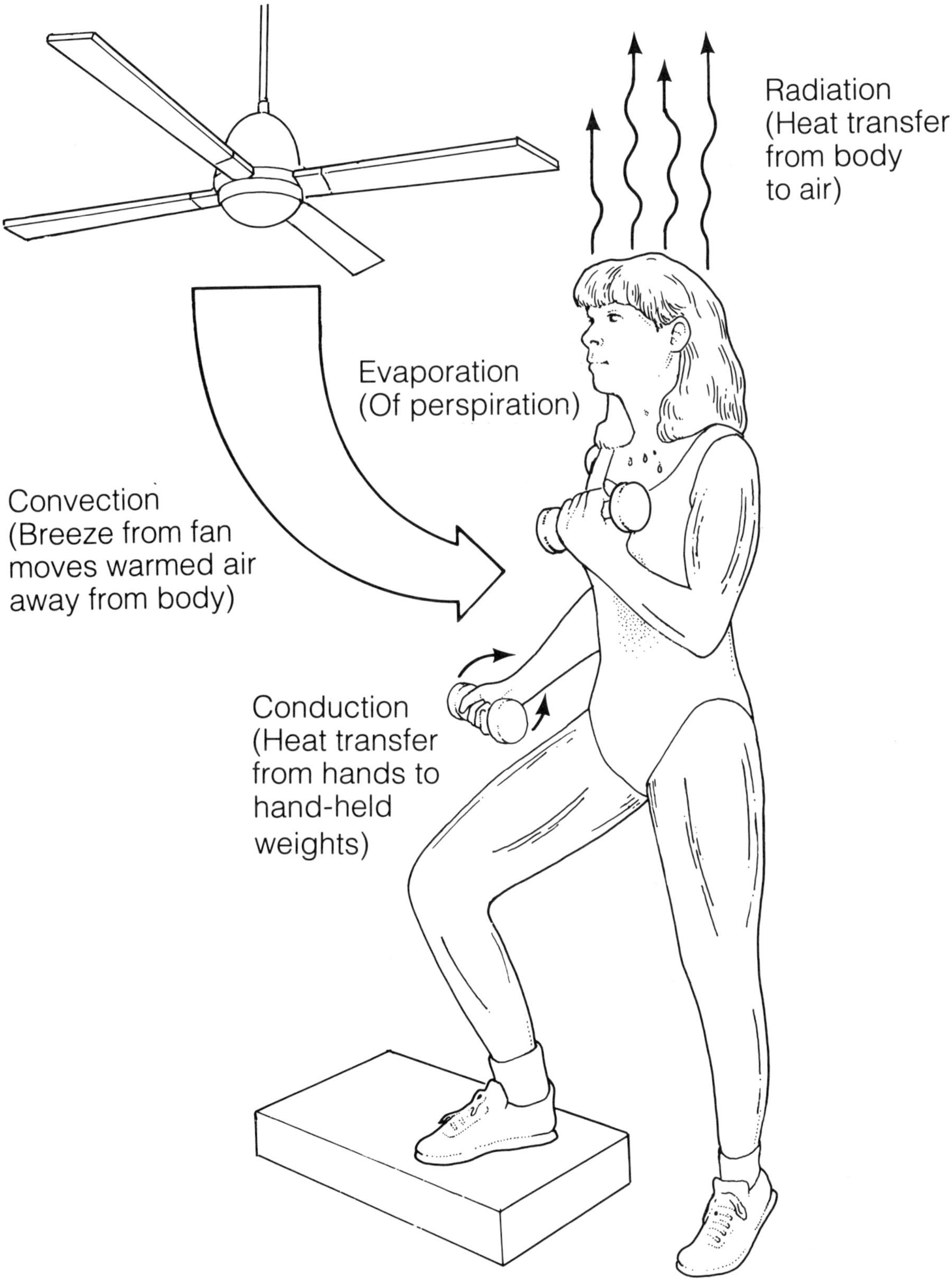

Body temperature regulation

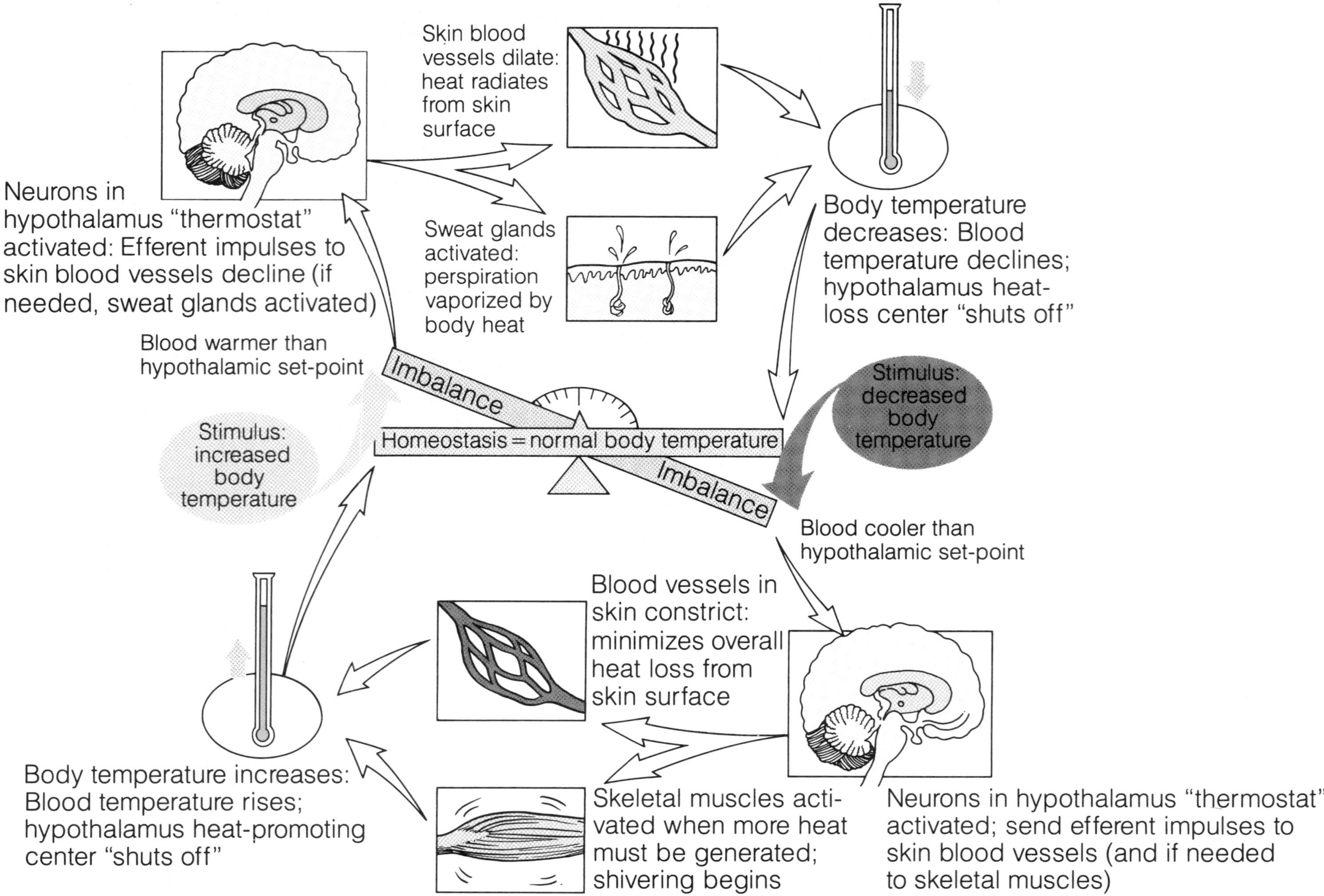

Location and structure of nephrons

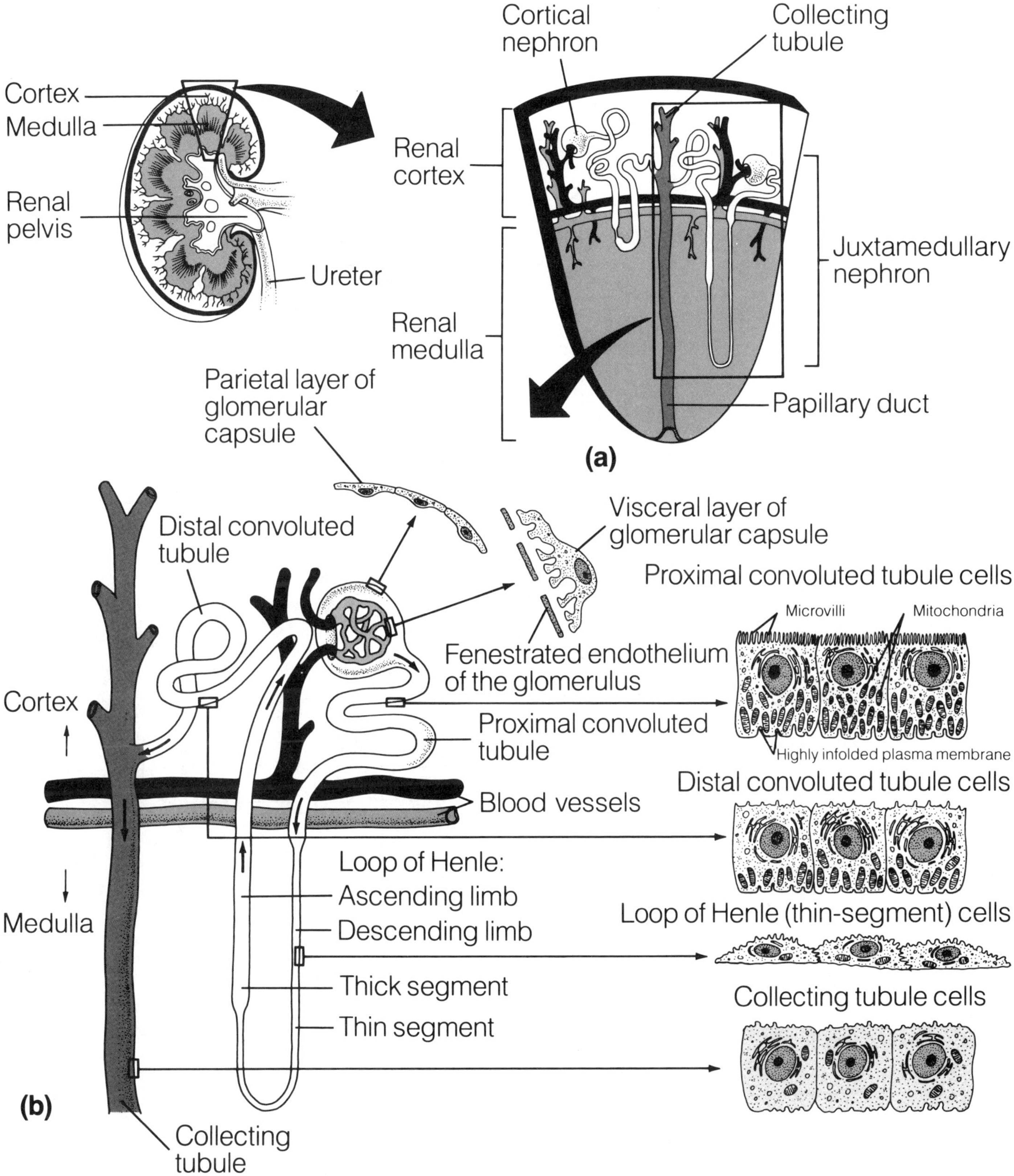

Detailed anatomy of a nephron and its vasculature

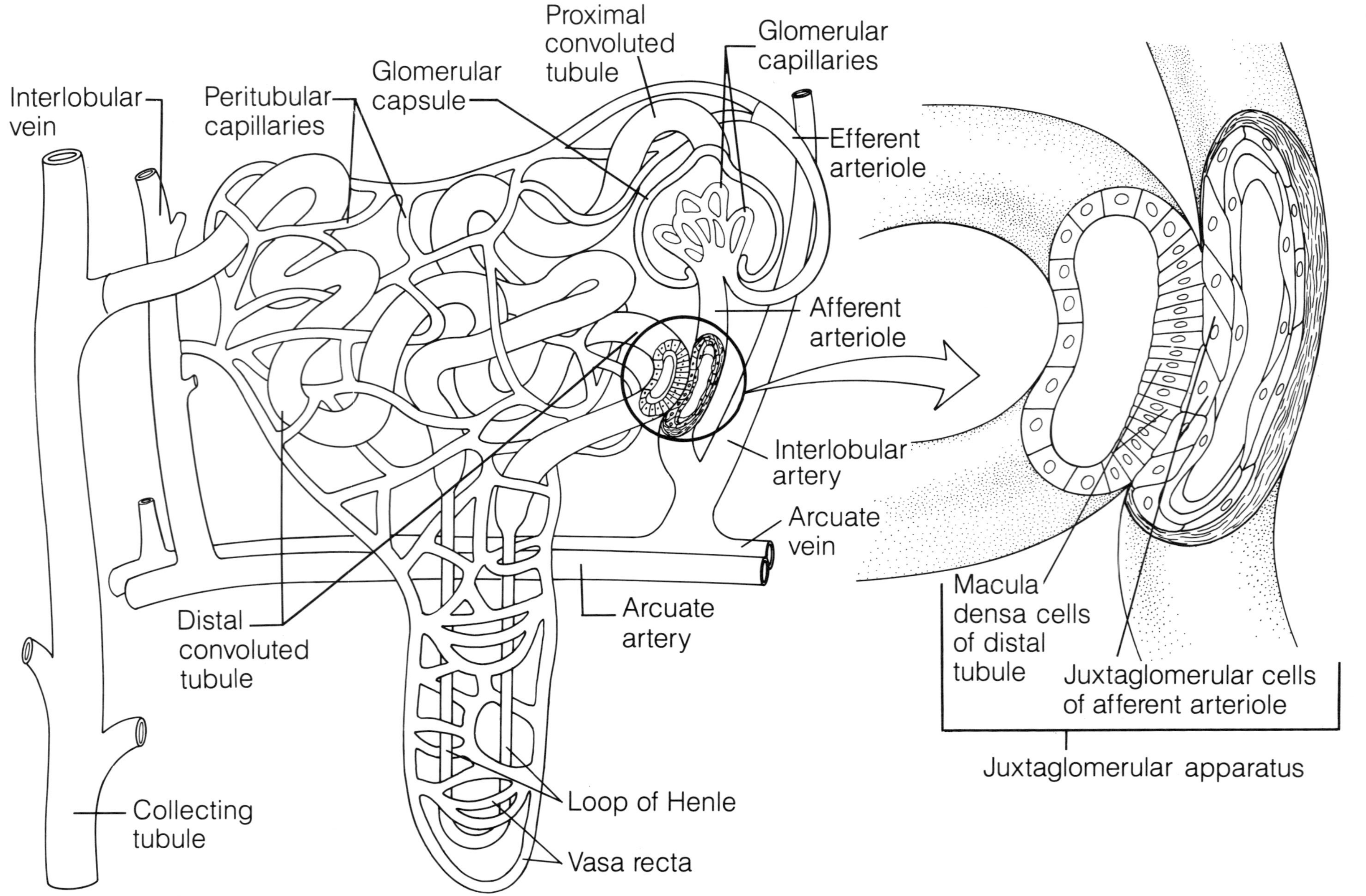

(a)

Relative blood pressures in the renal circulation

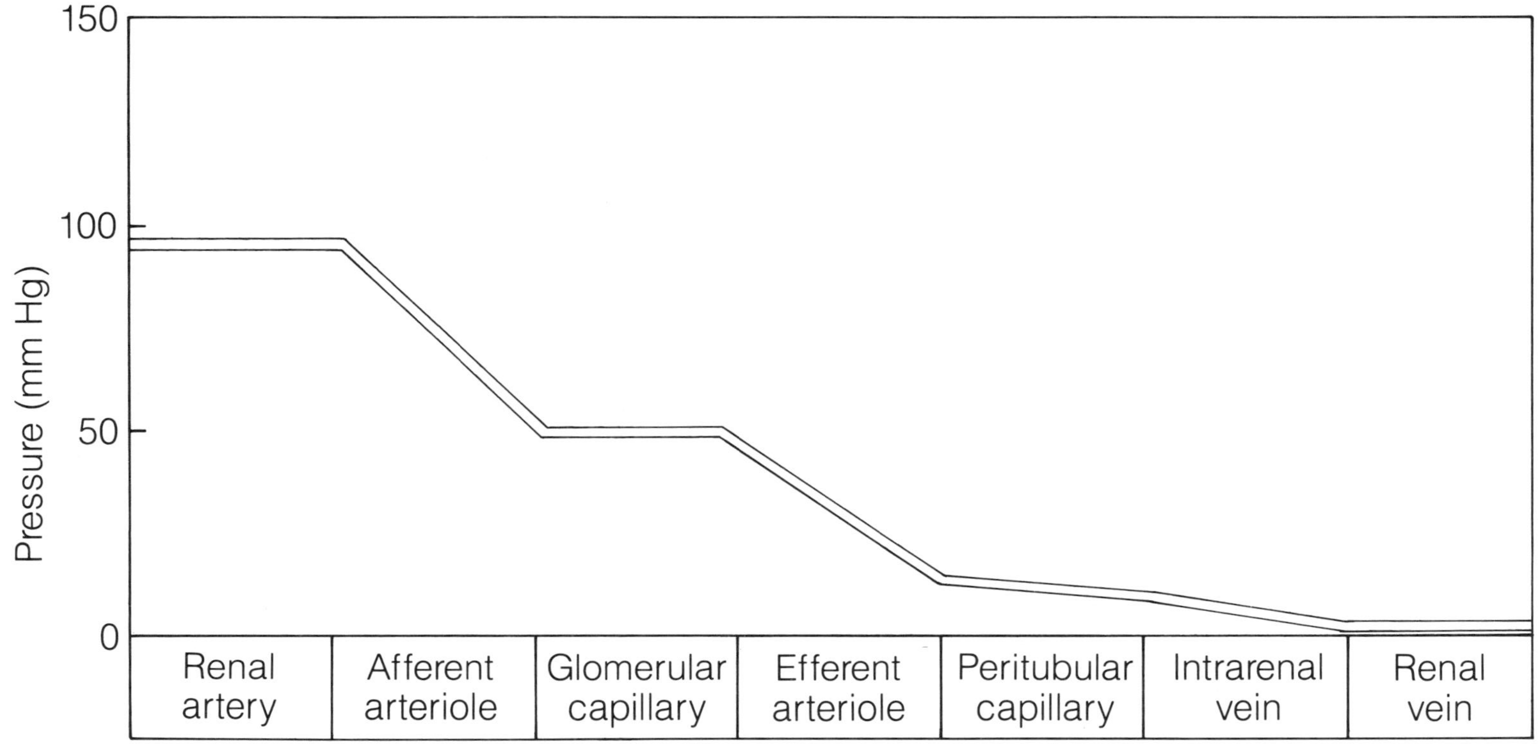

The kidney depicted as a single, large nephron

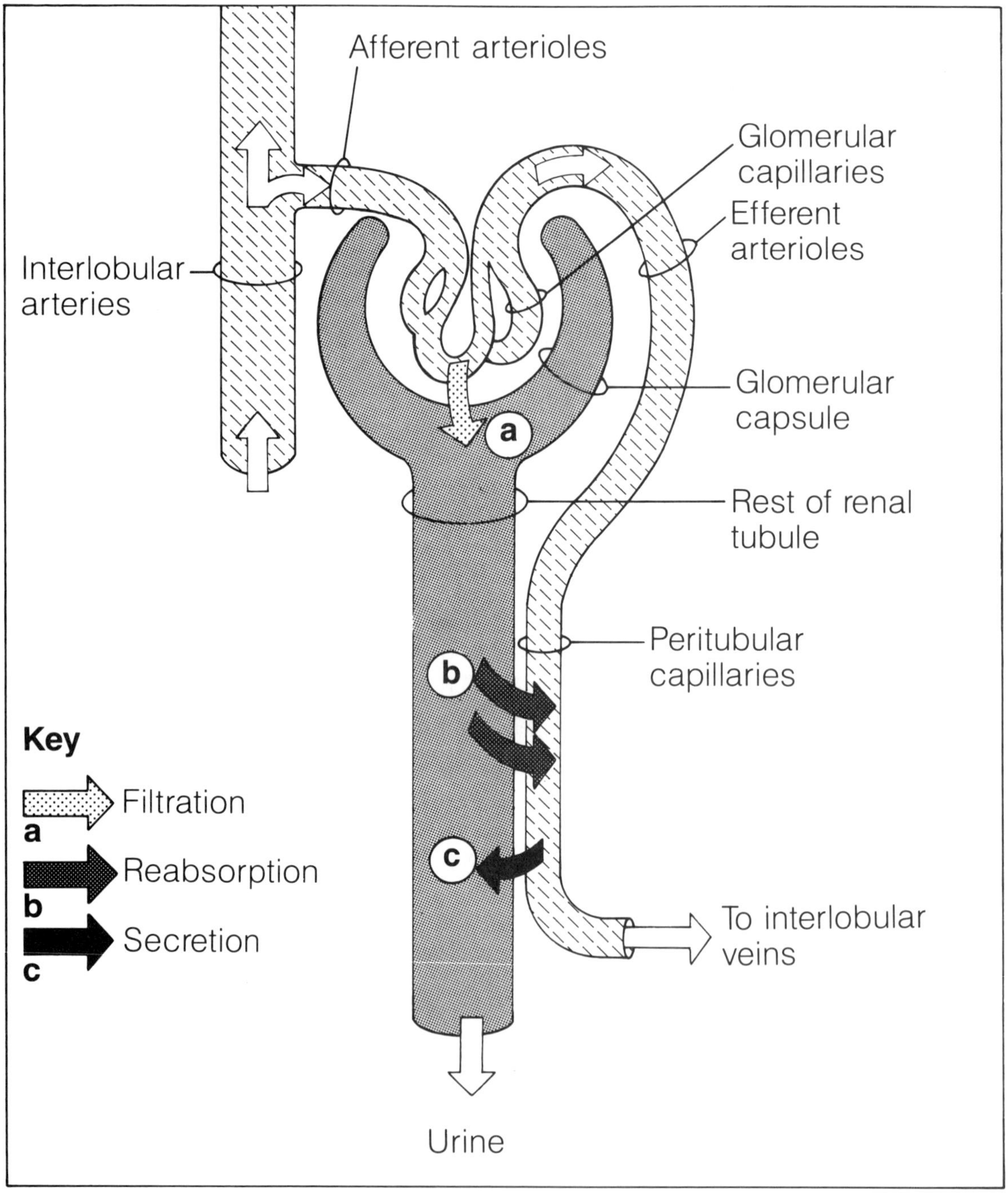

The filtration membrane

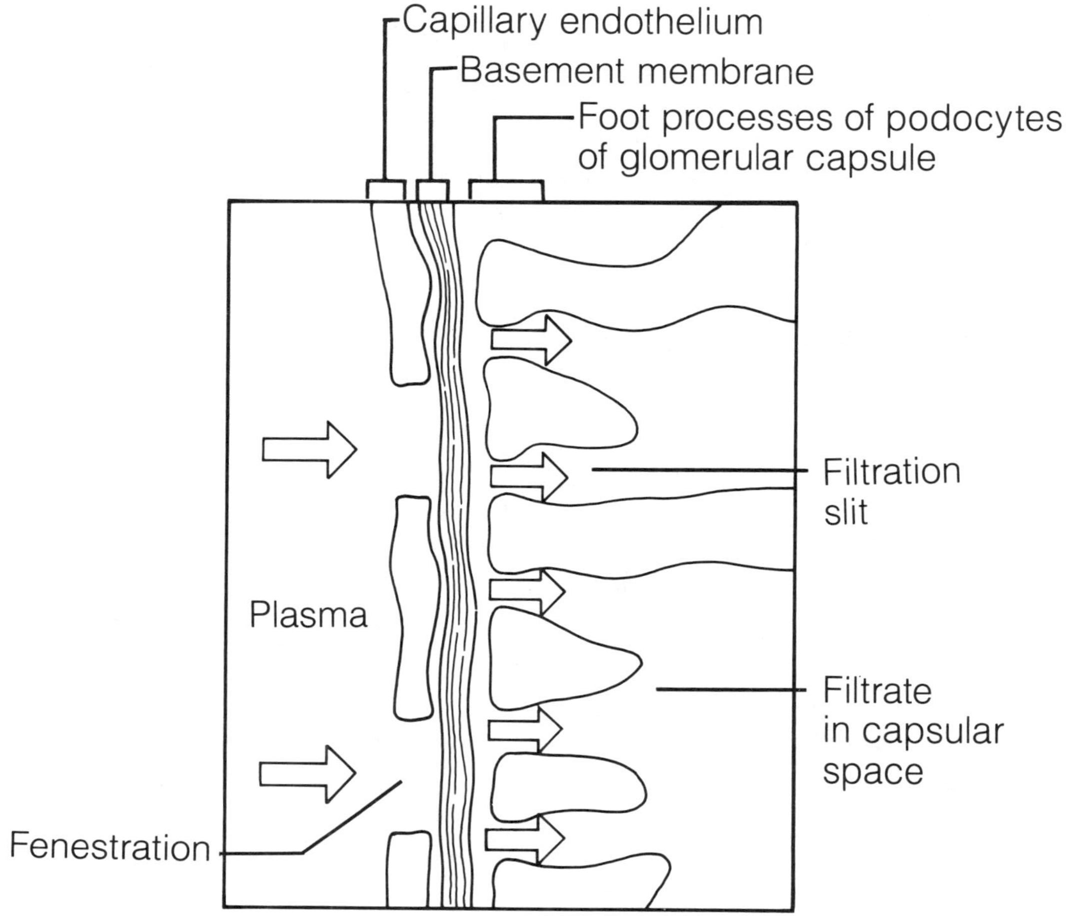

(c)

Forces that determine glomerular filtration and the effective filtration pressure

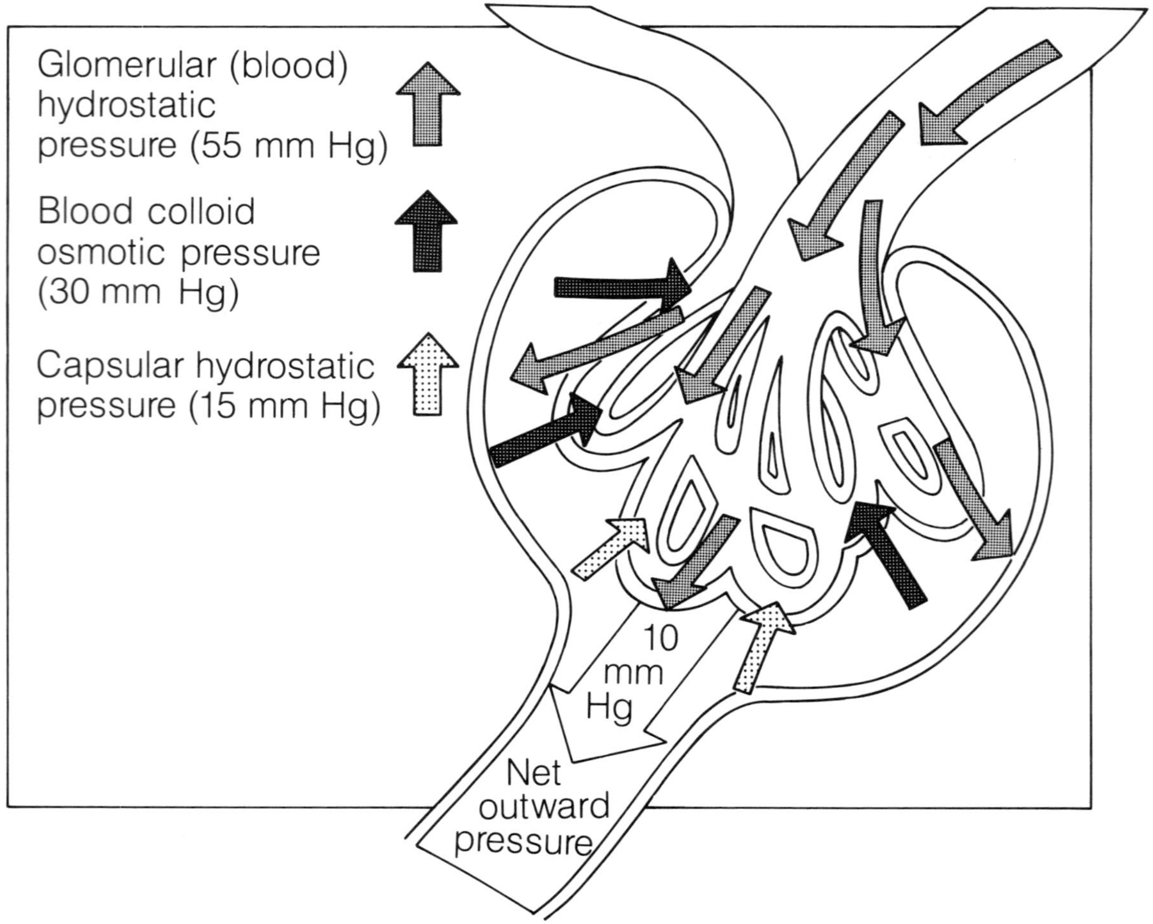

Flowchart indicating the mechanisms regulating the glomerular filtration rate (GFR) in the kidneys

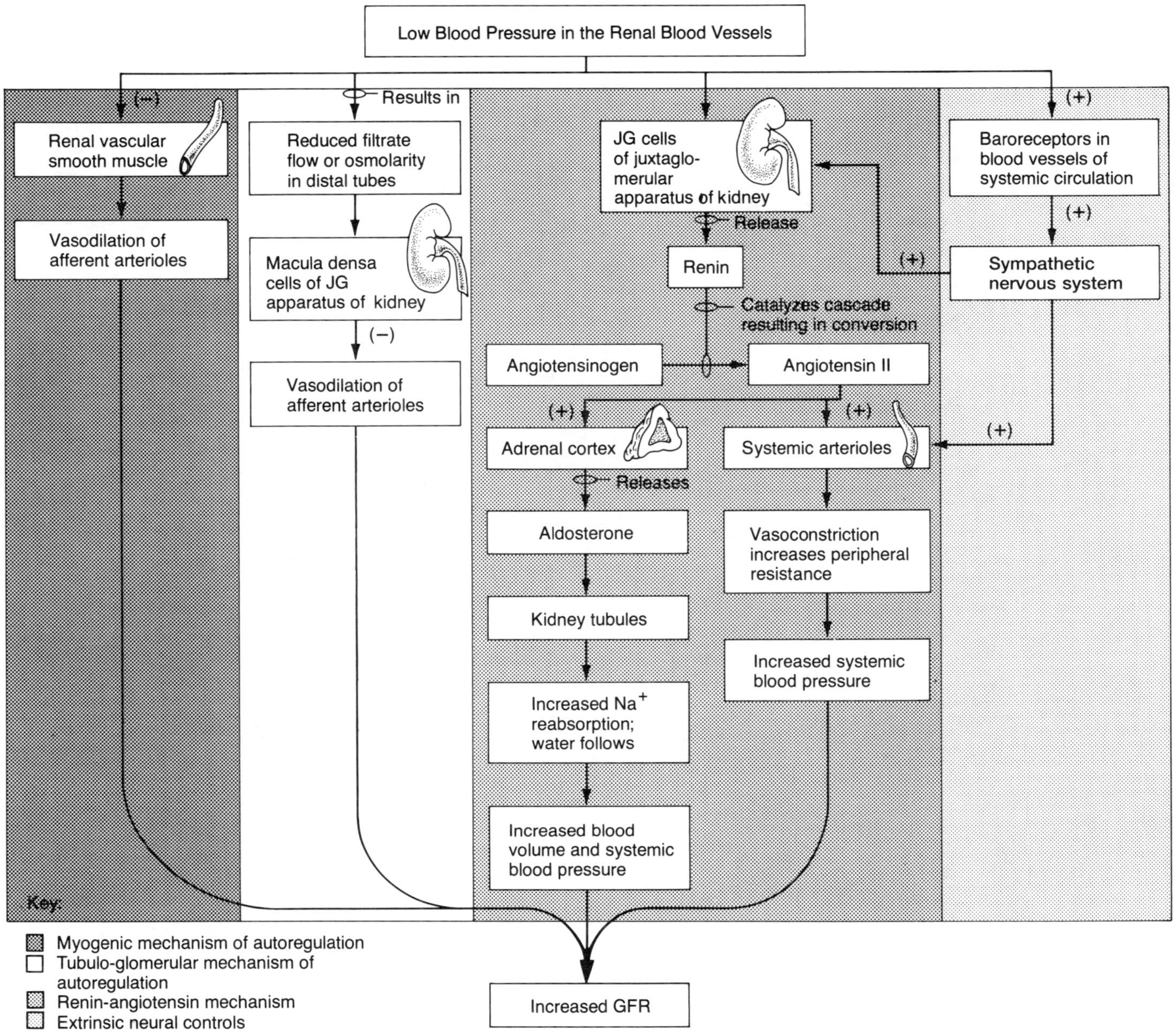

Hypothetical countercurrent mechanism for establishing and maintaining the medullary osmotic gradient

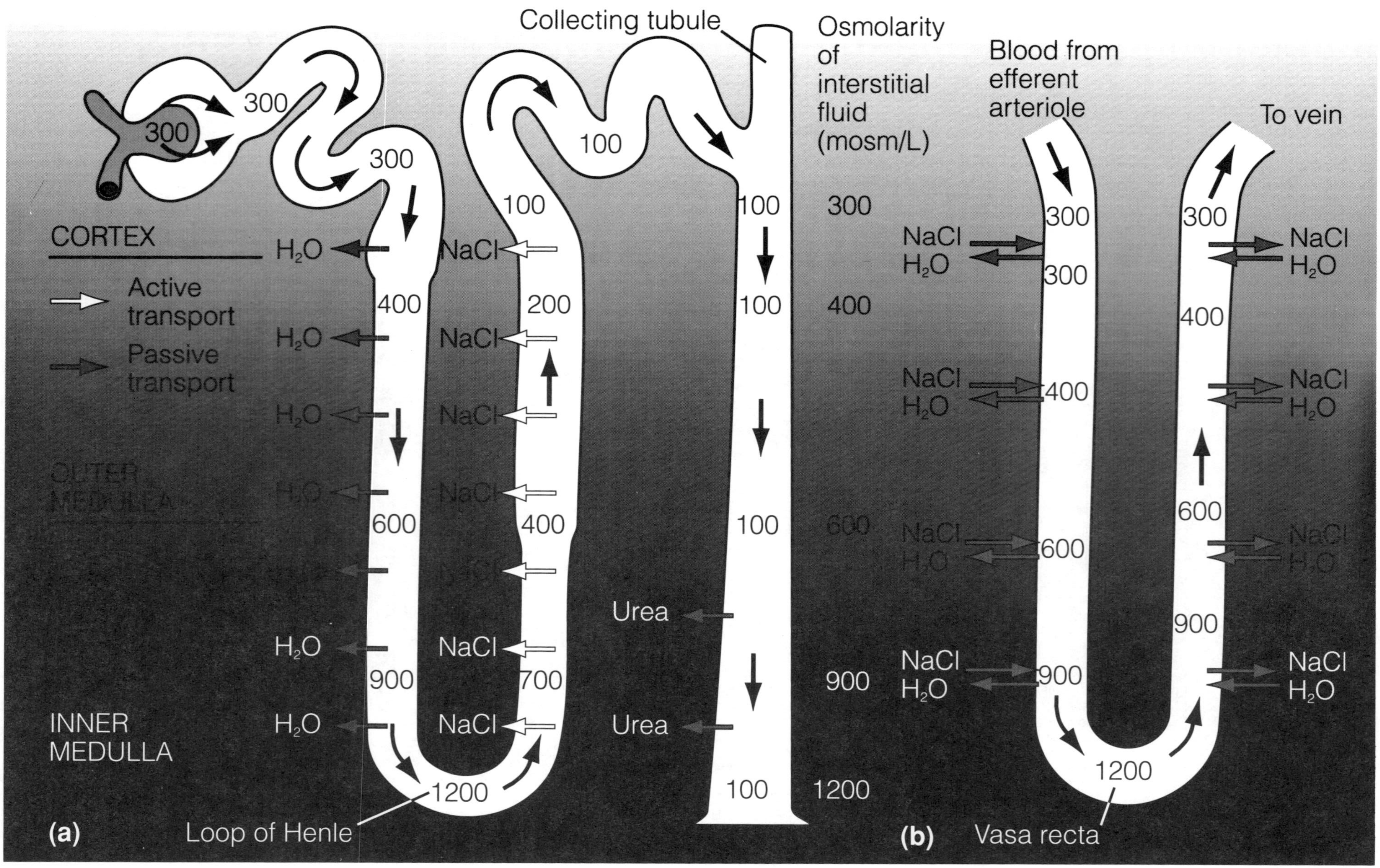

Summary of nephron functions

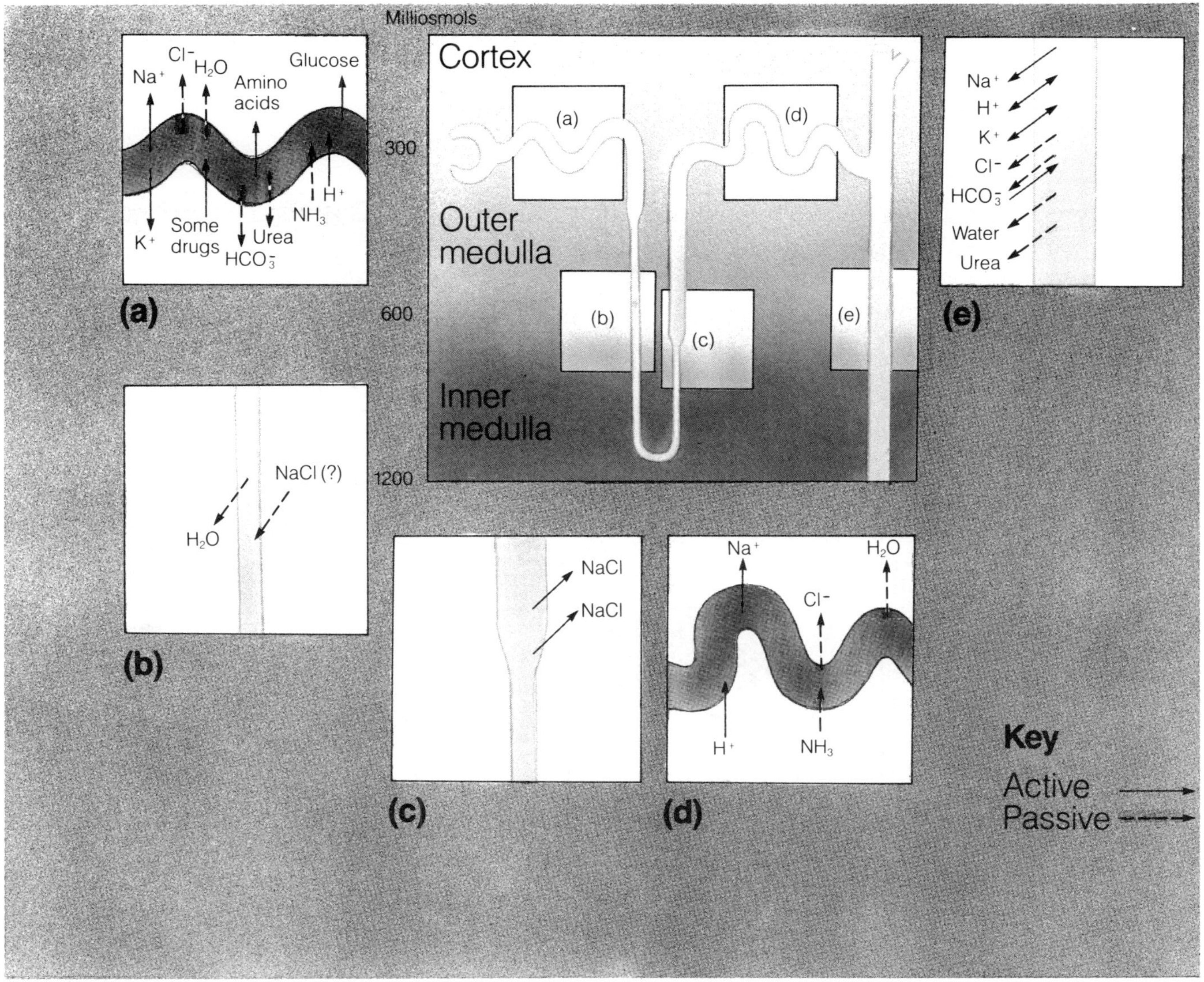

Micturition reflex arc pathway

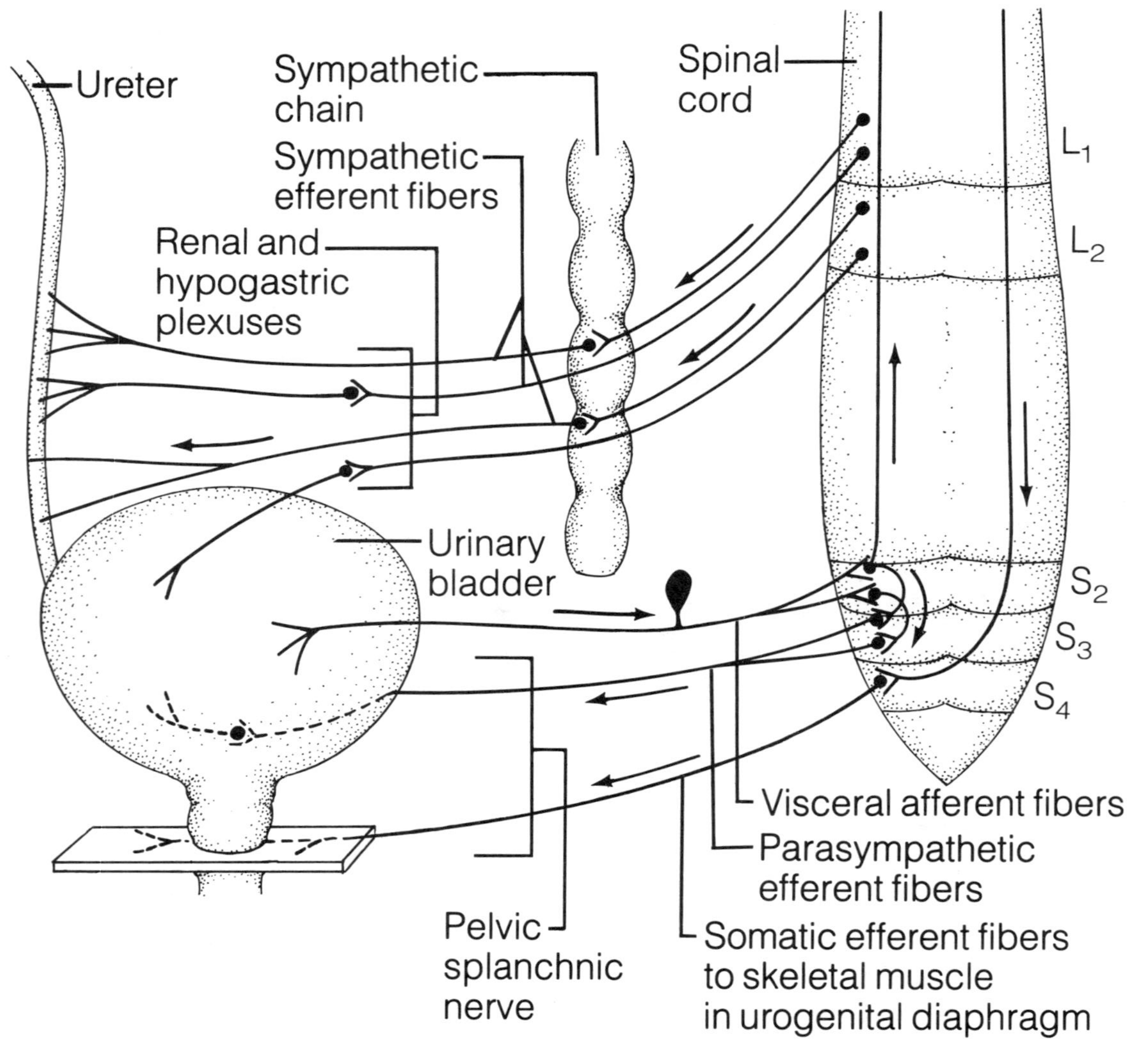

The major fluid compartments of the body

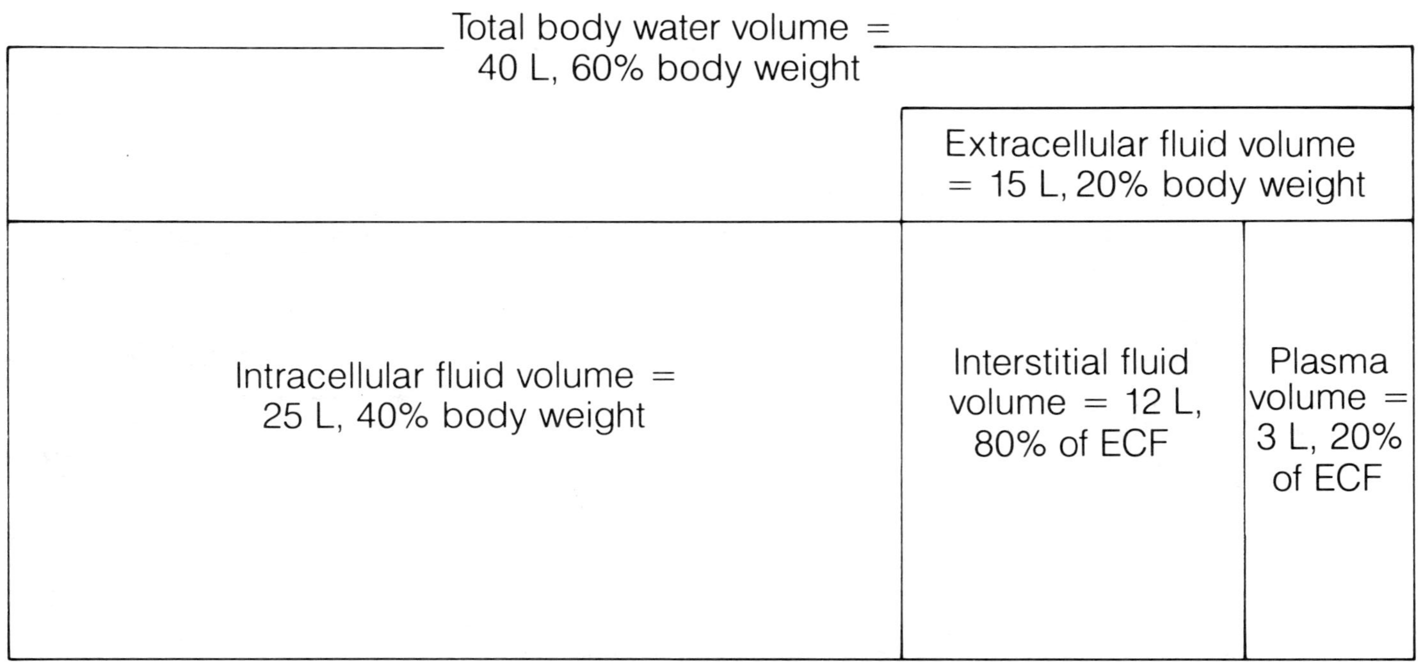

Comparison of the electrolyte composition of blood plasma, interstitial fluid, and intracellular fluid

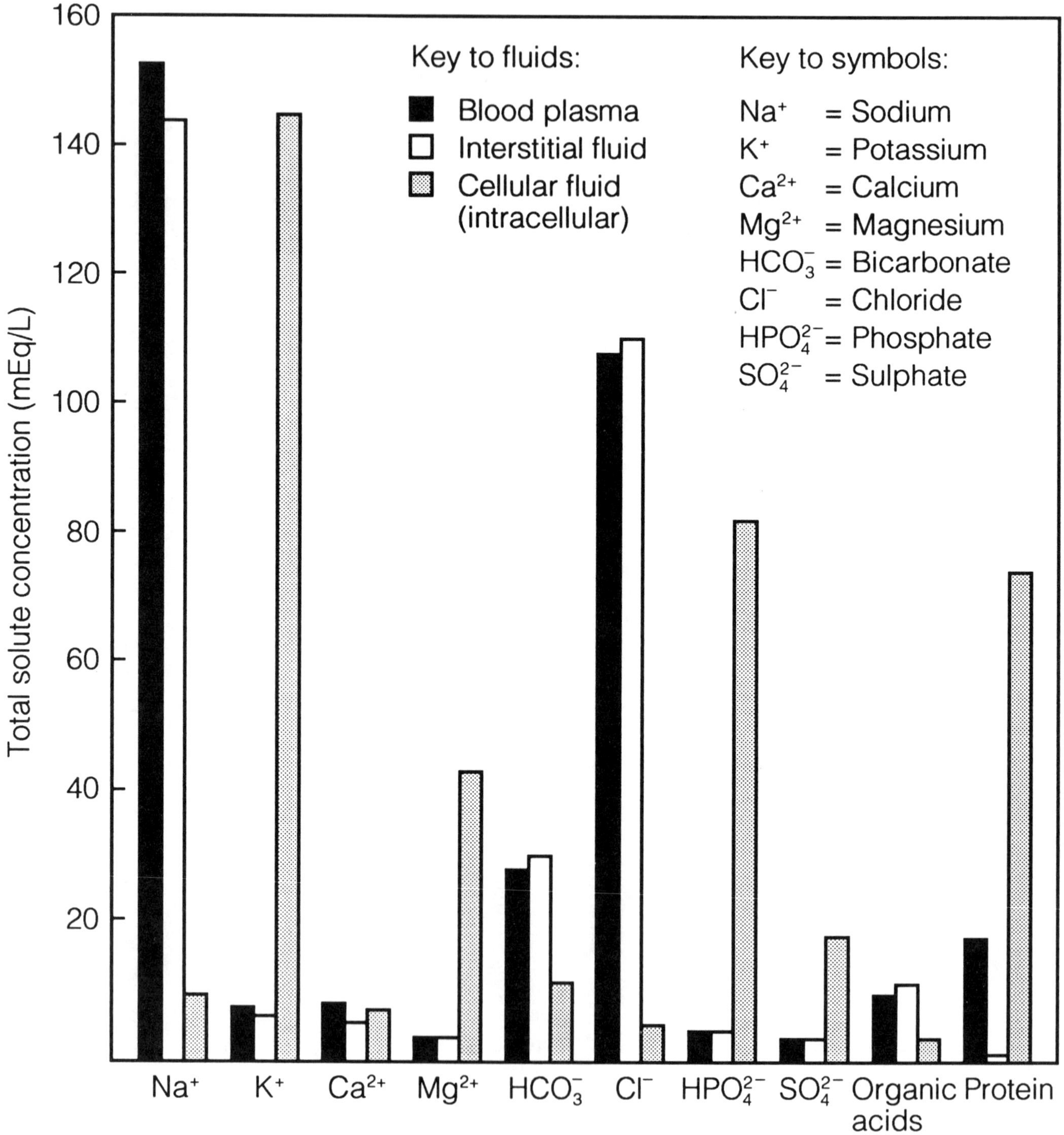

Fluid movement between compartments follows hydrostatic and osmotic pressure gradients

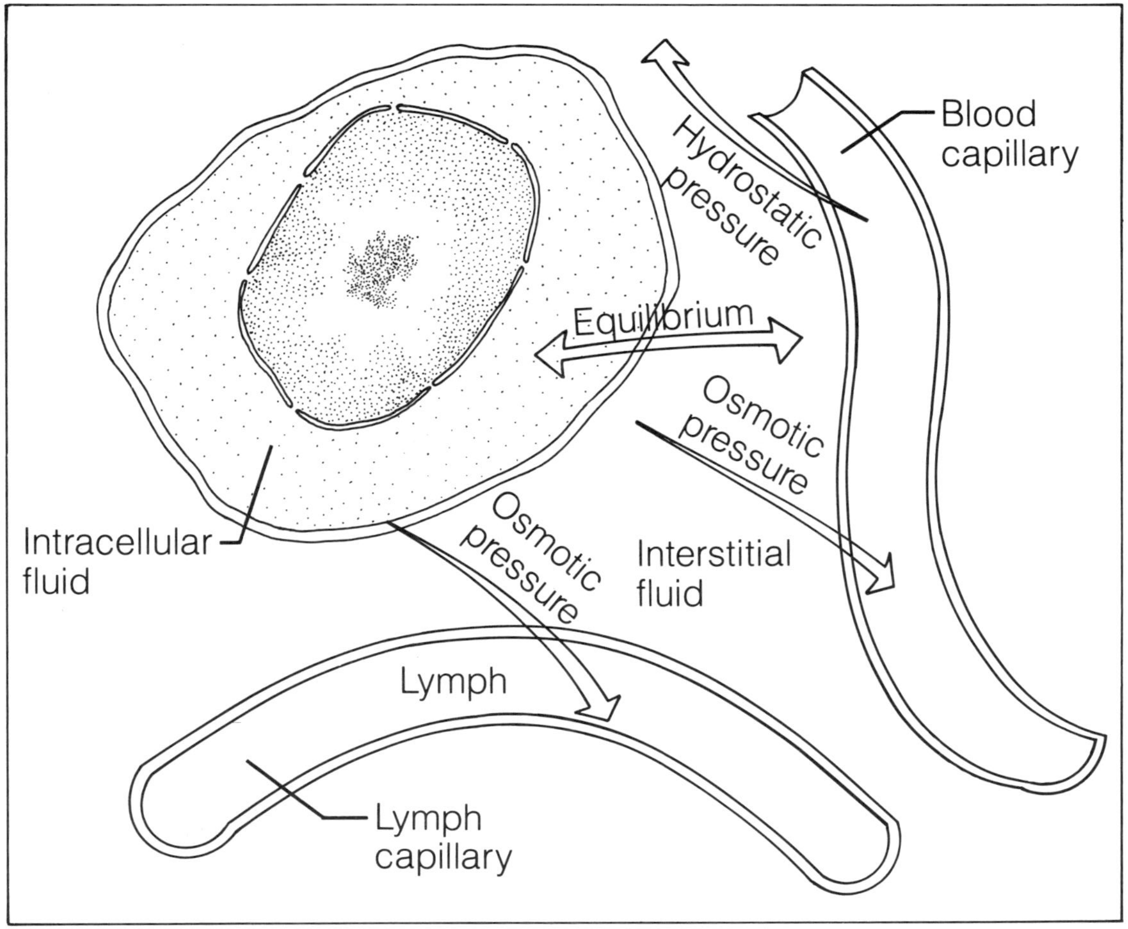

The continuous mixing of body fluids

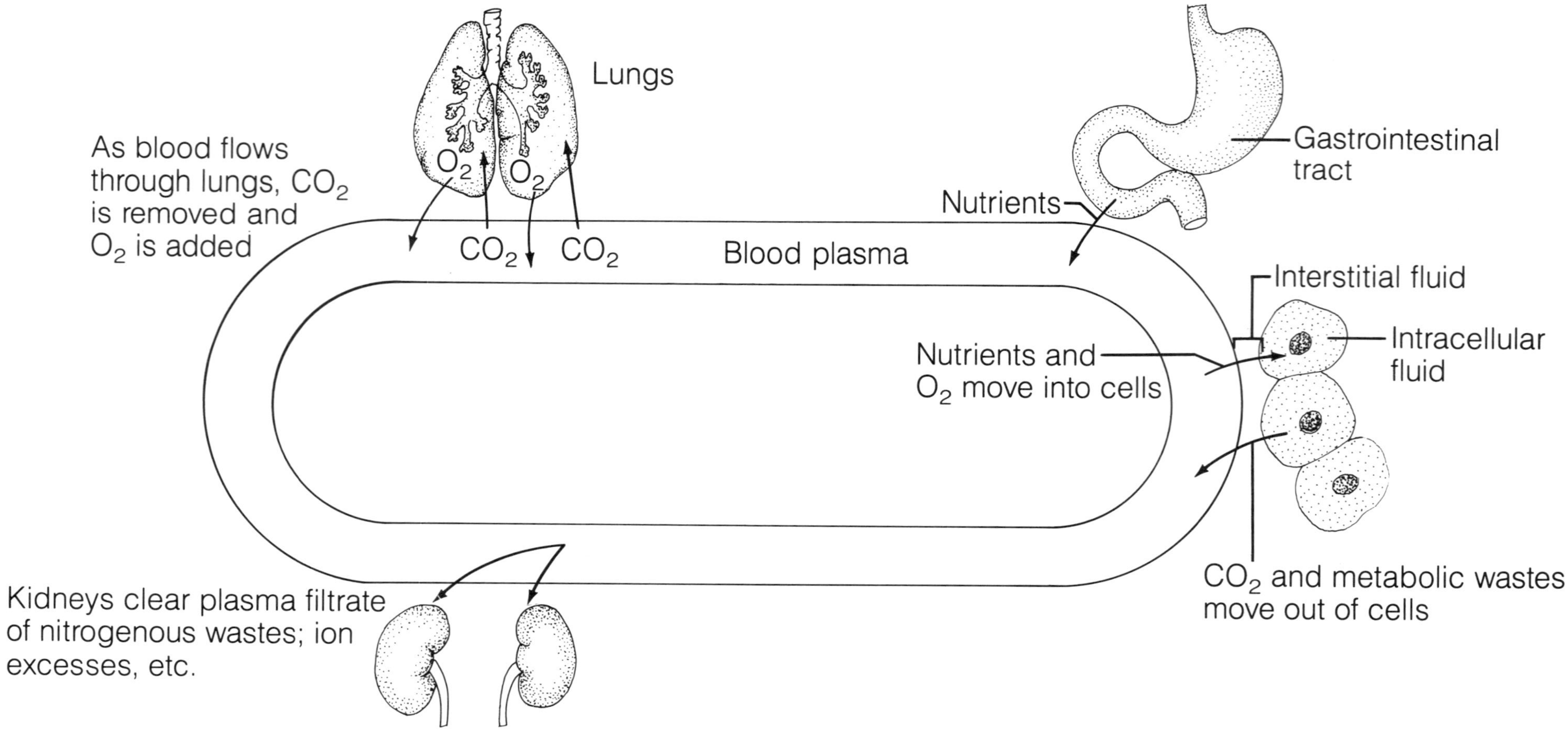

Major sources of body water (water intake) and routes of water loss from the body (water output)

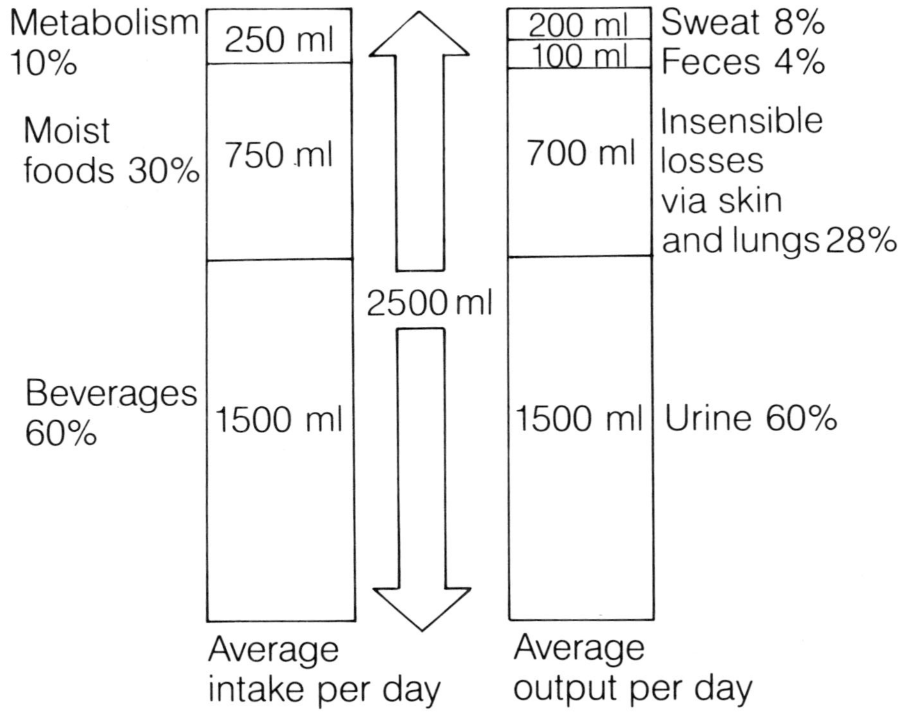

The thirst mechanism for regulating water intake

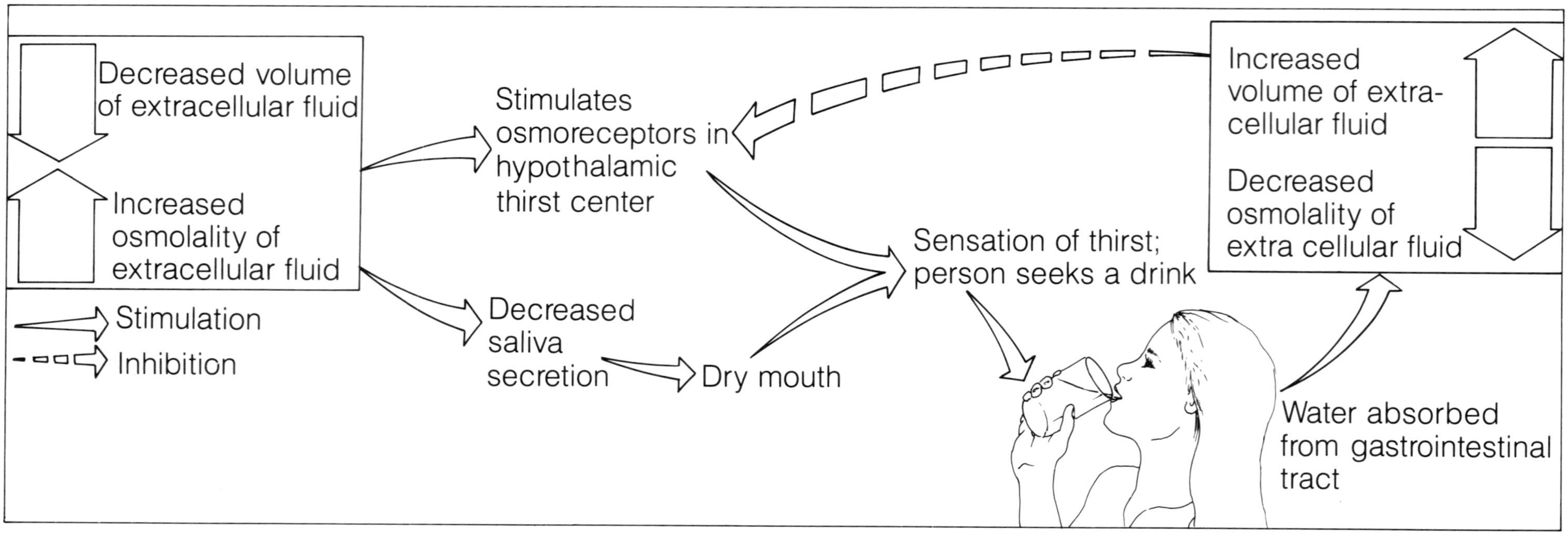

Disturbances in water balance

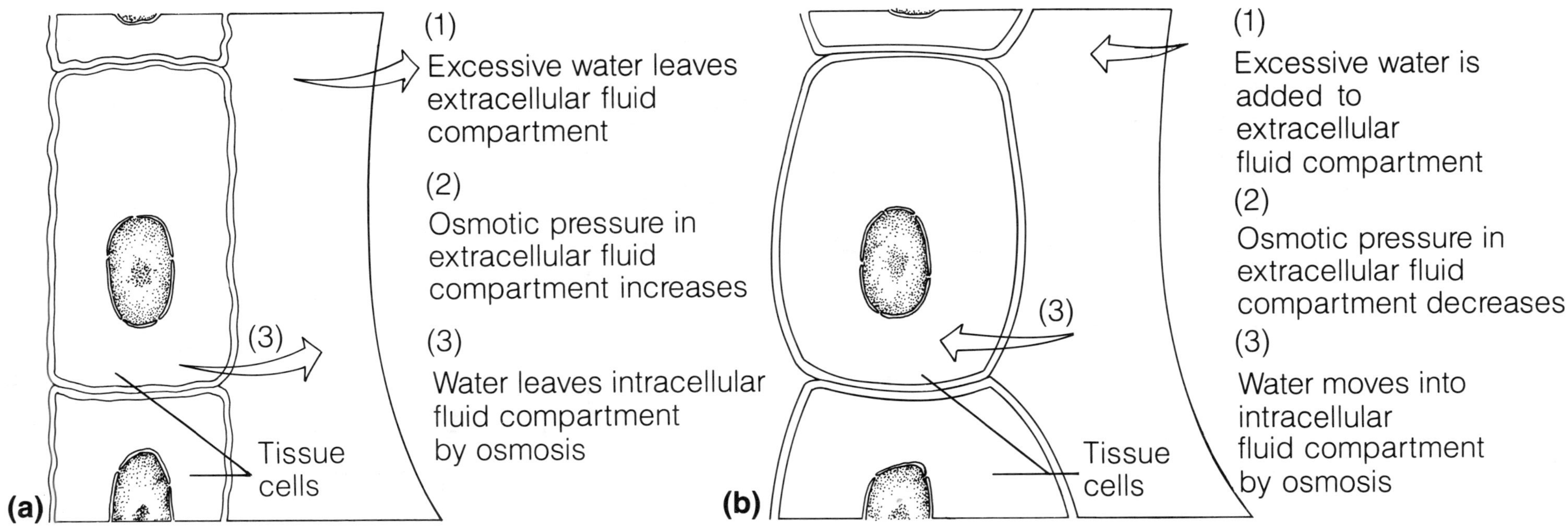

Mechanisms and consequences of aldosterone release

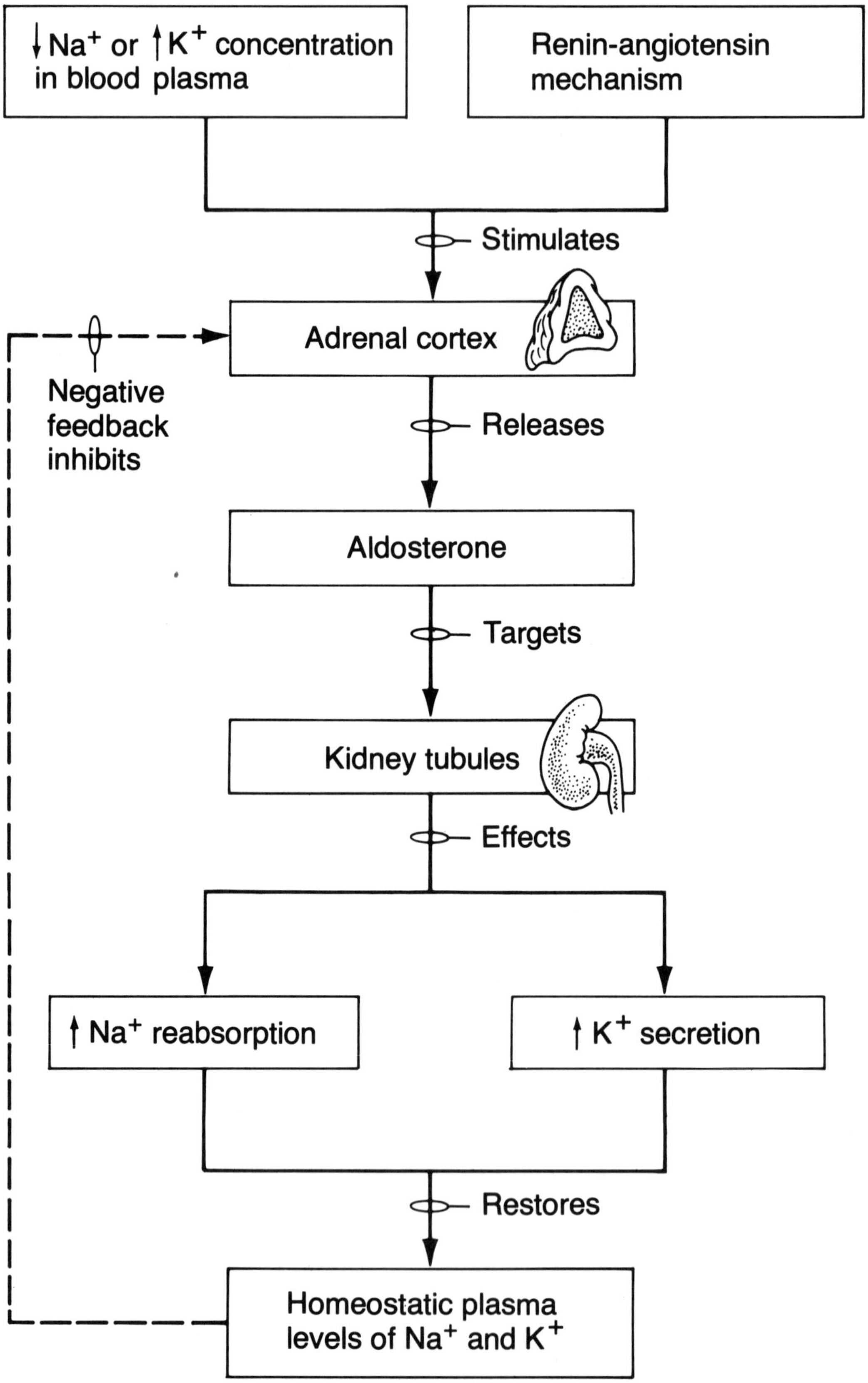

Mechanisms and consequences of ADH release

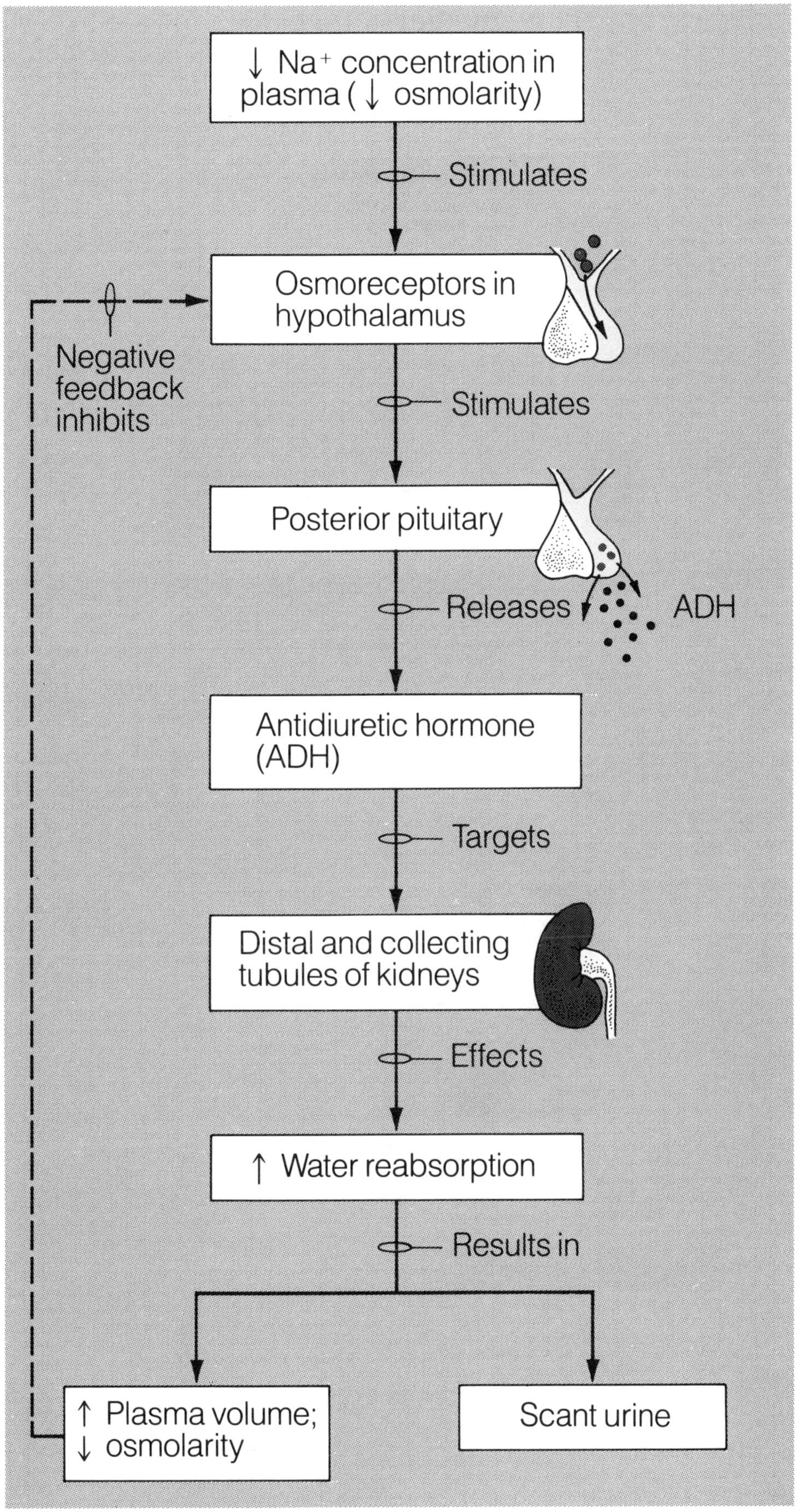

Flowchart illustrating the interaction of ADH, aldosterone, and blood pressure in the regulation of sodium and water balance mechanisms mediated through the kidneys

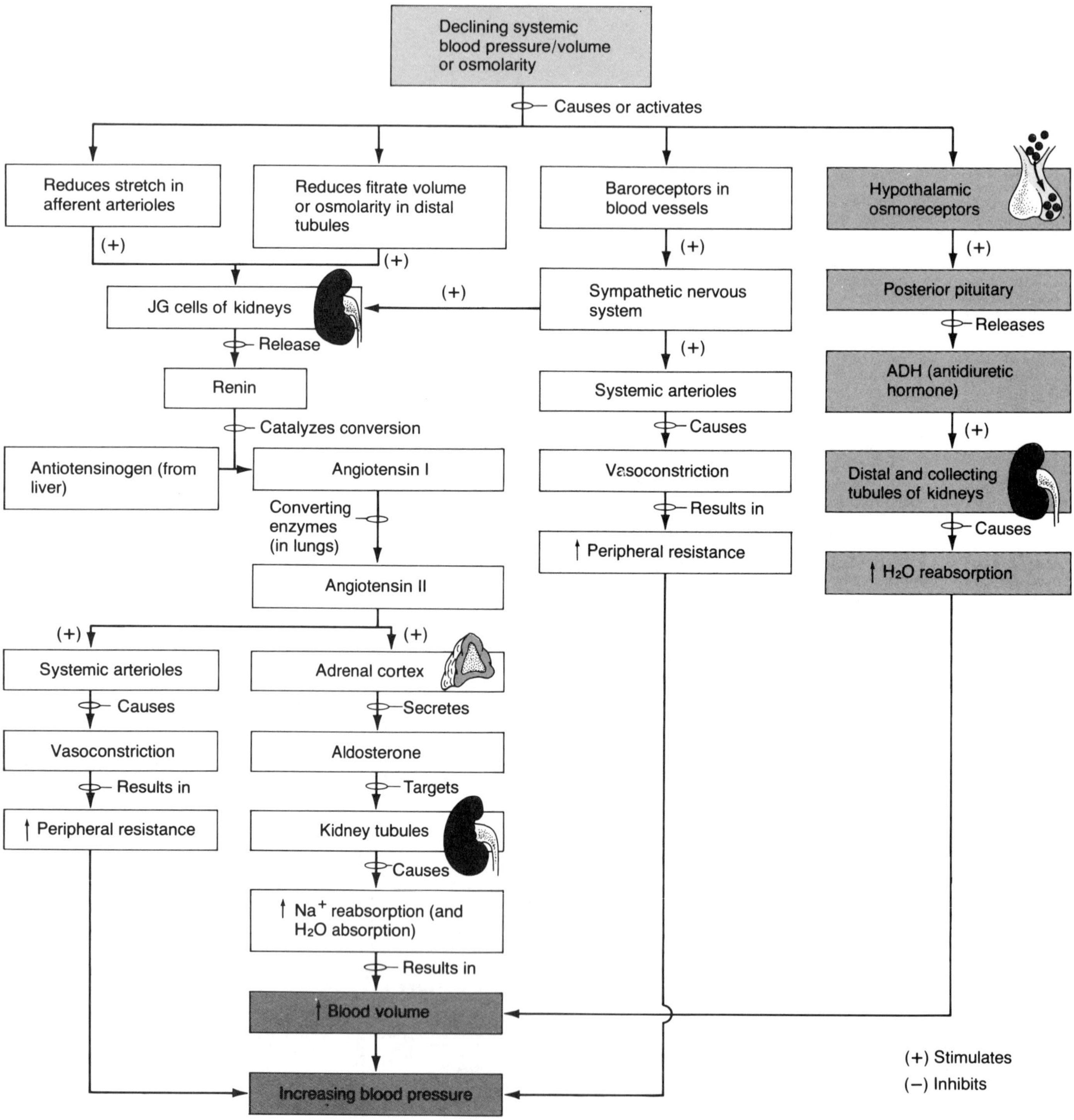

Sources of hydrogen ions (H^+) in the body

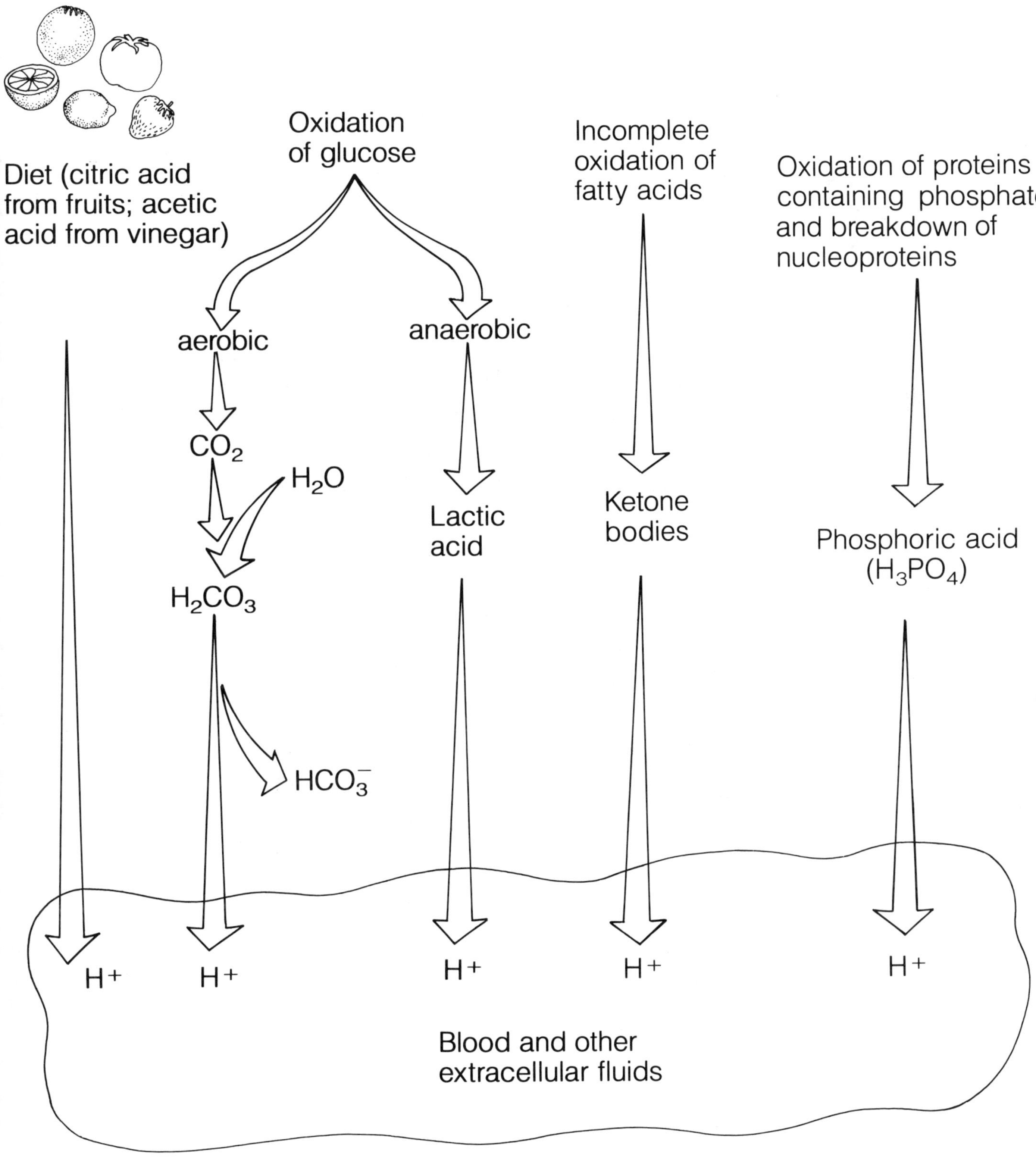

Comparison of dissociation of strong and weak acids

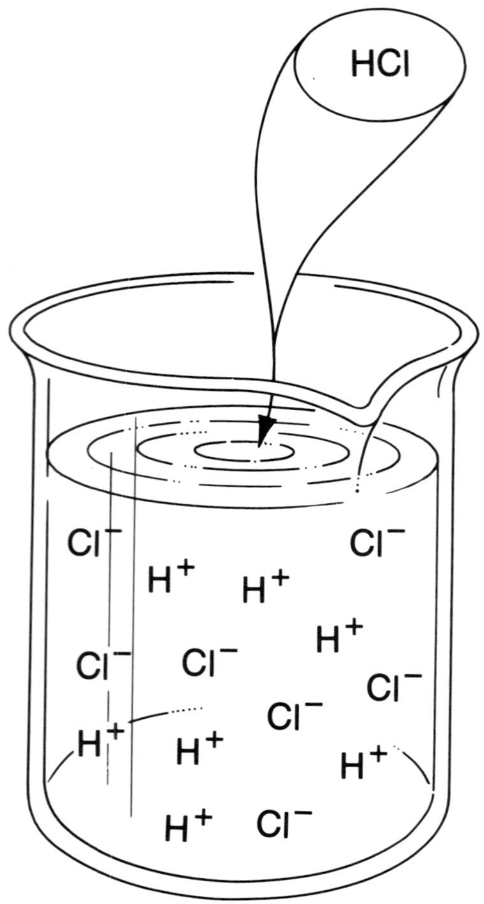

(a)

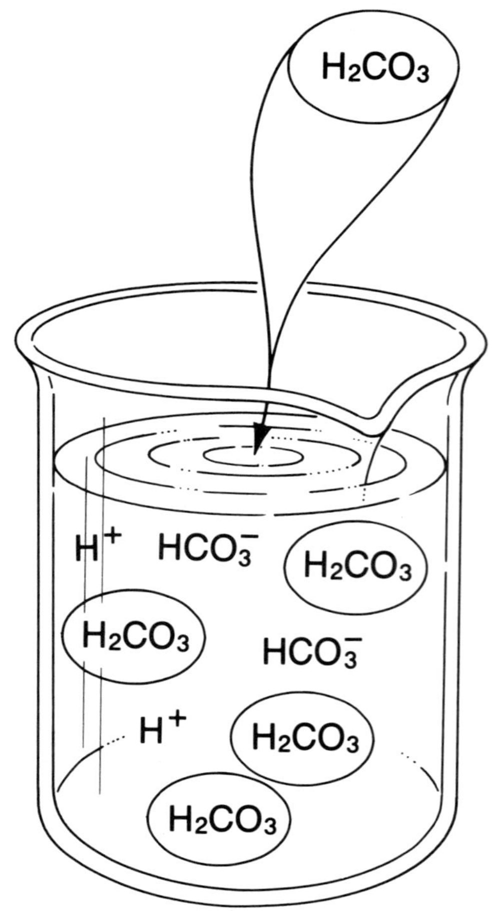

(b)

Reabsorption of filtered HCO_3^- is coupled to H^+ secretion

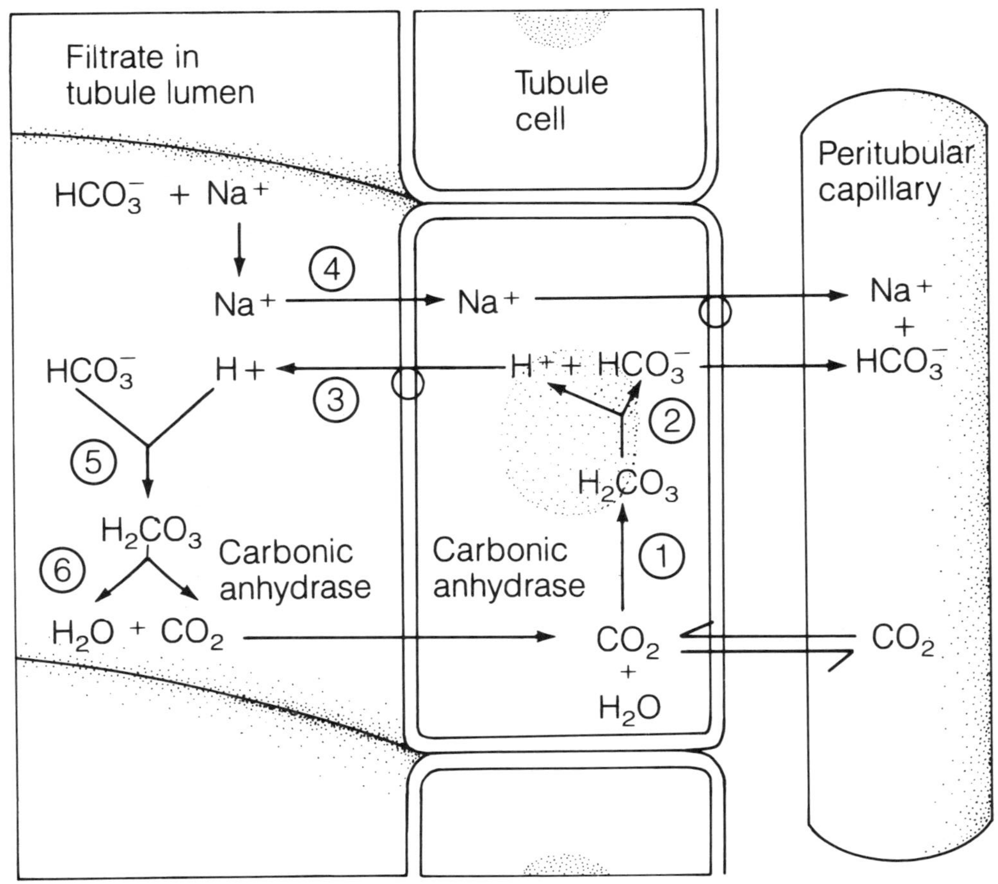

Buffering of excreted hydrogen ions in urine

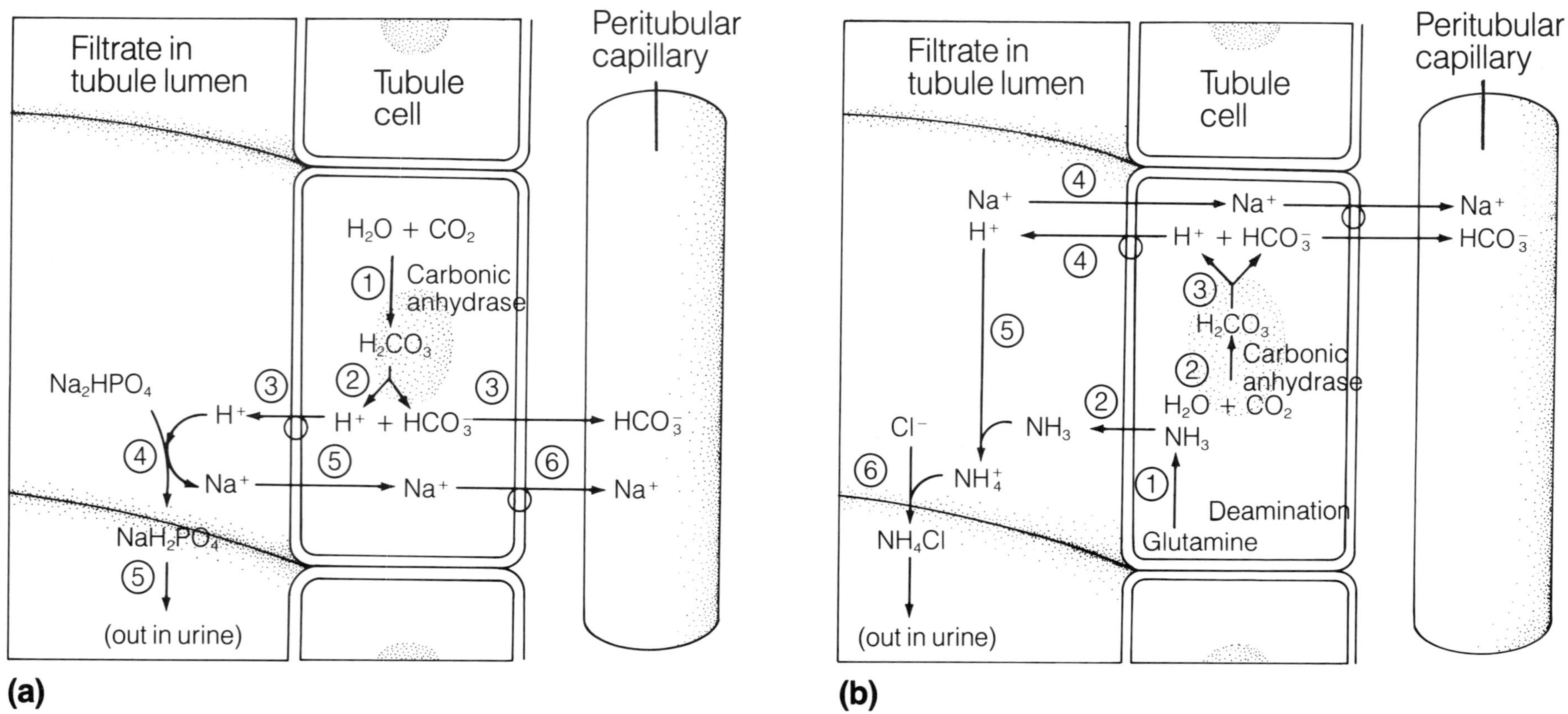

Internal structure of the testis

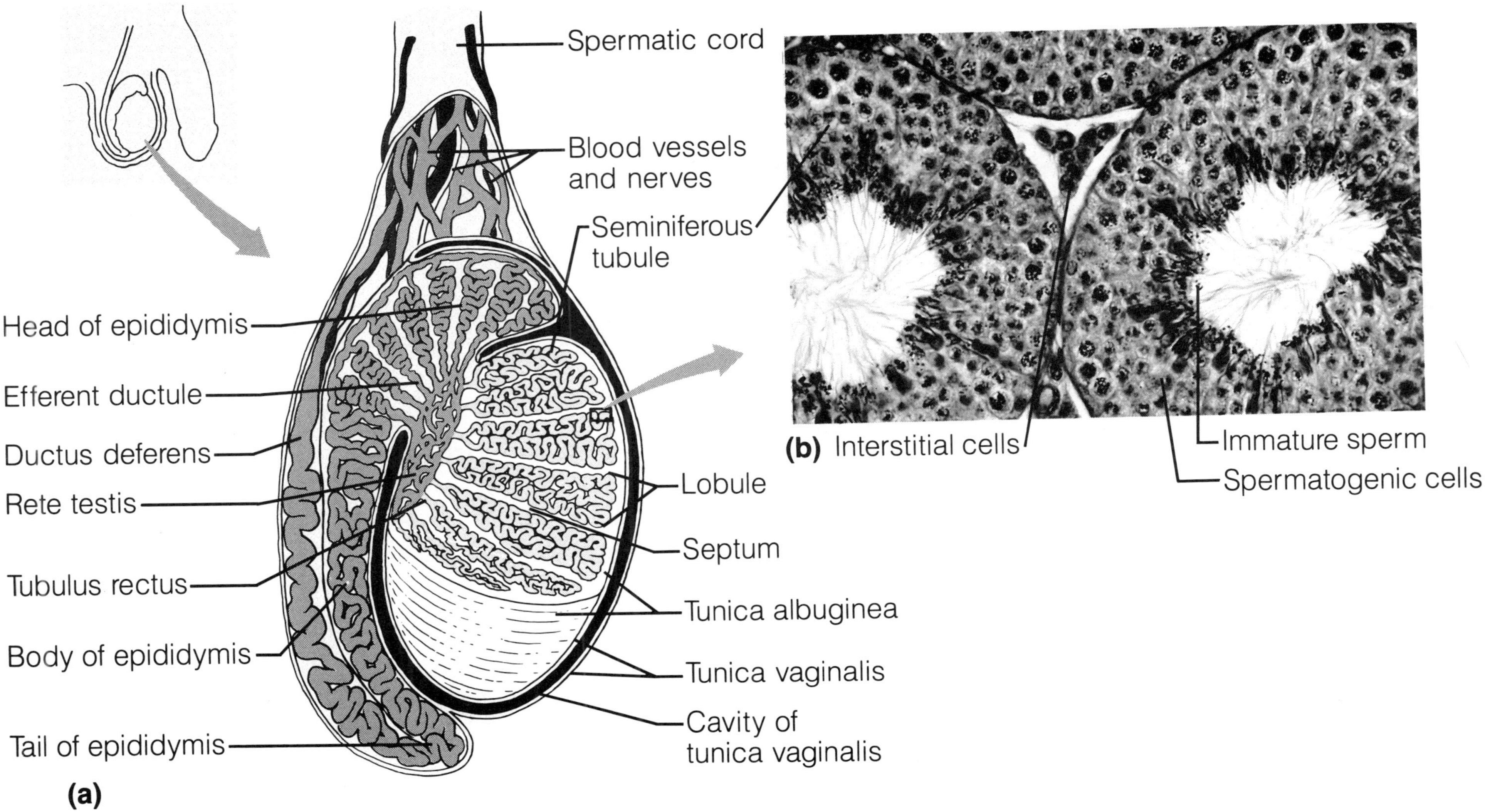

Comparison of mitosis and meiosis

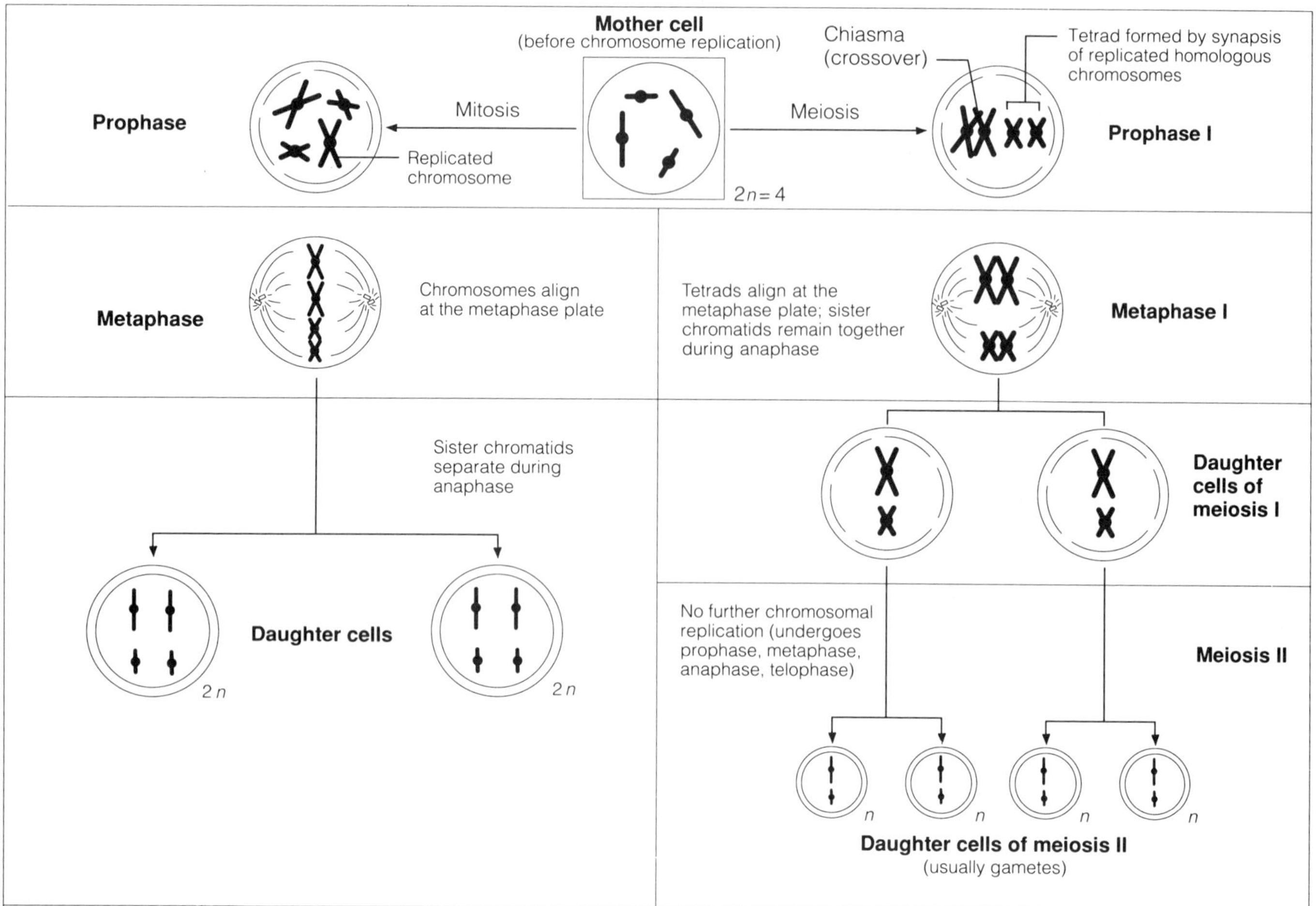

Event	Mitosis	Meiosis
DNA Replication	Occurs during interphase before the nuclear division process begins	Occurs in interphase before the nuclear division process begins
Number of divisions	One, consisting of prophase, metaphase, anaphase, and telophase	Two, each consisting of prophase, metaphase, anaphase, and telophase; DNA replication does not occur between the two nuclear divisions; an event unique to meiosis is that during meiosis I, the homologous chromosomes synapse (join along their length), forming tetrads (groups of four chromatids)
Number of daughter cells and genetic composition	Two; each diploid ($2n$) and identical to the mother cell	Four; each containing half as many chromosomes as the mother cell (hence haploid or n); nonidentical to mother cell
Importance in the body	To produce cells for growth and tissue repair; ensures constancy of genetic makeup of all body cells	To produce cells for reproduction (gametes); introduces variability in the gametes and reduces chromosomal number by half so that when fertilization occurs, the normal diploid chromosomal number of the species is restored; in humans, $2n = 46$

Spermatogenesis

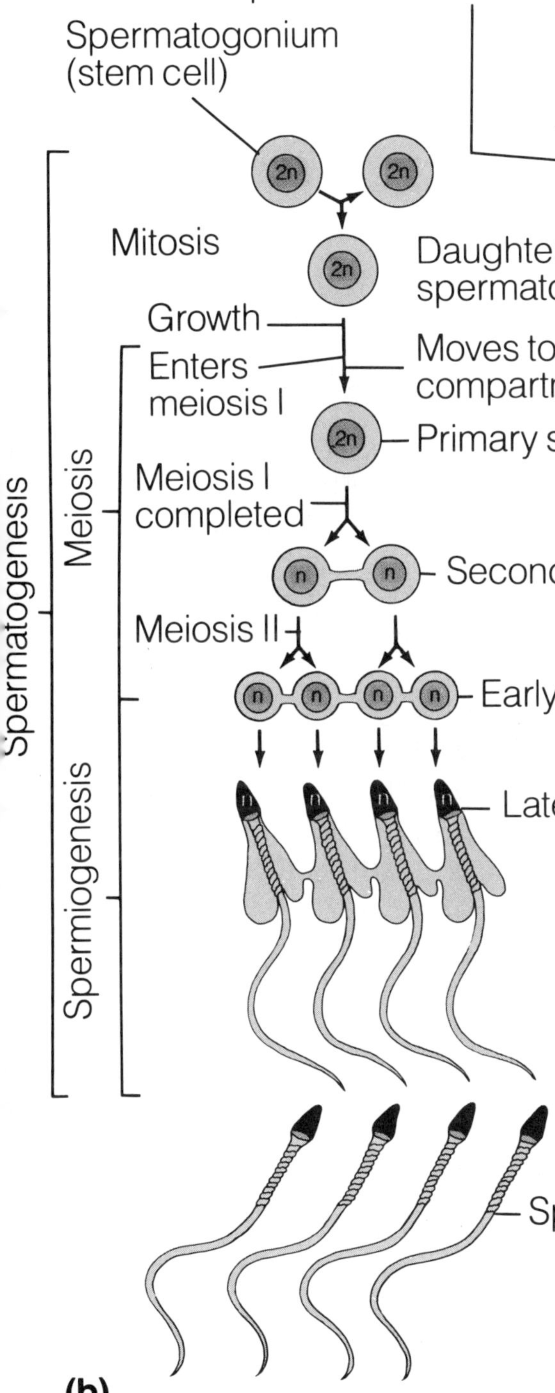

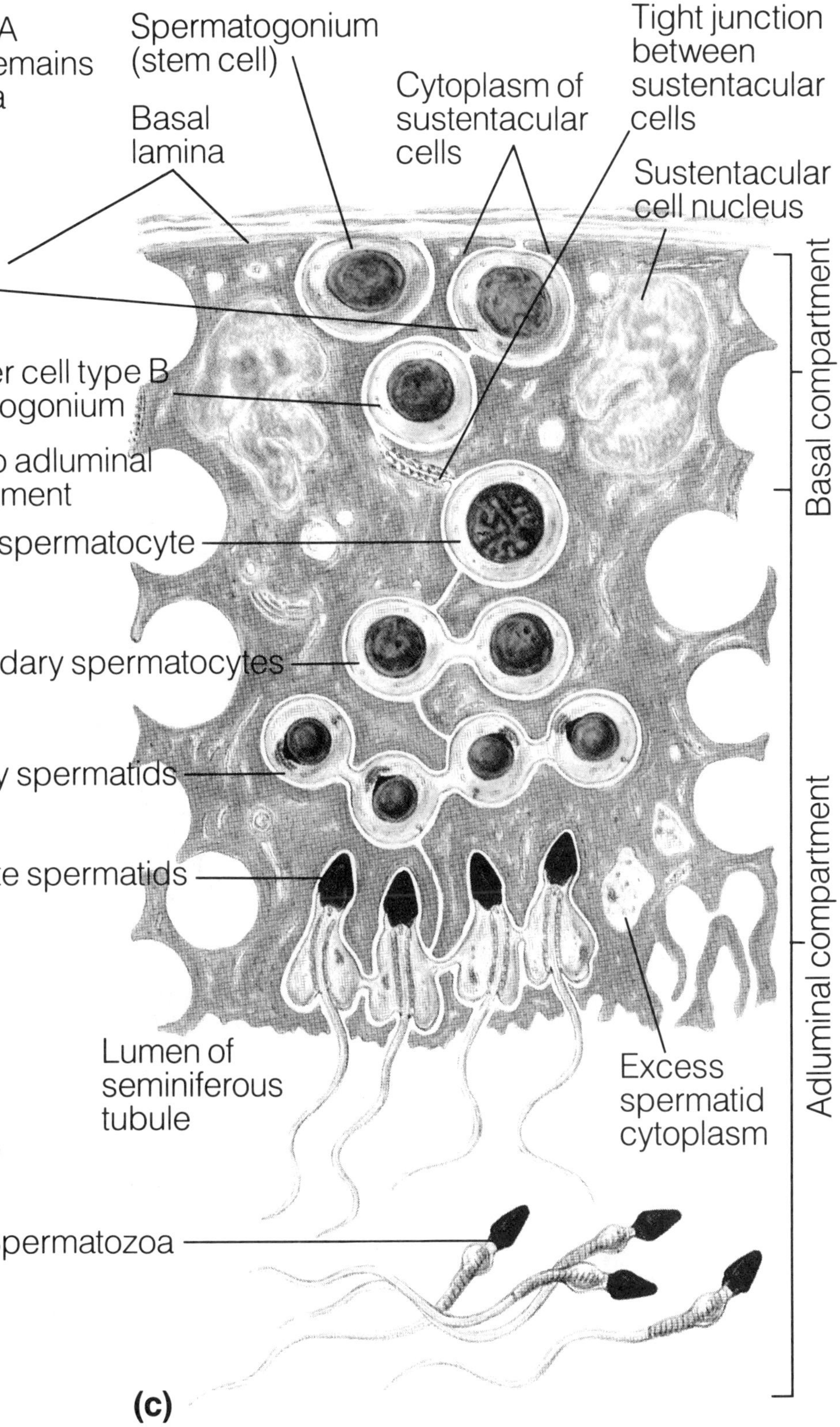

Schematic view of the relationship between a sustentacular cell and the spermatogenic cells and of the sustentacular cell's role in maintaining the blood-testis barrier

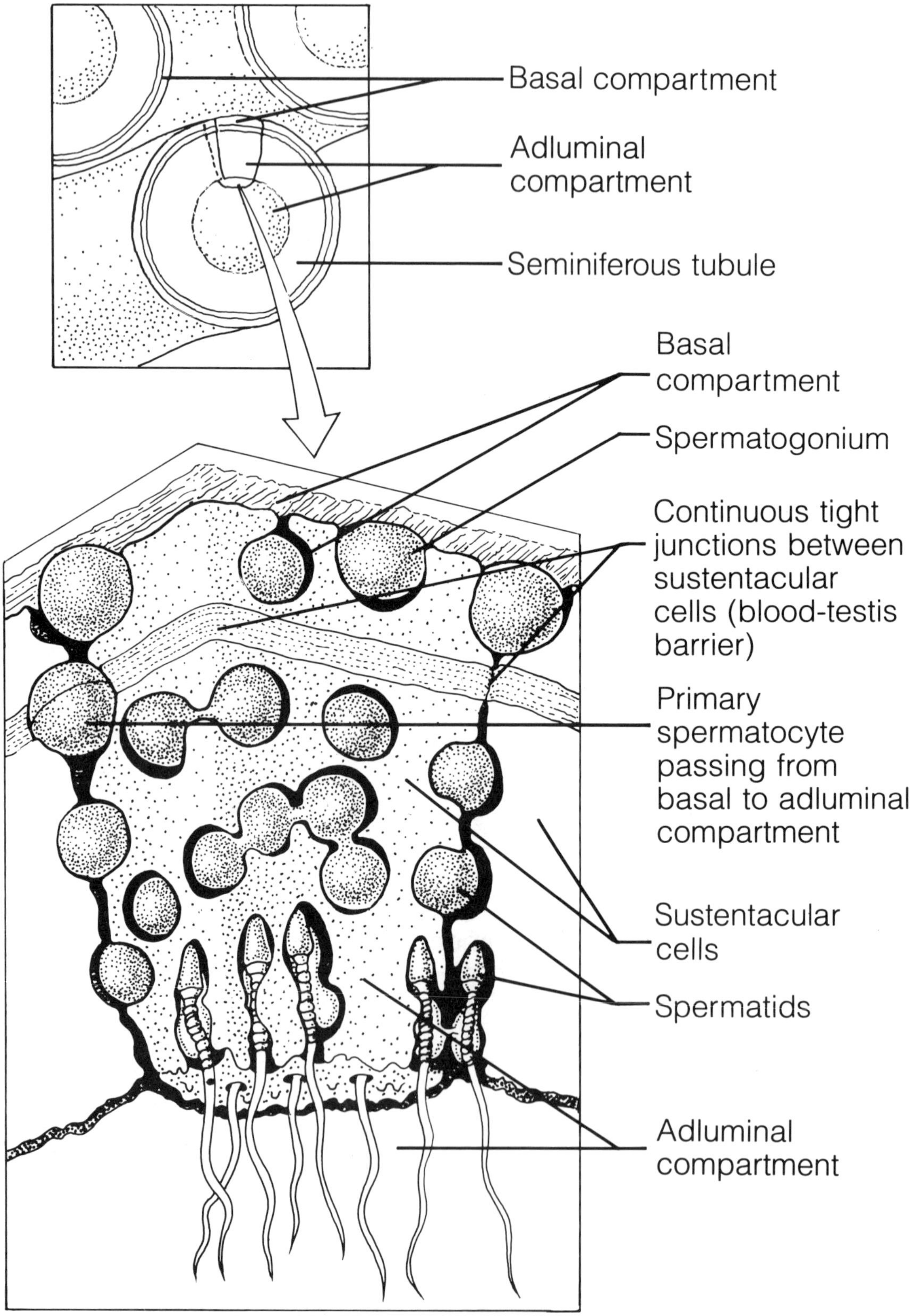

Hormonal regulation of testicular function, the brain-testicular axis

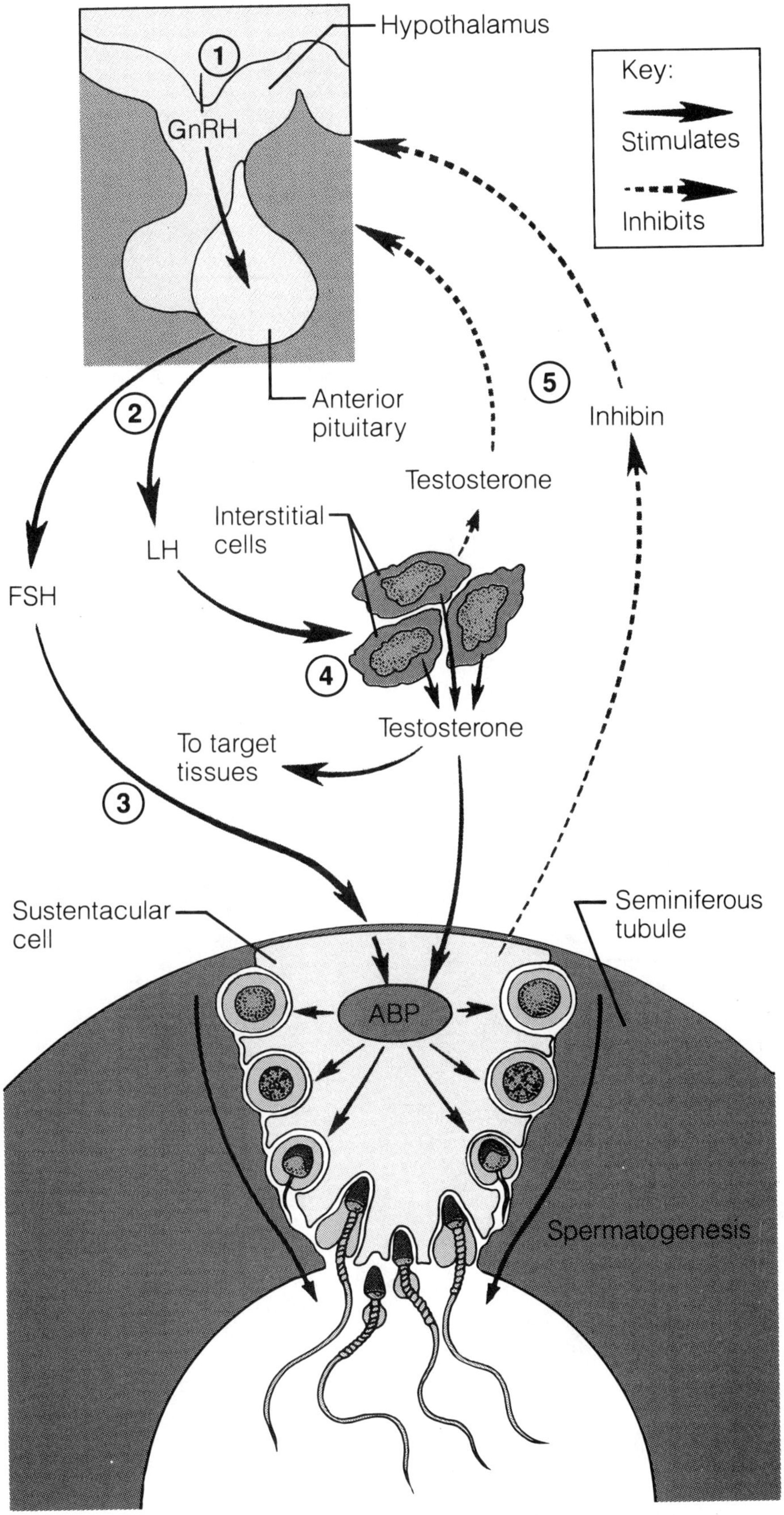

Structure of the endometrium and its blood supply

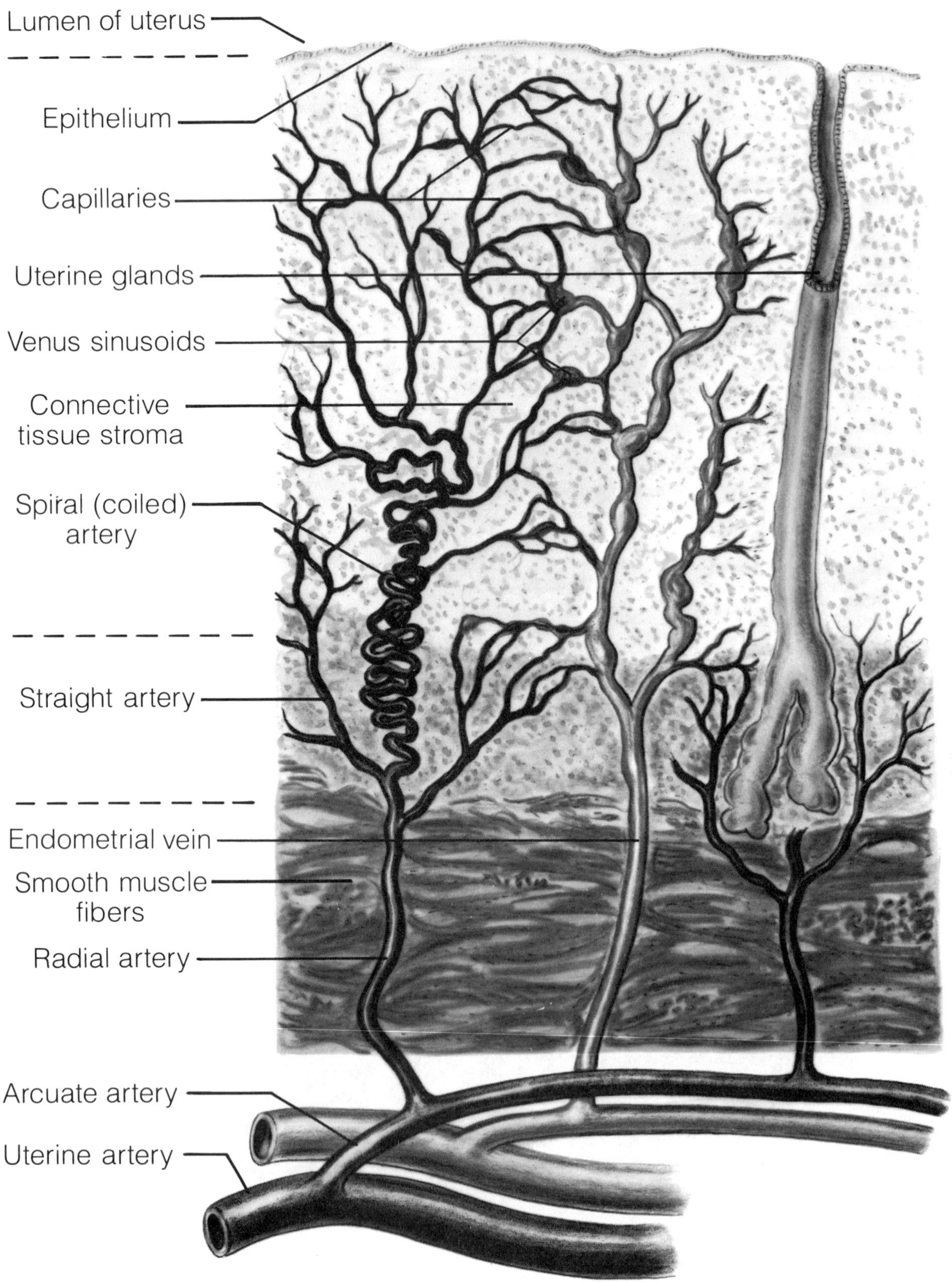

(b)

Events of oogenesis

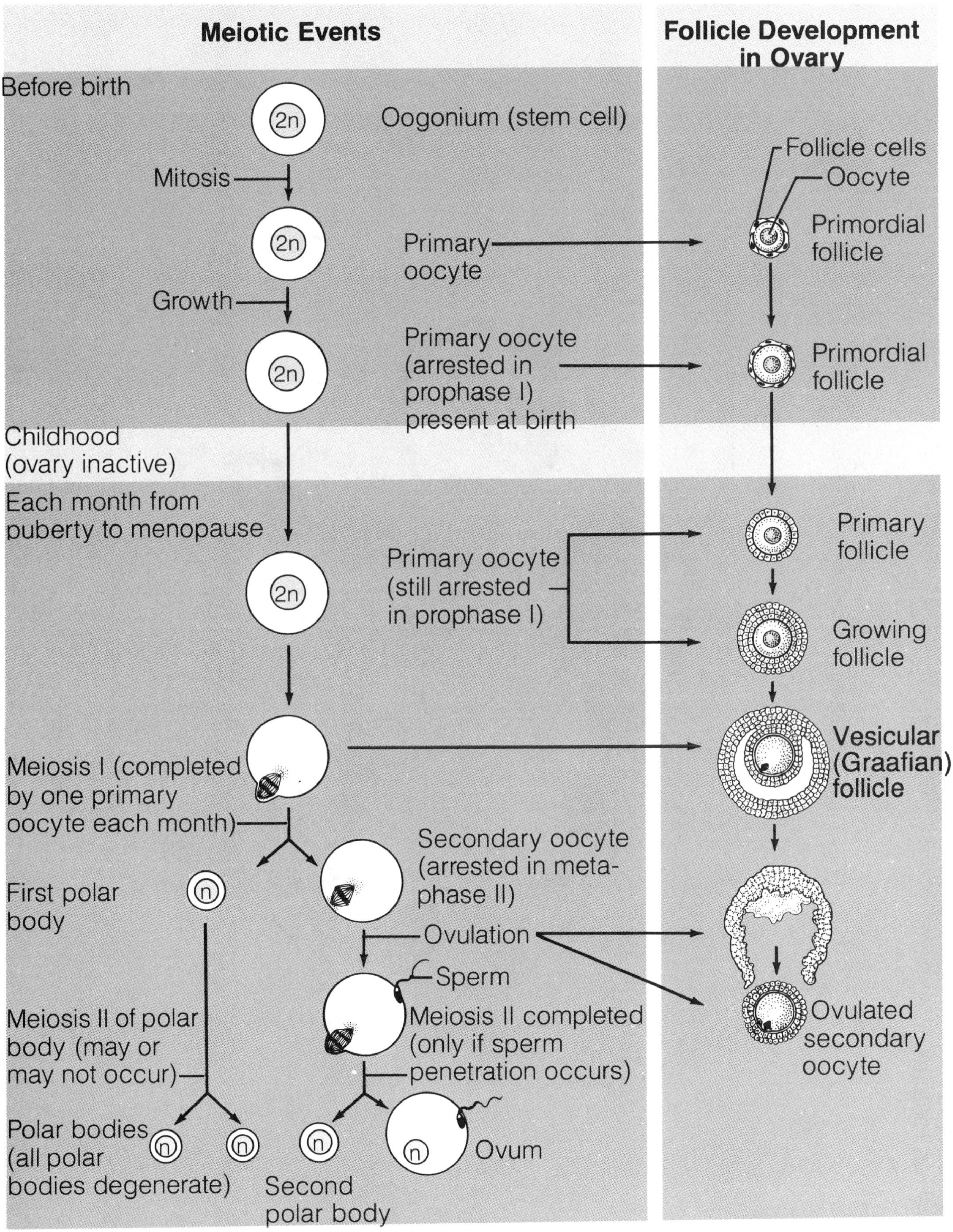

The ovarian cycle: development and fate of ovarian follicles

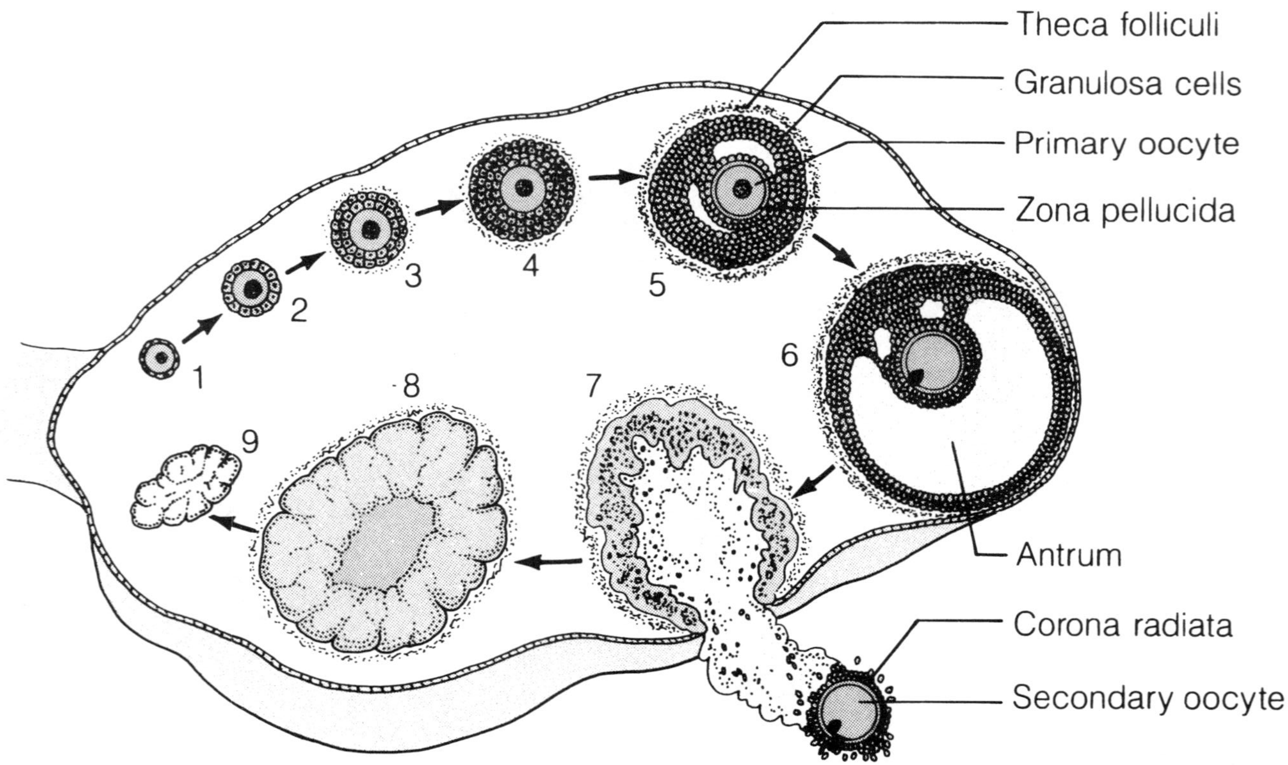

The regulation of ovarian function

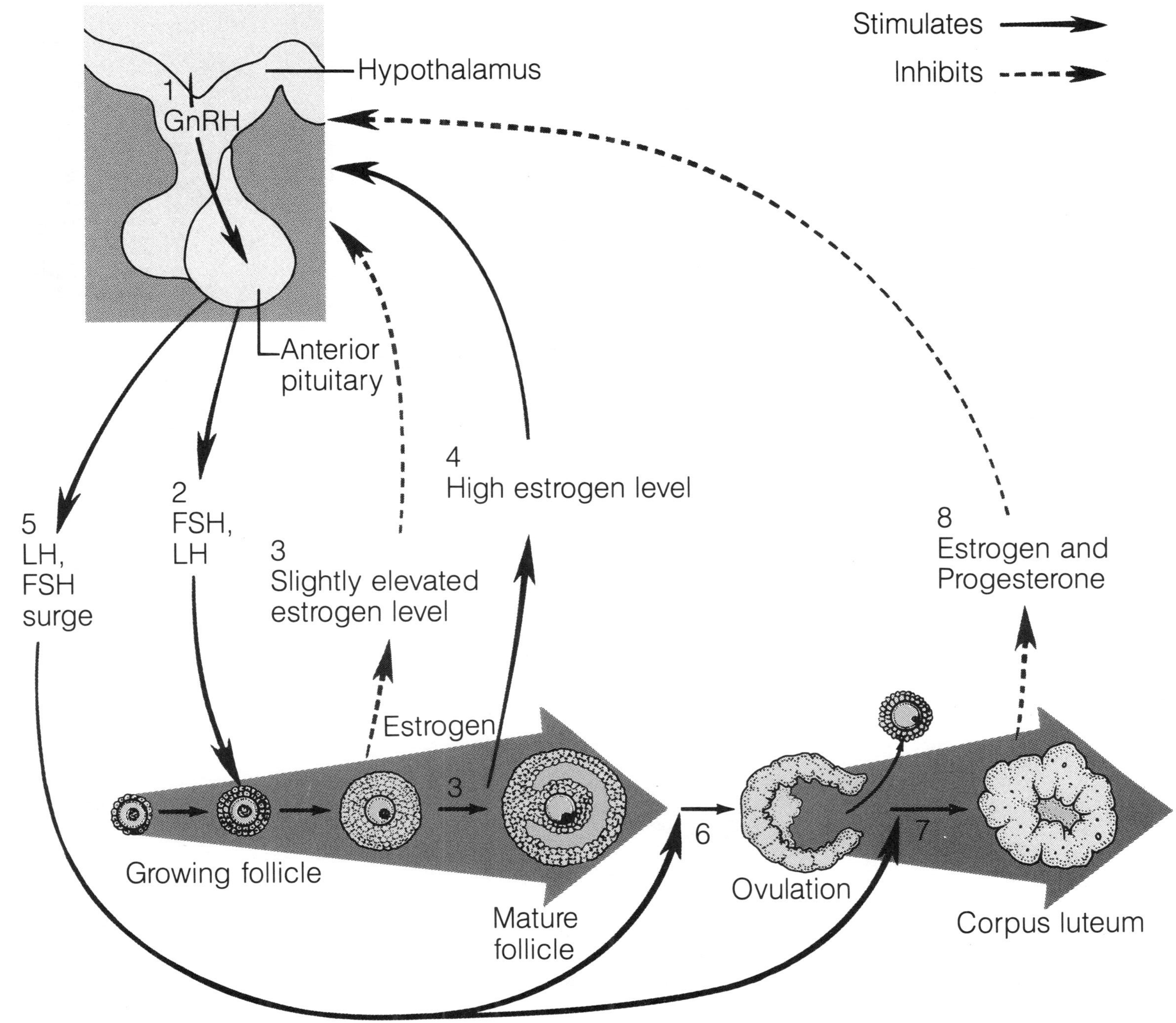

The uterine cycle and its correlation to the ovarian cycle

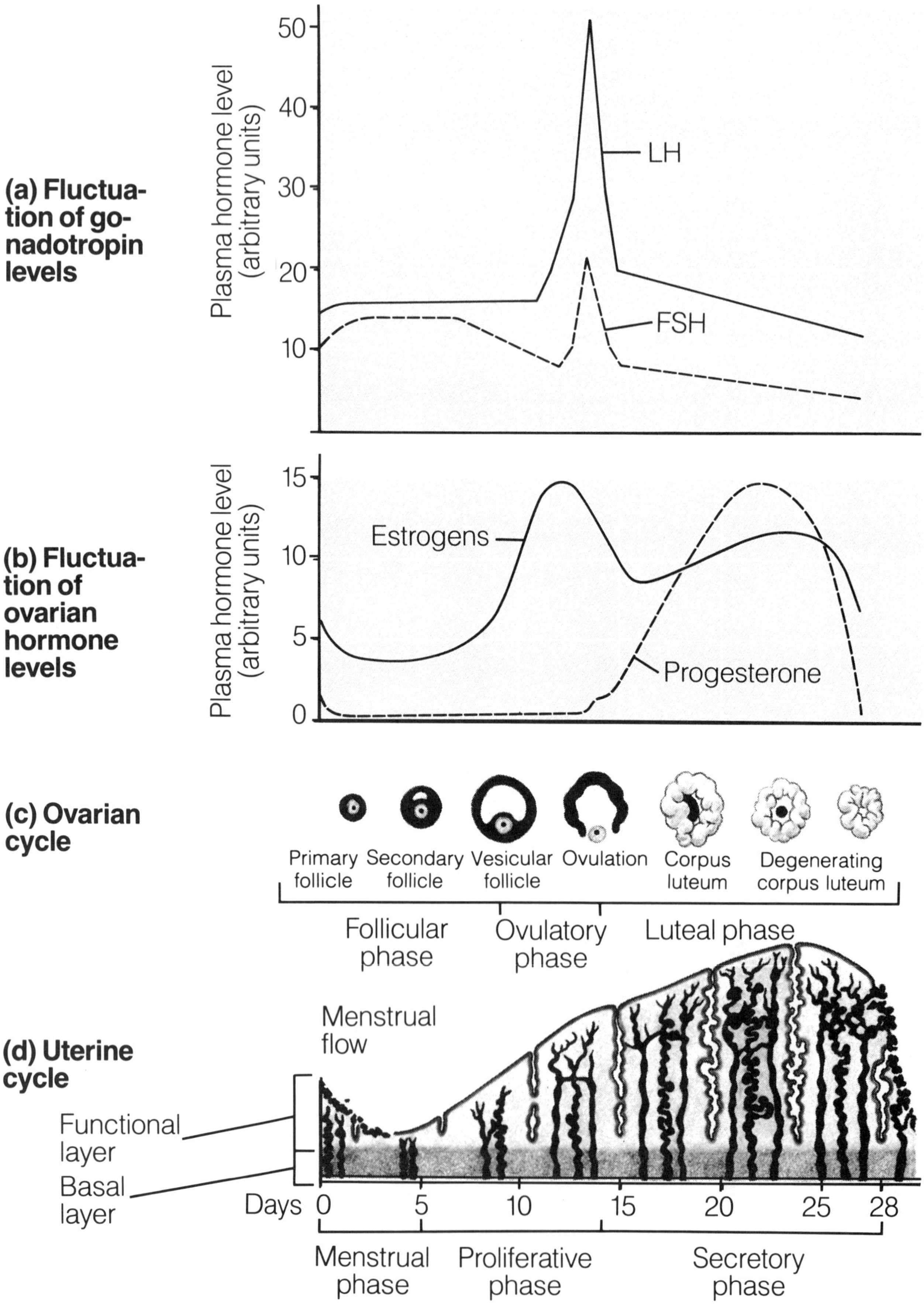

Development of the internal reproductive organs

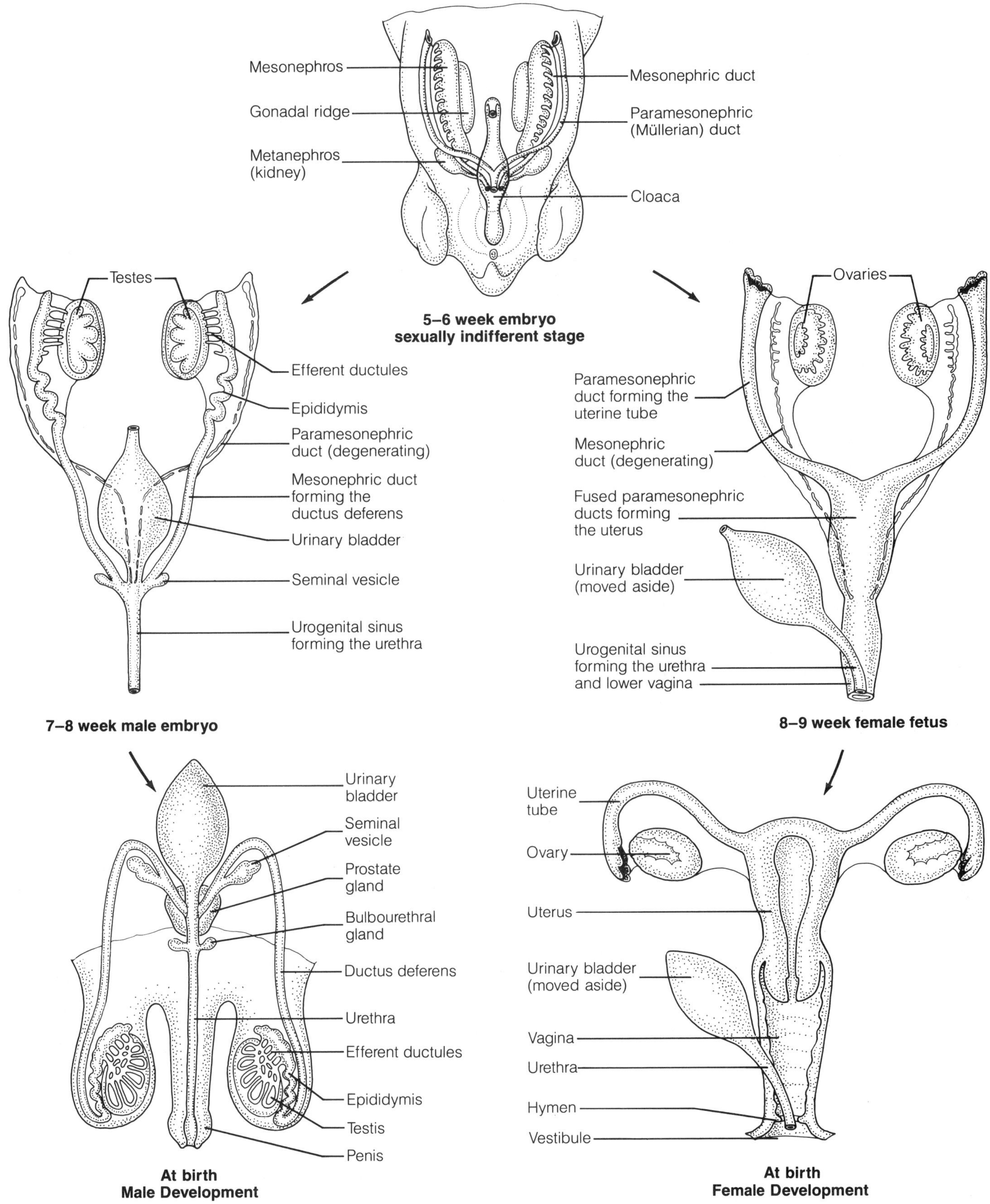

Development of external genitalia in humans

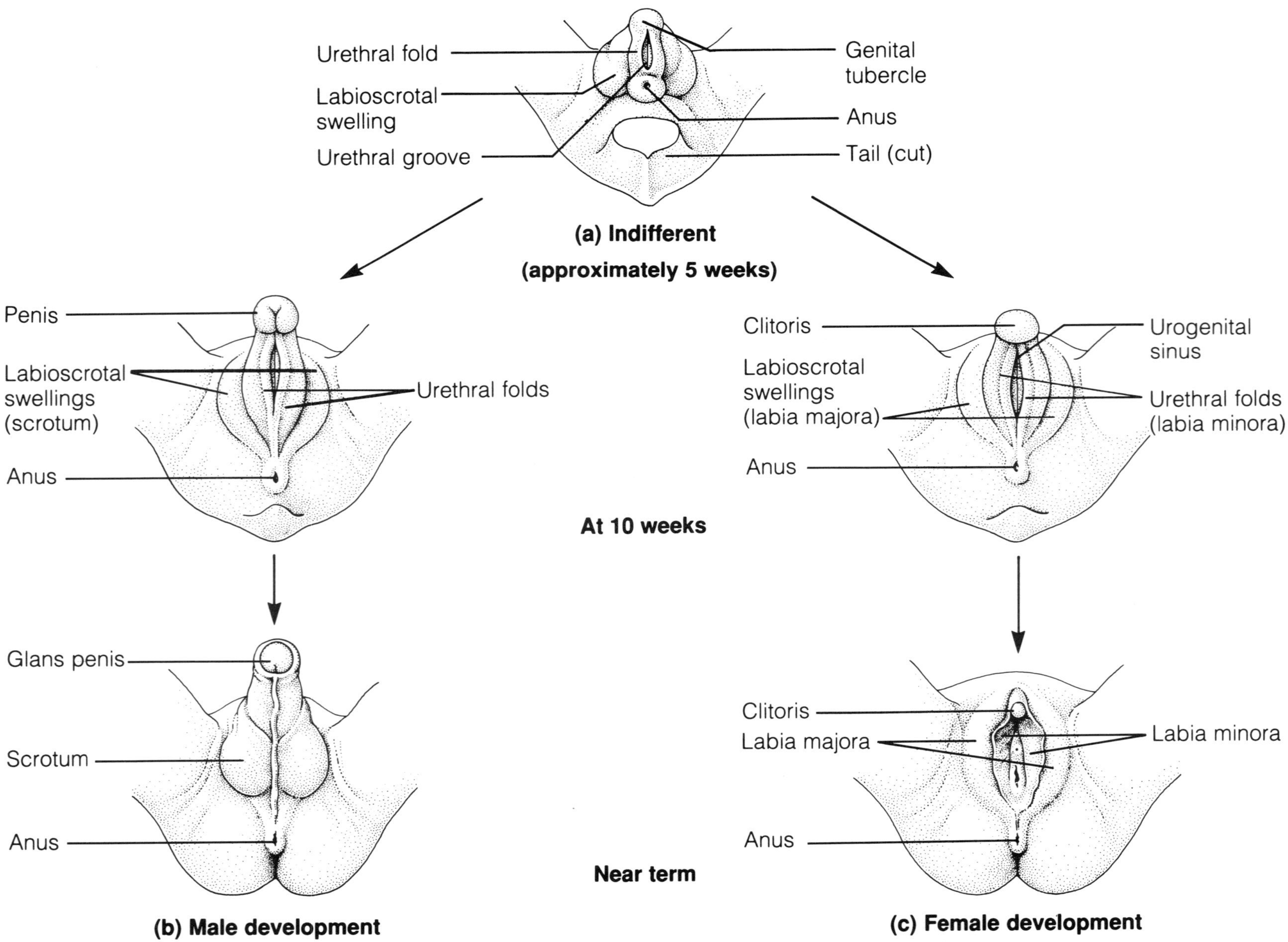

Descent of the testes

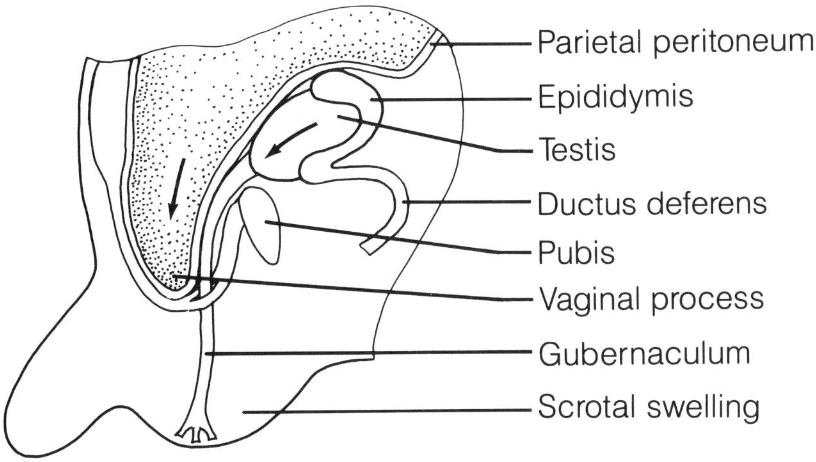

(a) 7-month fetus

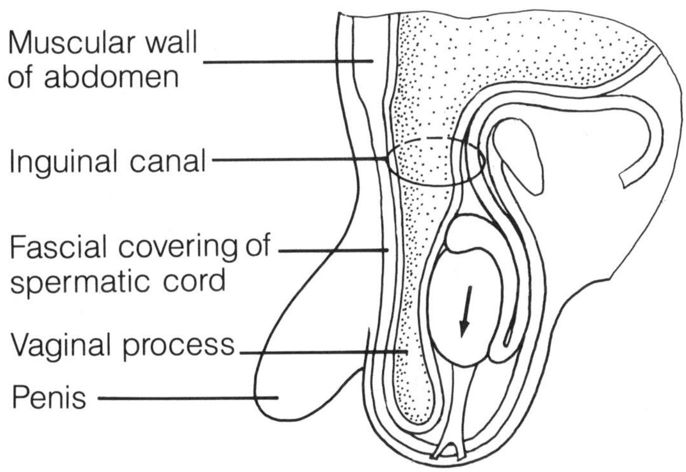

(b) 8-month fetus

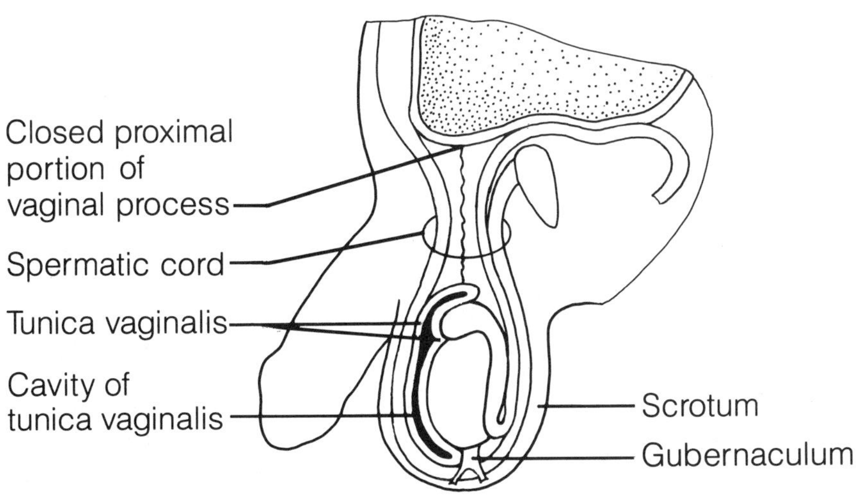

(c) 1-month-old infant

A CLOSER LOOK: Contraception

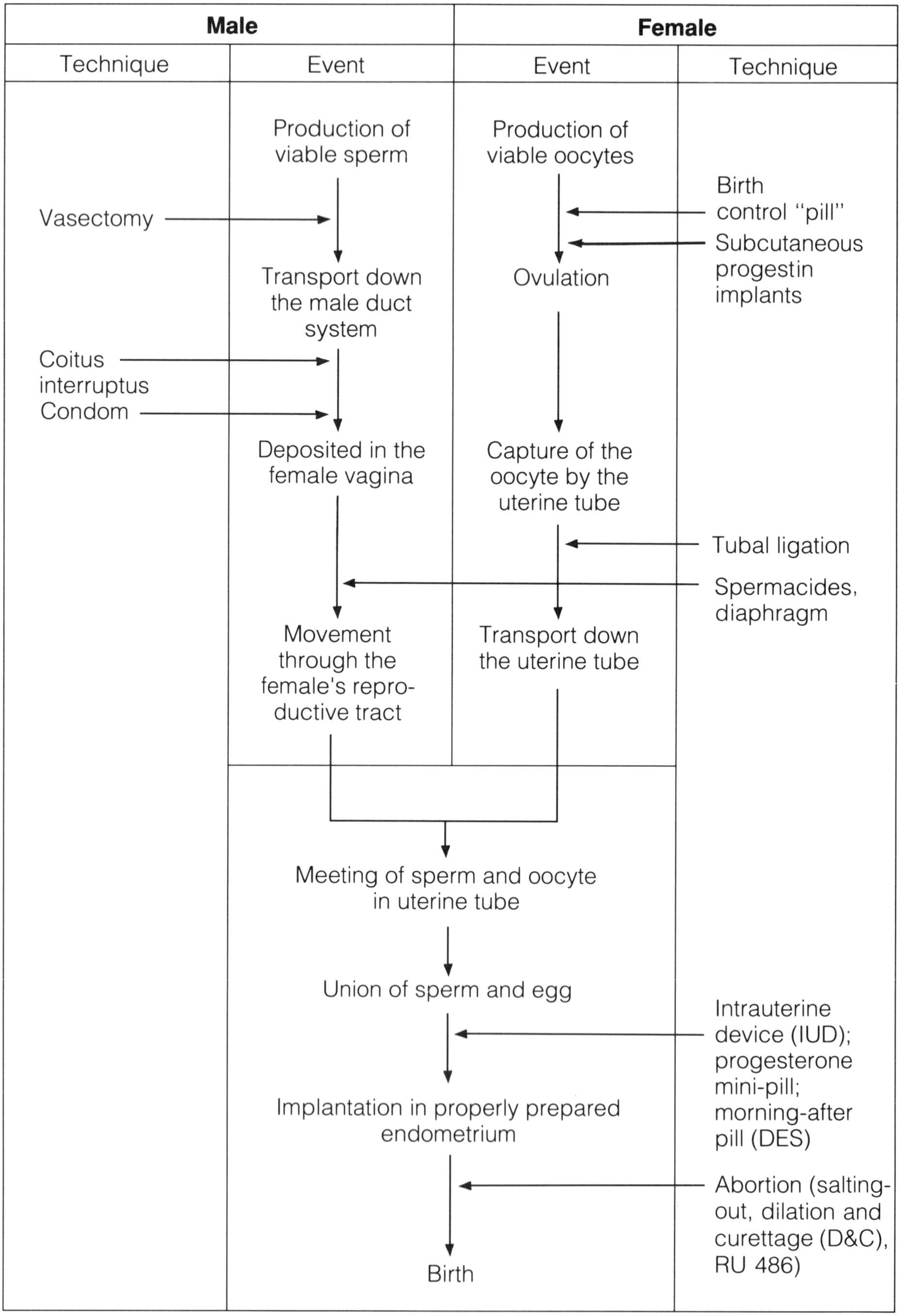

Sperm penetration and the cortical reaction (slow block to polyspermy) (Figure 29.2a)

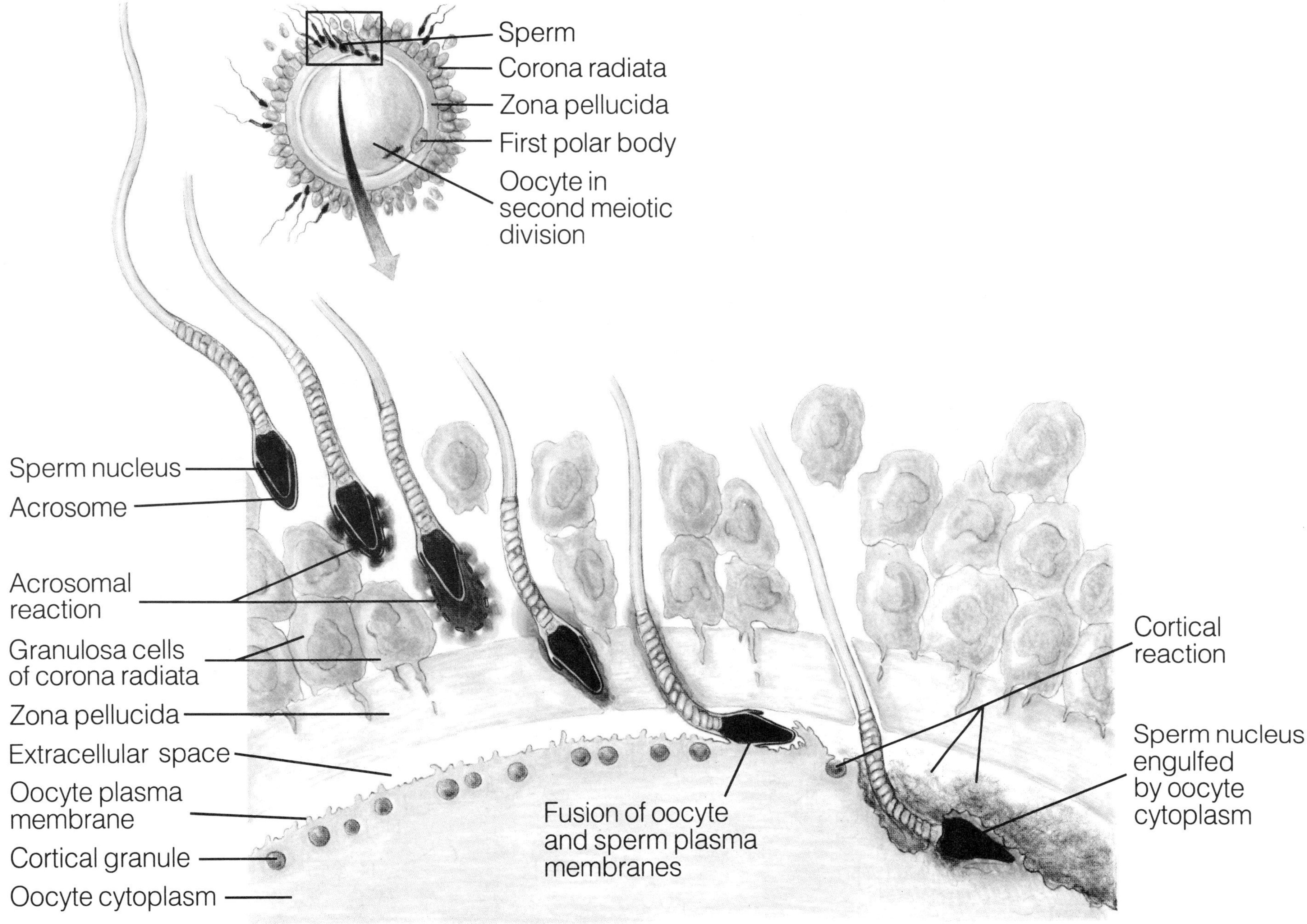

(a)

Events immediately following sperm penetration

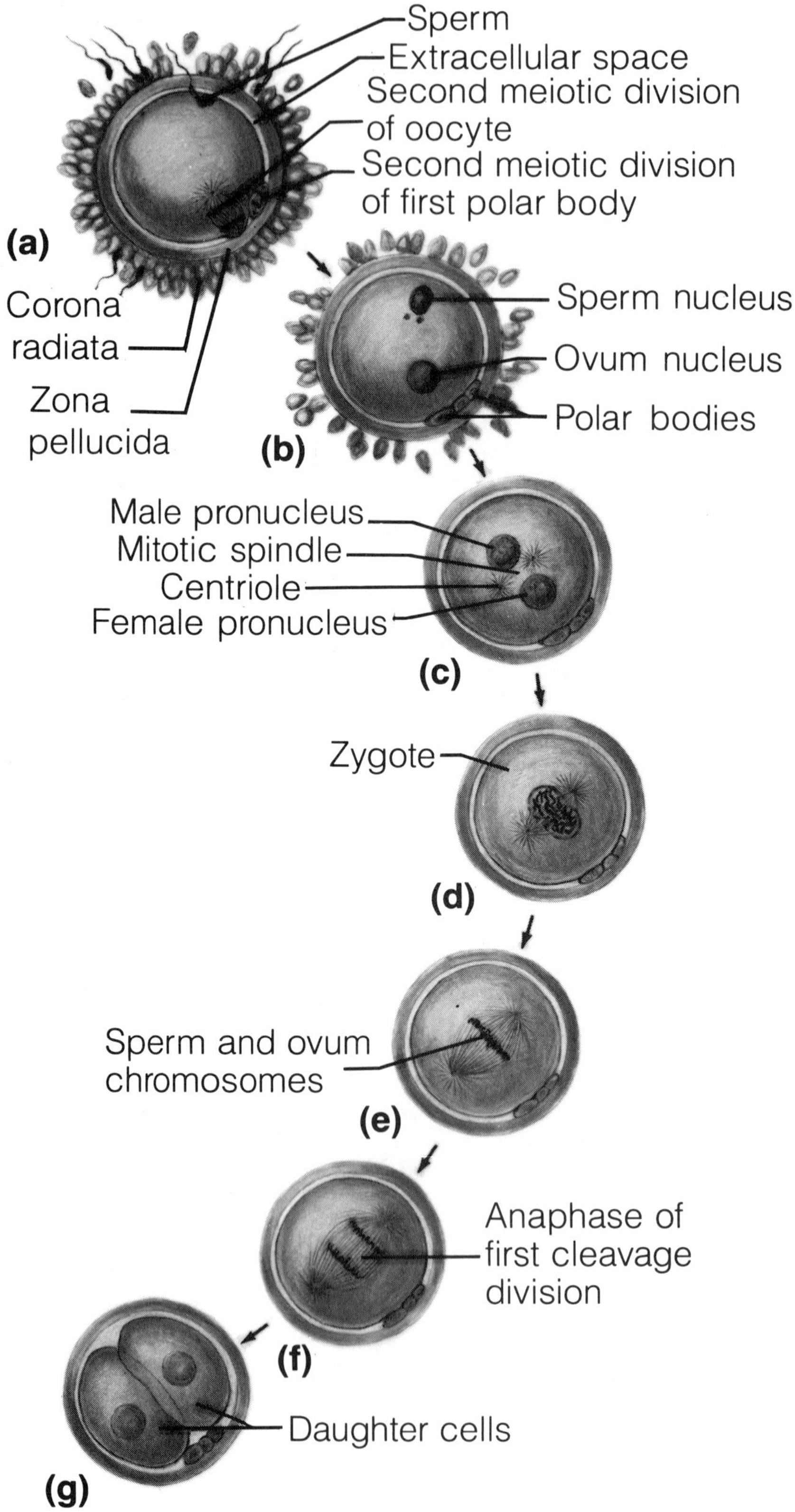

Cleavage

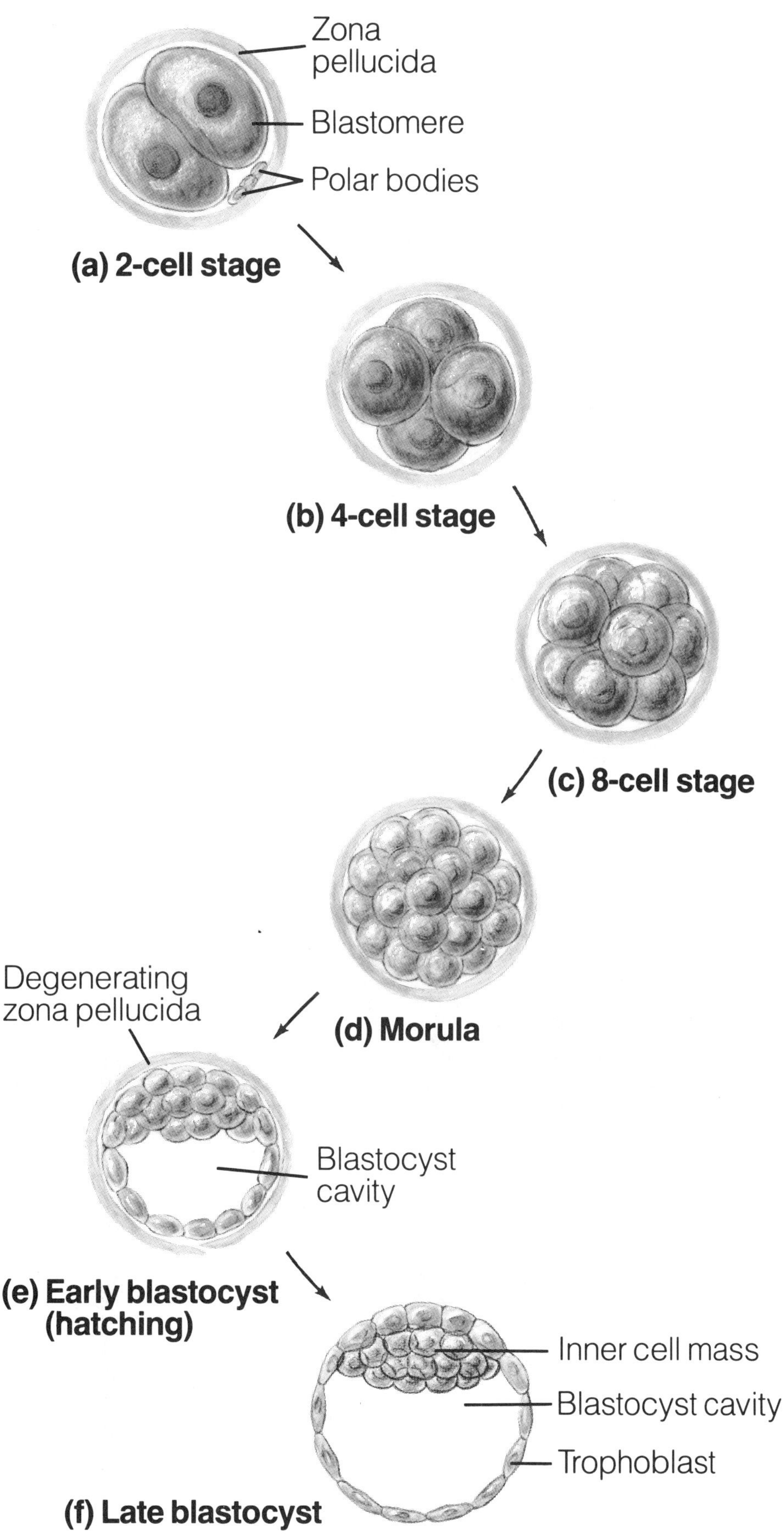
Zona pellucida
Blastomere
Polar bodies
(a) 2-cell stage
(b) 4-cell stage
(c) 8-cell stage
(d) Morula
Degenerating zona pellucida
Blastocyst cavity
(e) Early blastocyst (hatching)
Inner cell mass
Blastocyst cavity
Trophoblast
(f) Late blastocyst

Implantation of the blastocyst

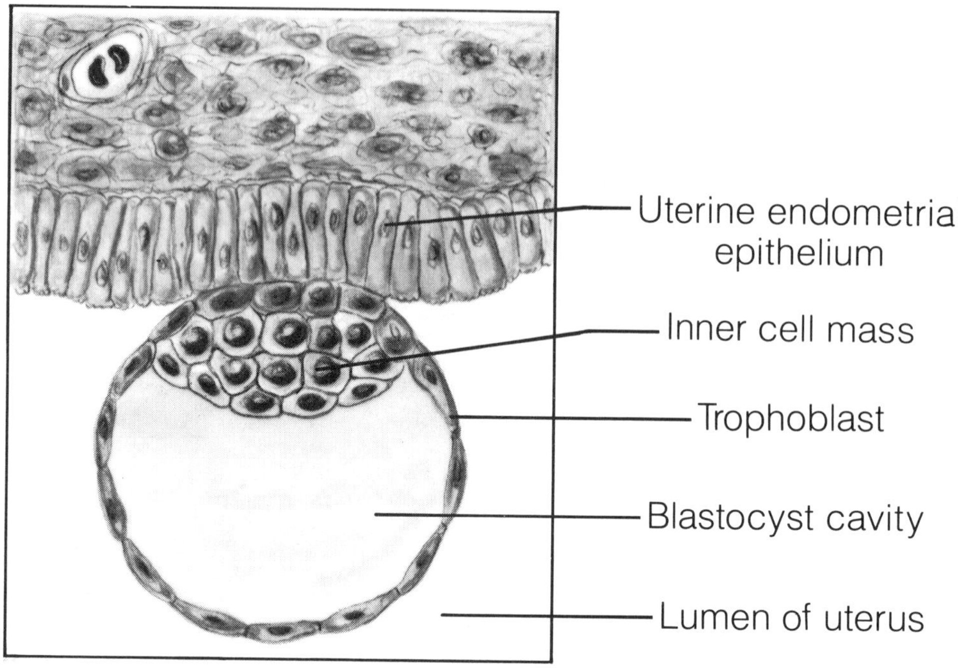

(a)

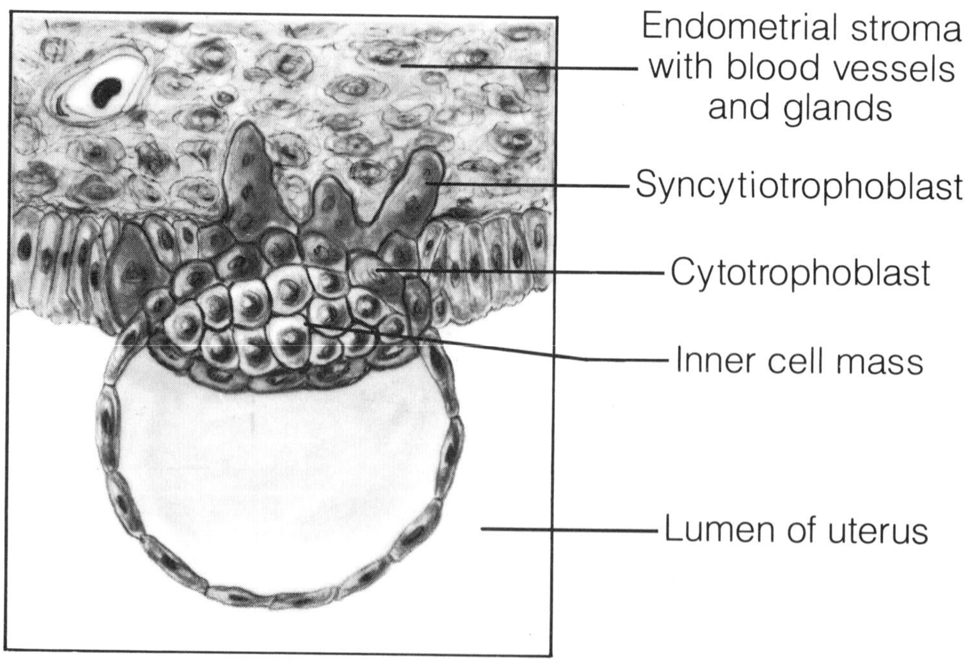

(c)

Relative change in maternal blood levels of human chorionic gonadotropin, estrogen, and progesterone during pregnancy

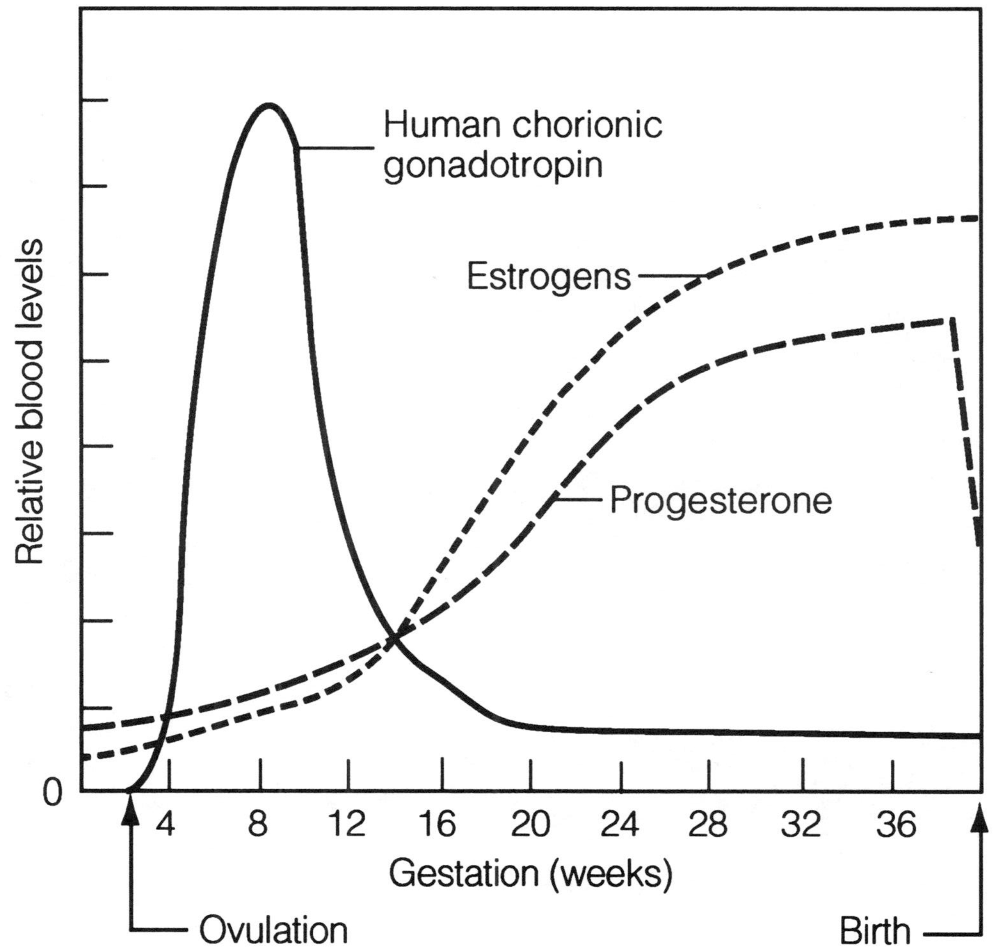

Fetal circulation

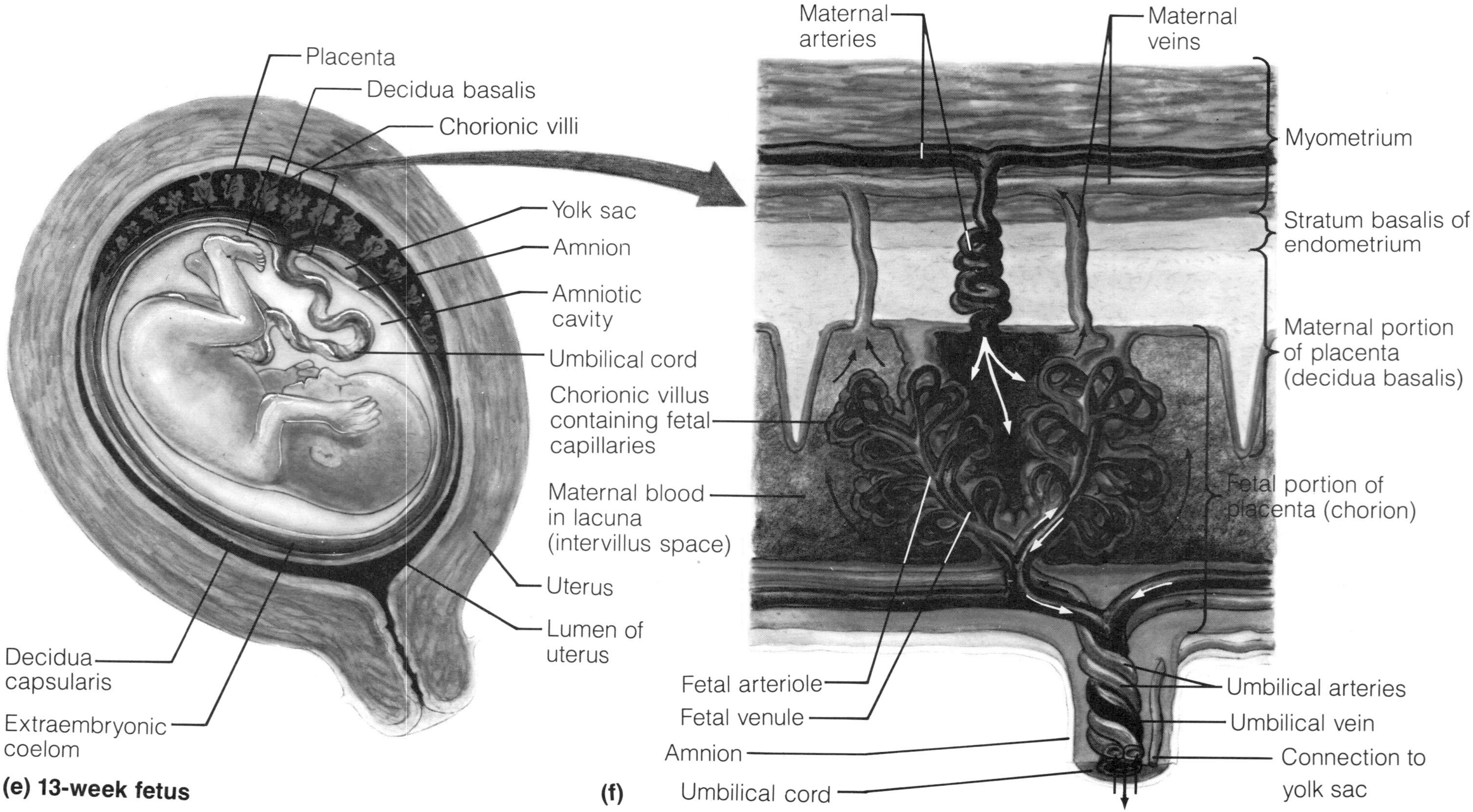

(e) 13-week fetus

(f)

Gastrulation: formation of the three primary germ layers

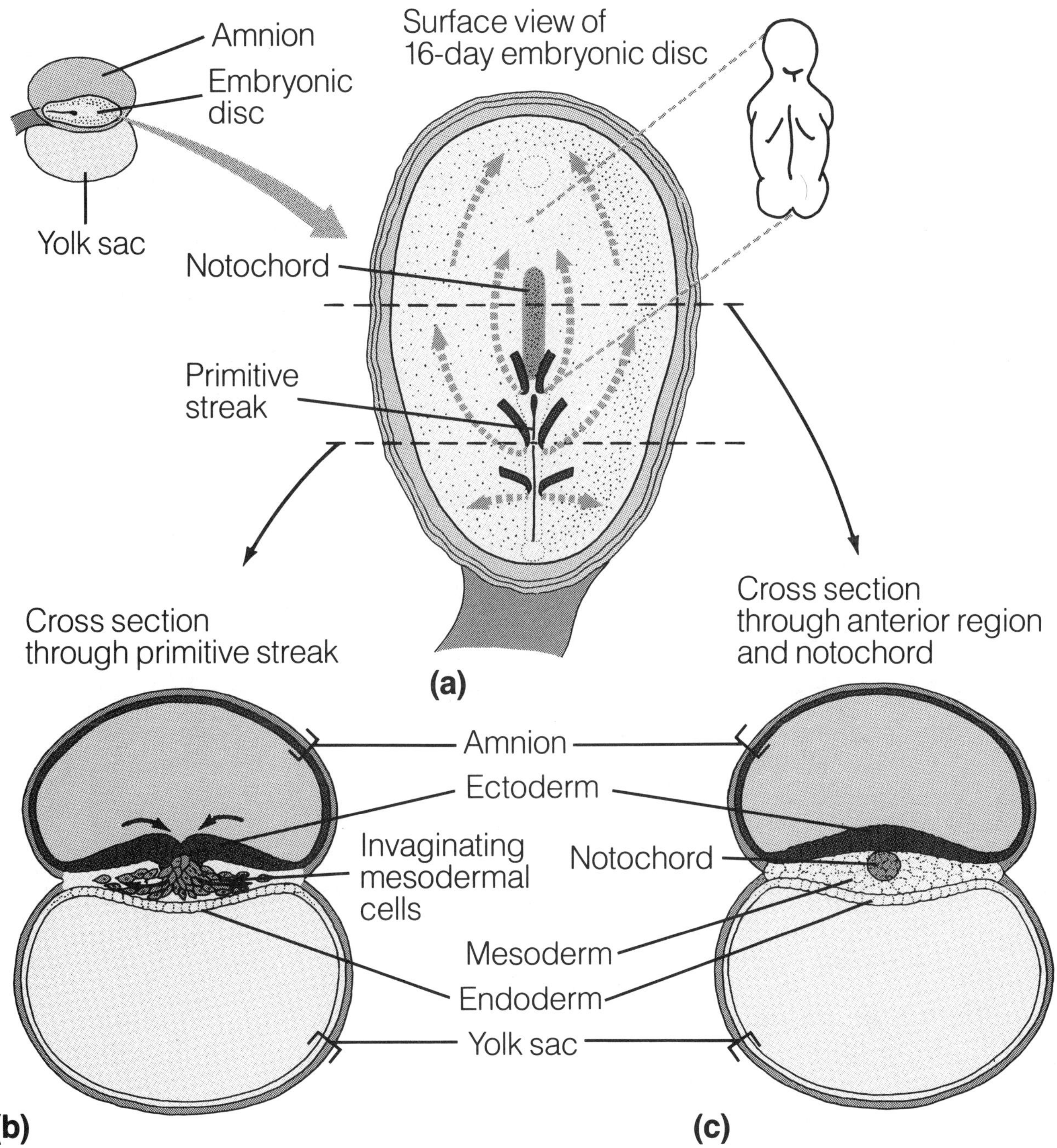

Circulation in fetus and newborn

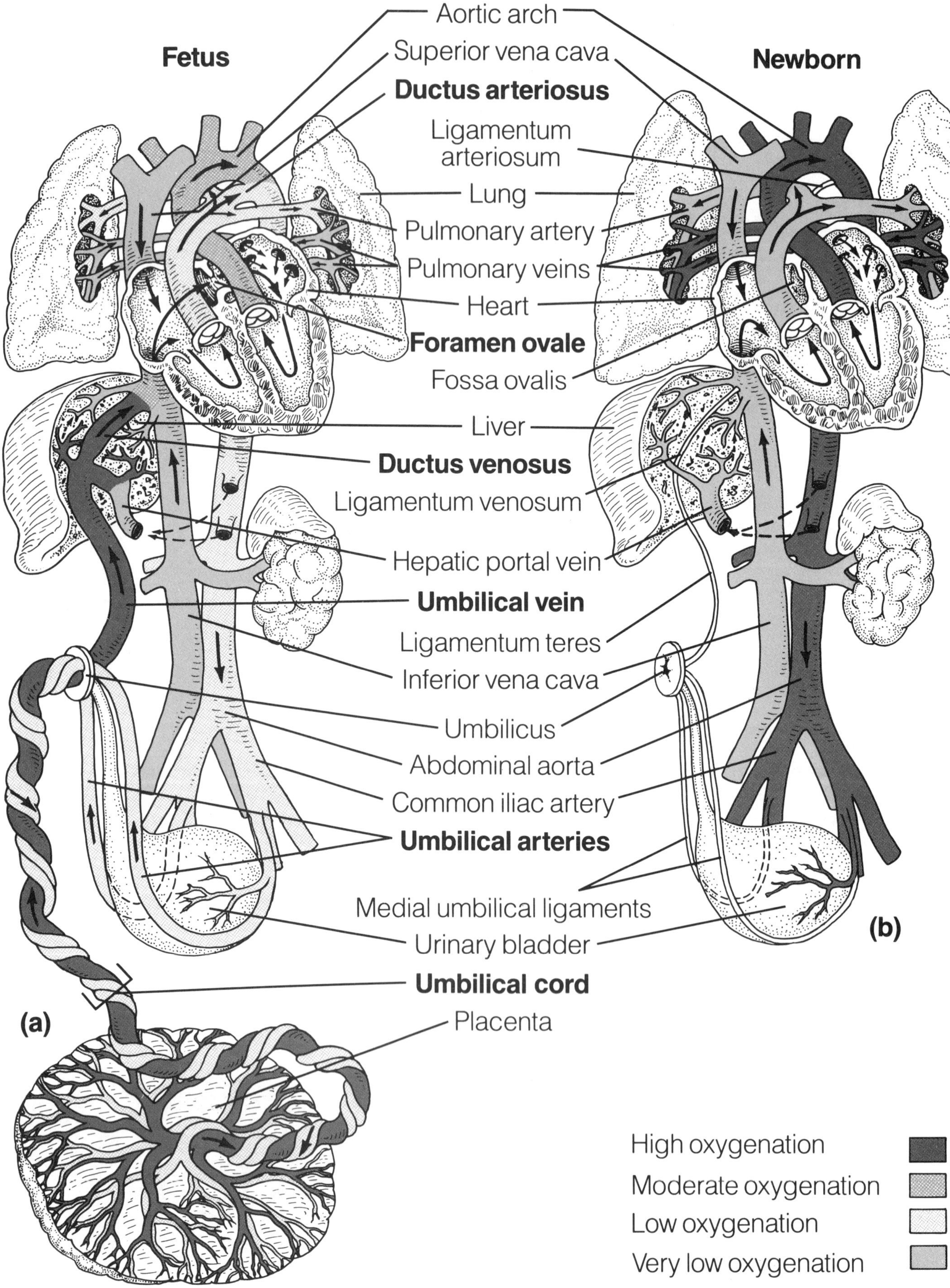

The positive-feedback mechanism by which oxytocin promotes labor contractions during birth

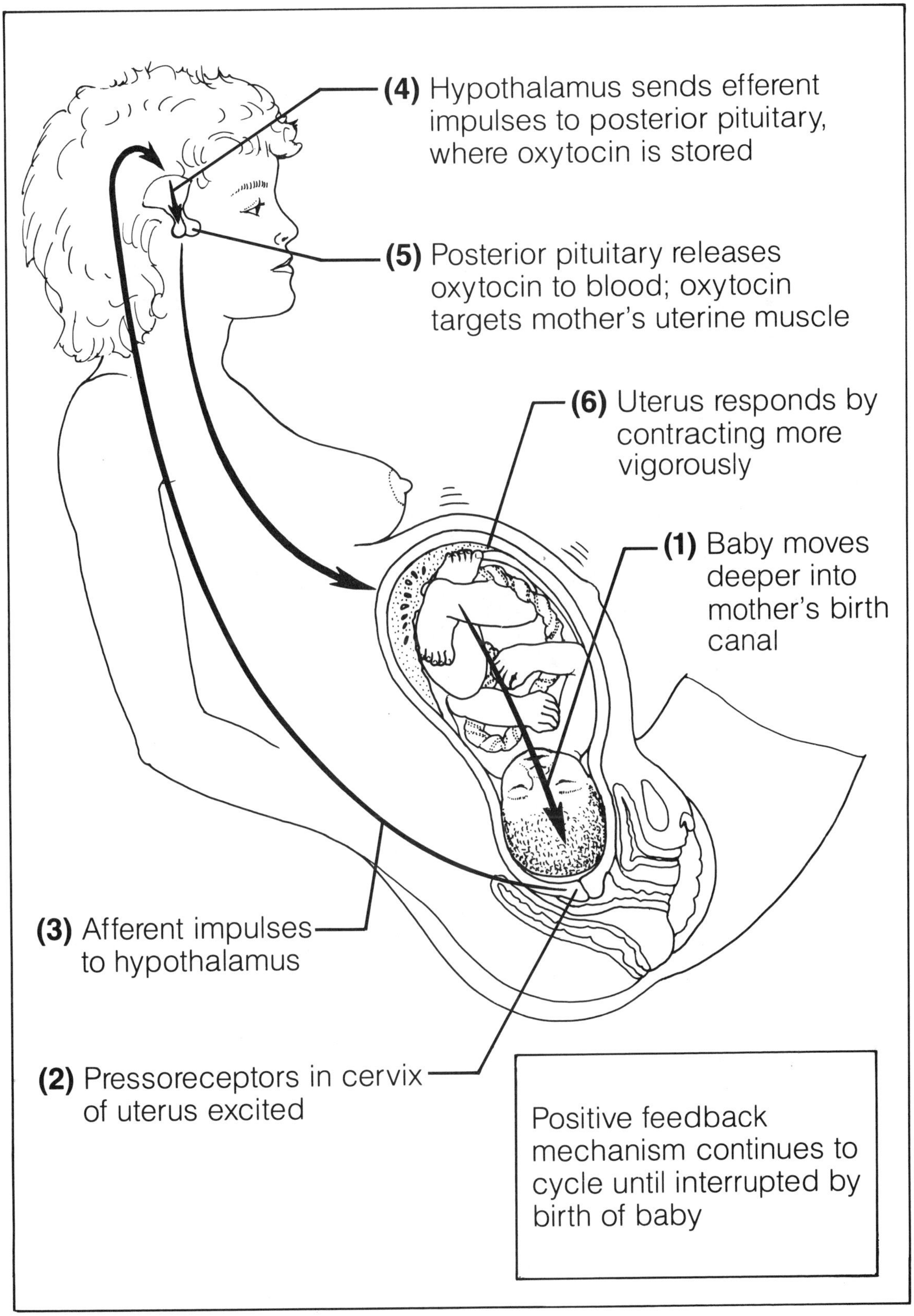

Positive feedback mechanism of the milk let-down reflex

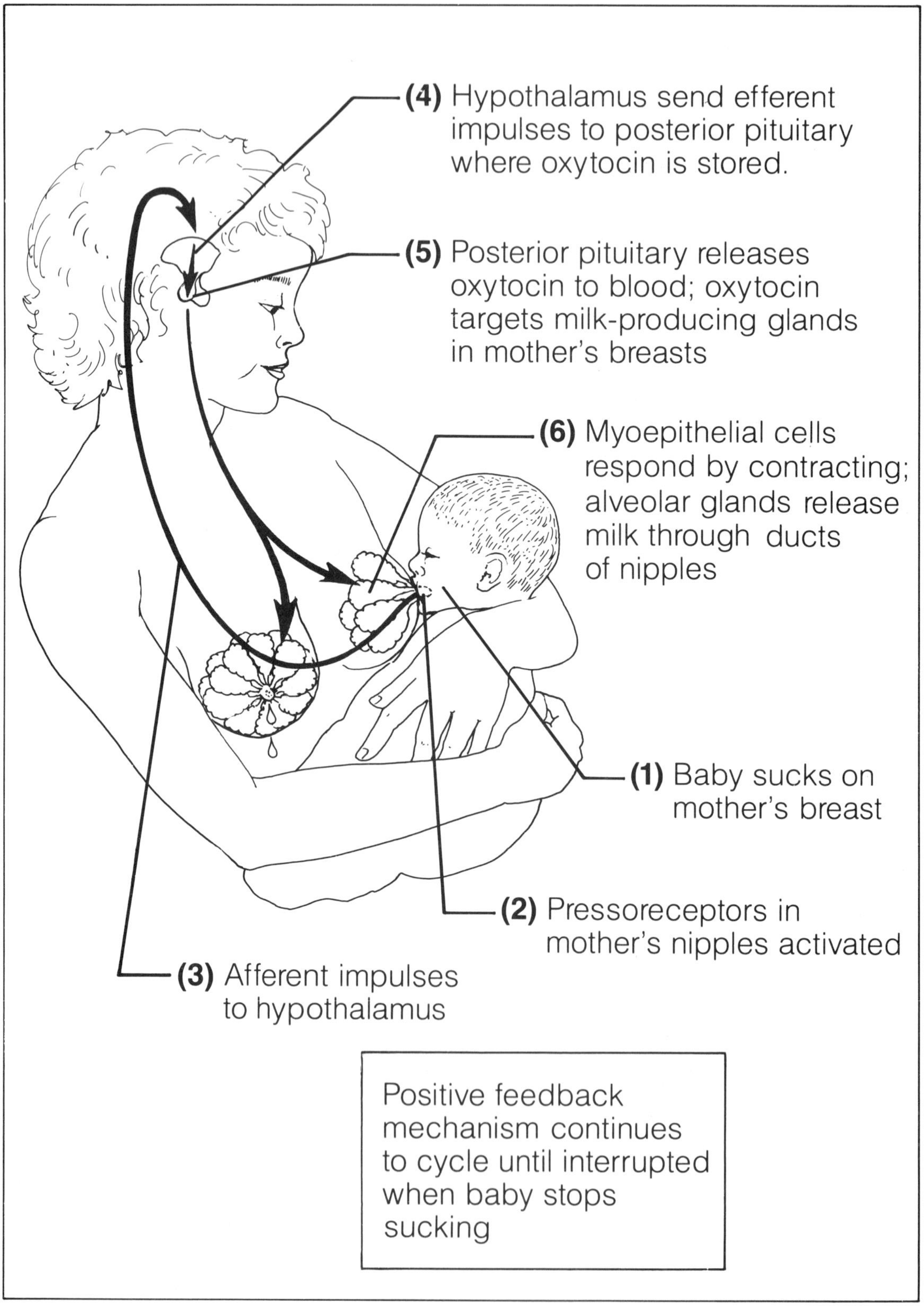

Fetal testing

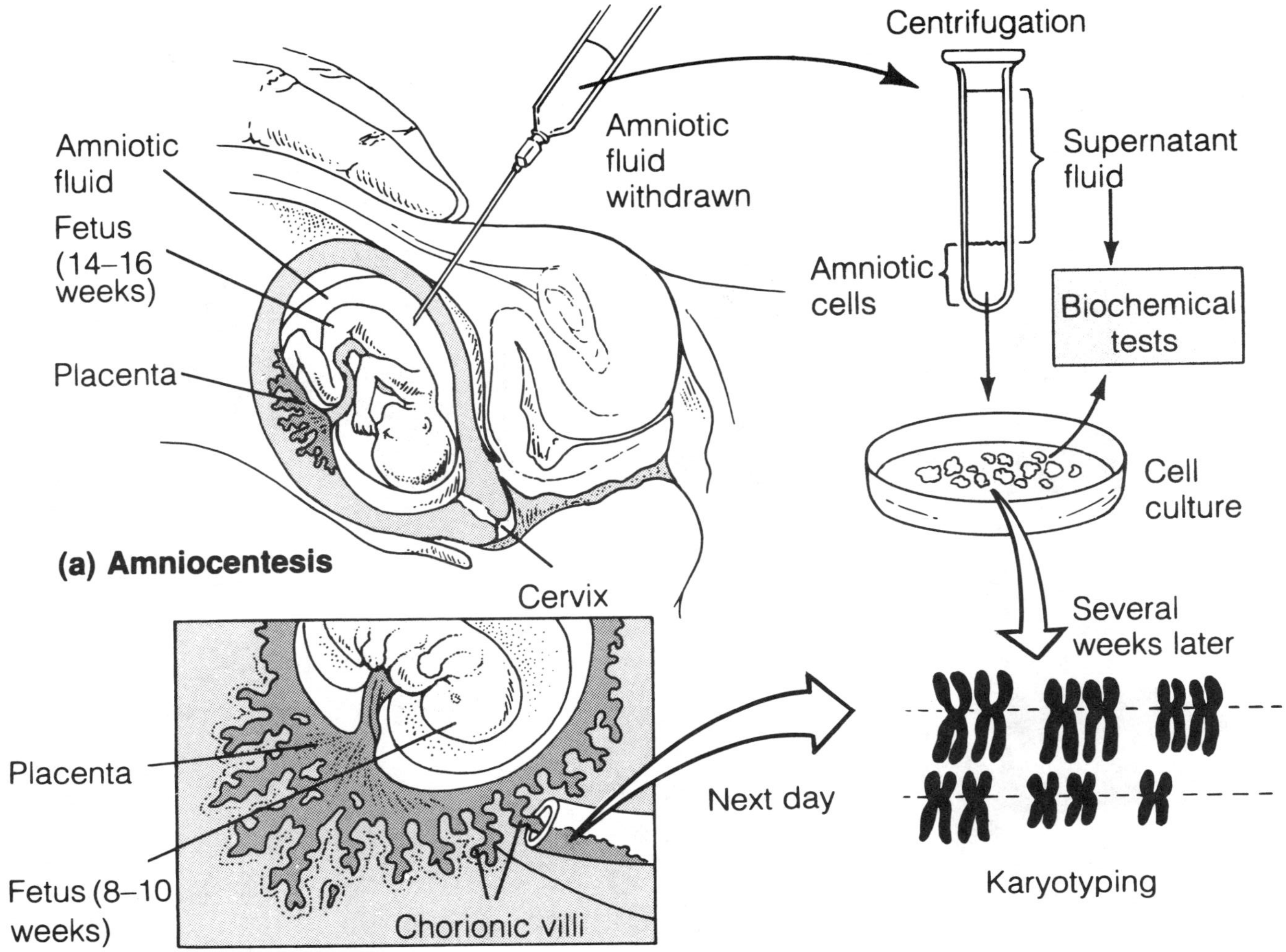

(a) **Amniocentesis**

(b) **Chorionic villi sampling**

Levels of structural complexity

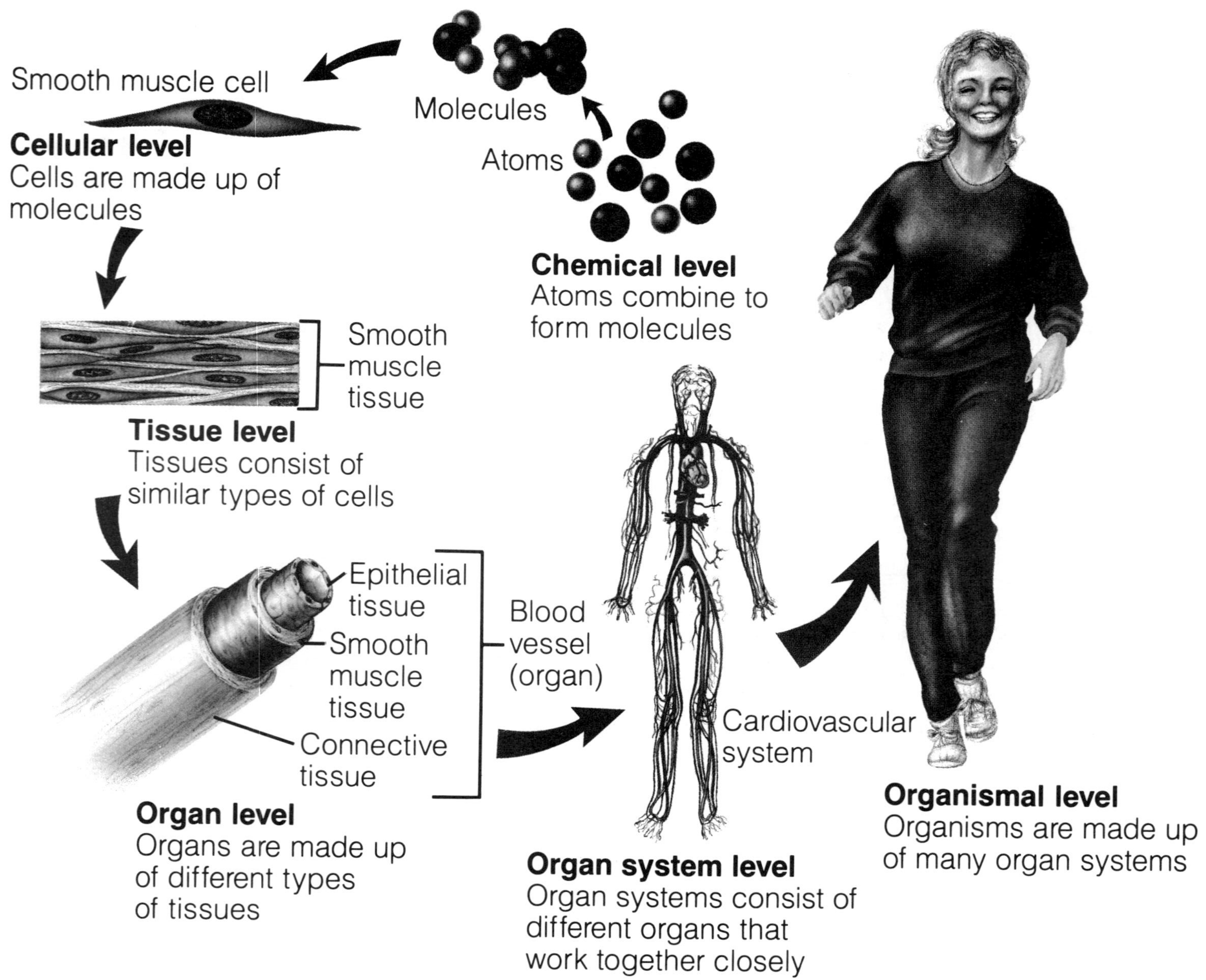

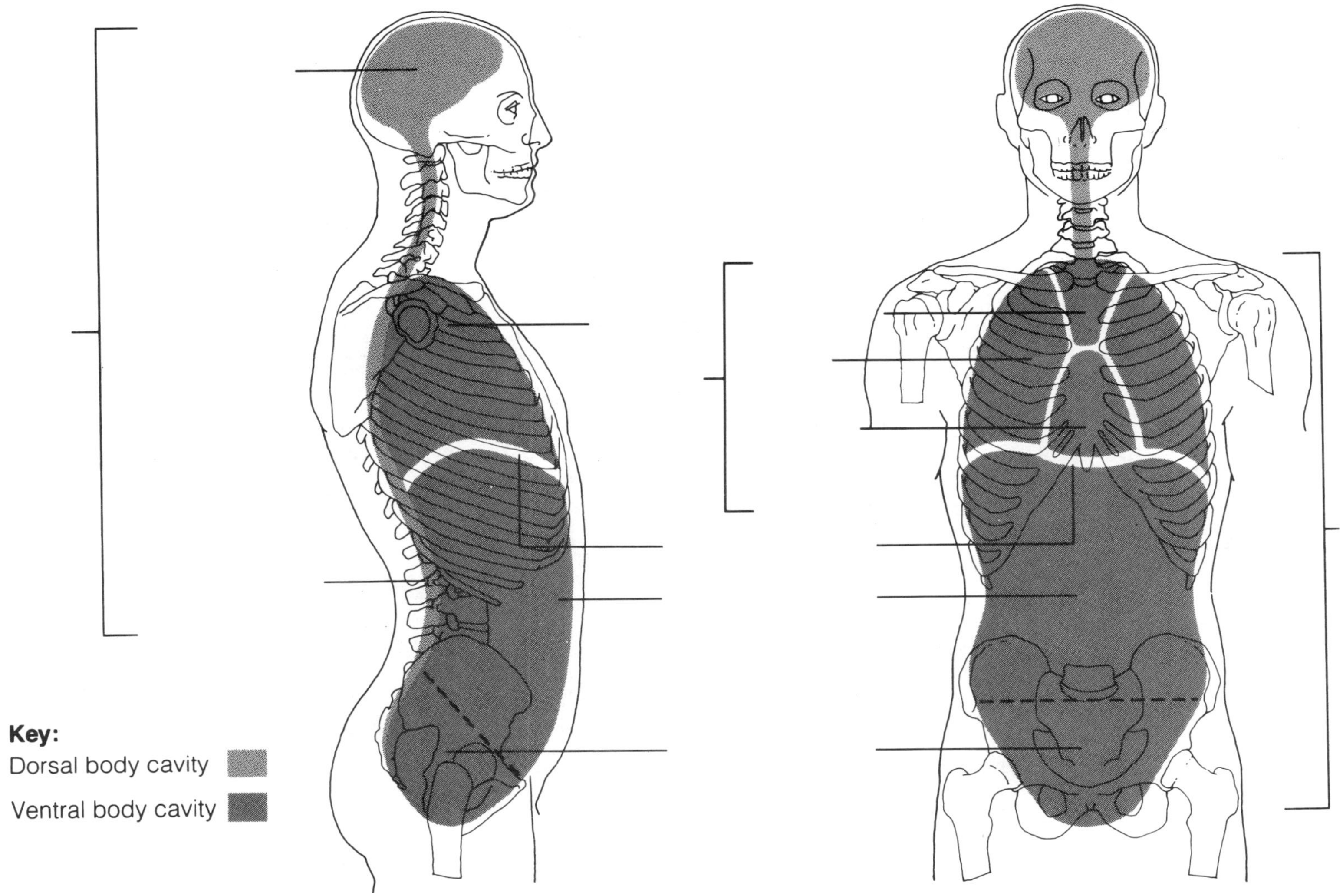
Key:
Dorsal body cavity
Ventral body cavity

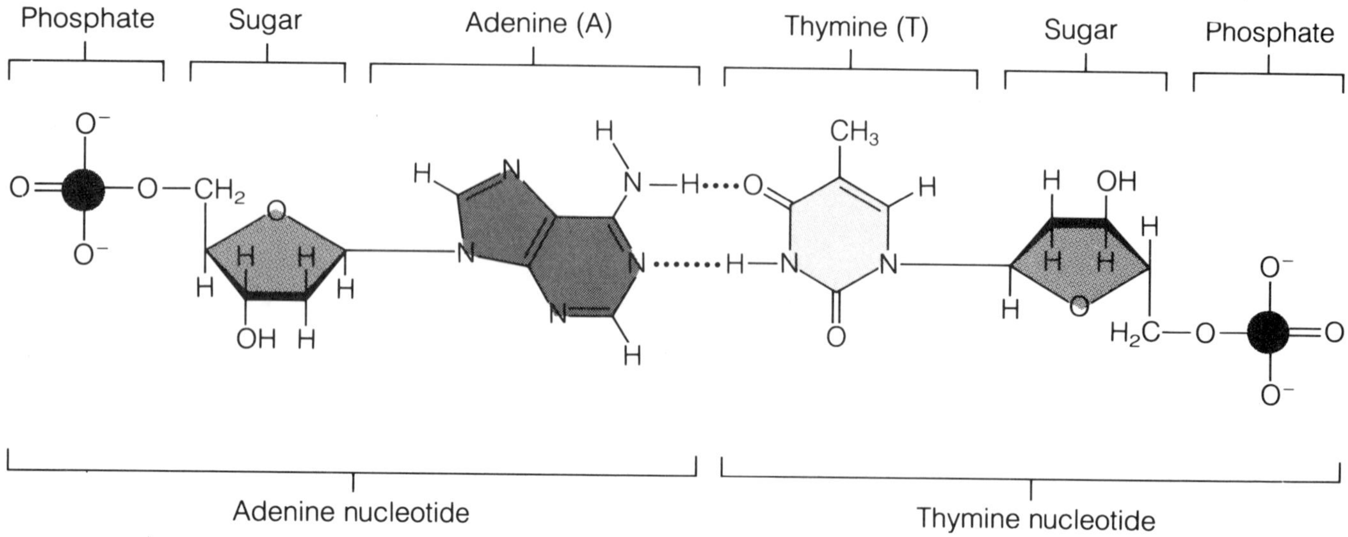
Phosphate
Sugar
Adenine (A)
Thymine (T)
Sugar
Phosphate
Adenine nucleotide
Thymine nucleotide

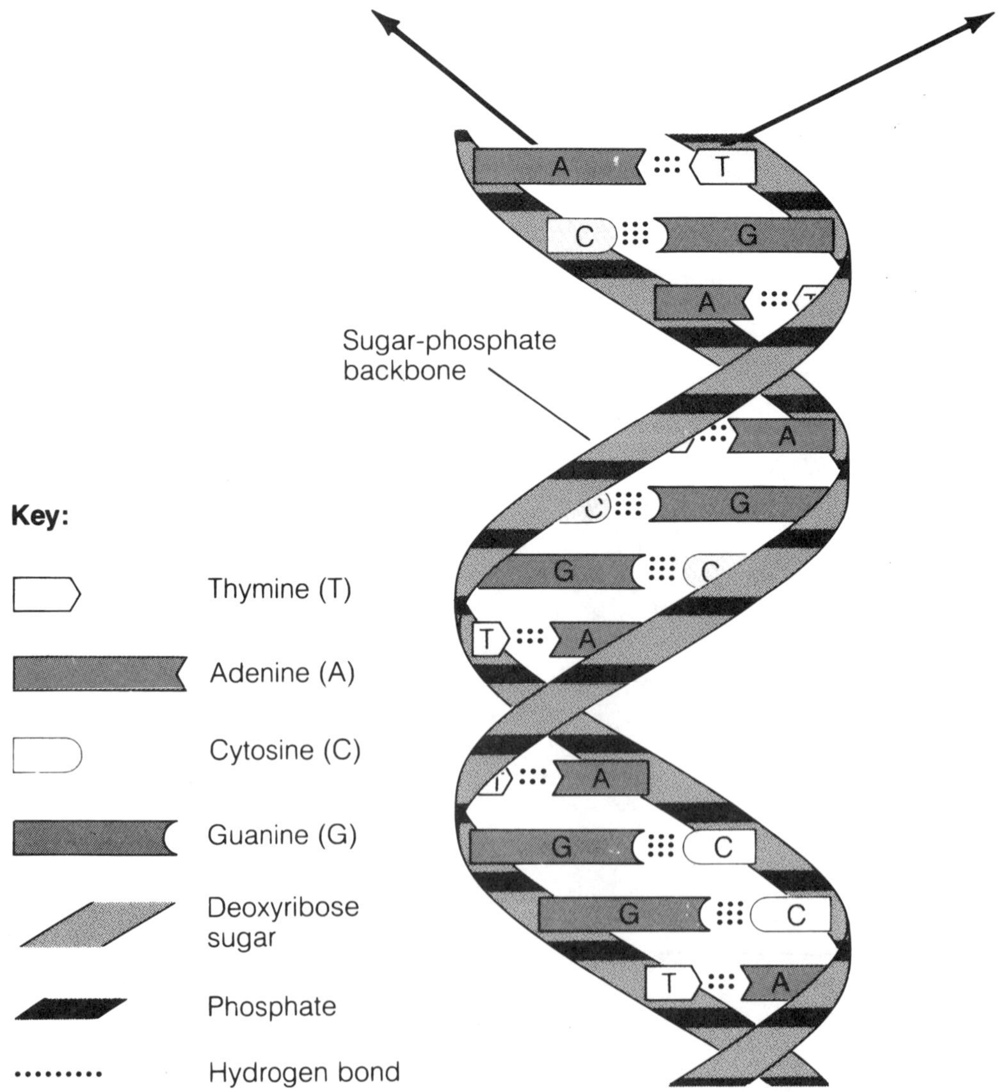
Sugar-phosphate backbone
Key:
Thymine (T)
Adenine (A)
Cytosine (C)
Guanine (G)
Deoxyribose sugar
Phosphate
Hydrogen bond

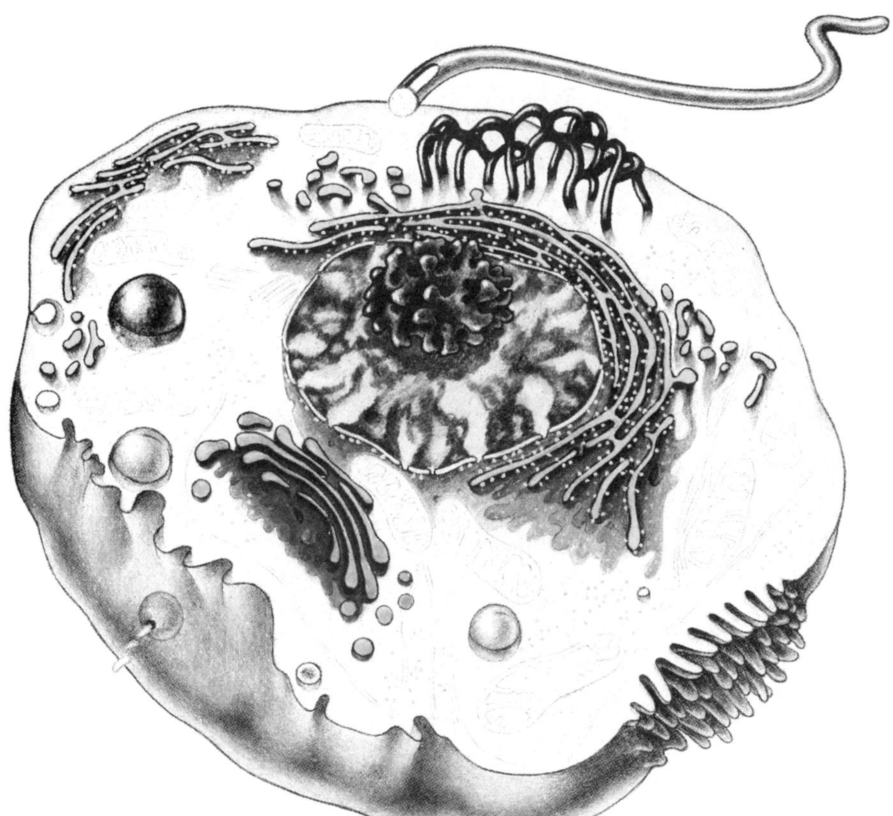

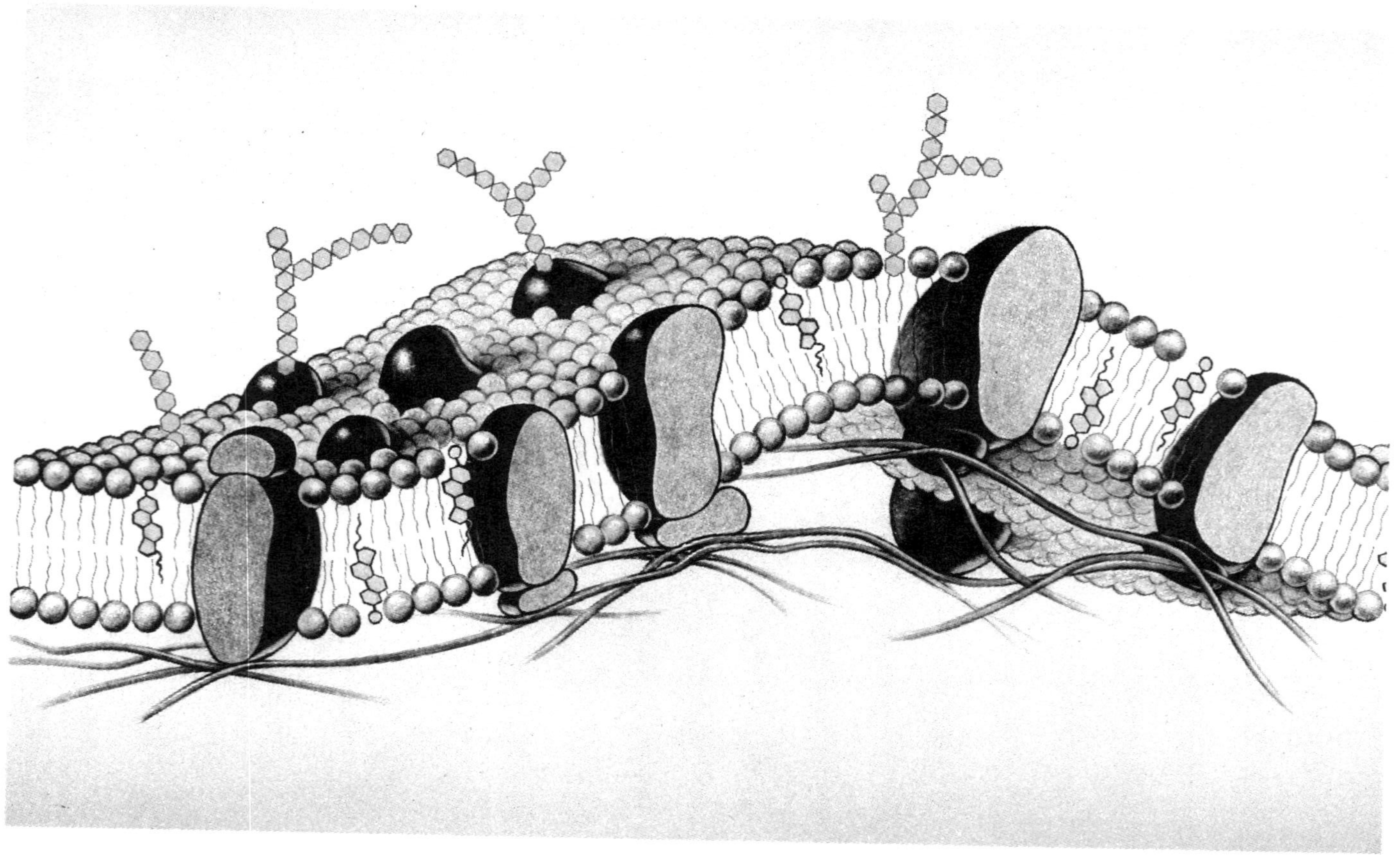

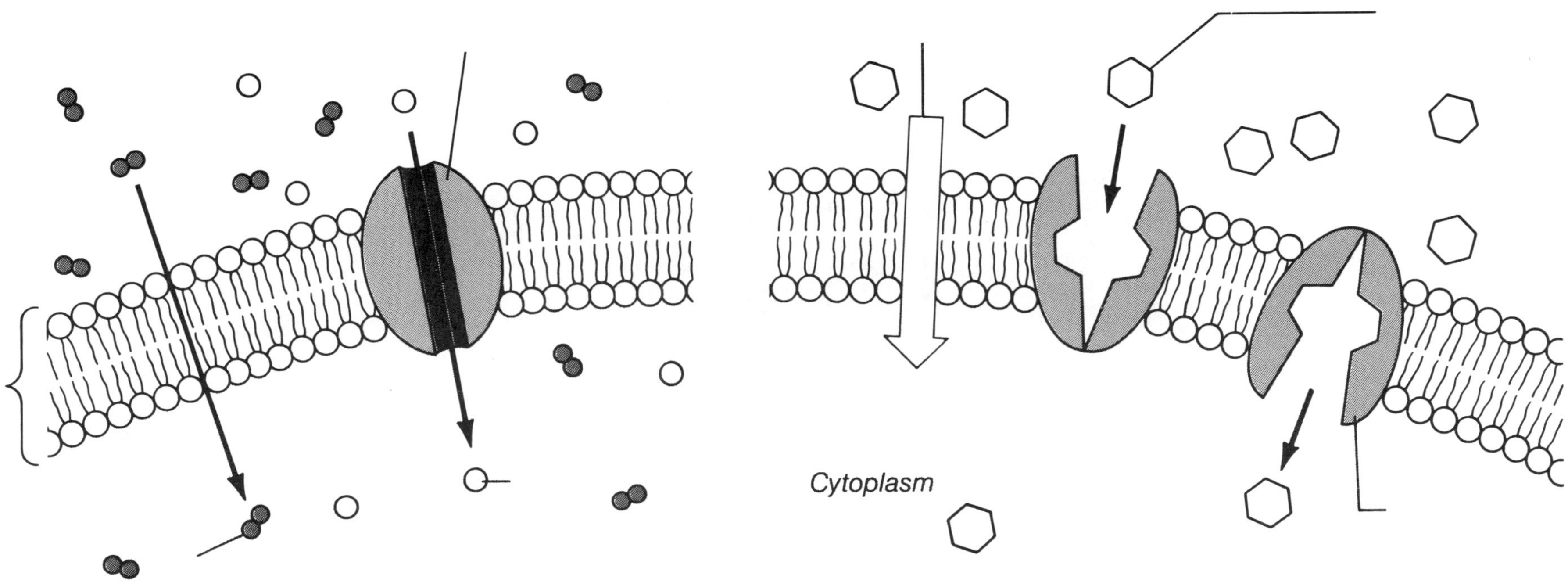
Cytoplasm

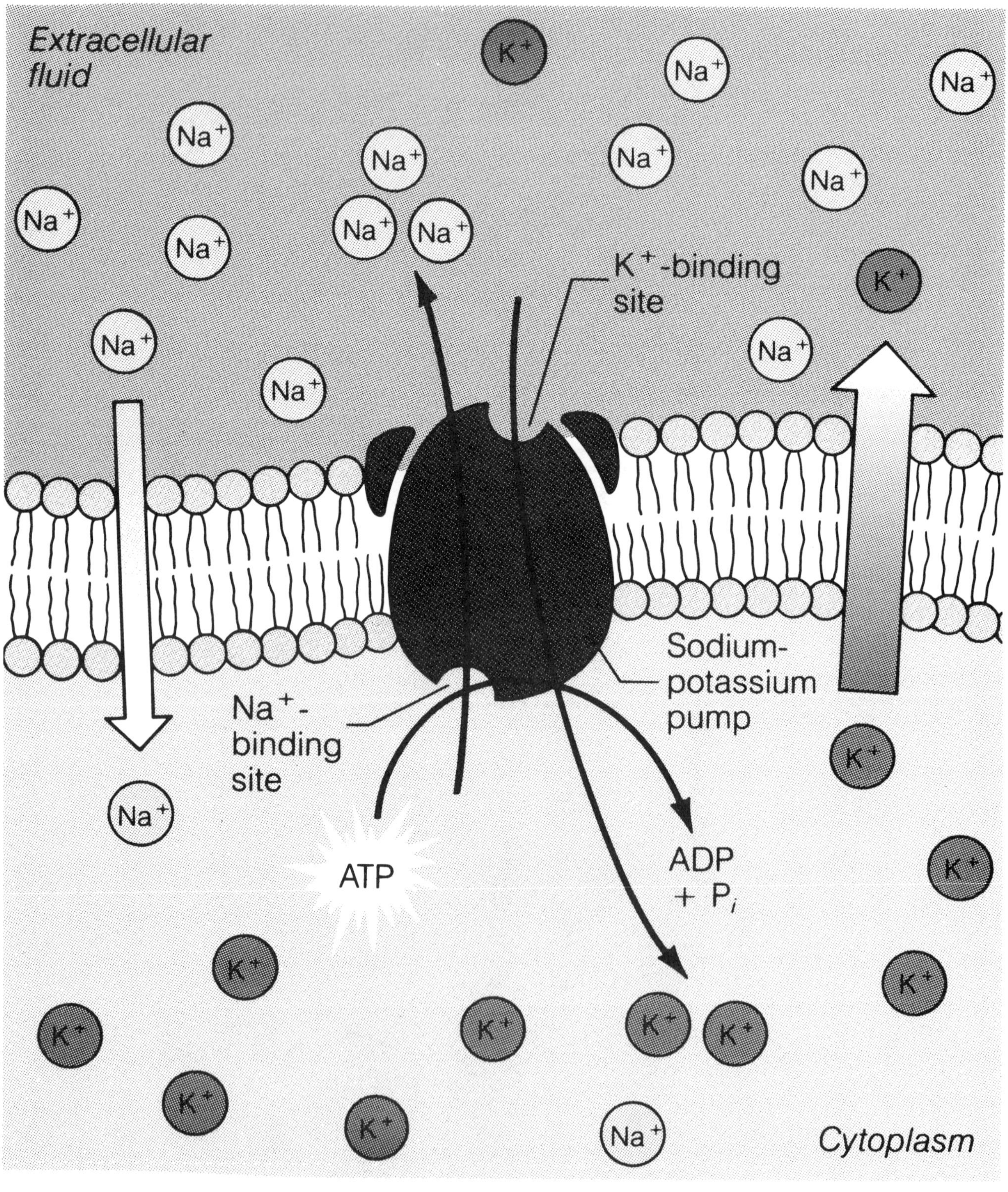
Extracellular fluid
K+-binding site
Sodium-potassium pump
Na+-binding site
ATP
ADP + Pi
Cytoplasm

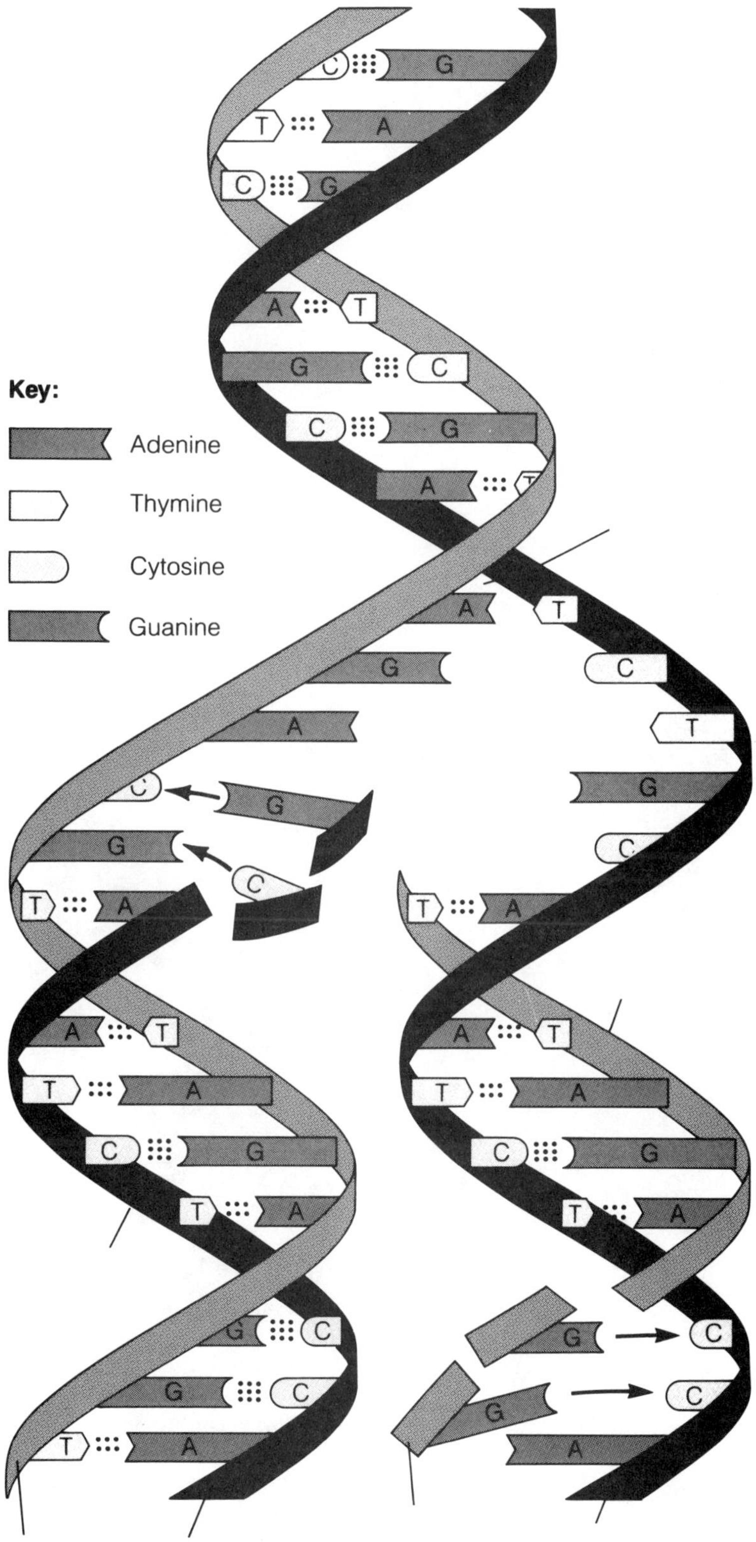
Key:
Adenine
Thymine
Cytosine
Guanine

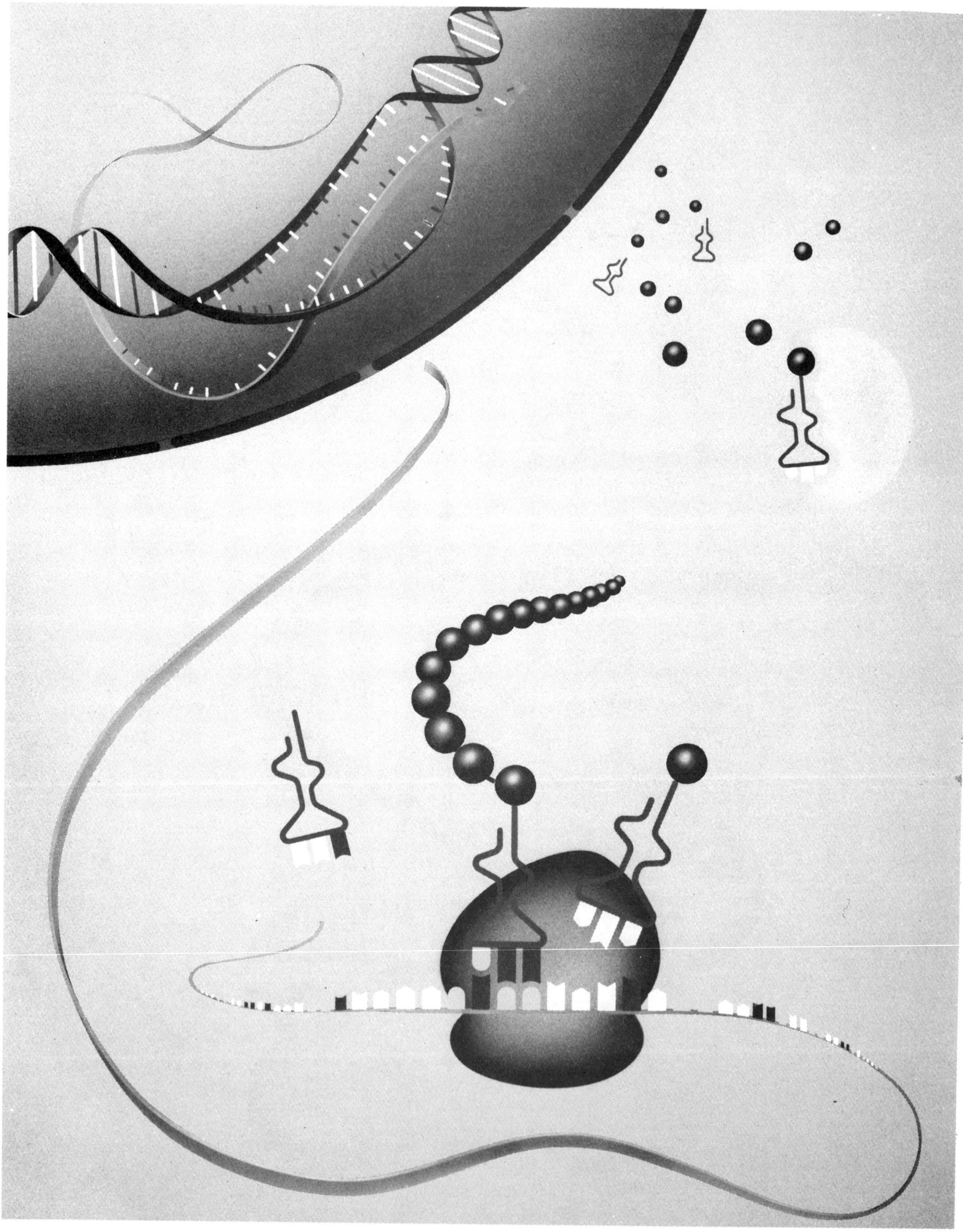

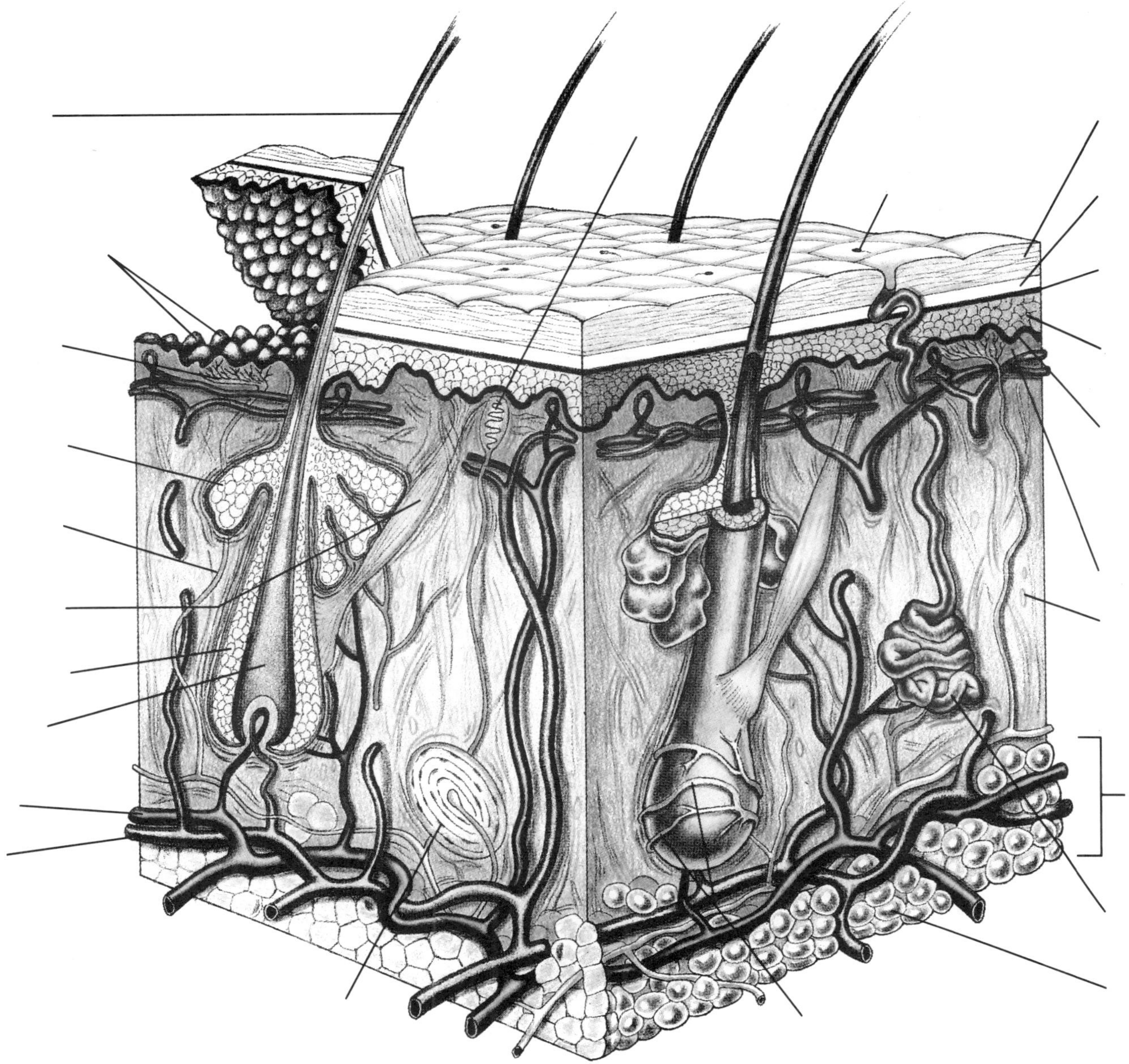

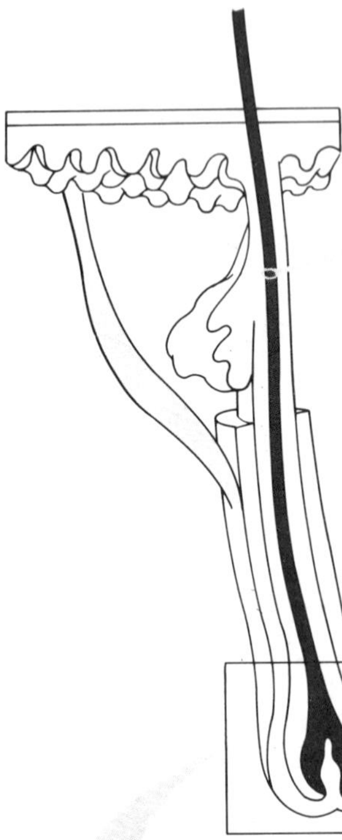

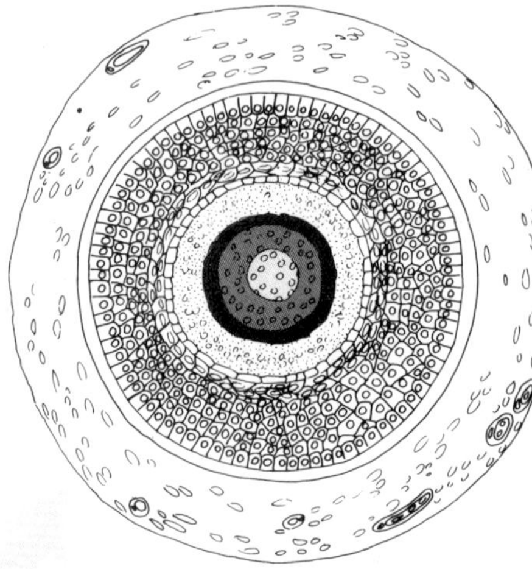

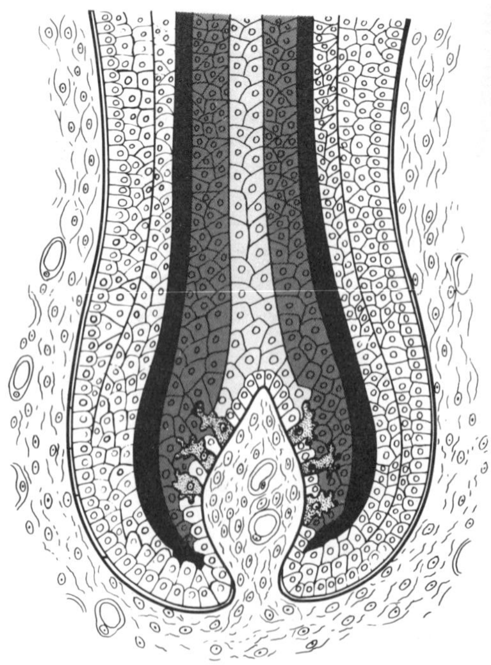

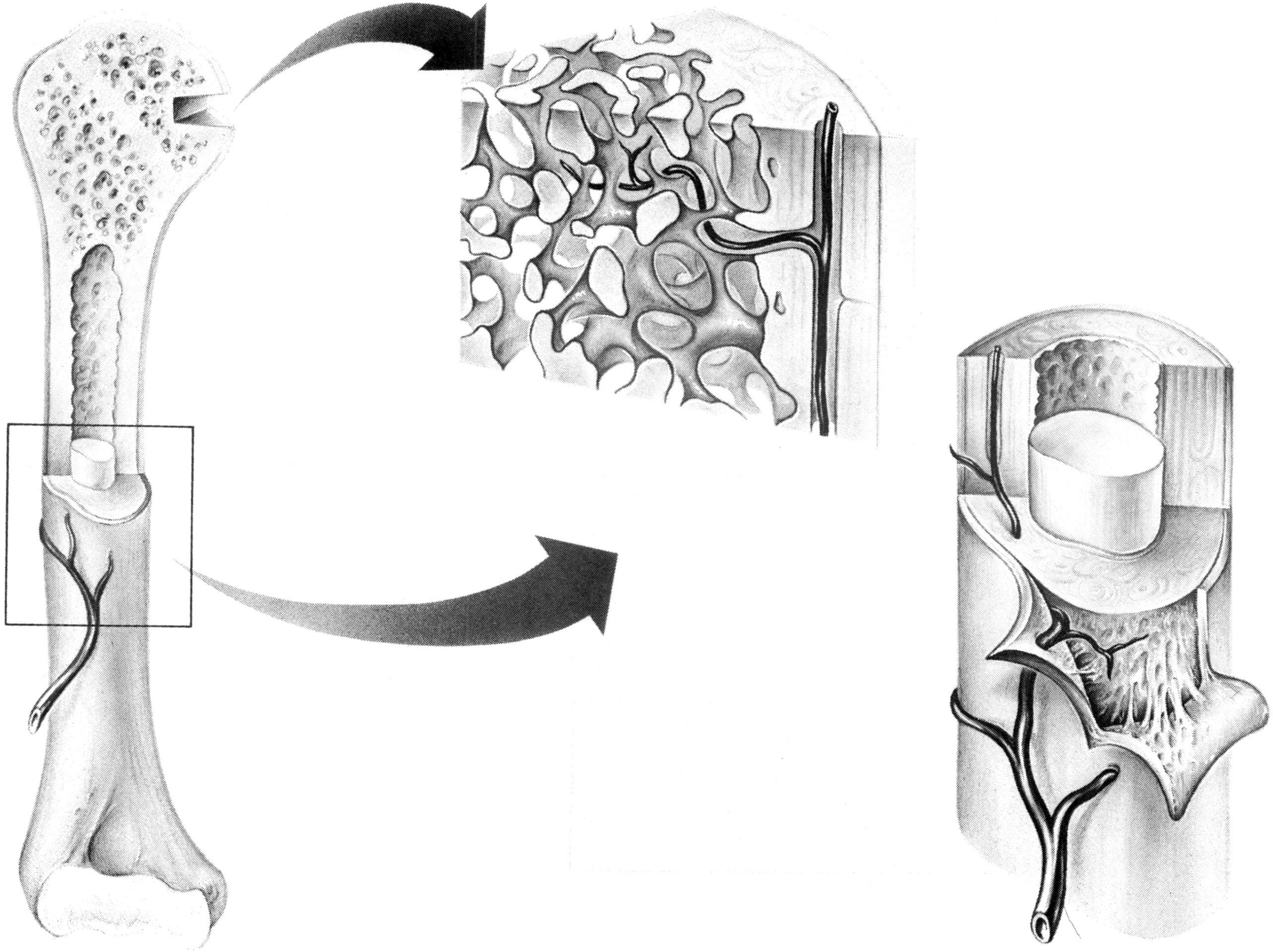

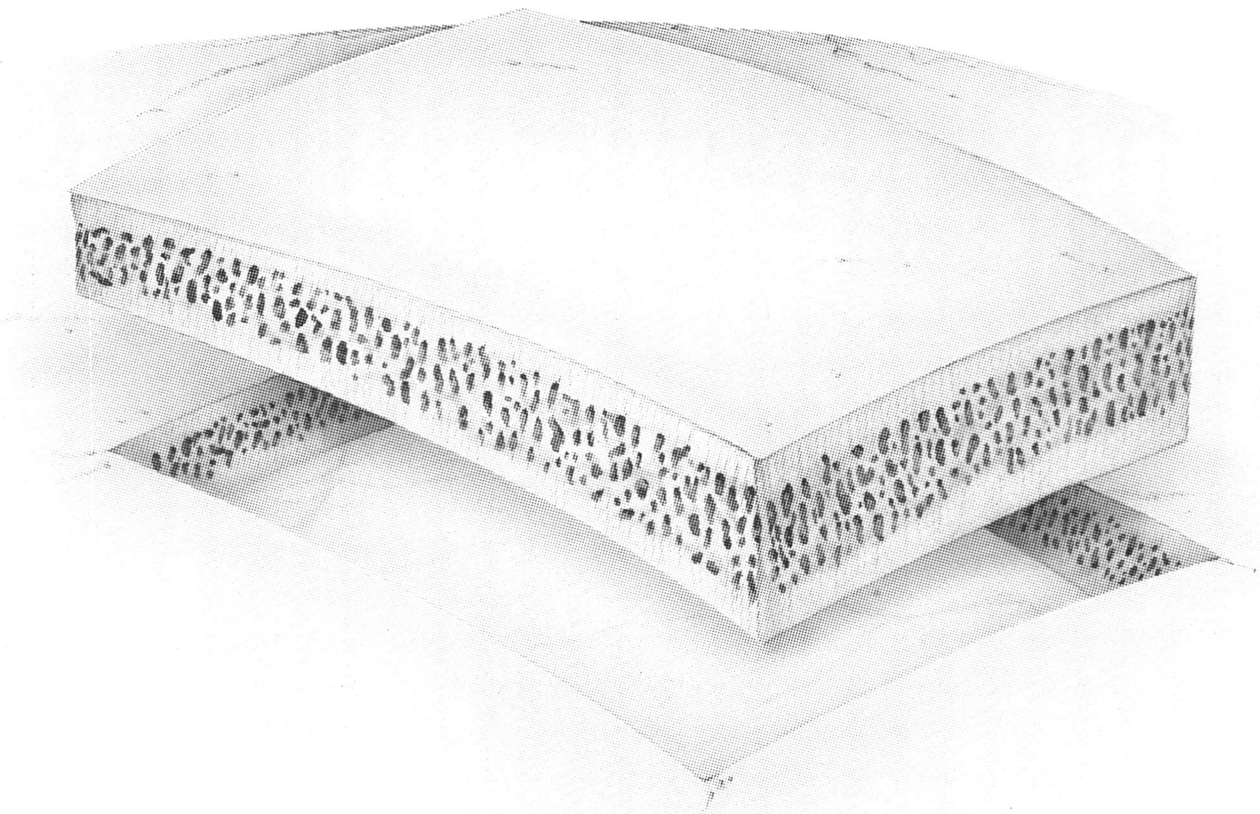

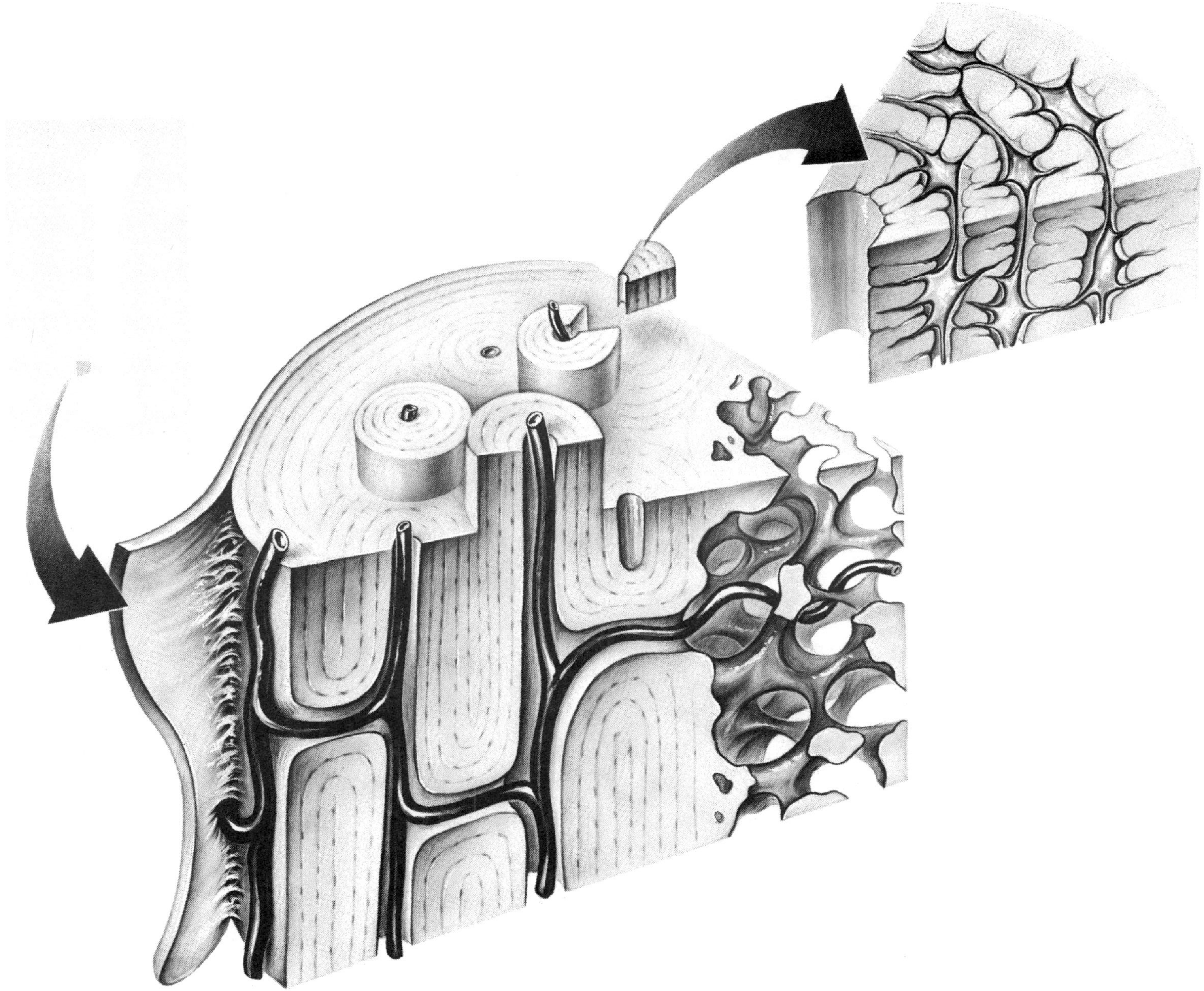

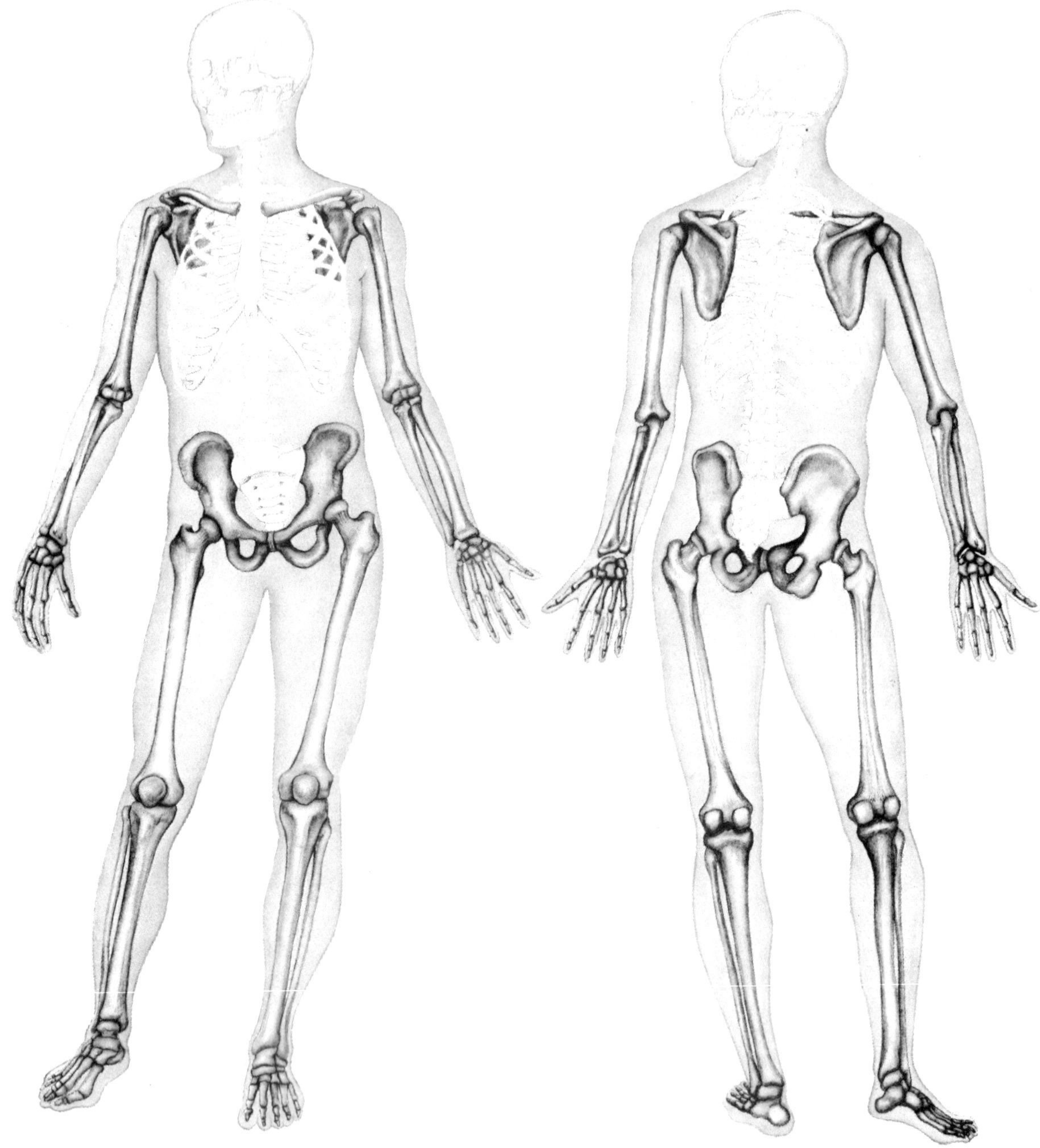

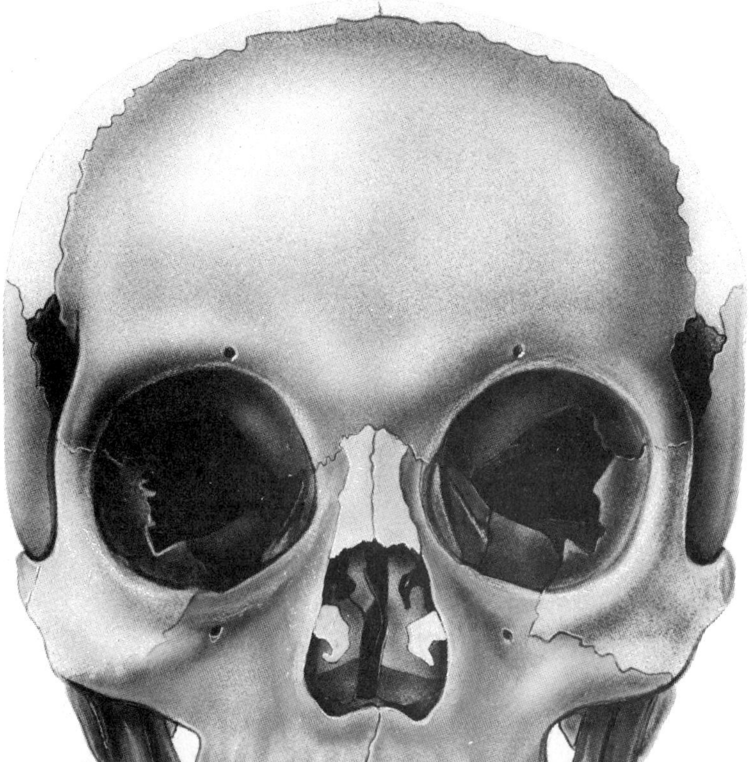

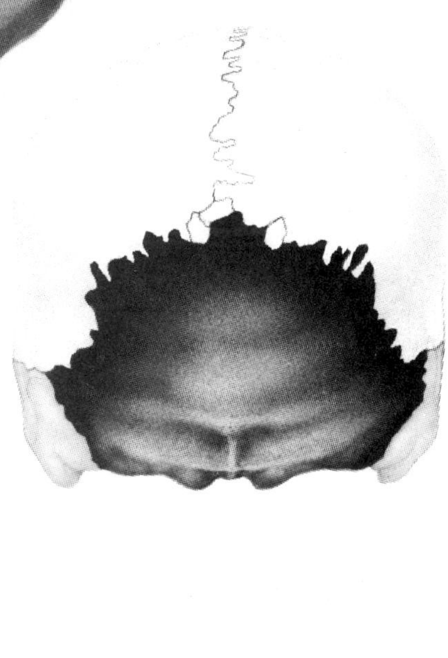

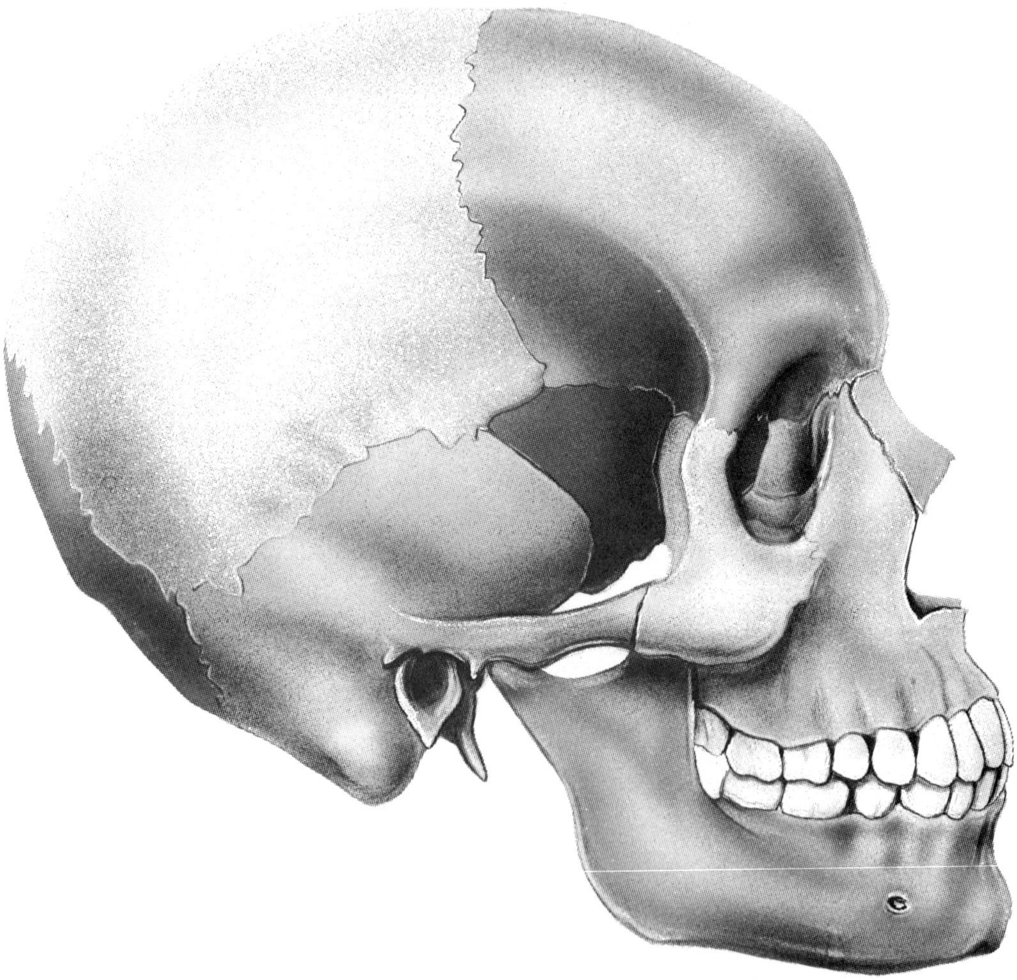

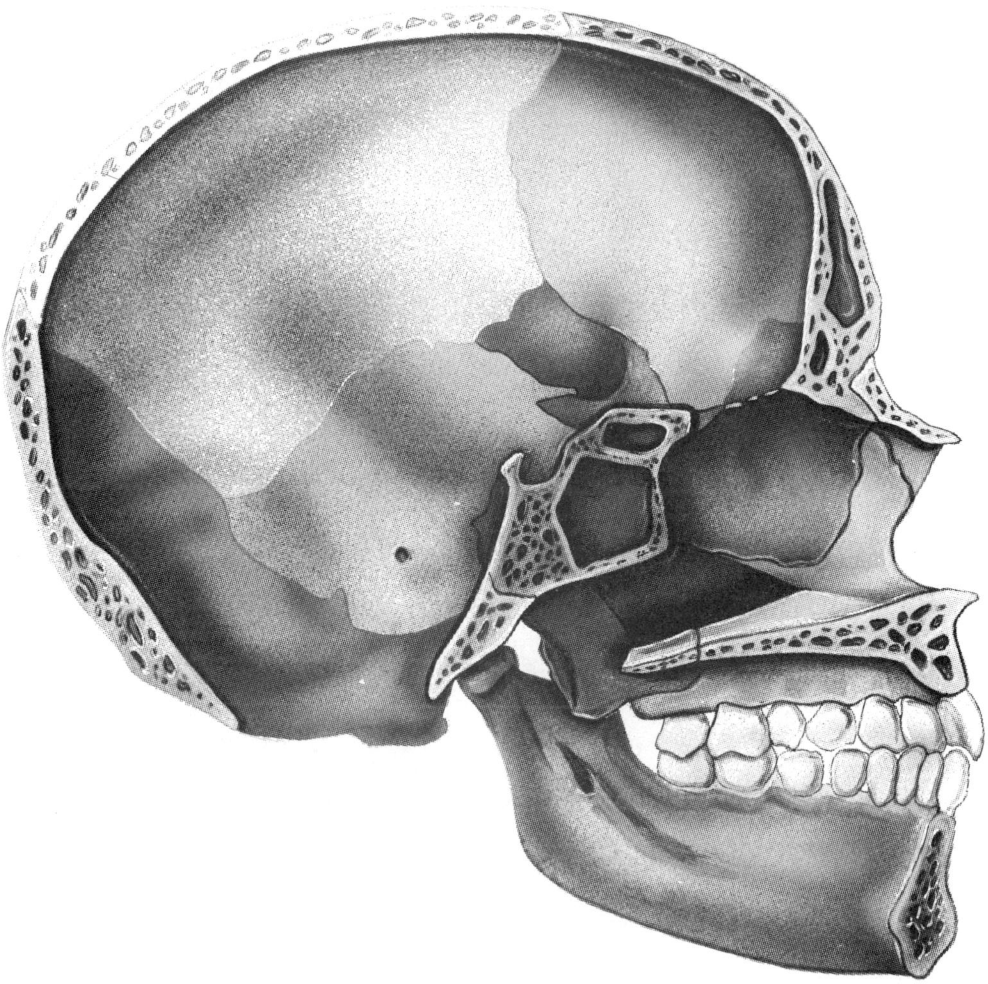

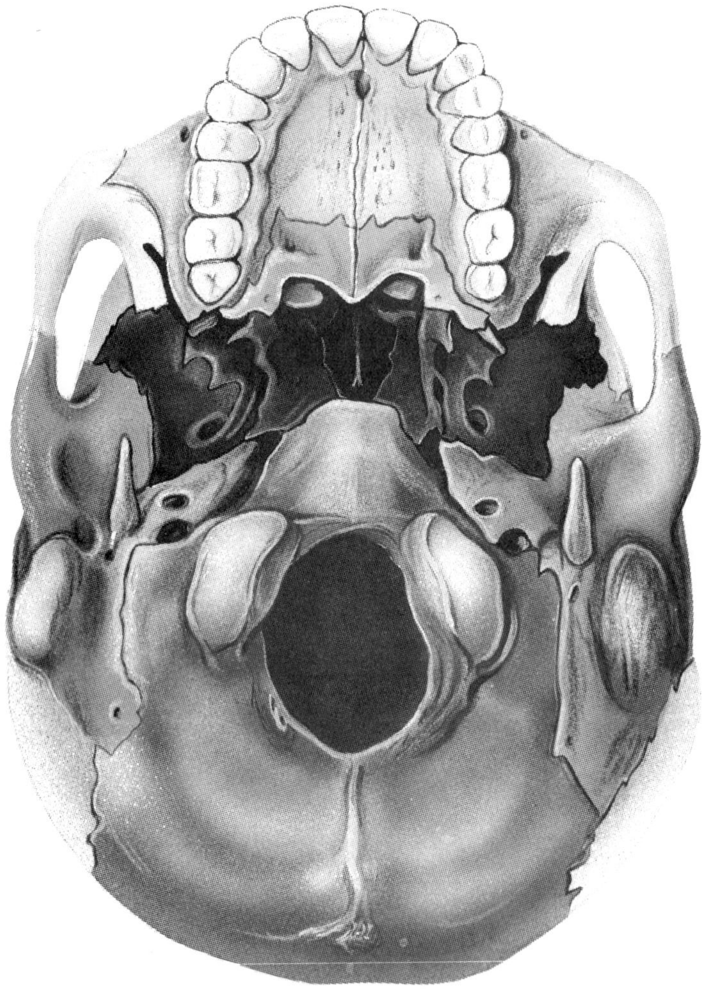

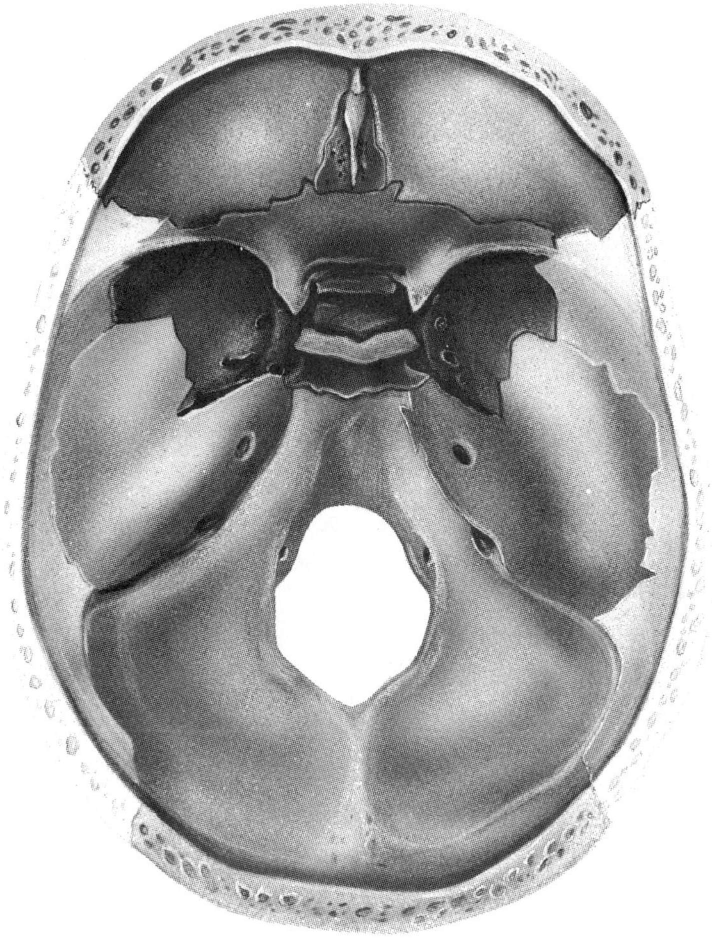

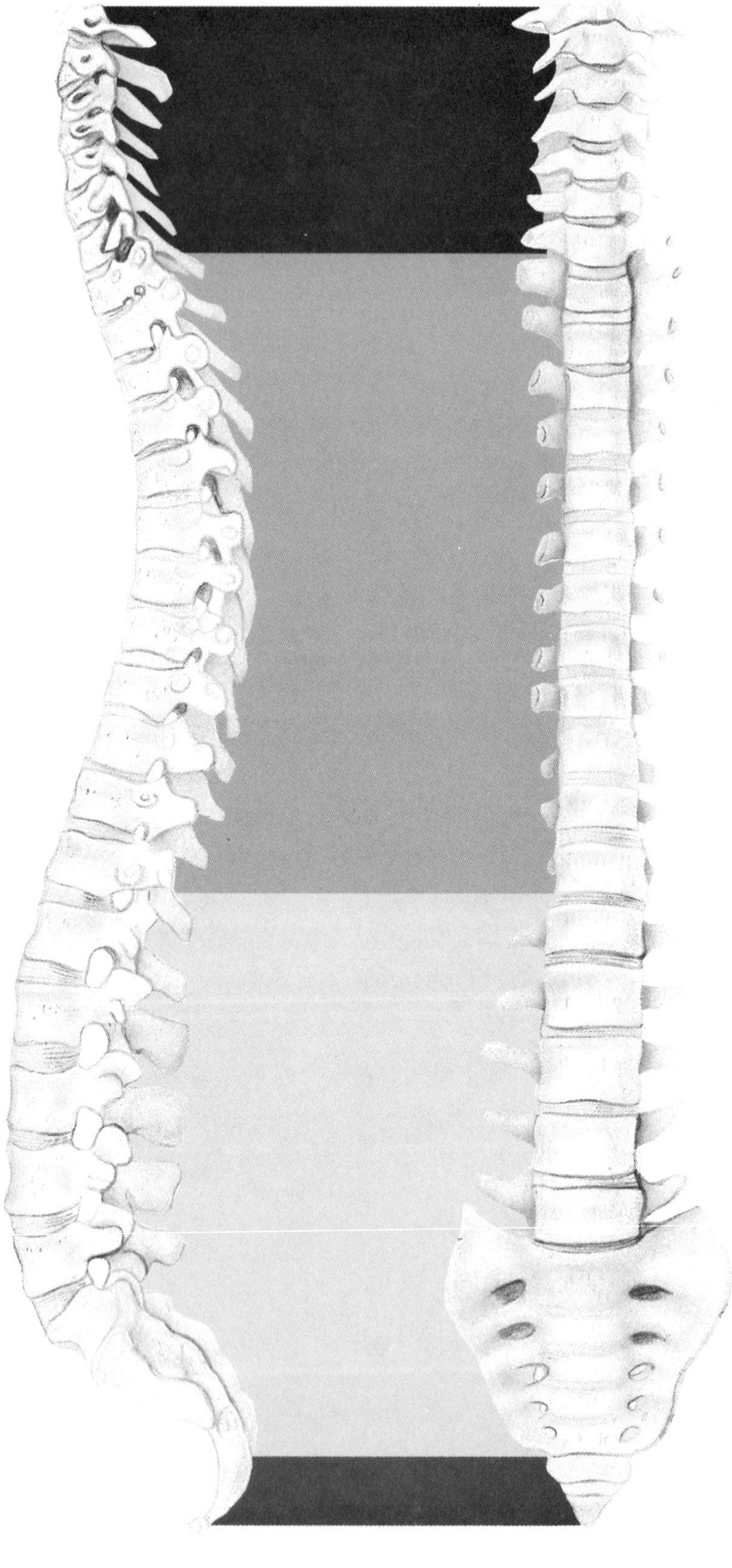

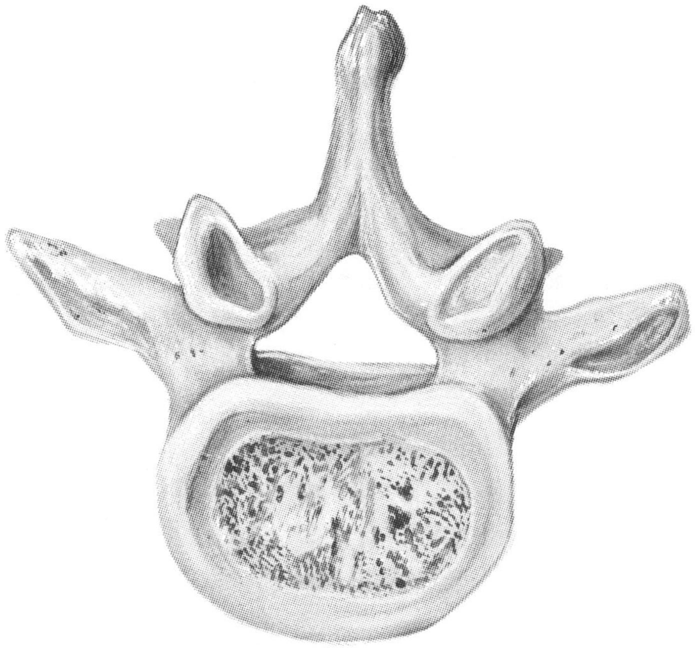

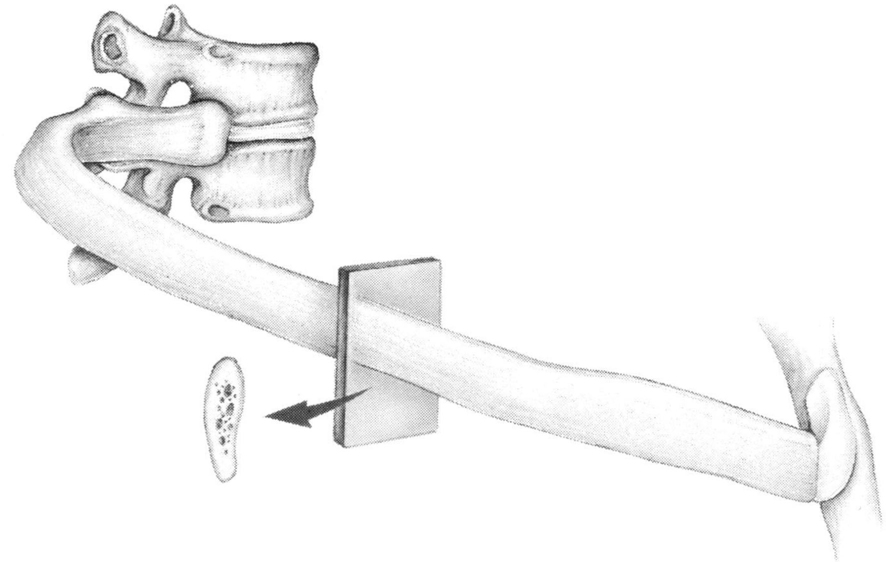

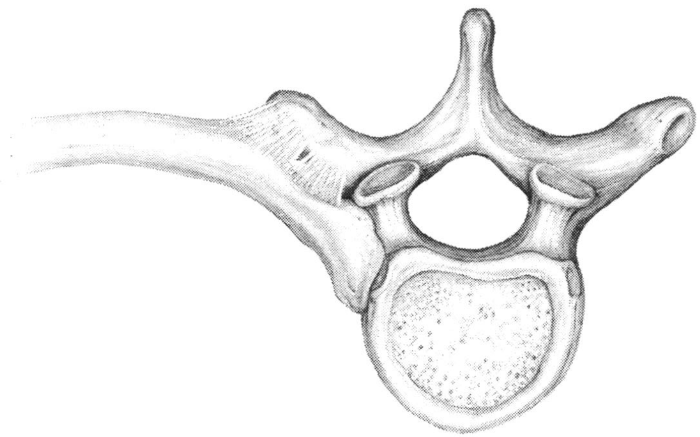

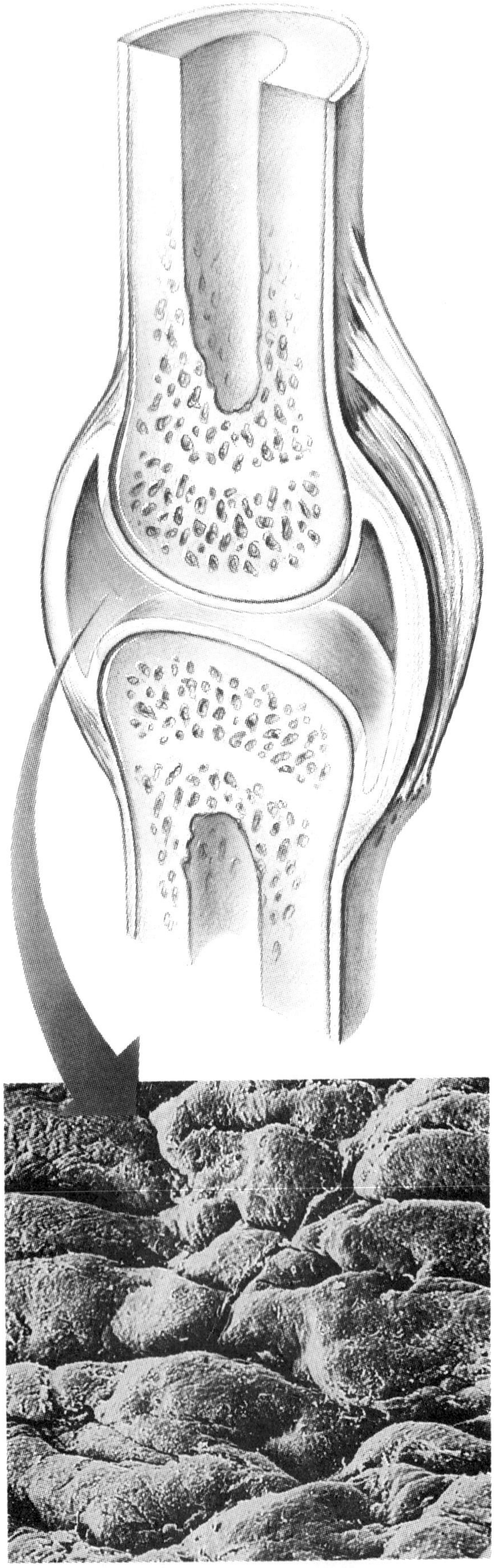

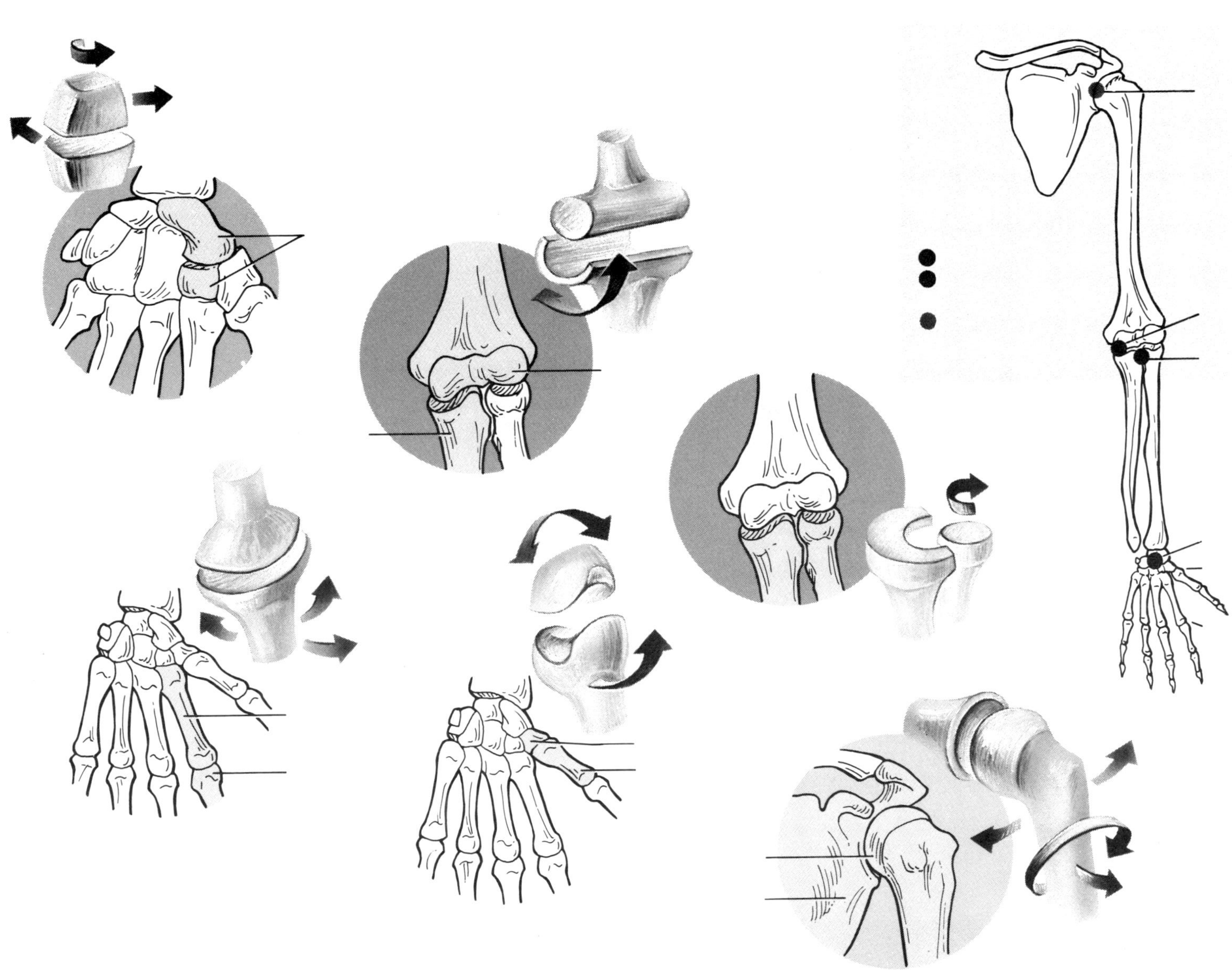

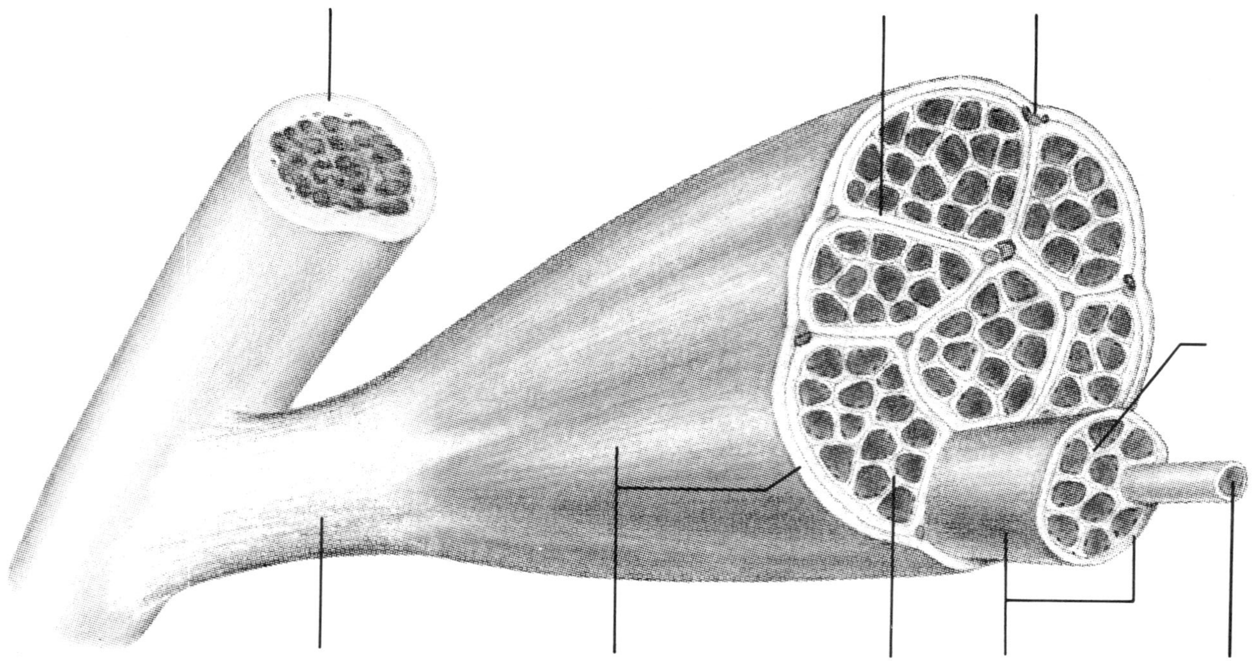

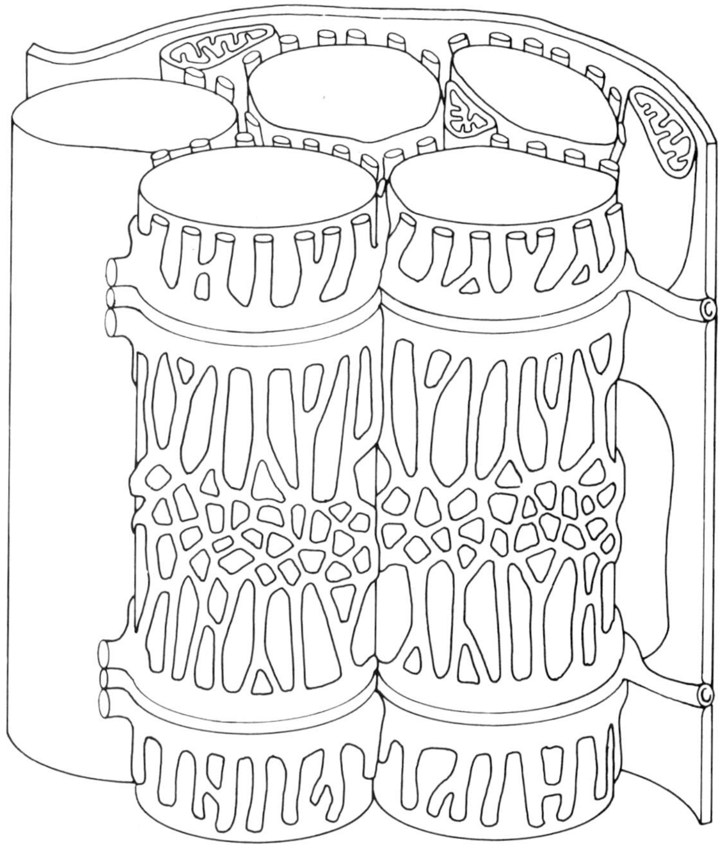

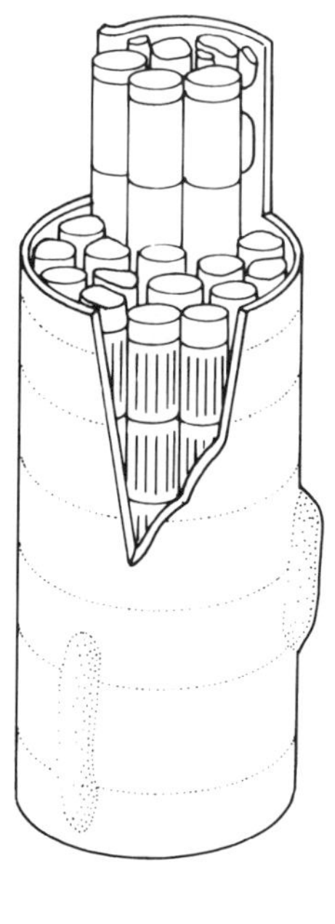

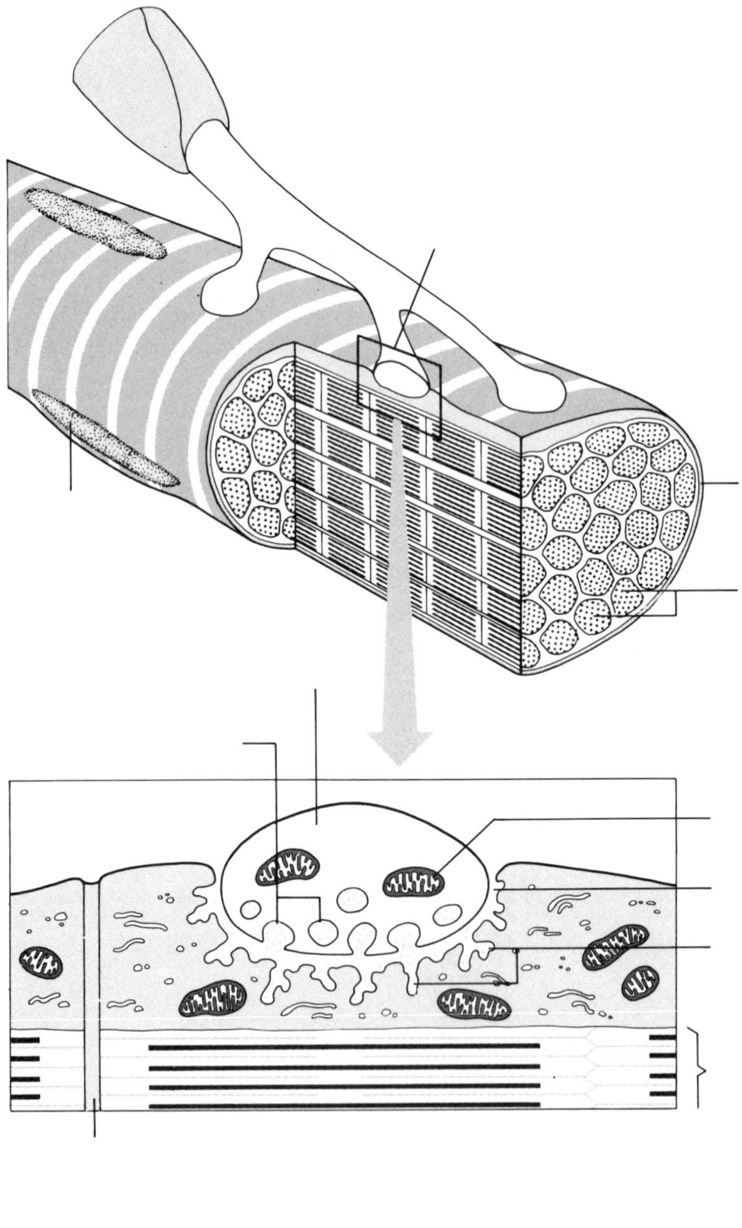

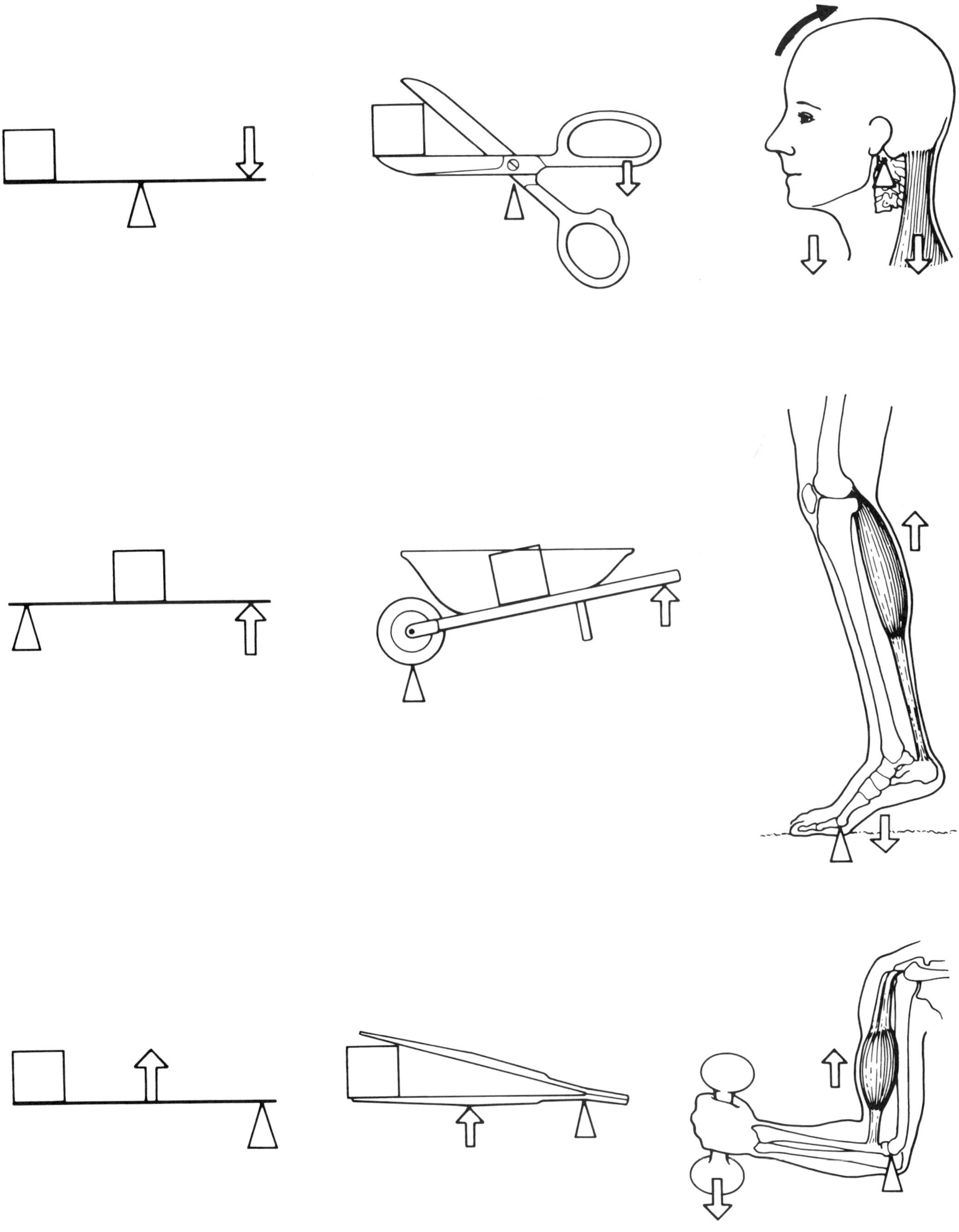

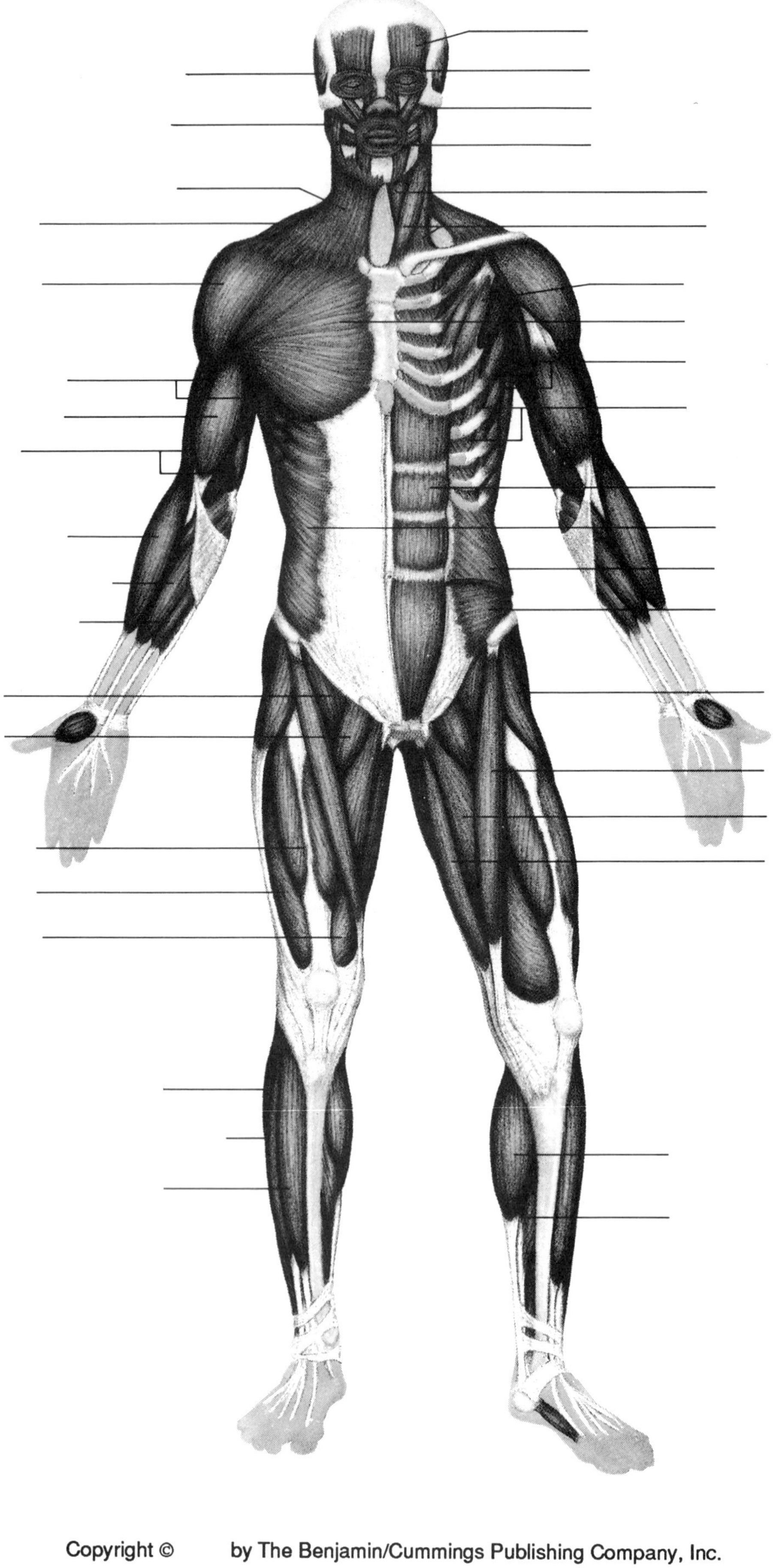

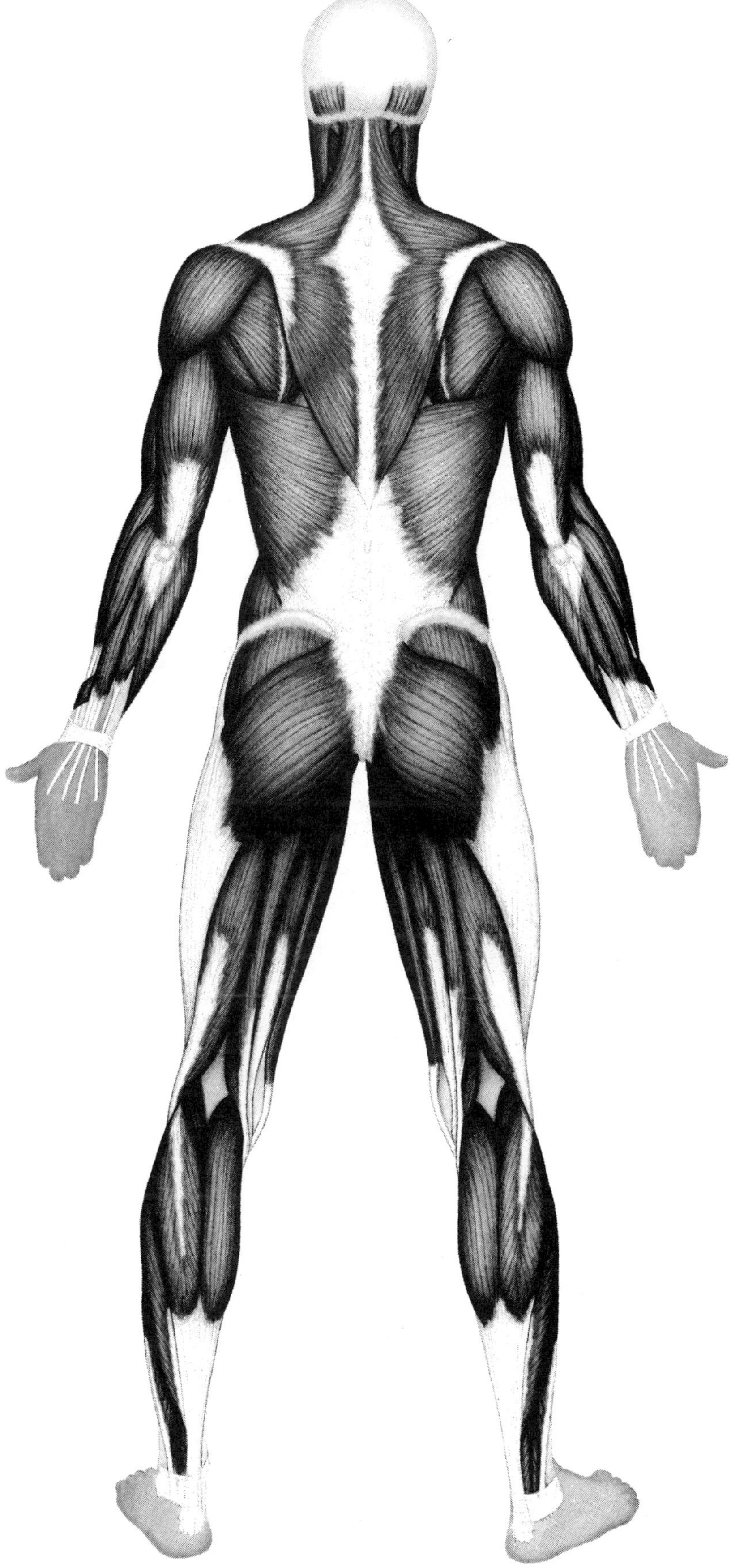

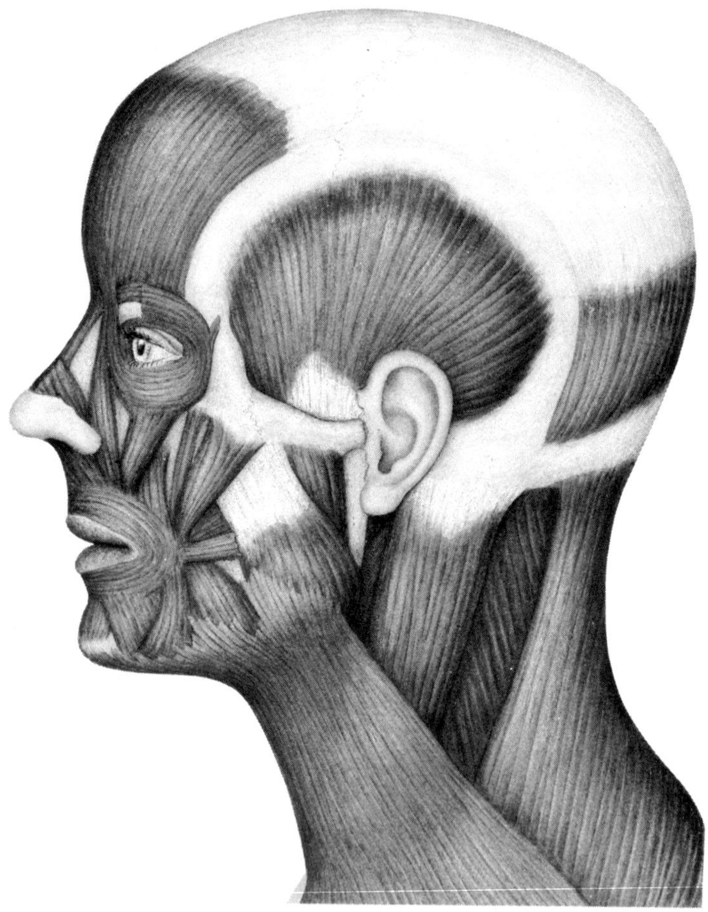

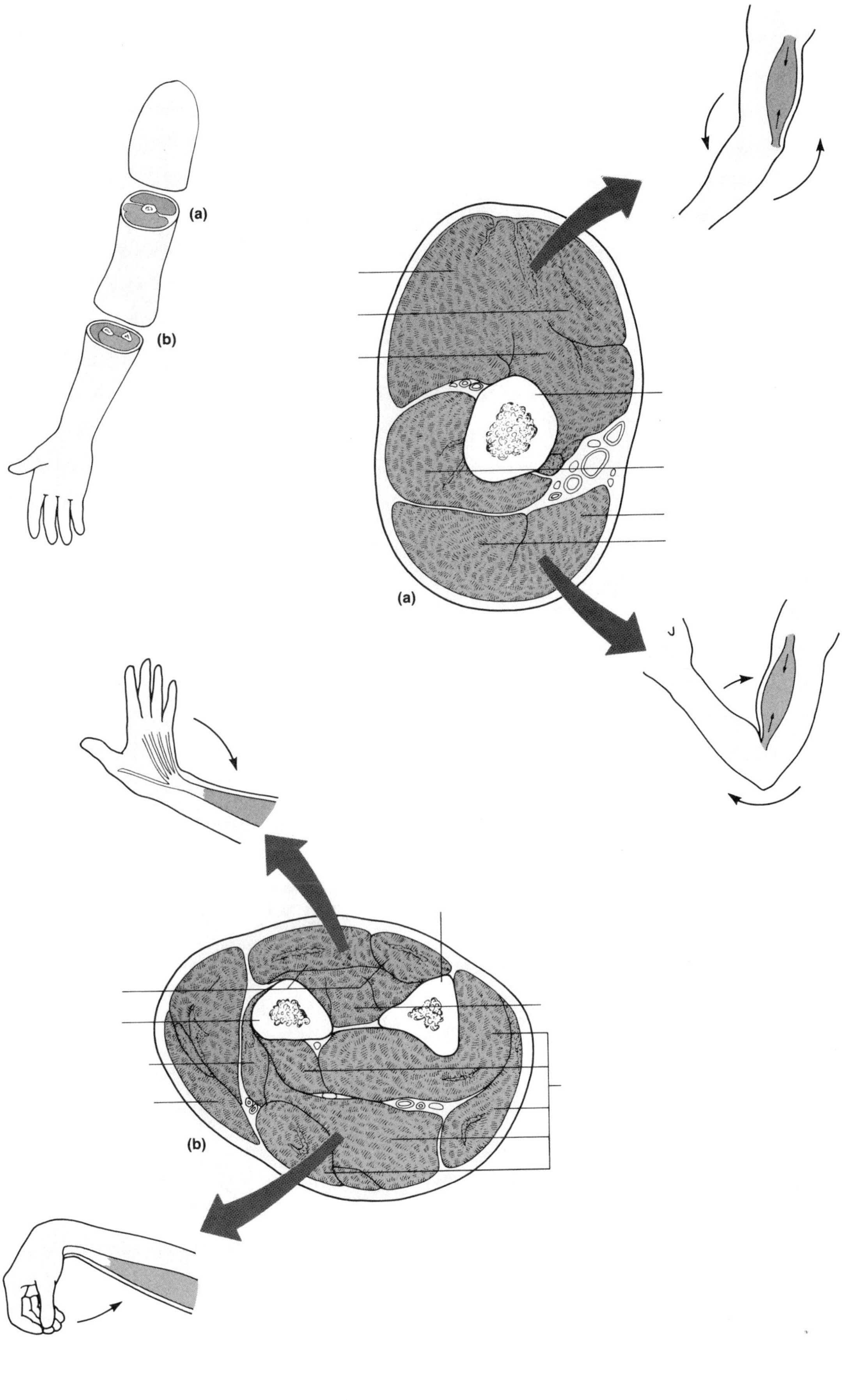

(a)
(b)
(a)
(b)

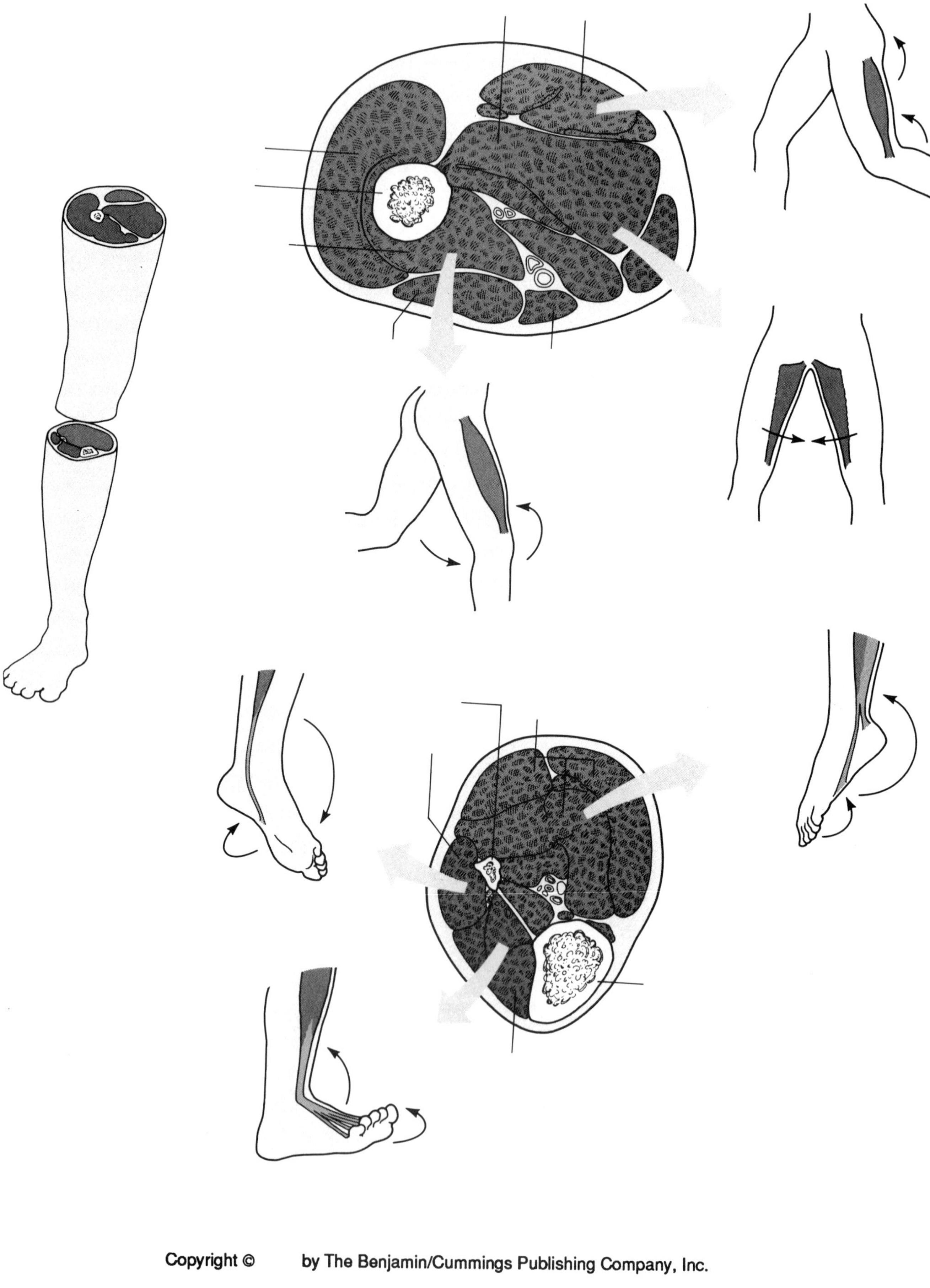

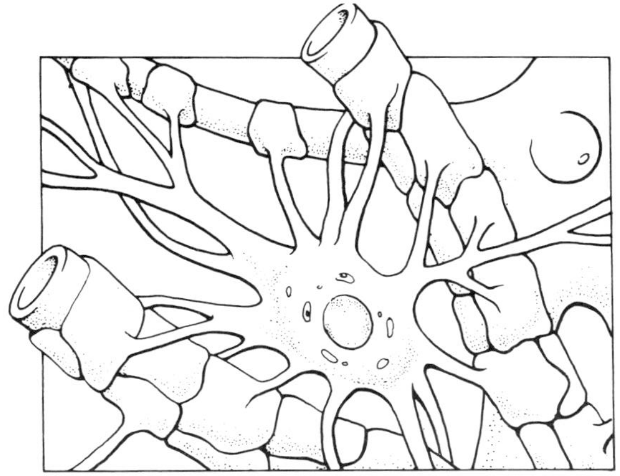

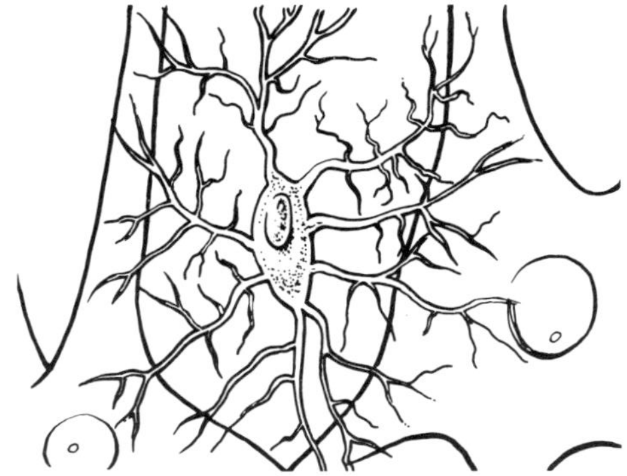

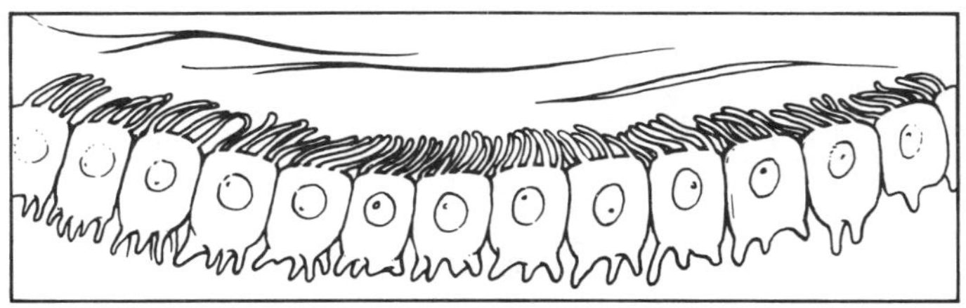

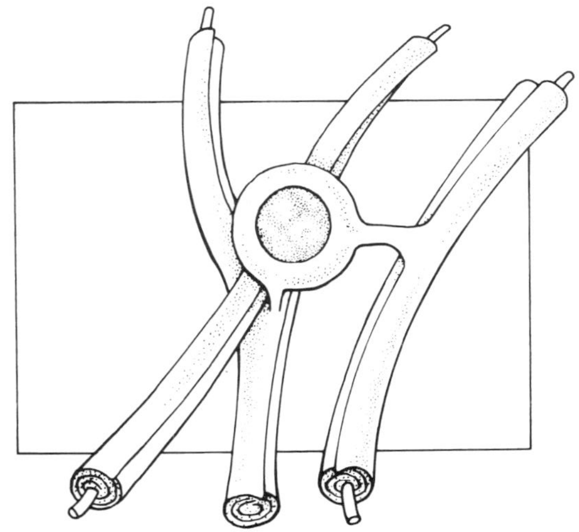

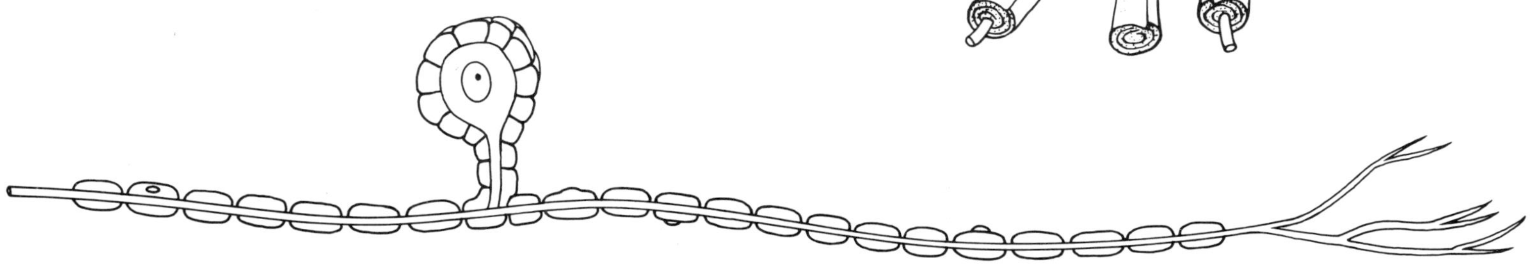

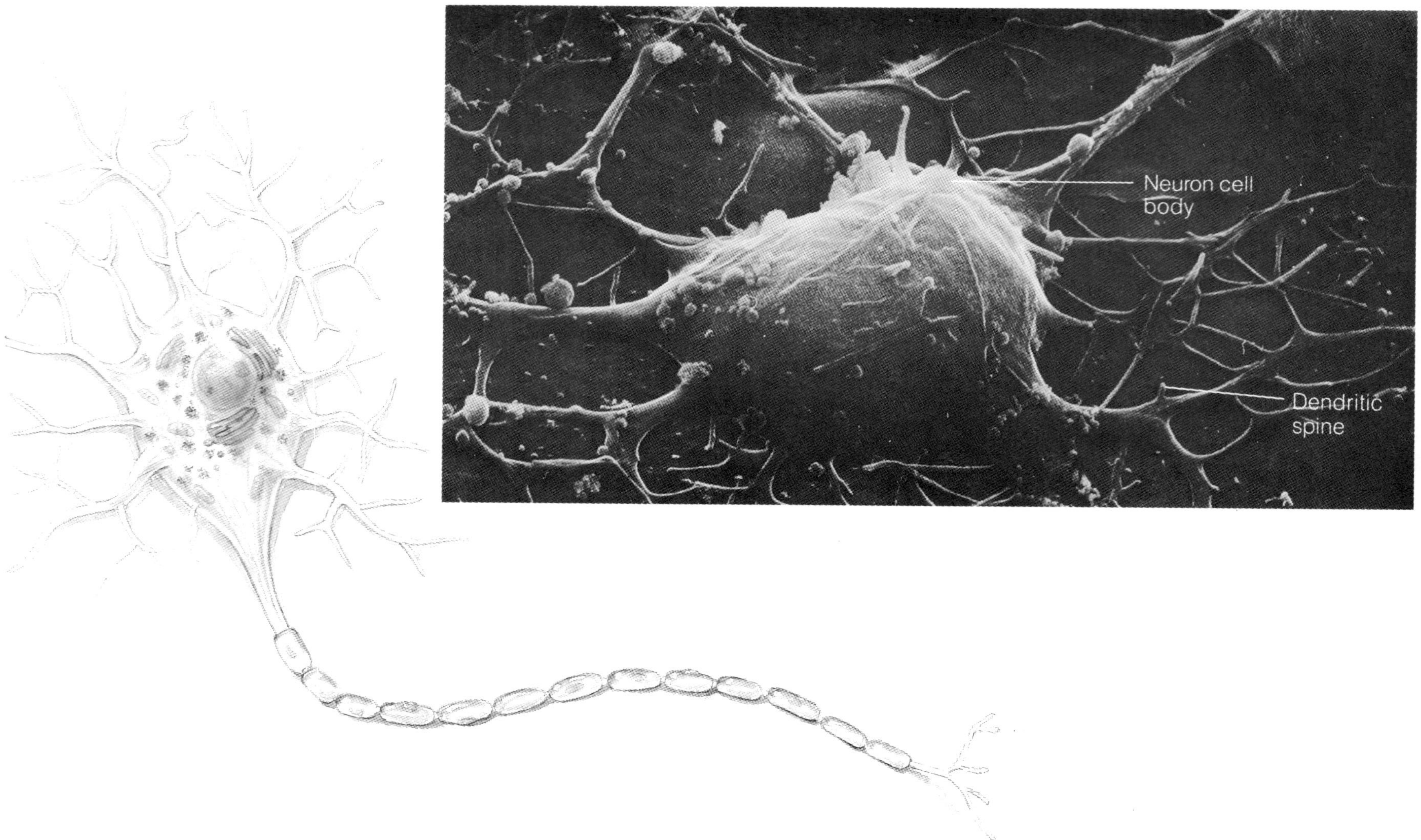
Neuron cell body
Dendritic spine

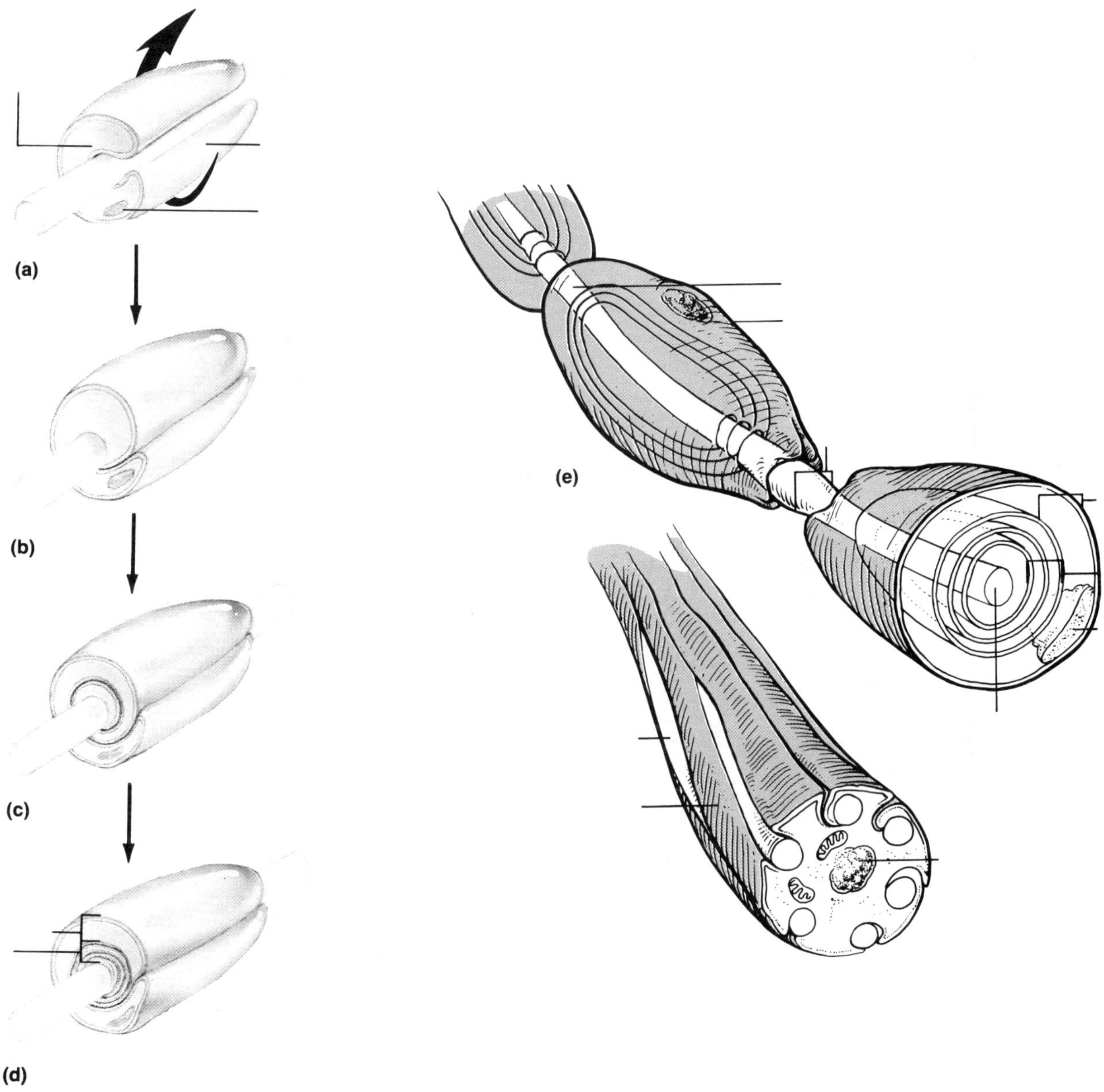
(a)
(b)
(c)
(d)
(e)

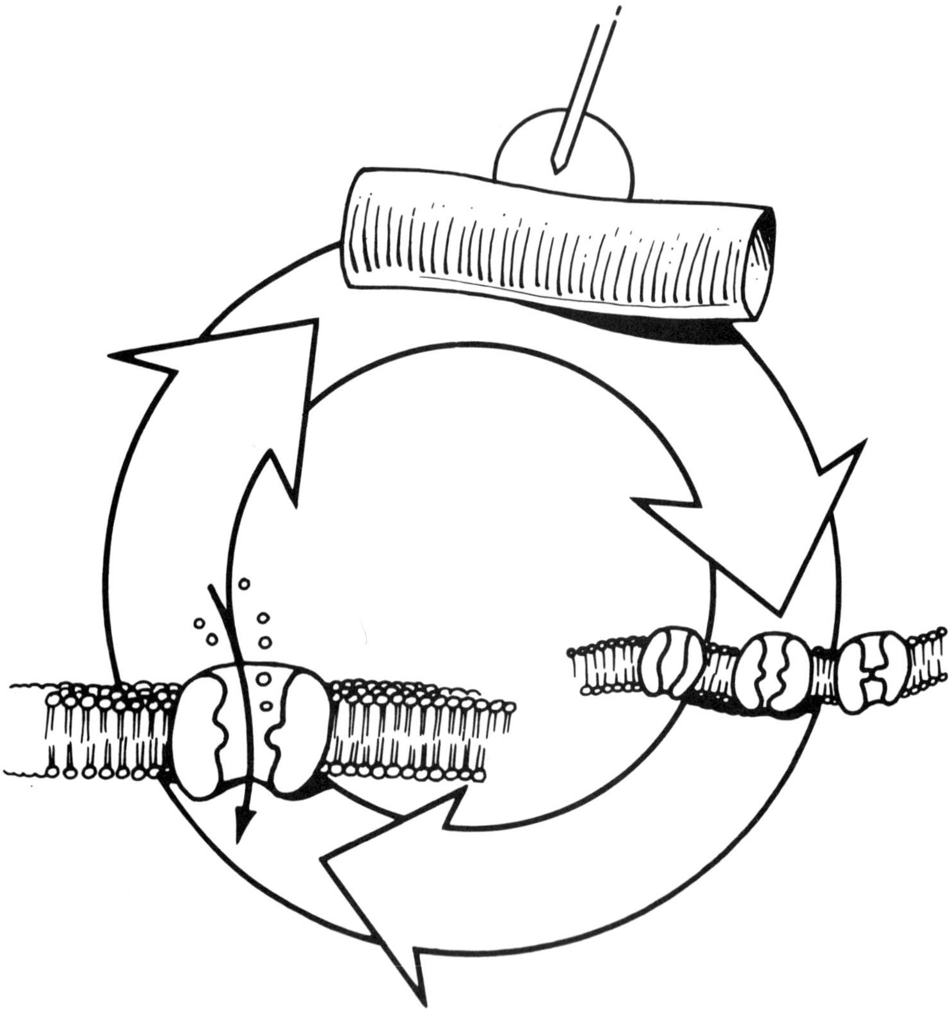

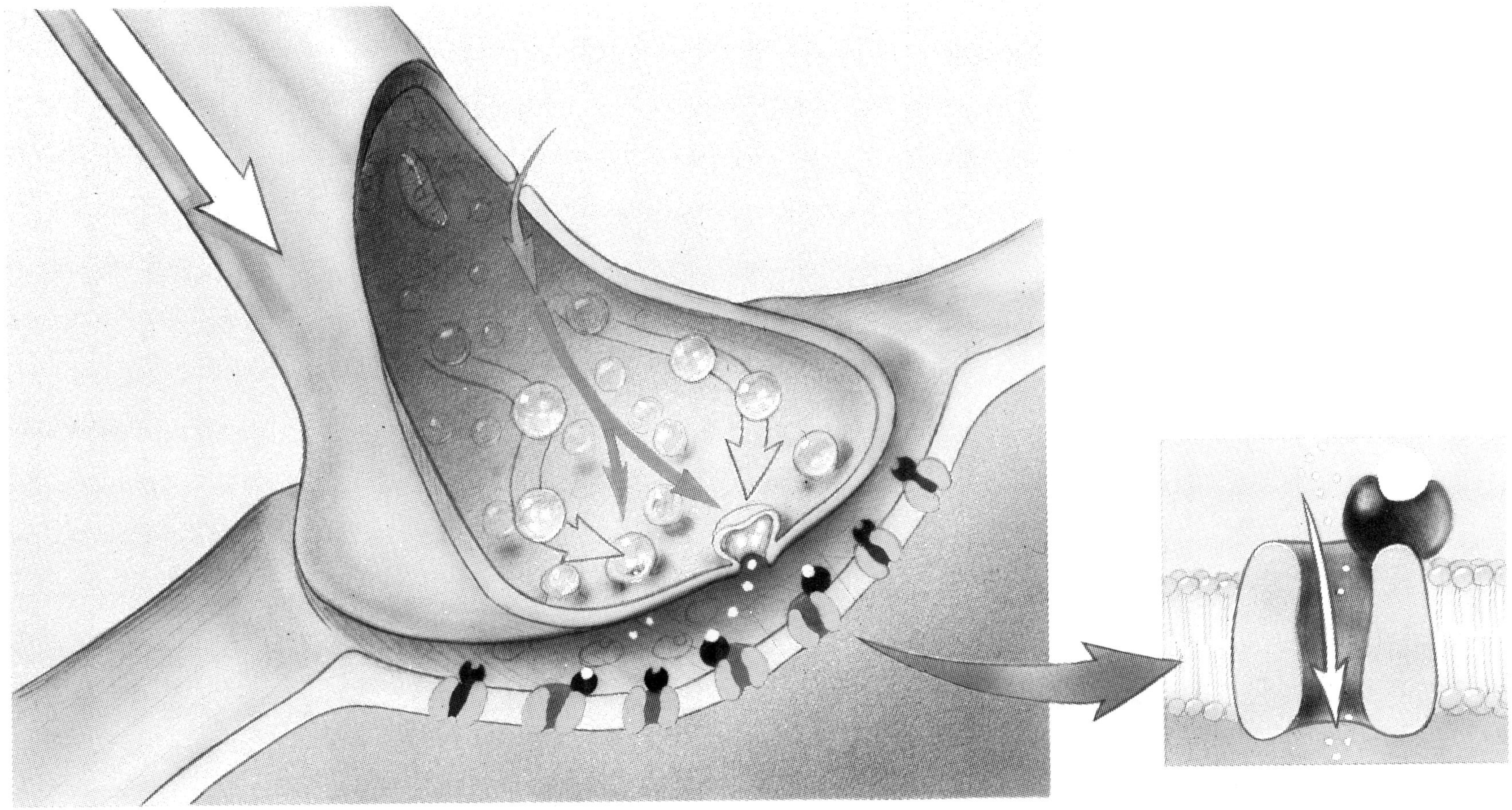

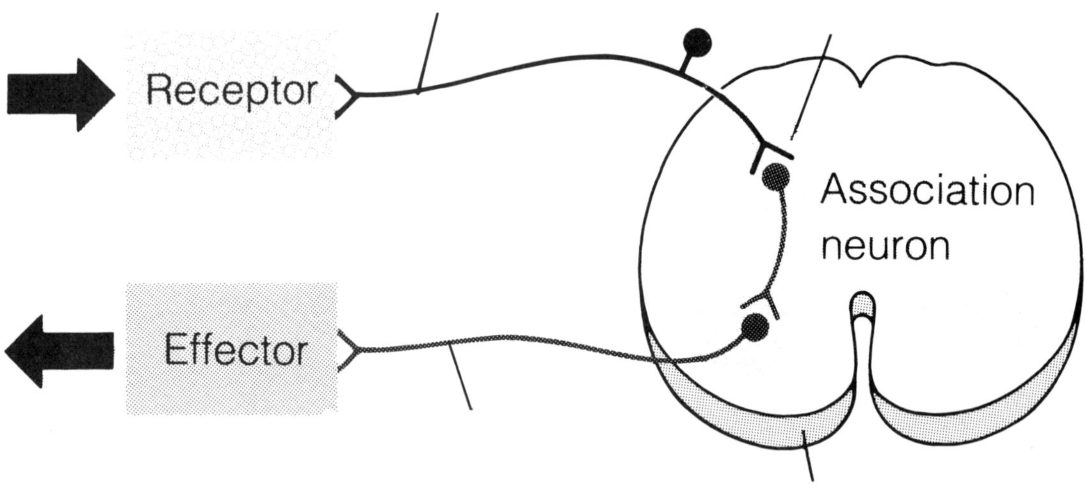
Receptor
Association
neuron
Effector

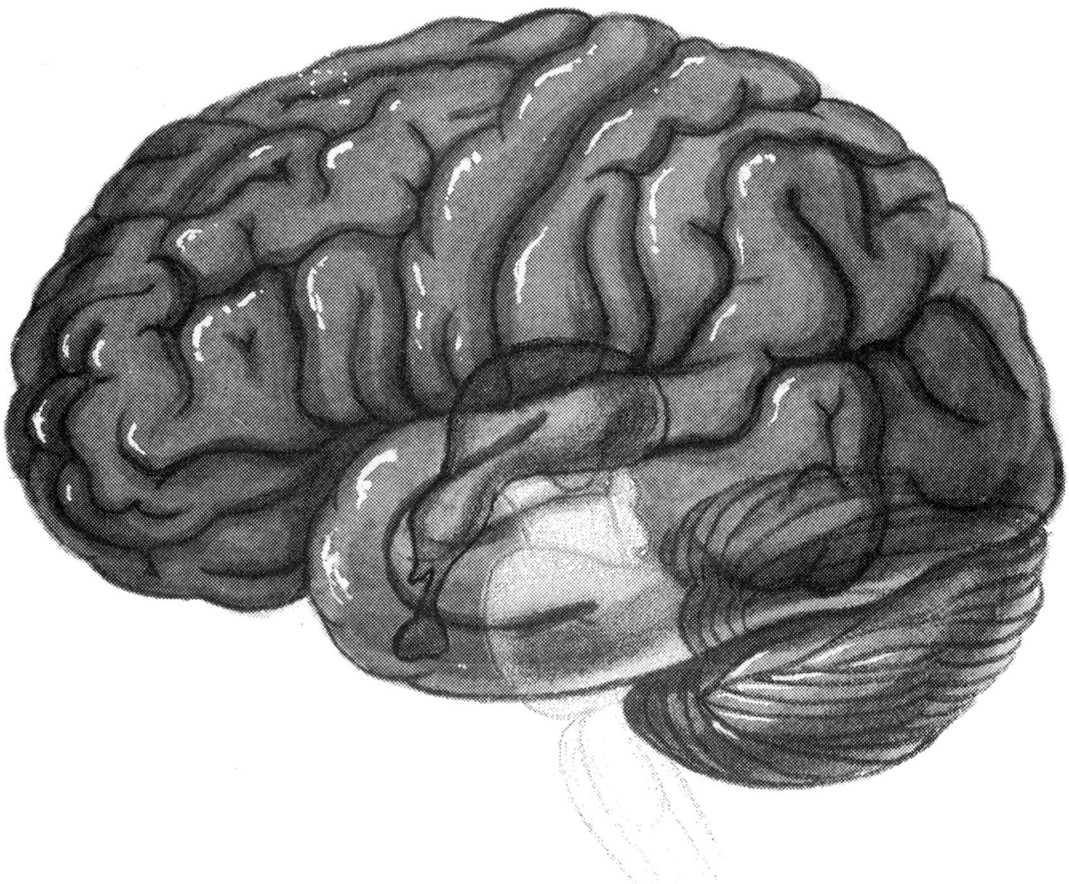

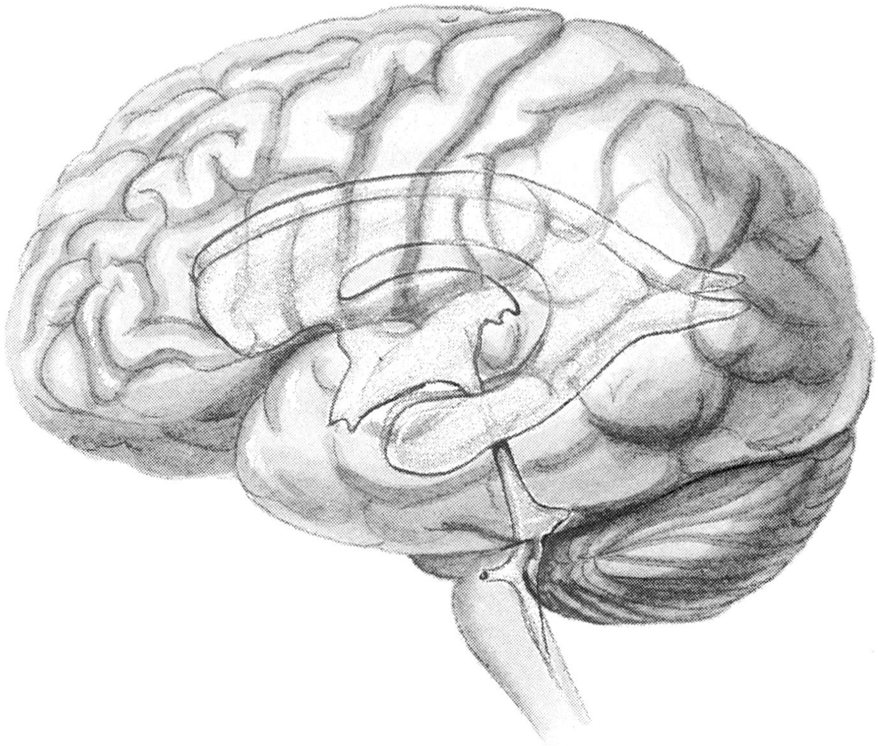

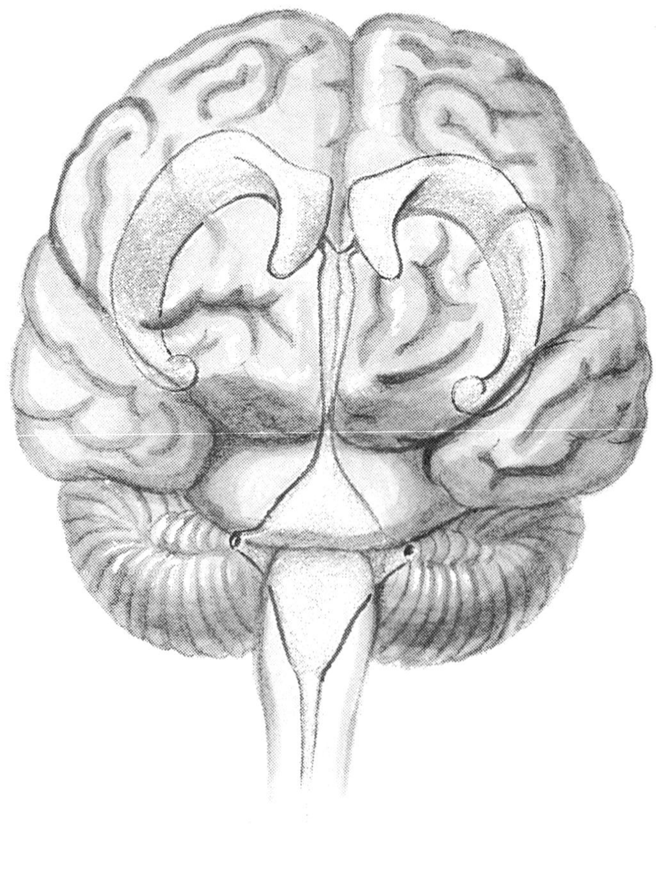

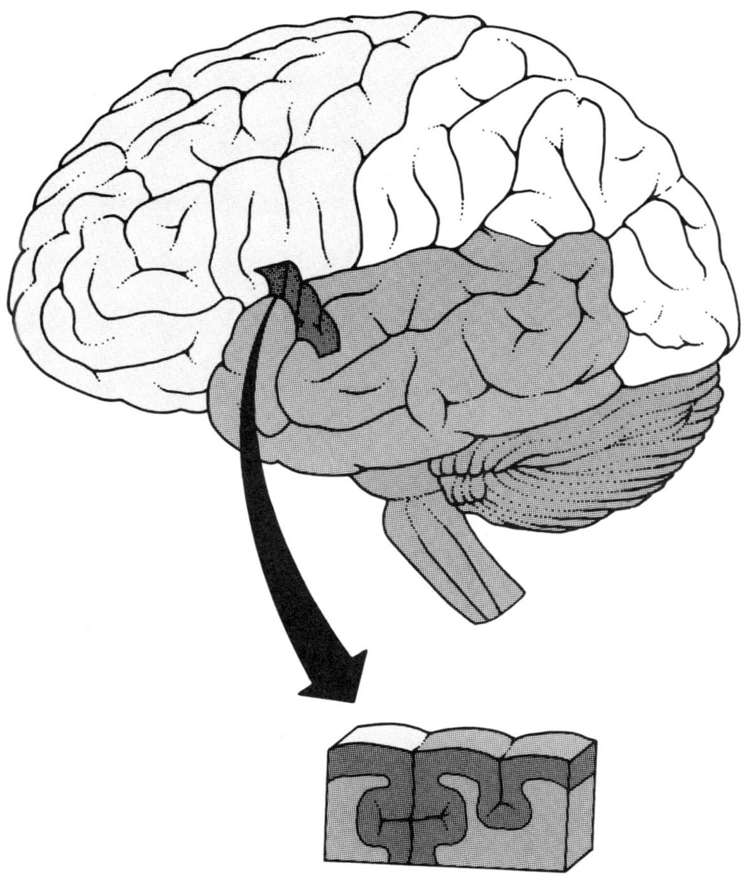

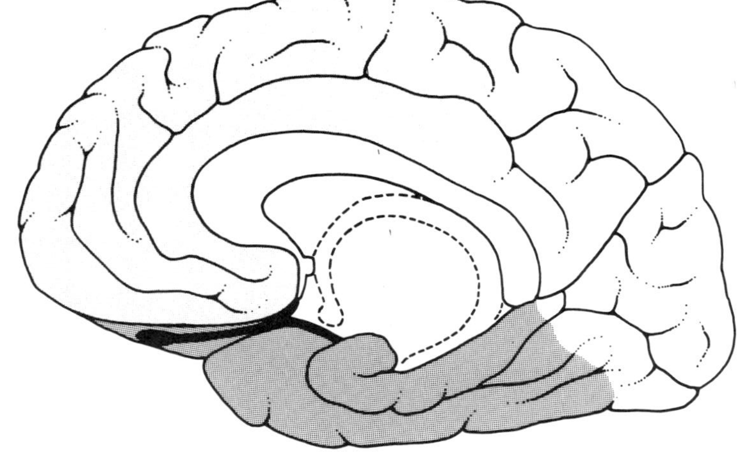

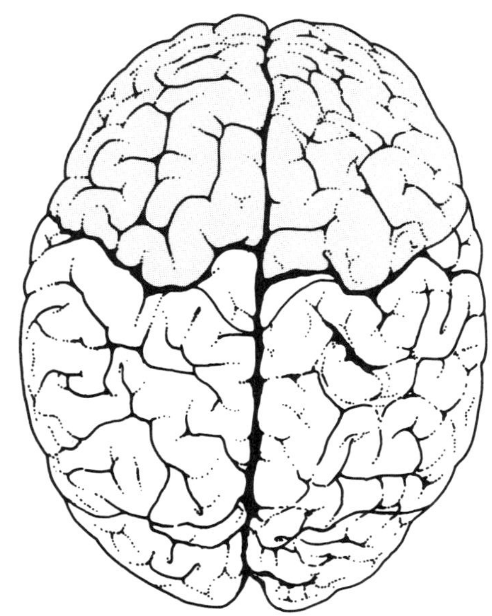

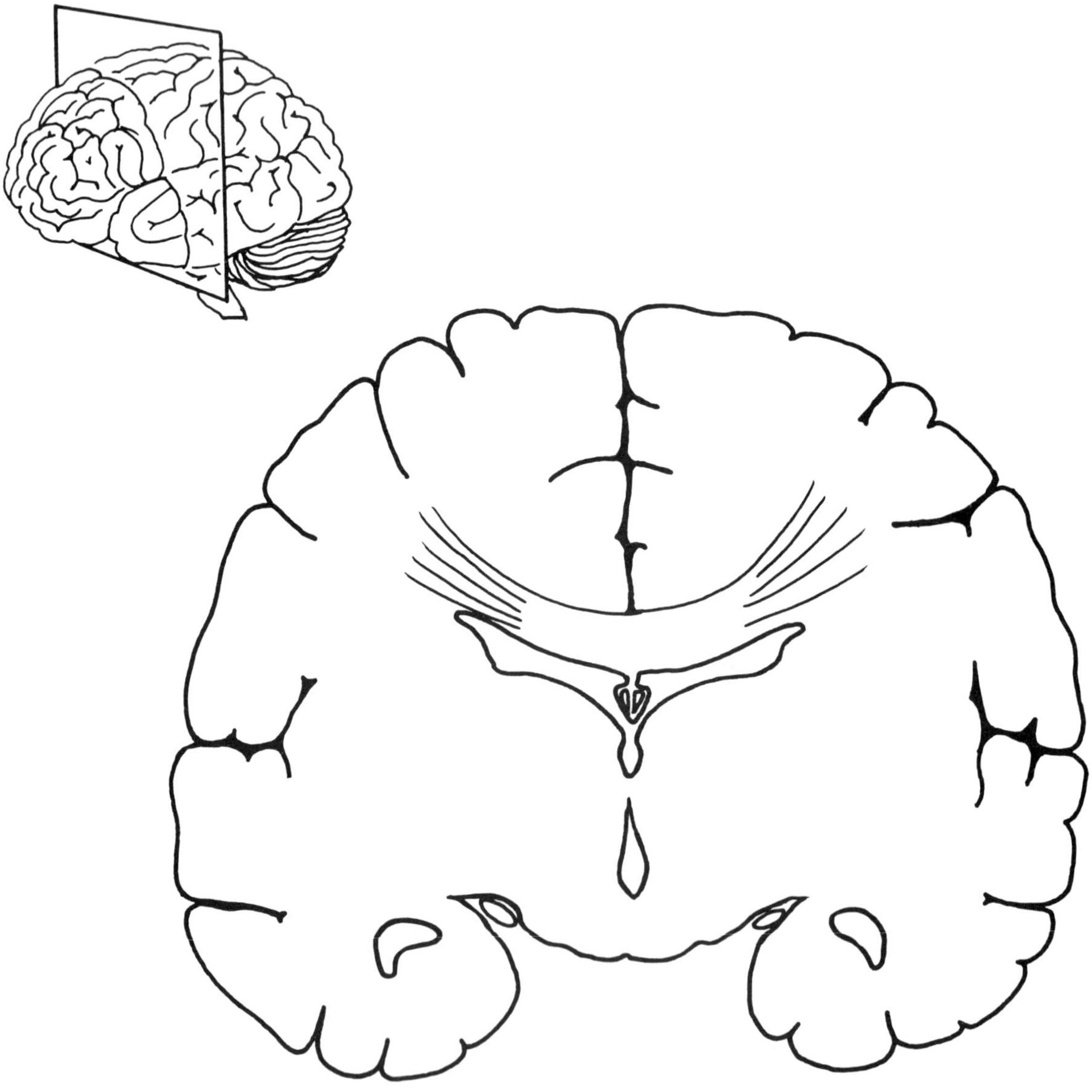

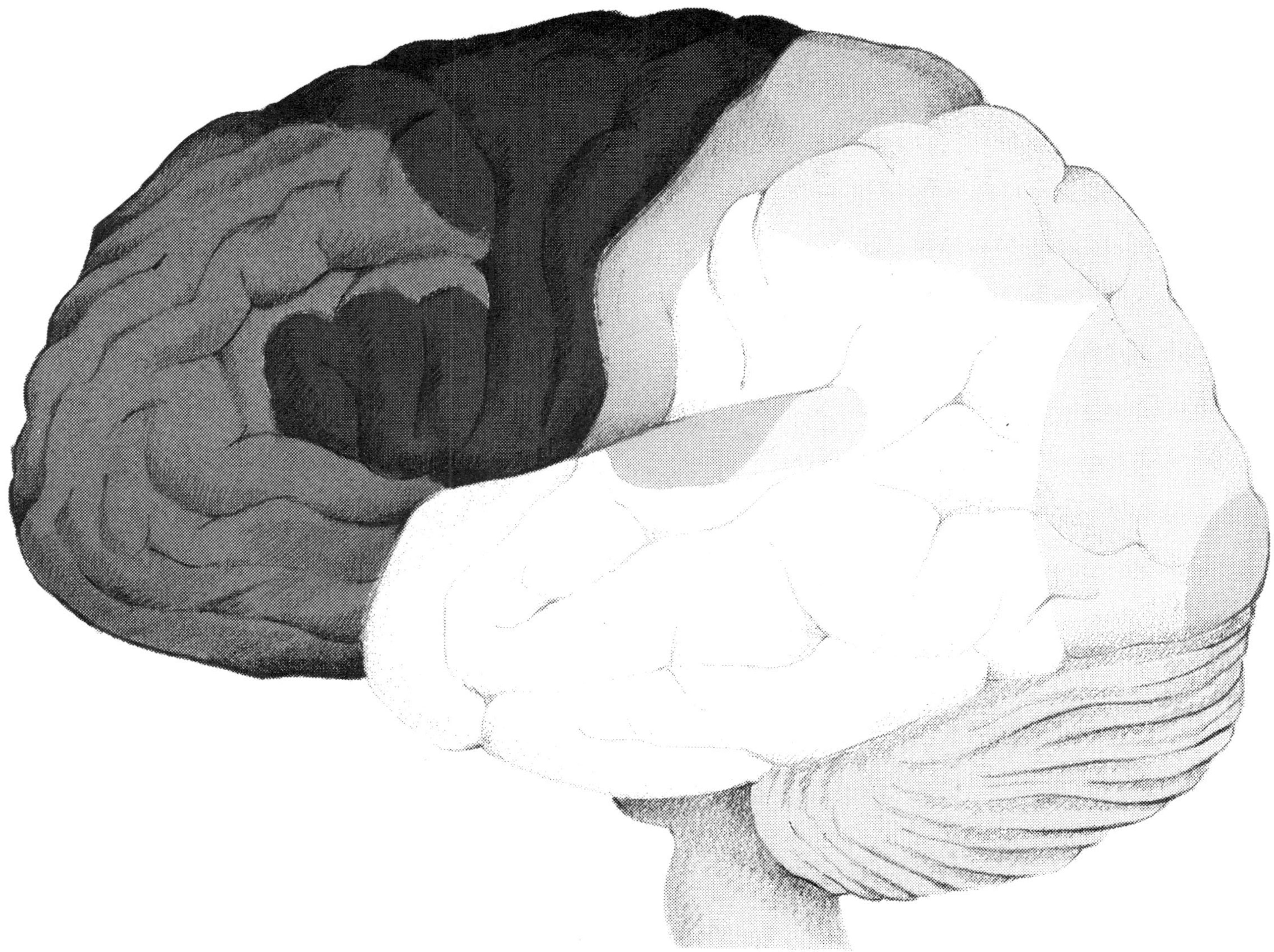

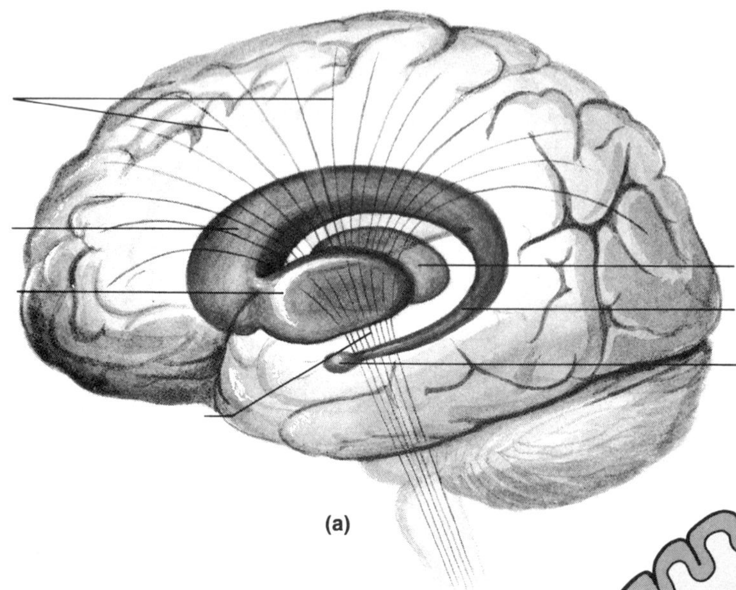

(a)

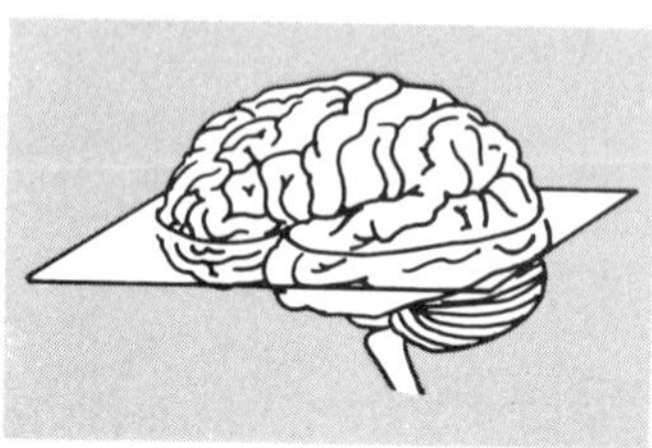

Anterior

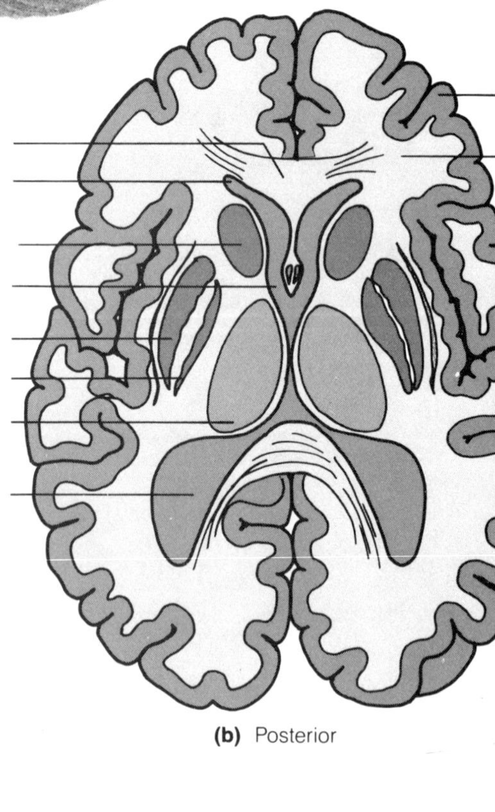

(b) Posterior

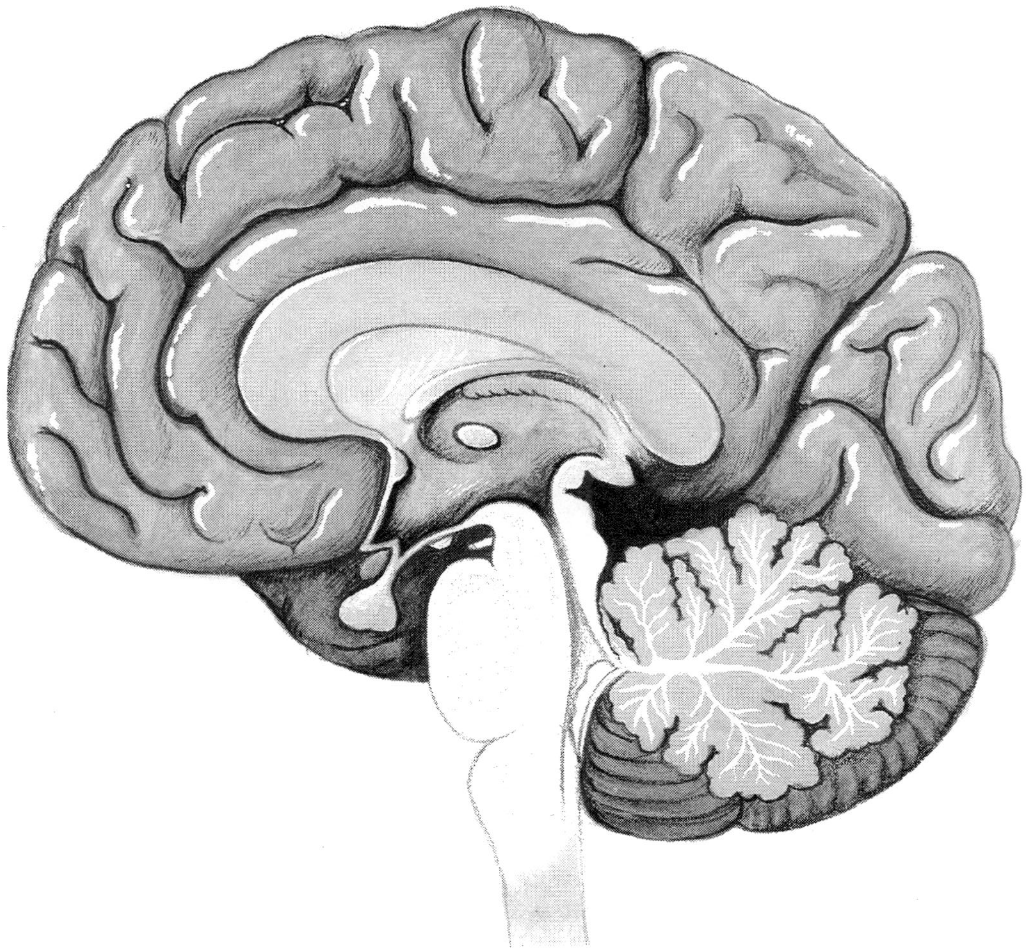

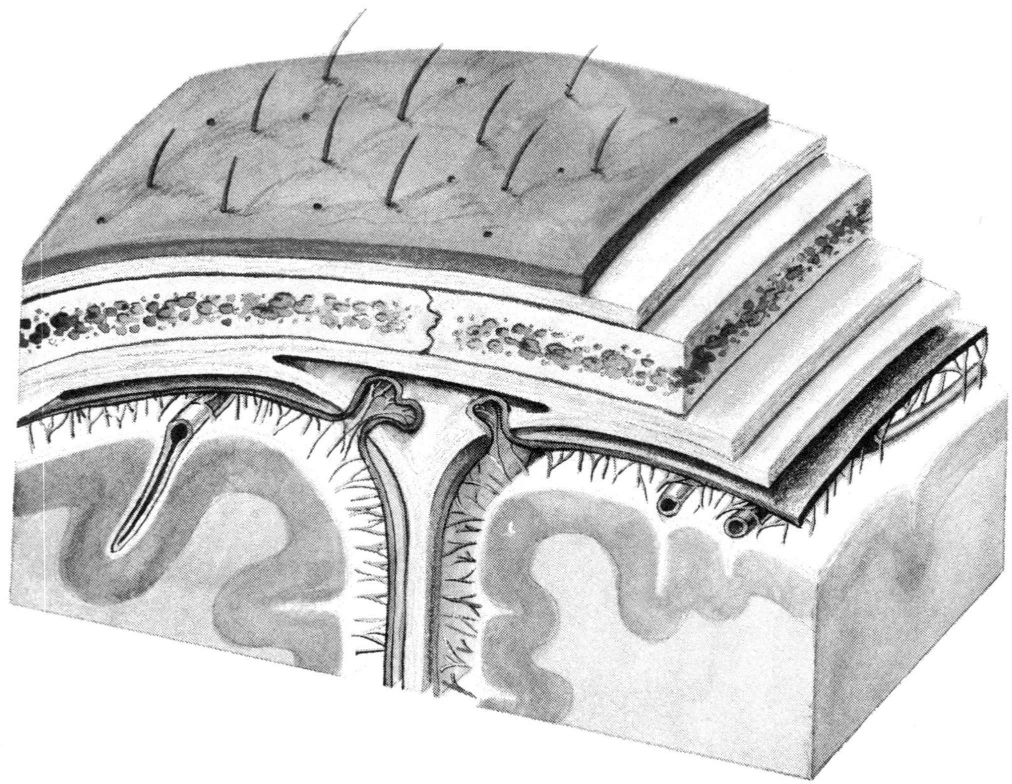

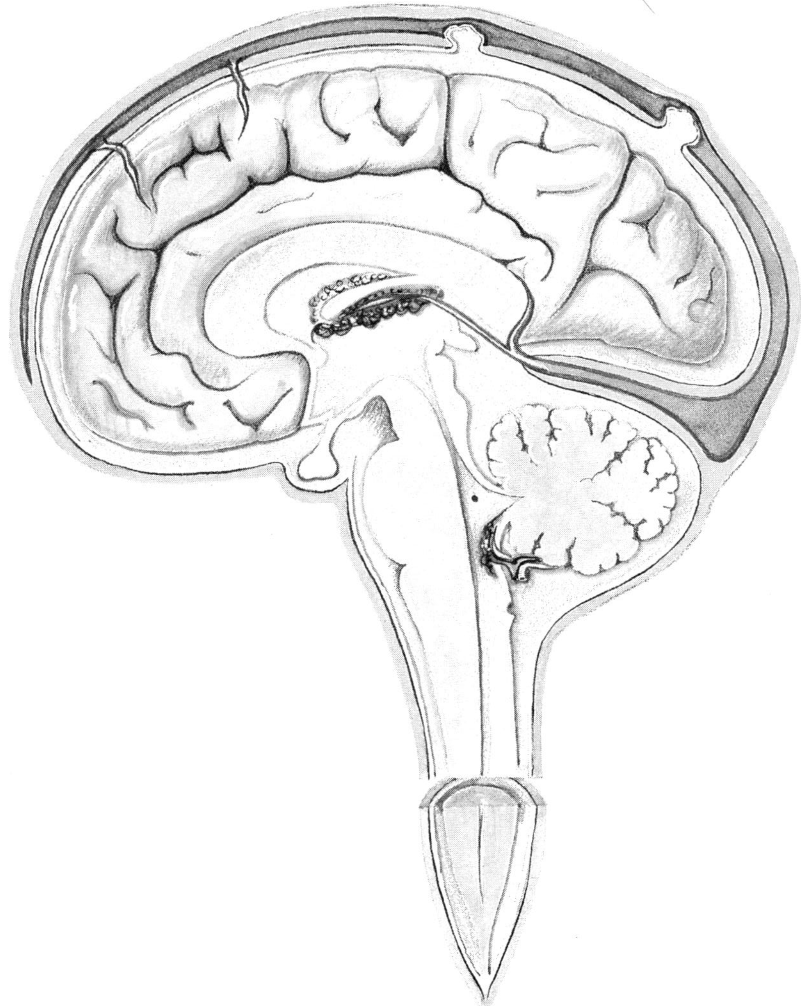

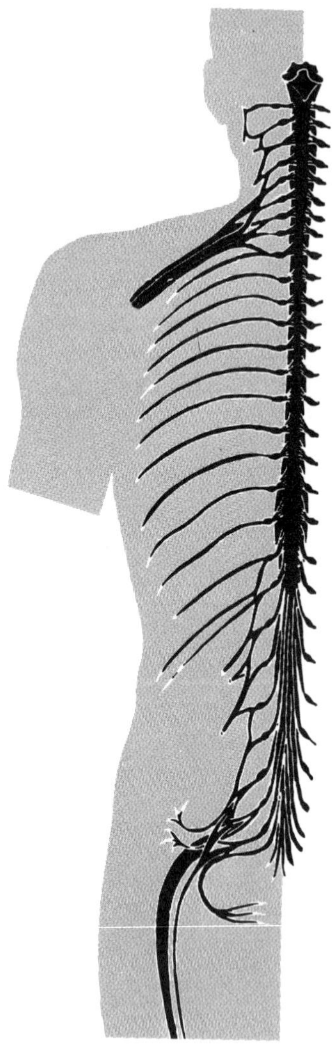

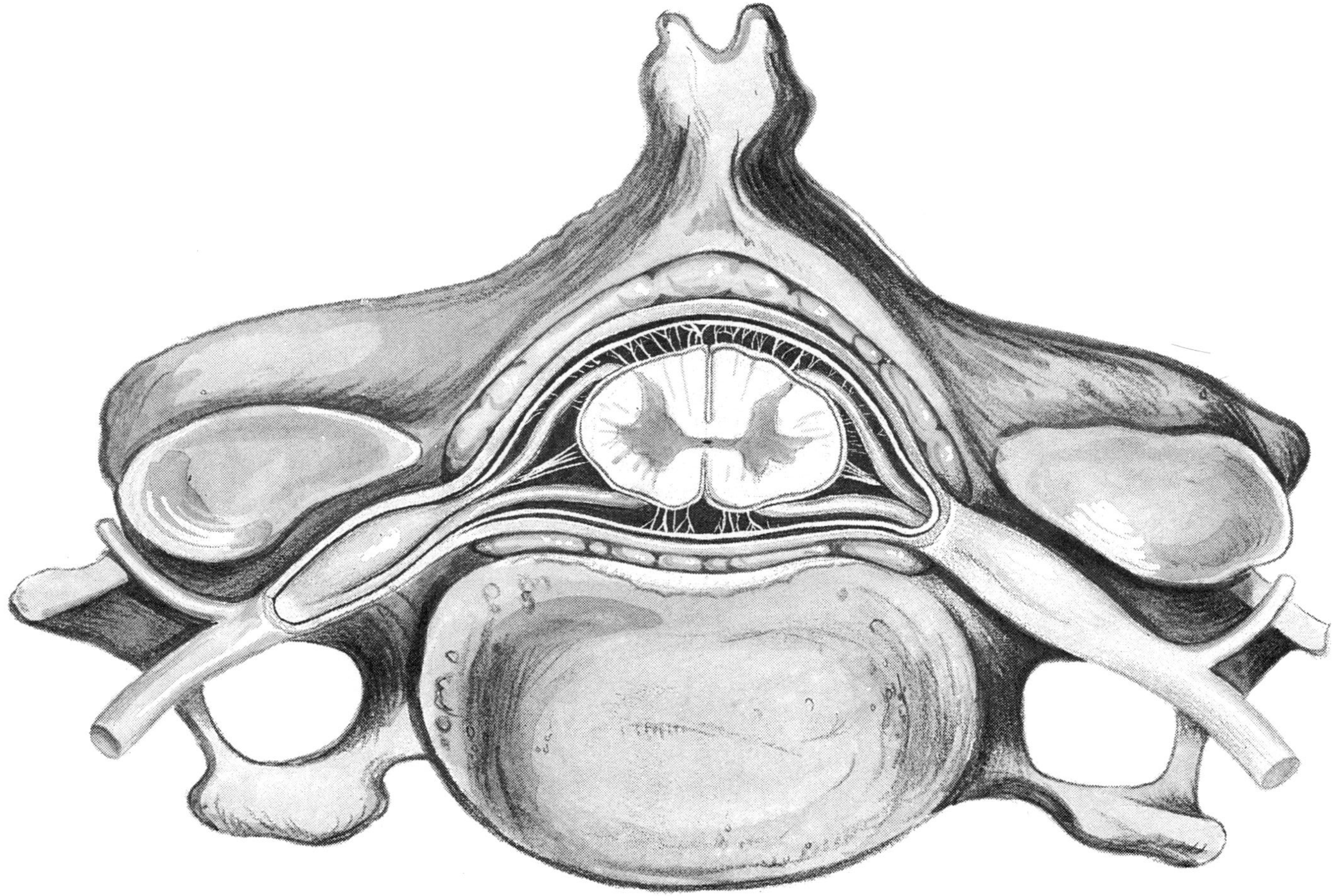

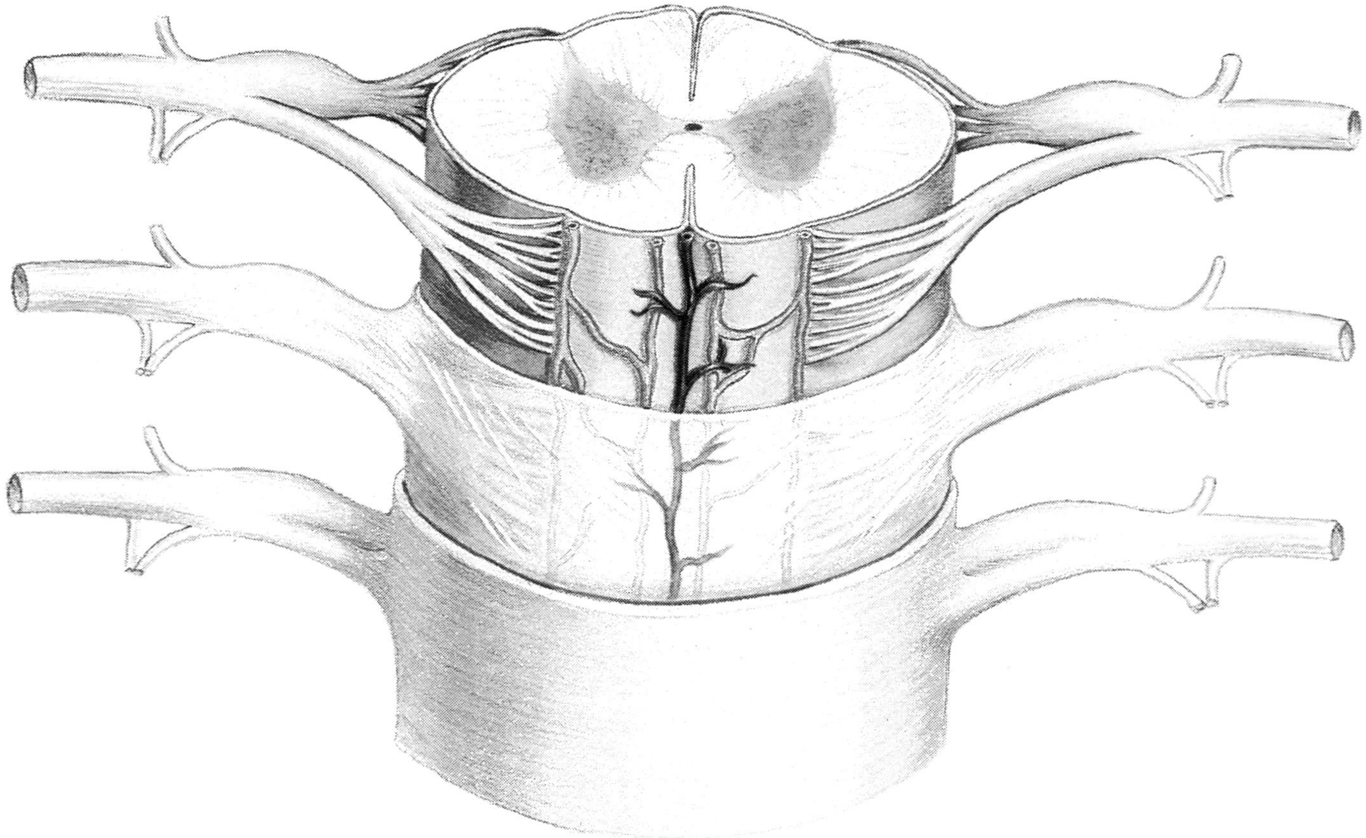

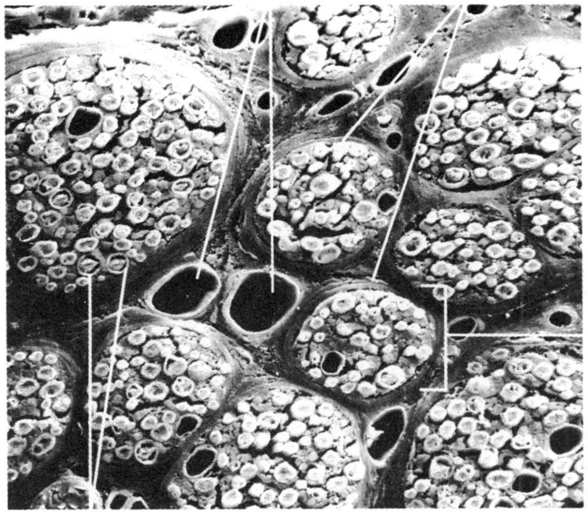

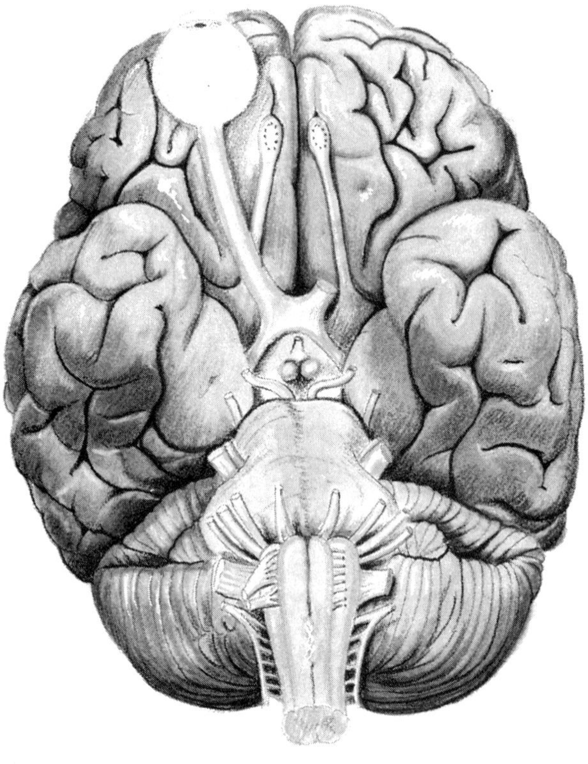

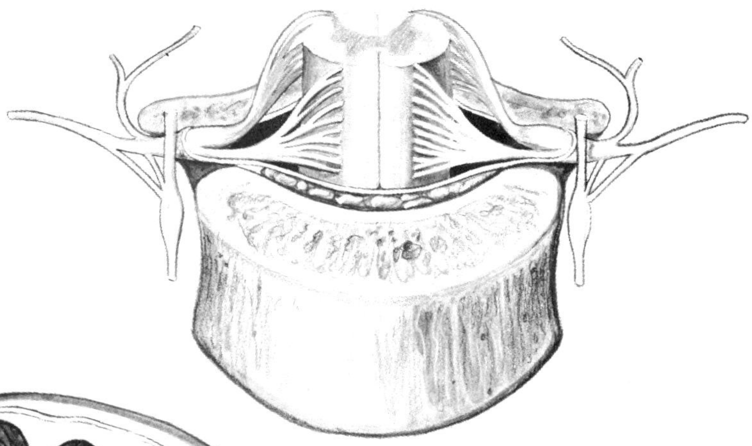

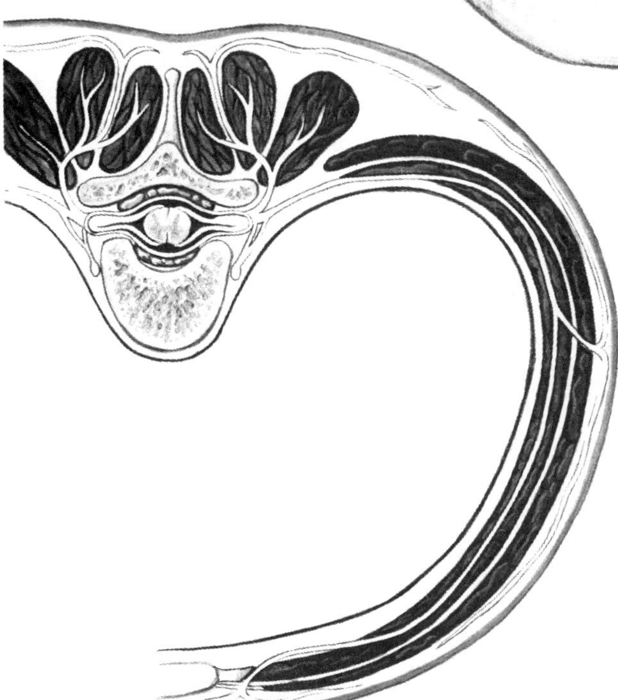

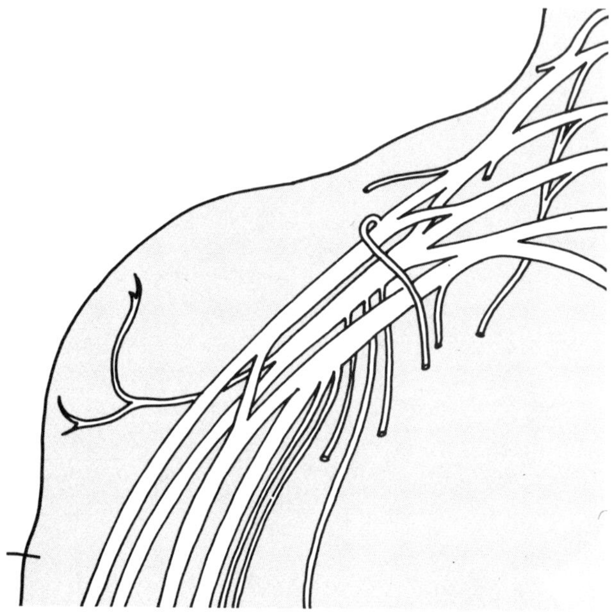

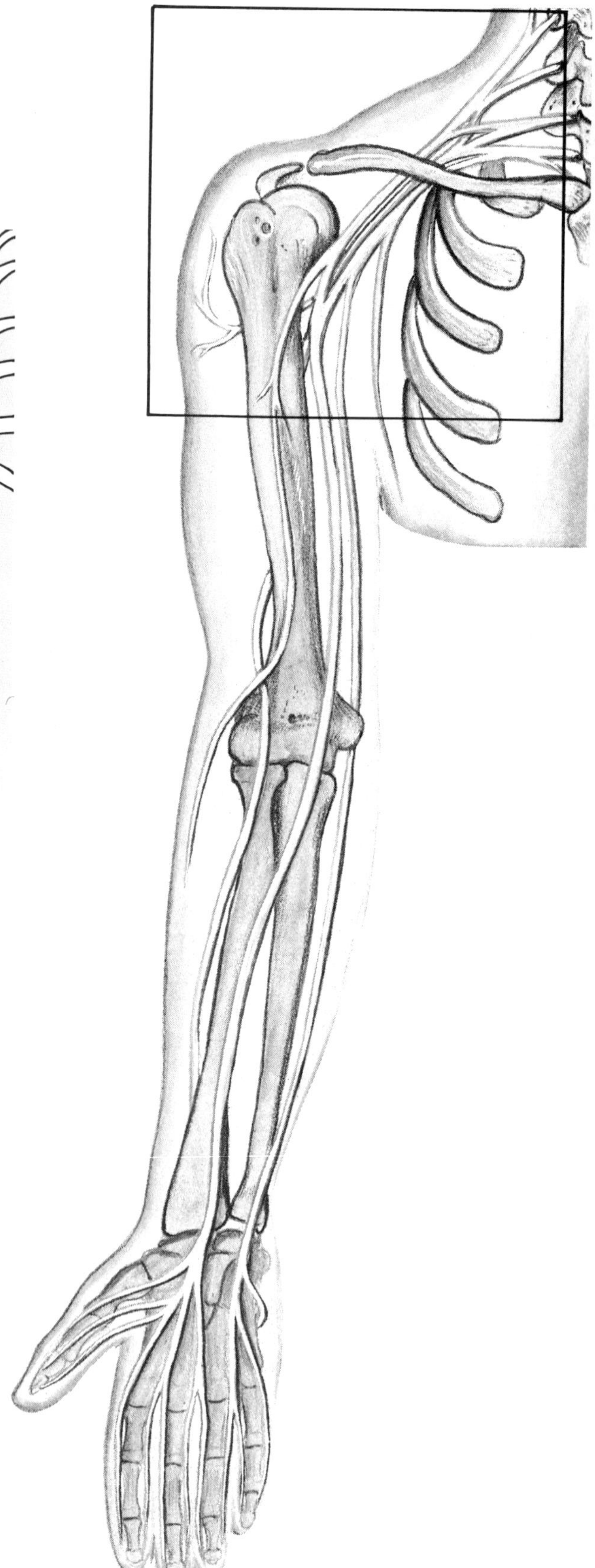

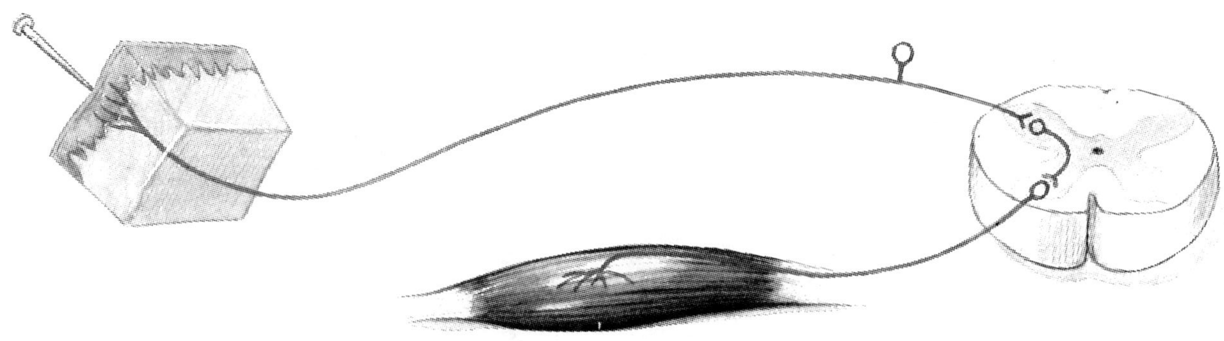

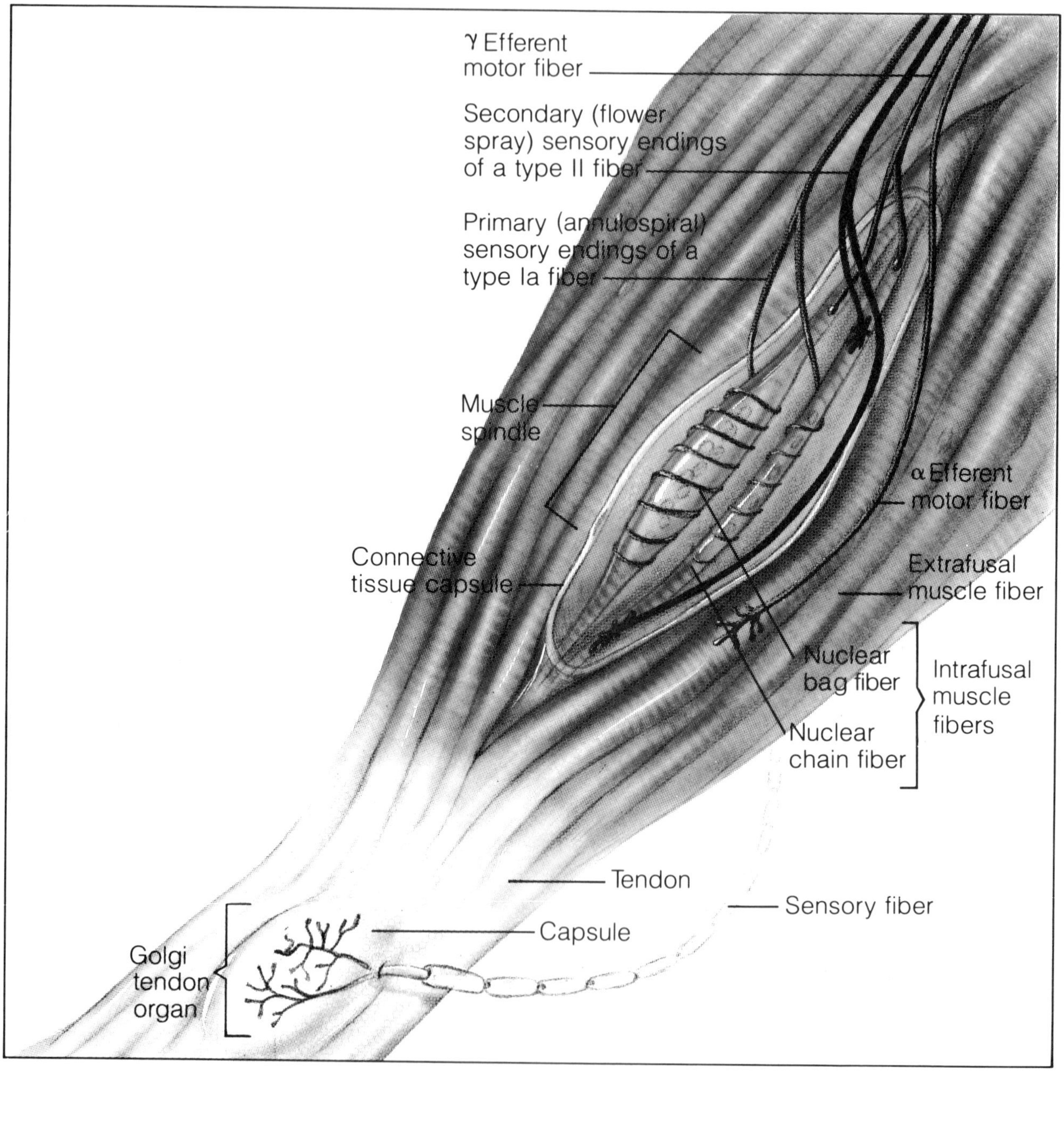
γ Efferent motor fiber
Secondary (flower spray) sensory endings of a type II fiber
Primary (annulospiral) sensory endings of a type Ia fiber
Muscle spindle
α Efferent motor fiber
Connective tissue capsule
Extrafusal muscle fiber
Nuclear bag fiber
Nuclear chain fiber
Intrafusal muscle fibers
Tendon
Sensory fiber
Capsule
Golgi tendon organ

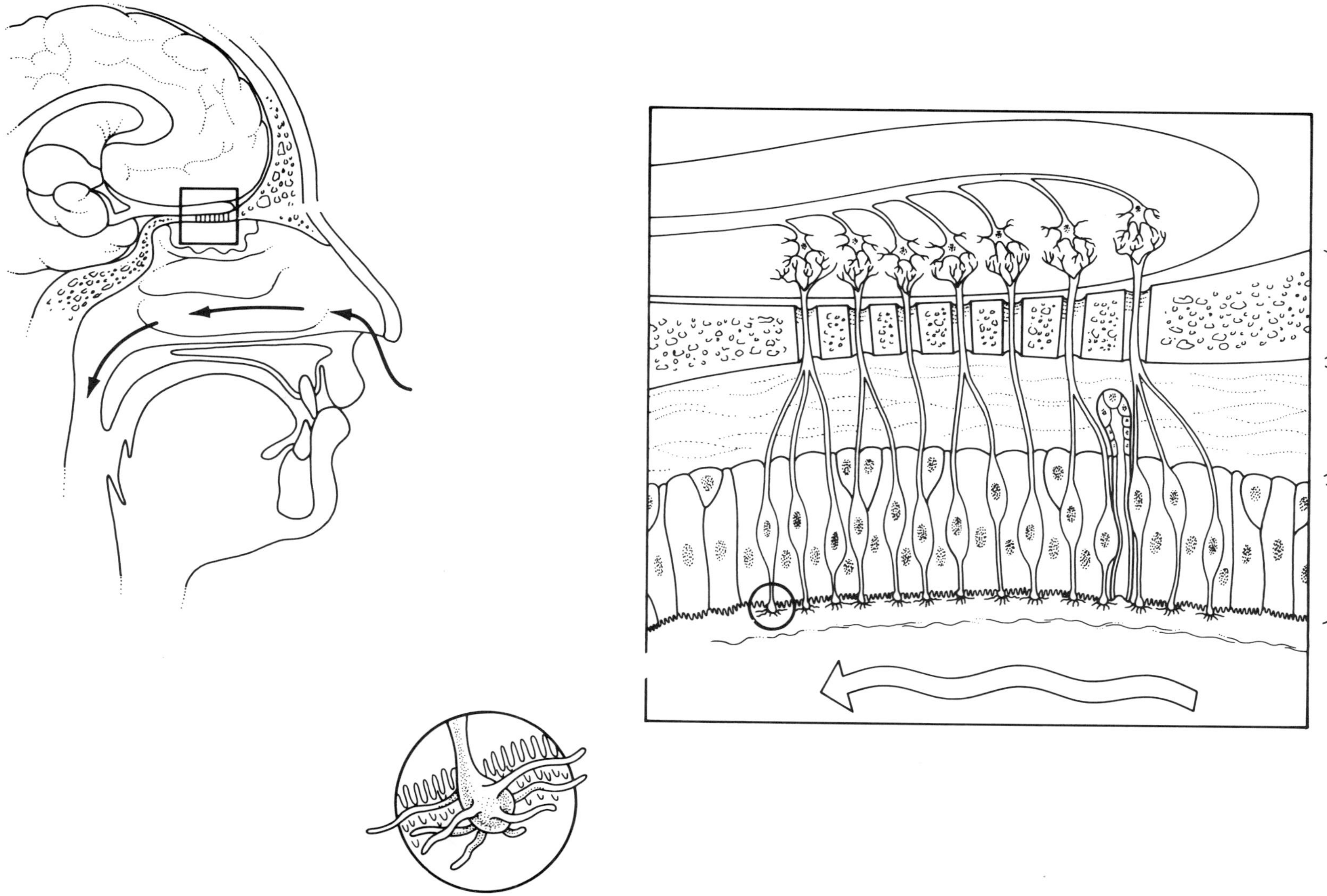

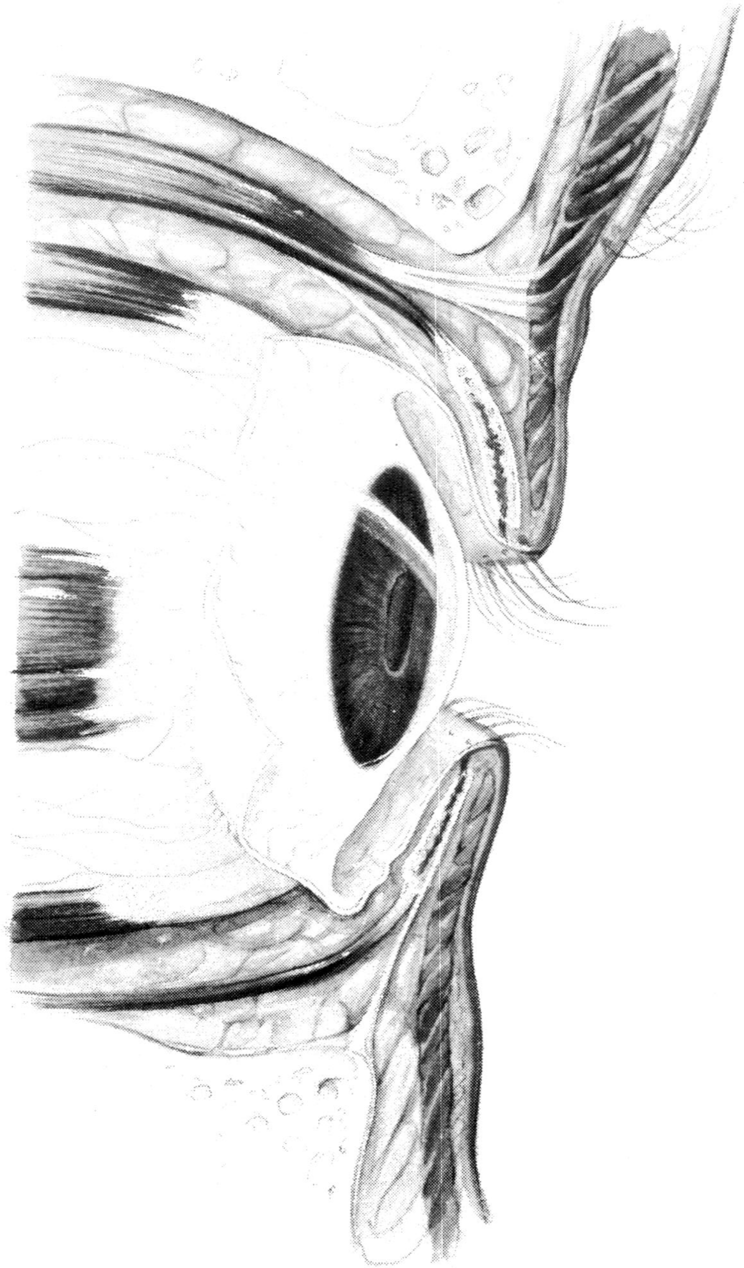

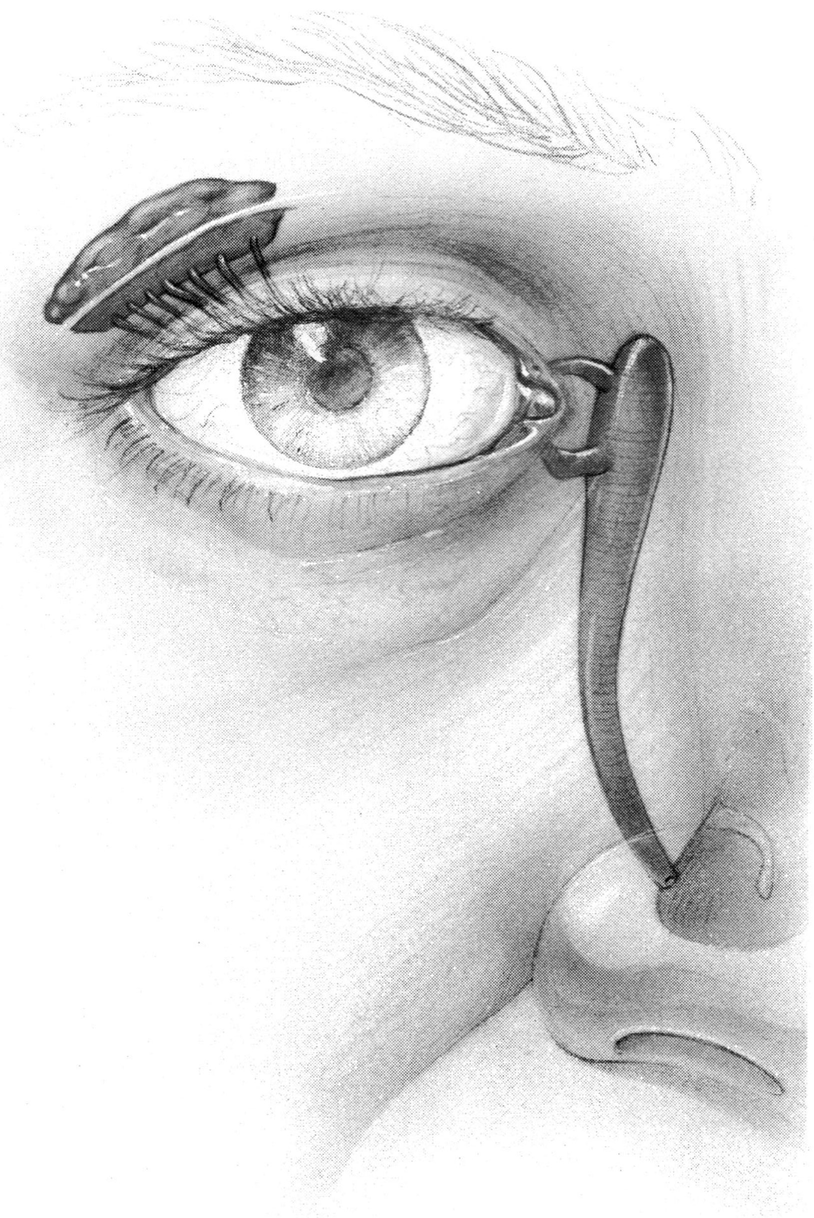

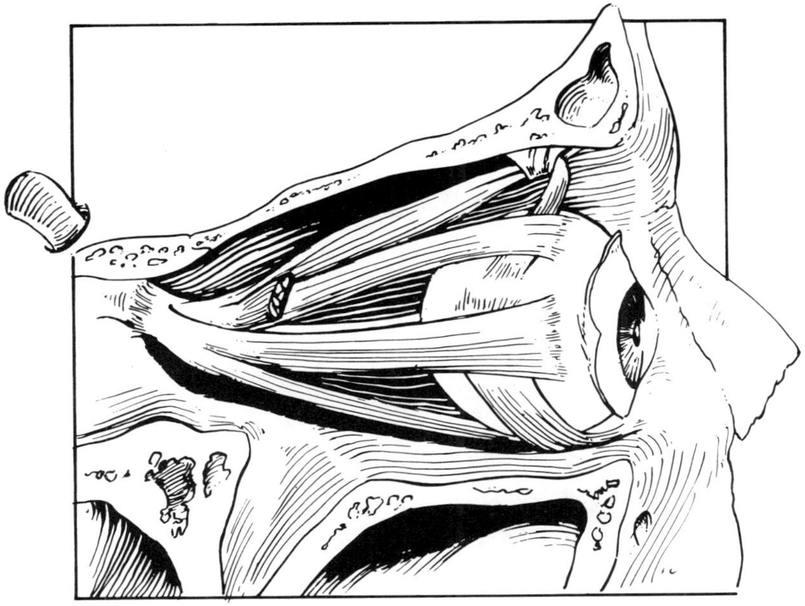

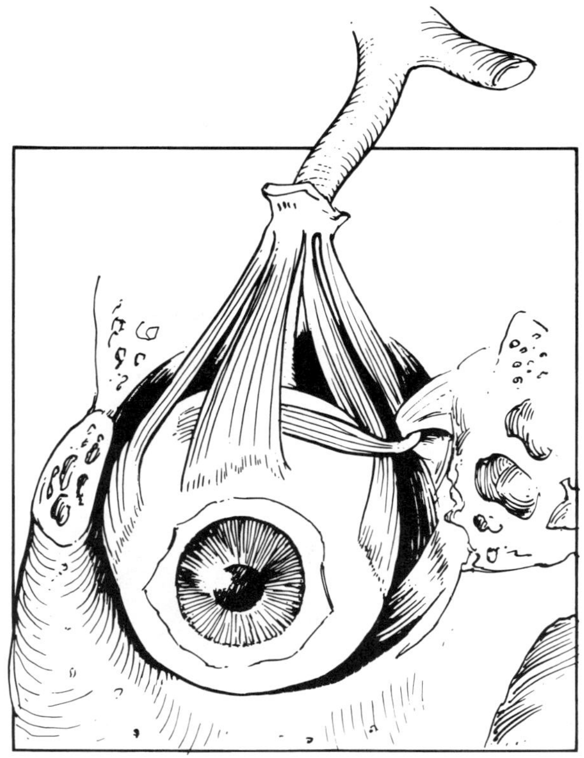

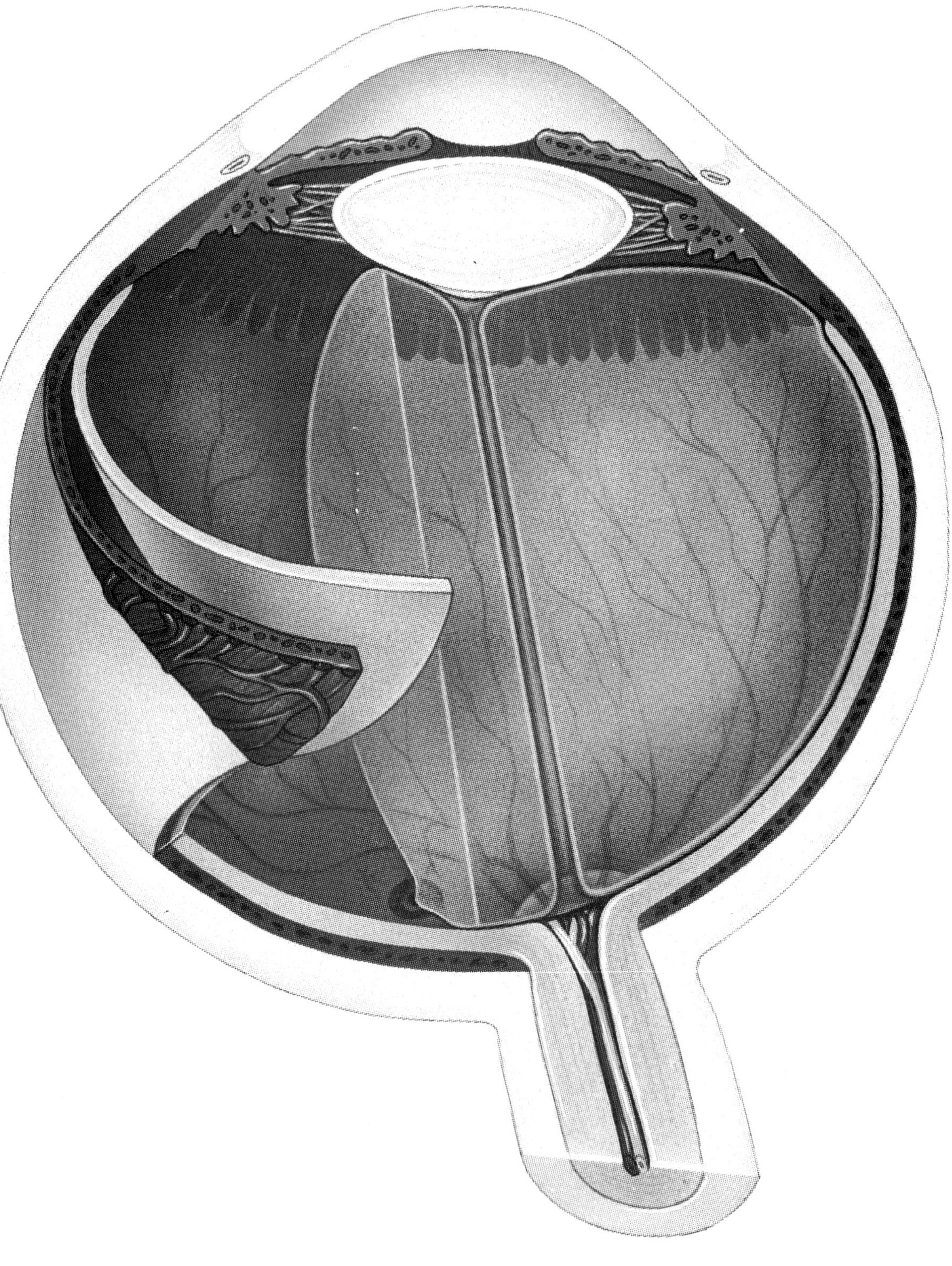

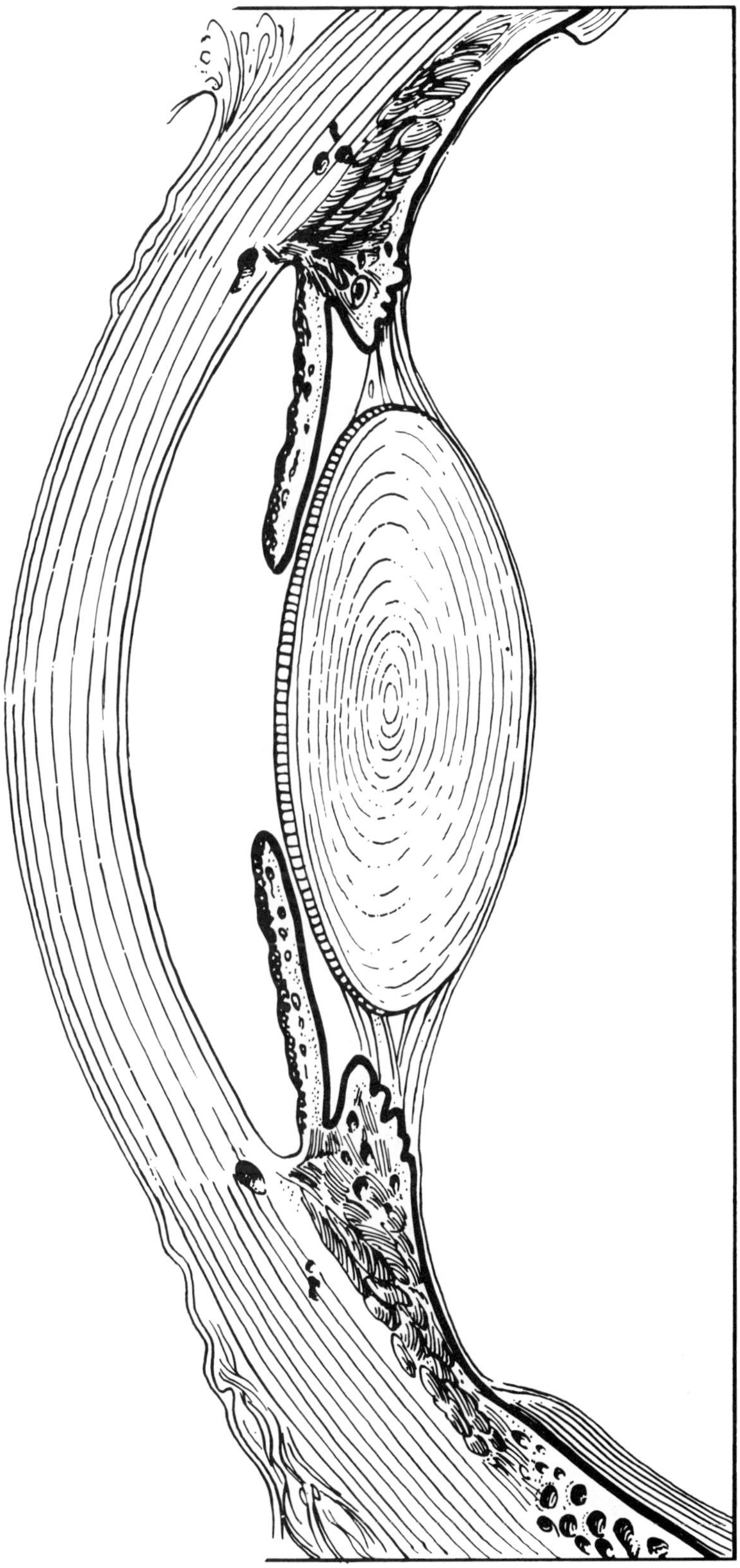

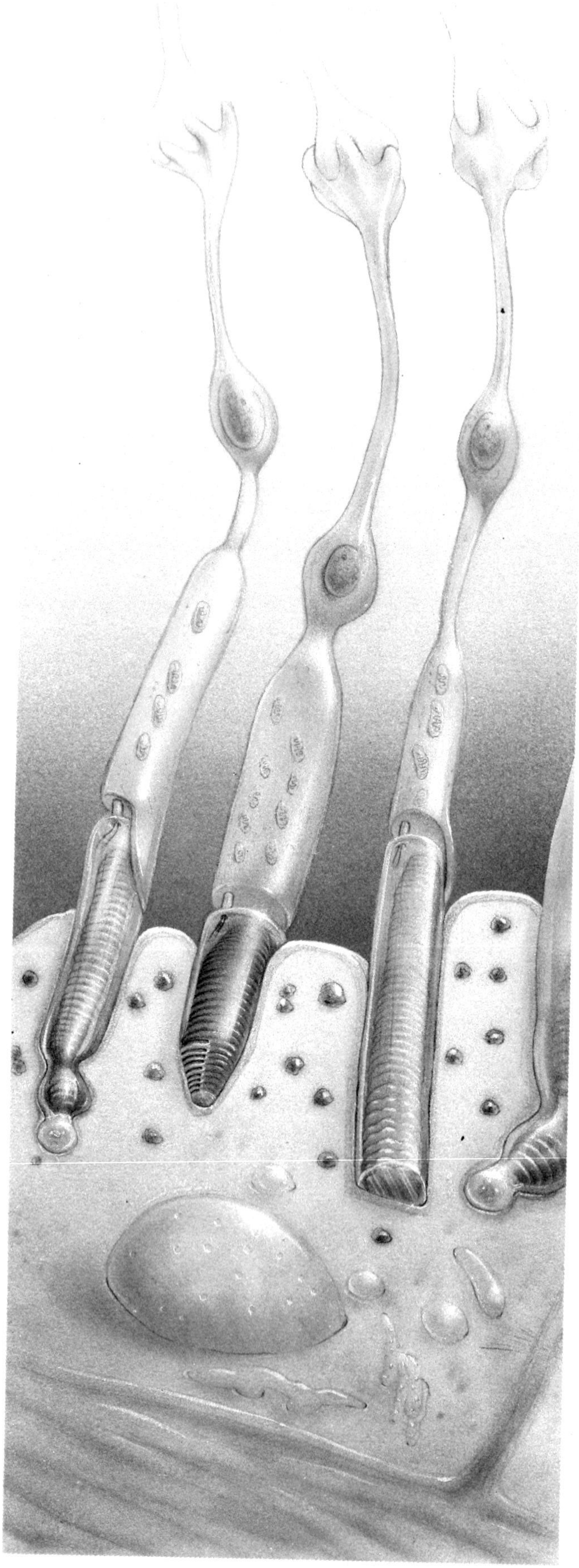

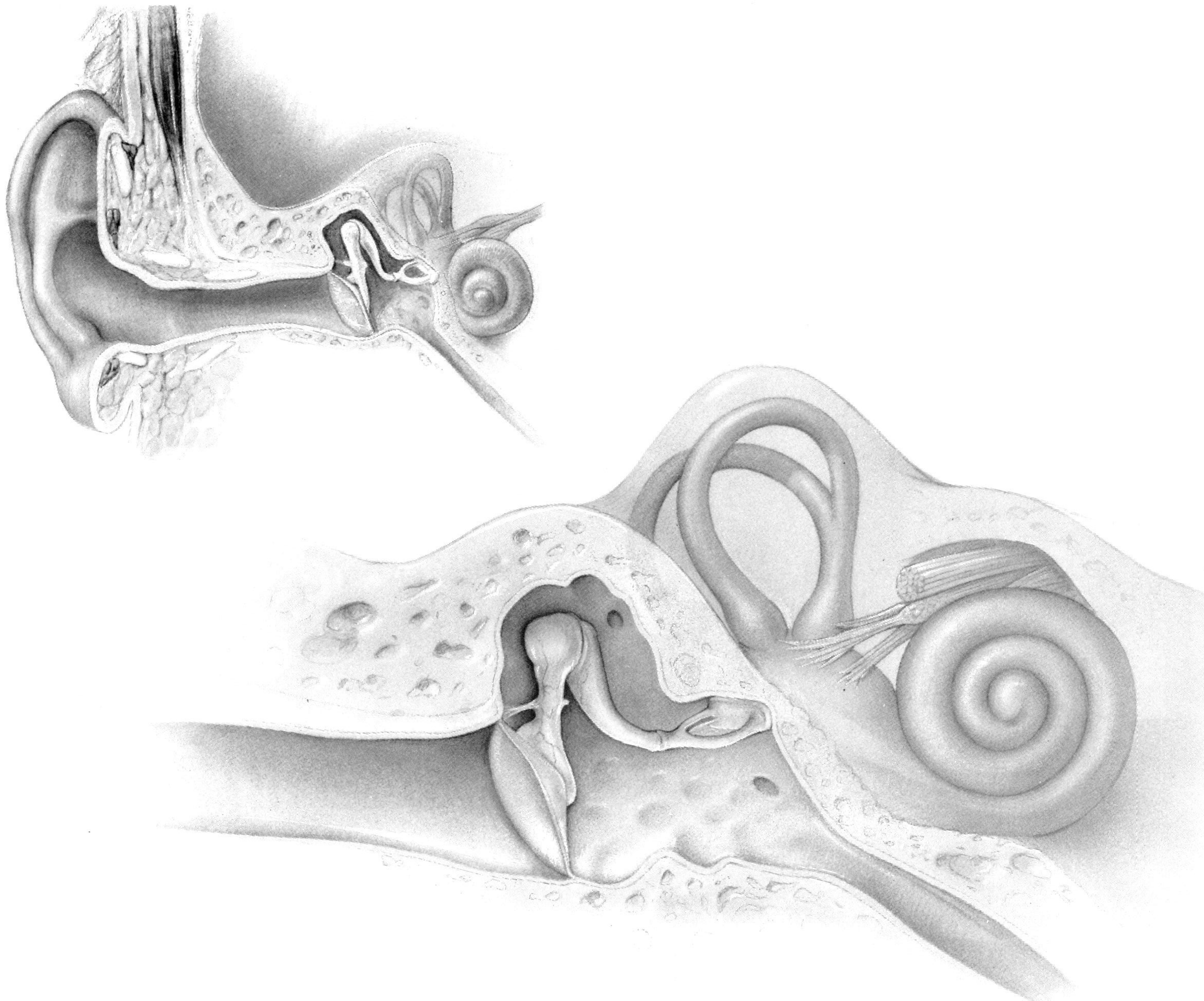

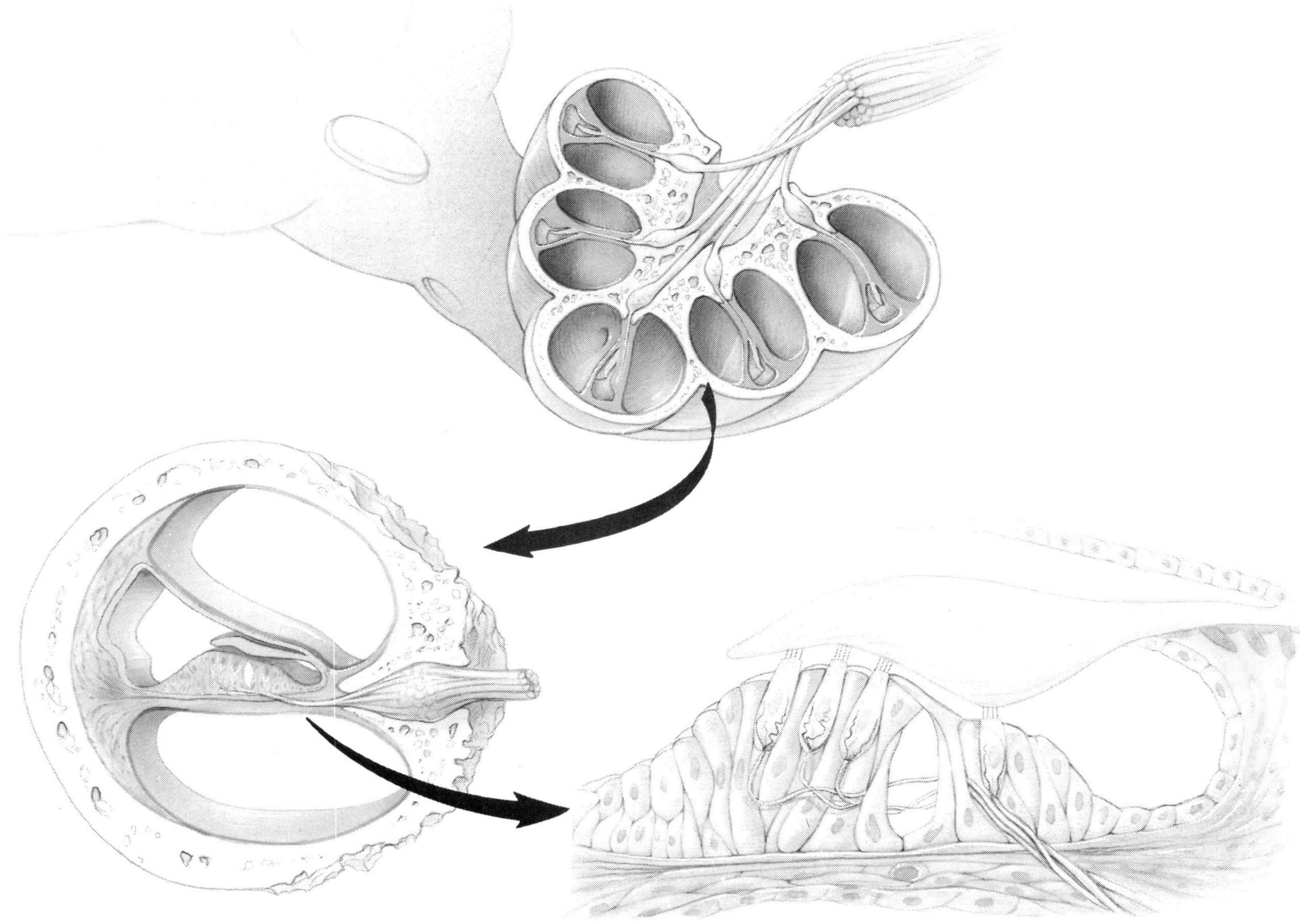

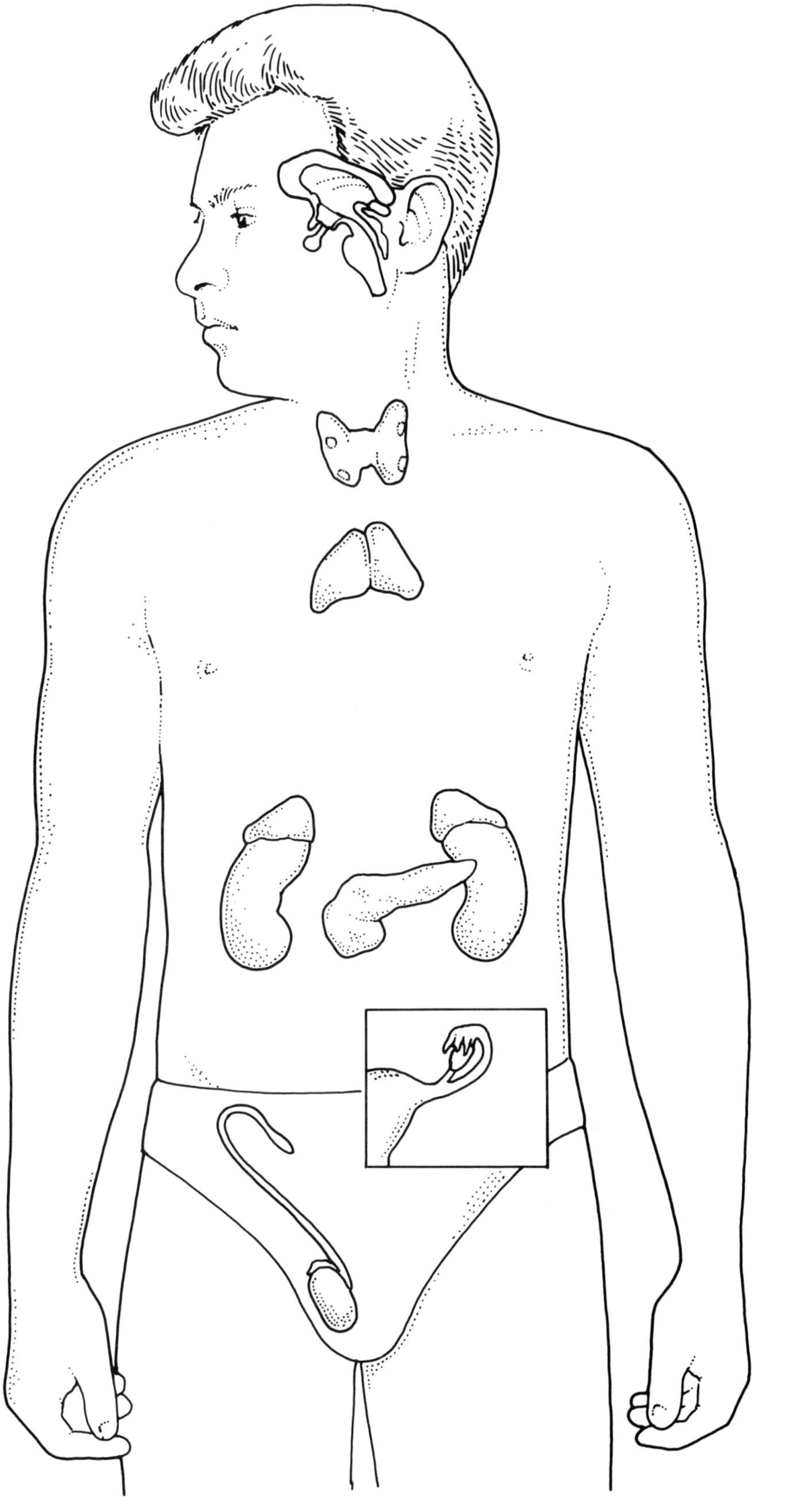

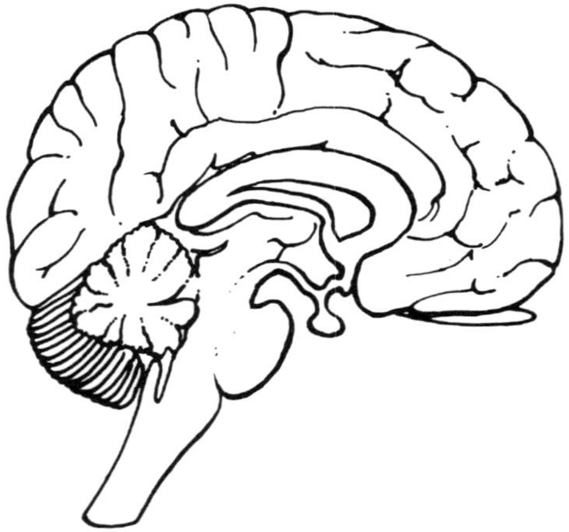

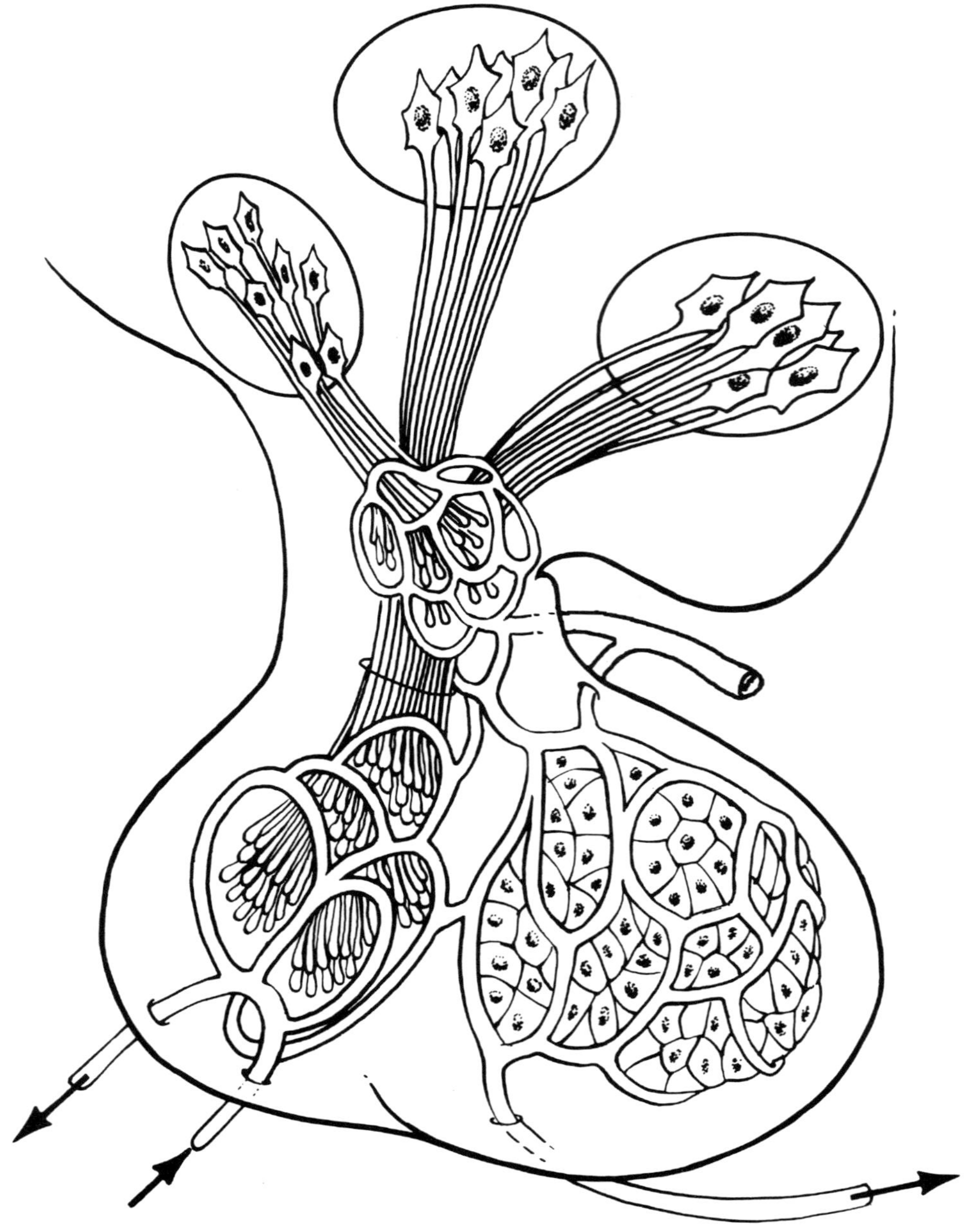

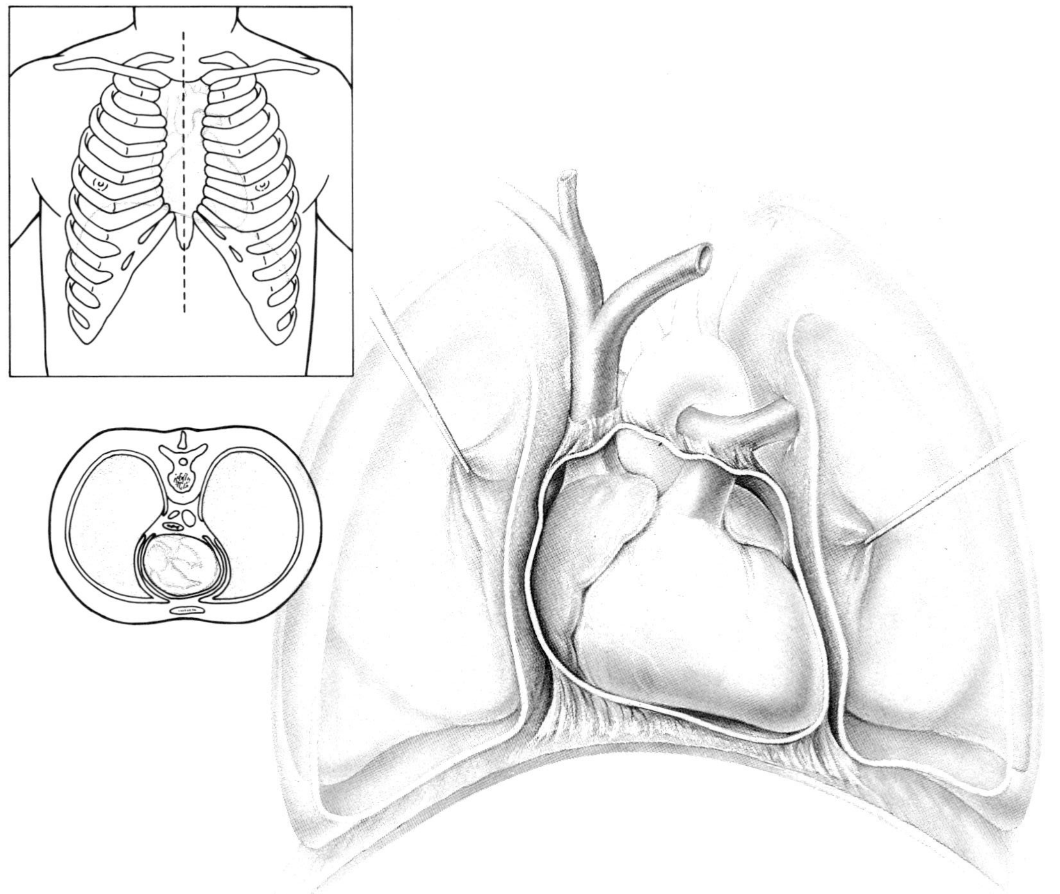

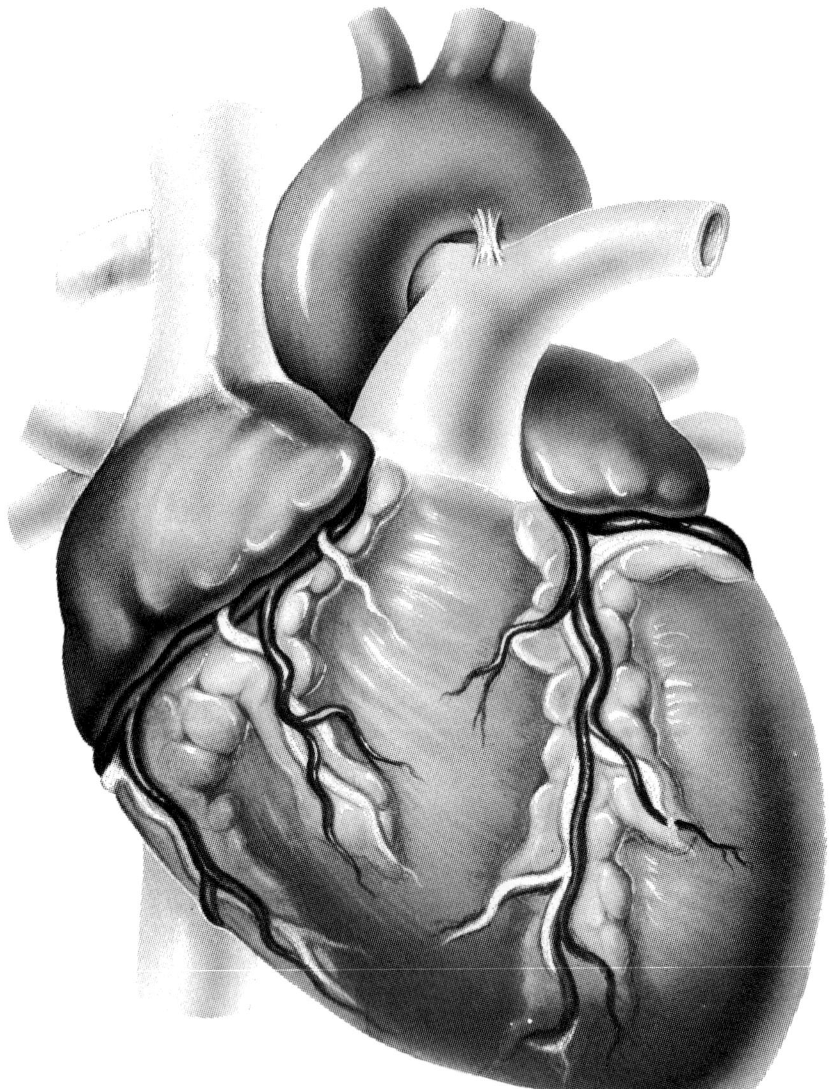

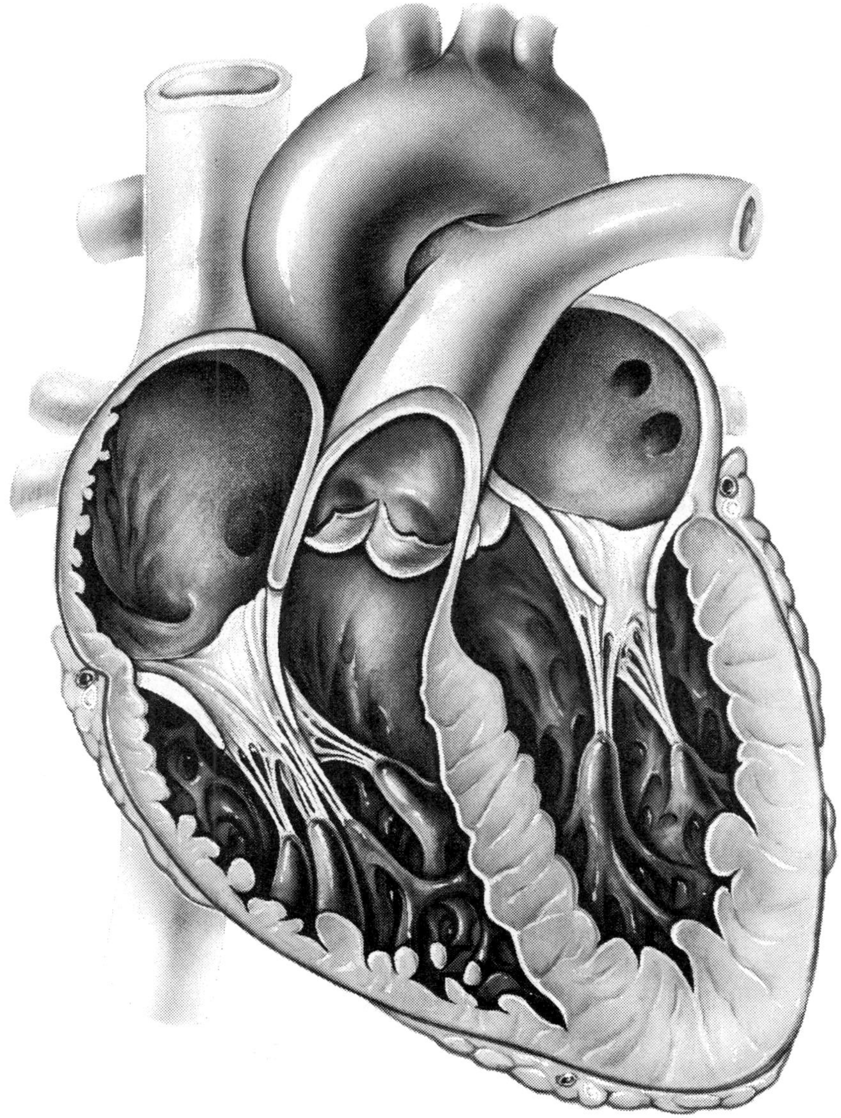

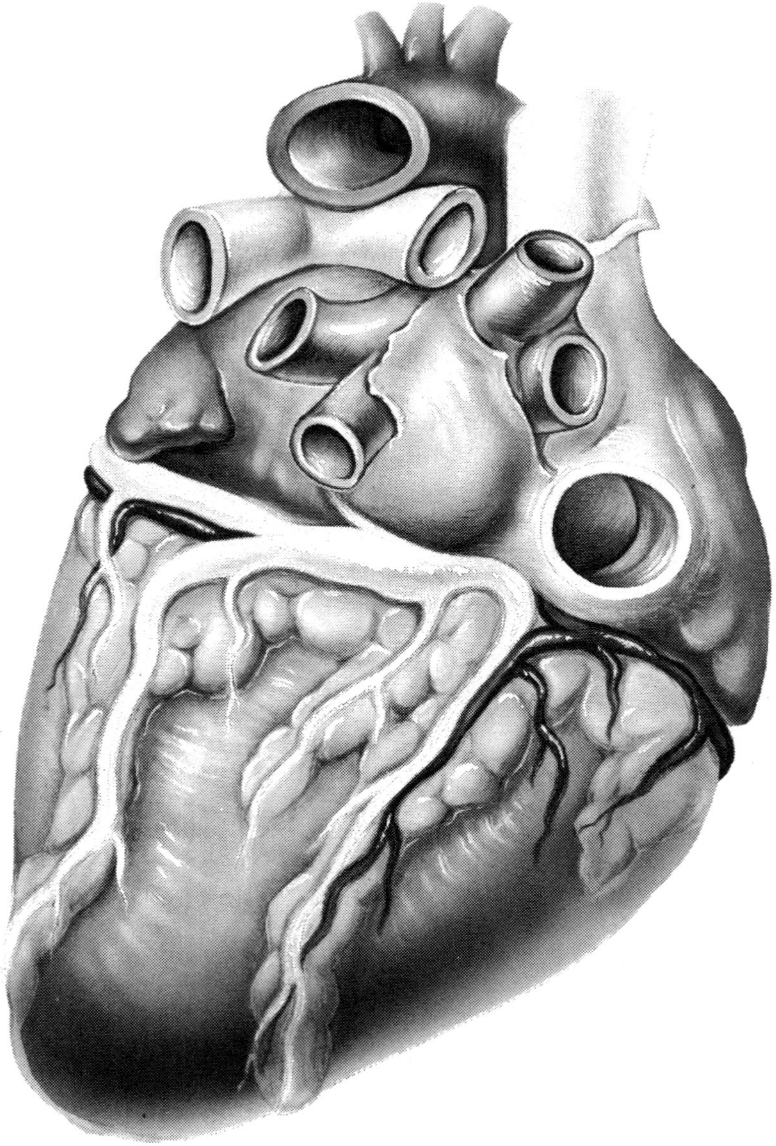

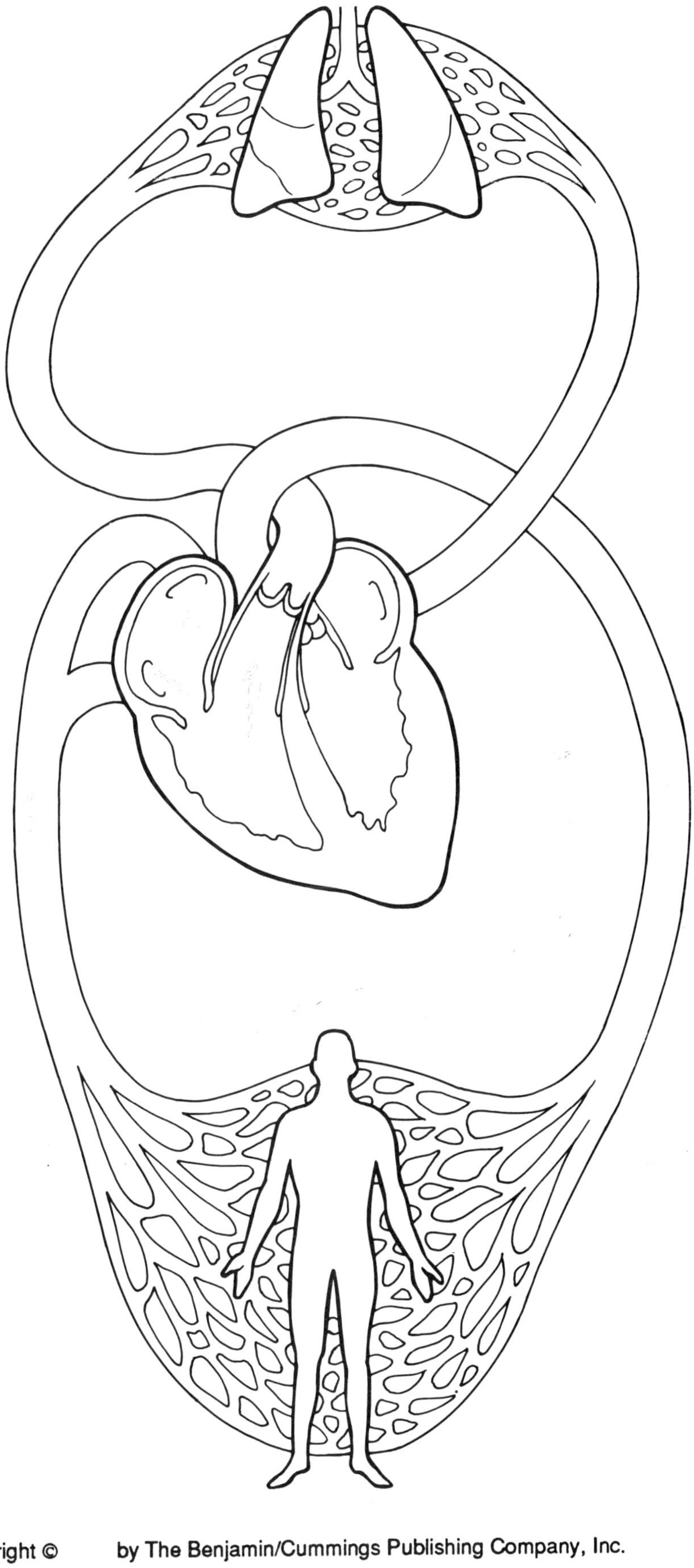

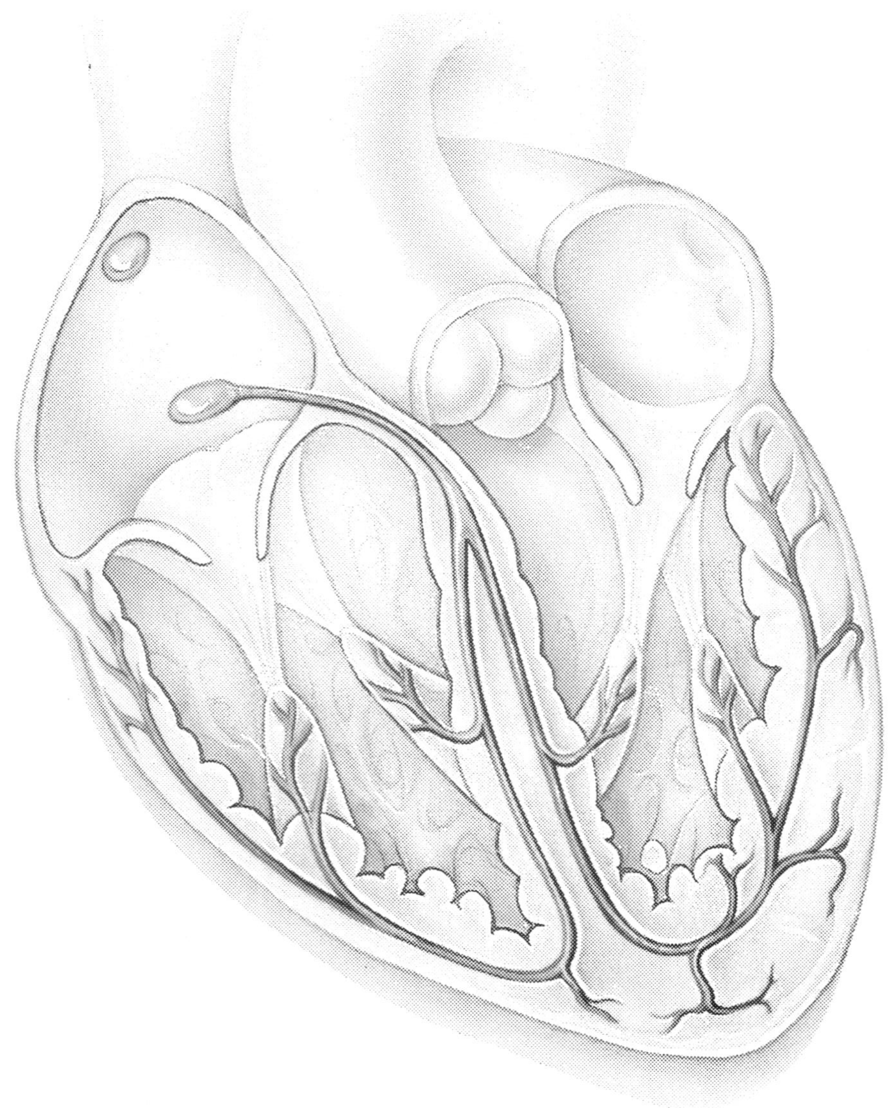

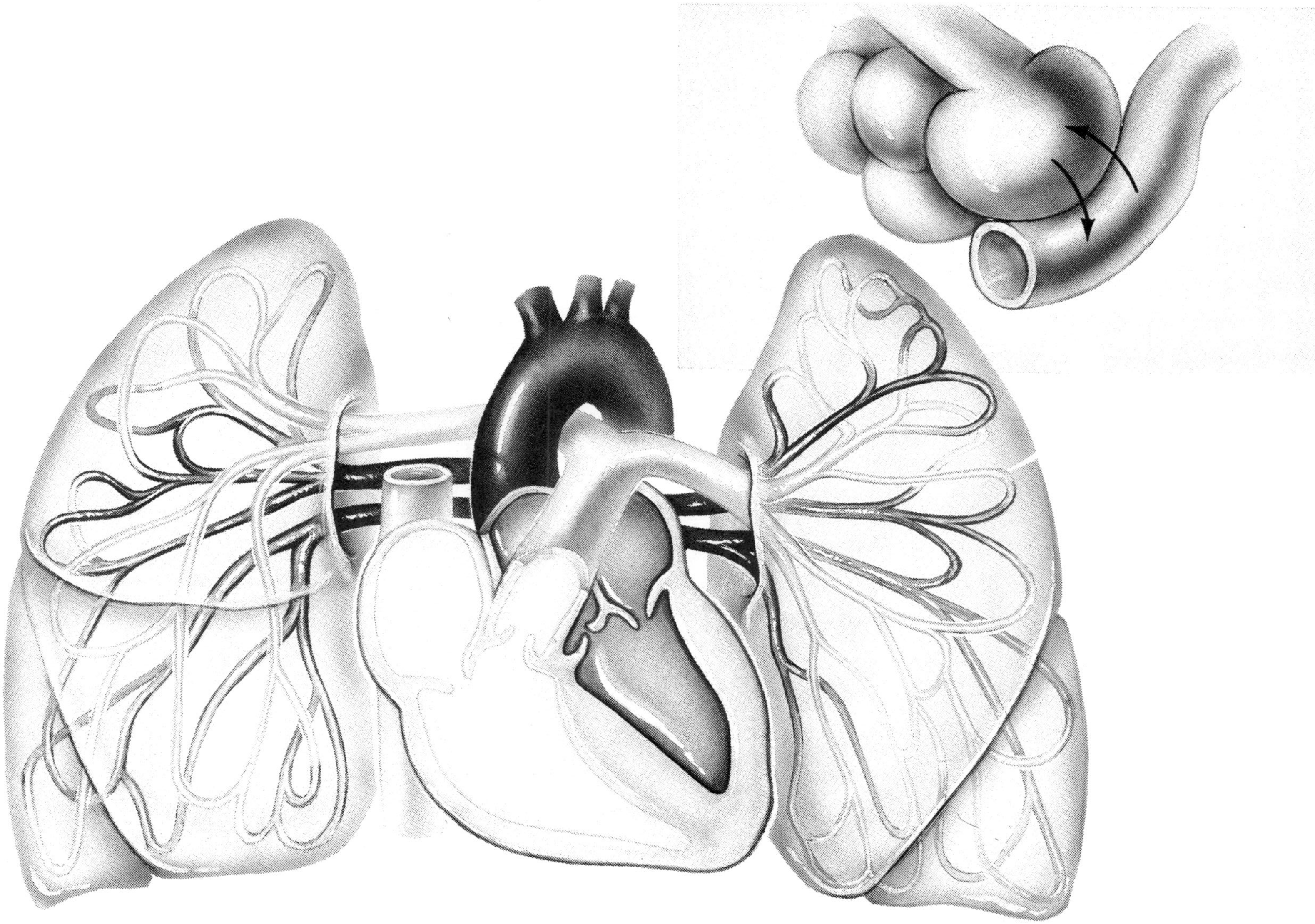

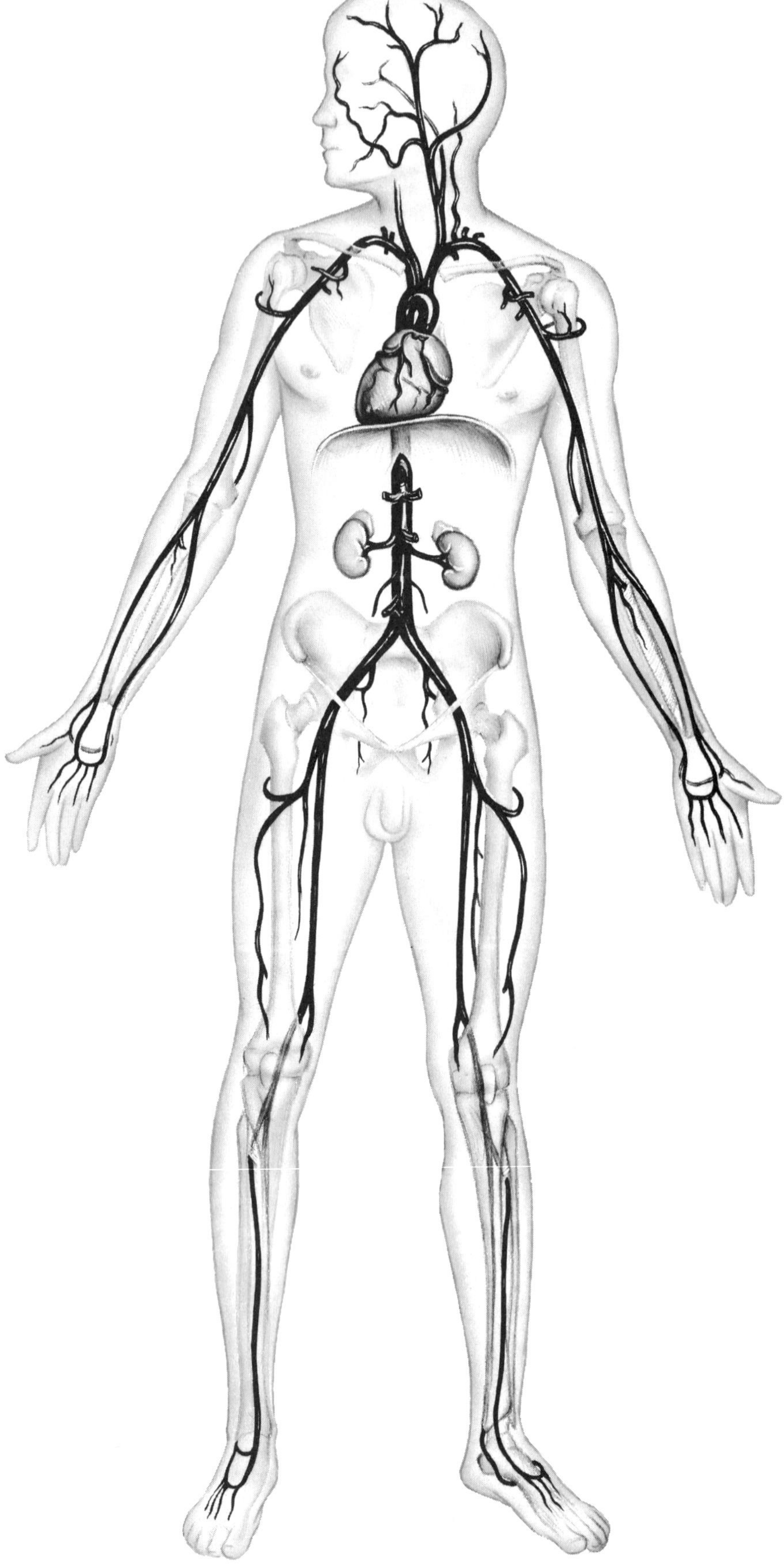

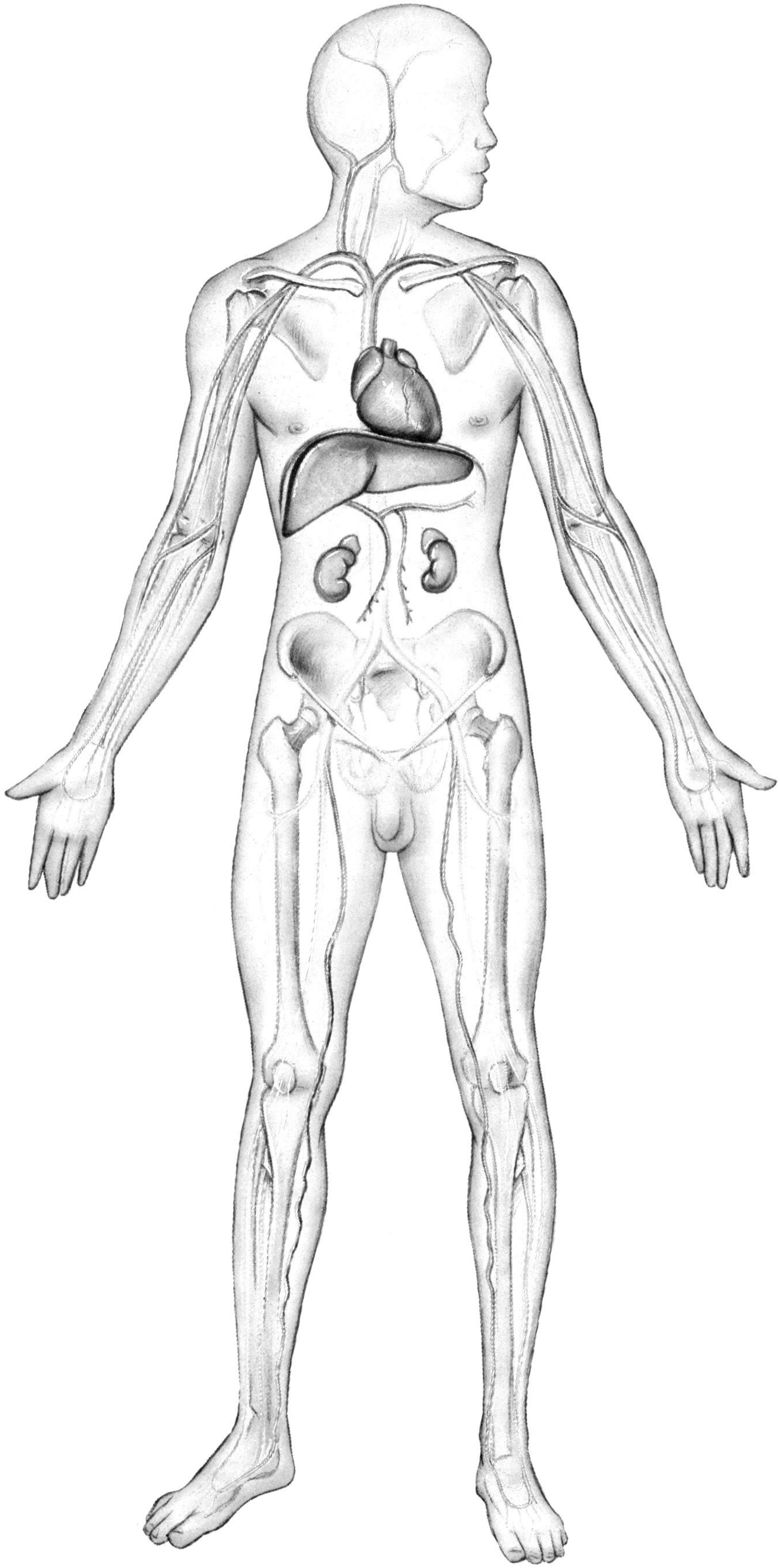

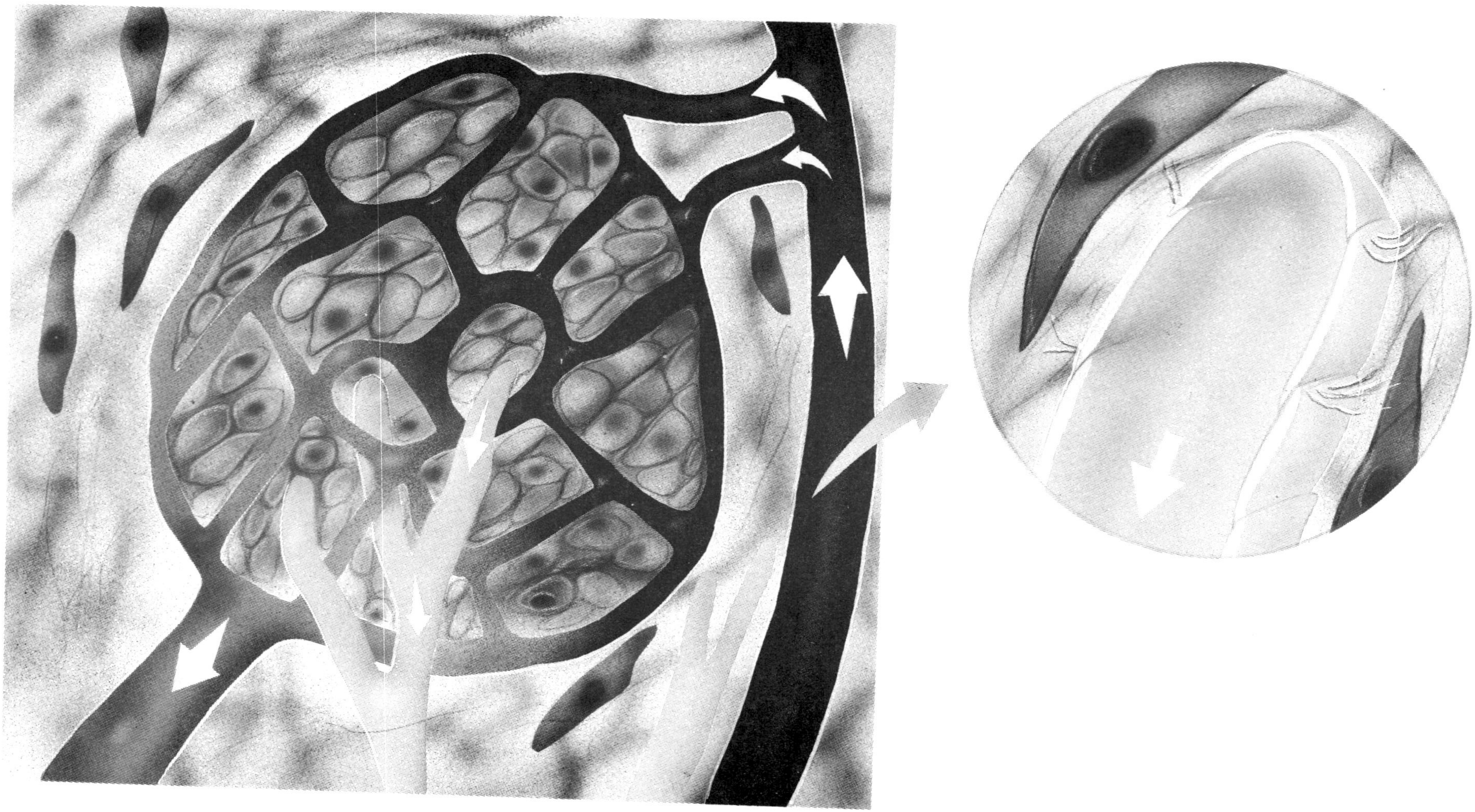

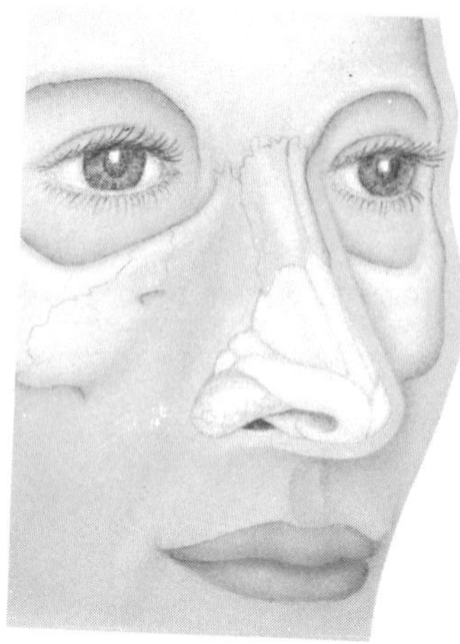

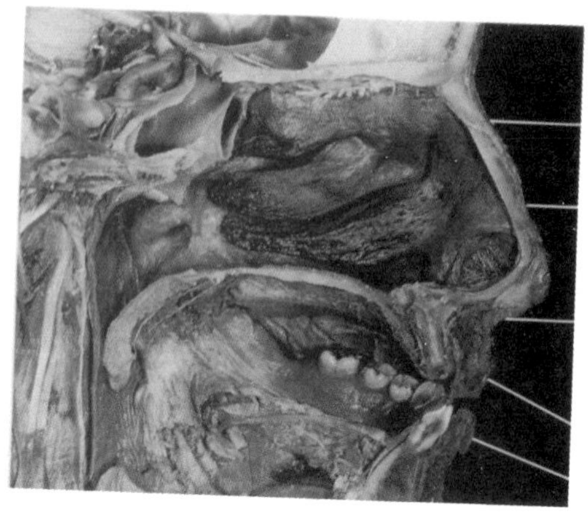

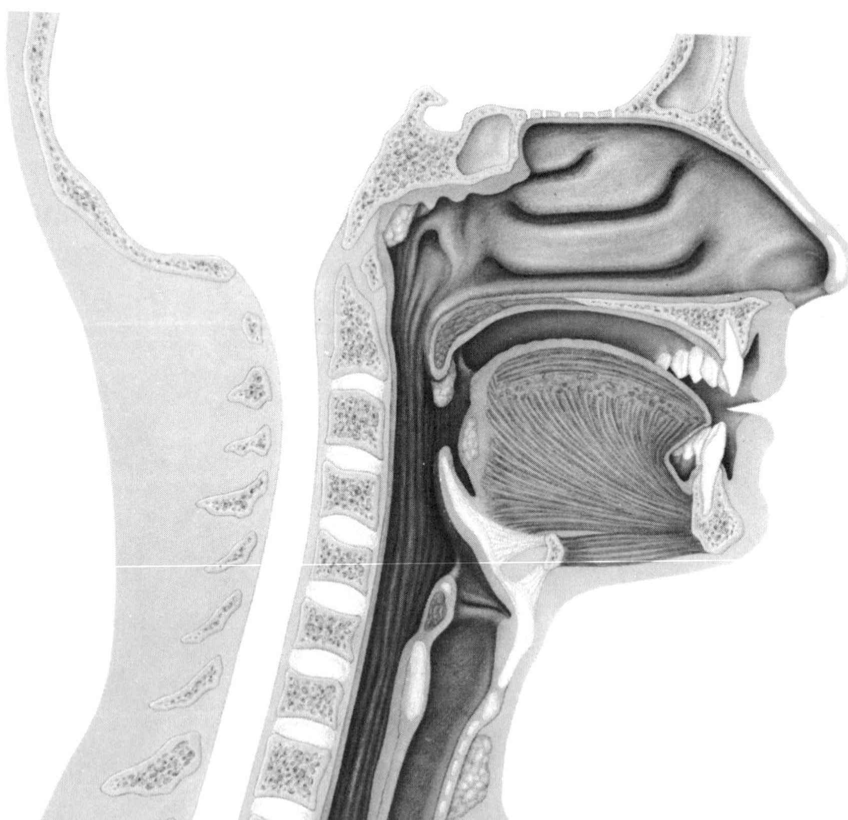

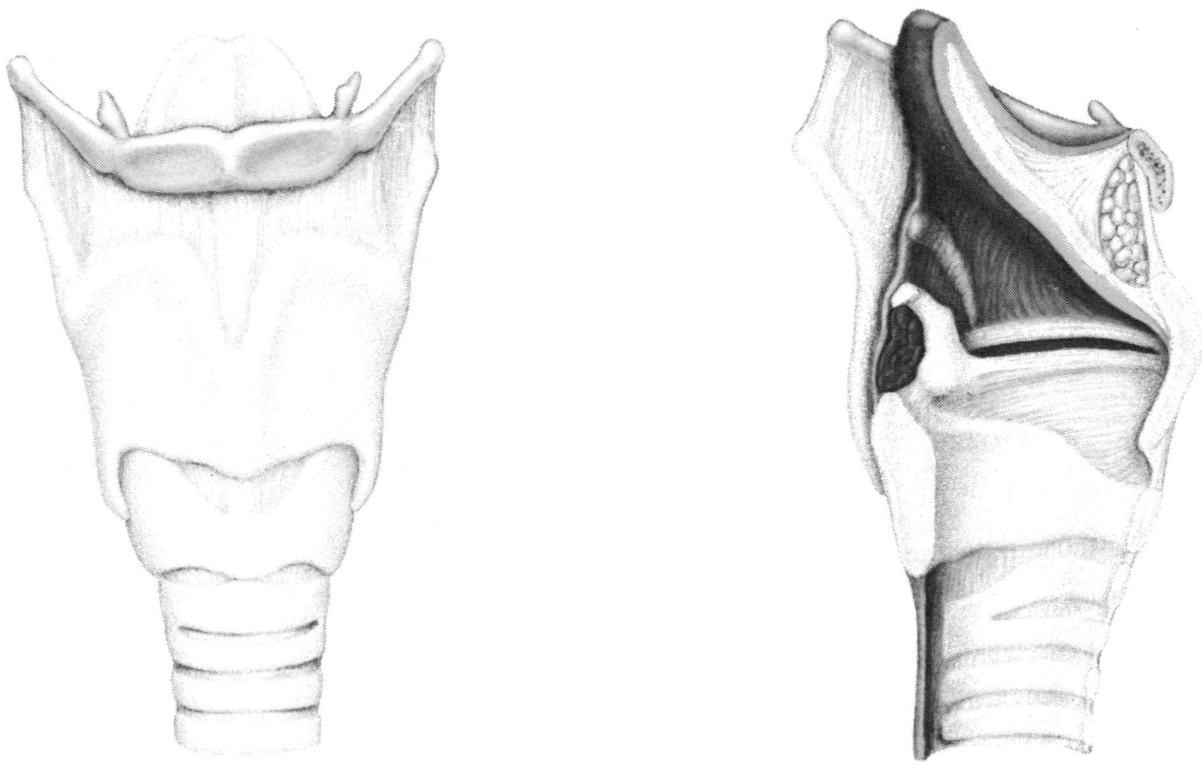

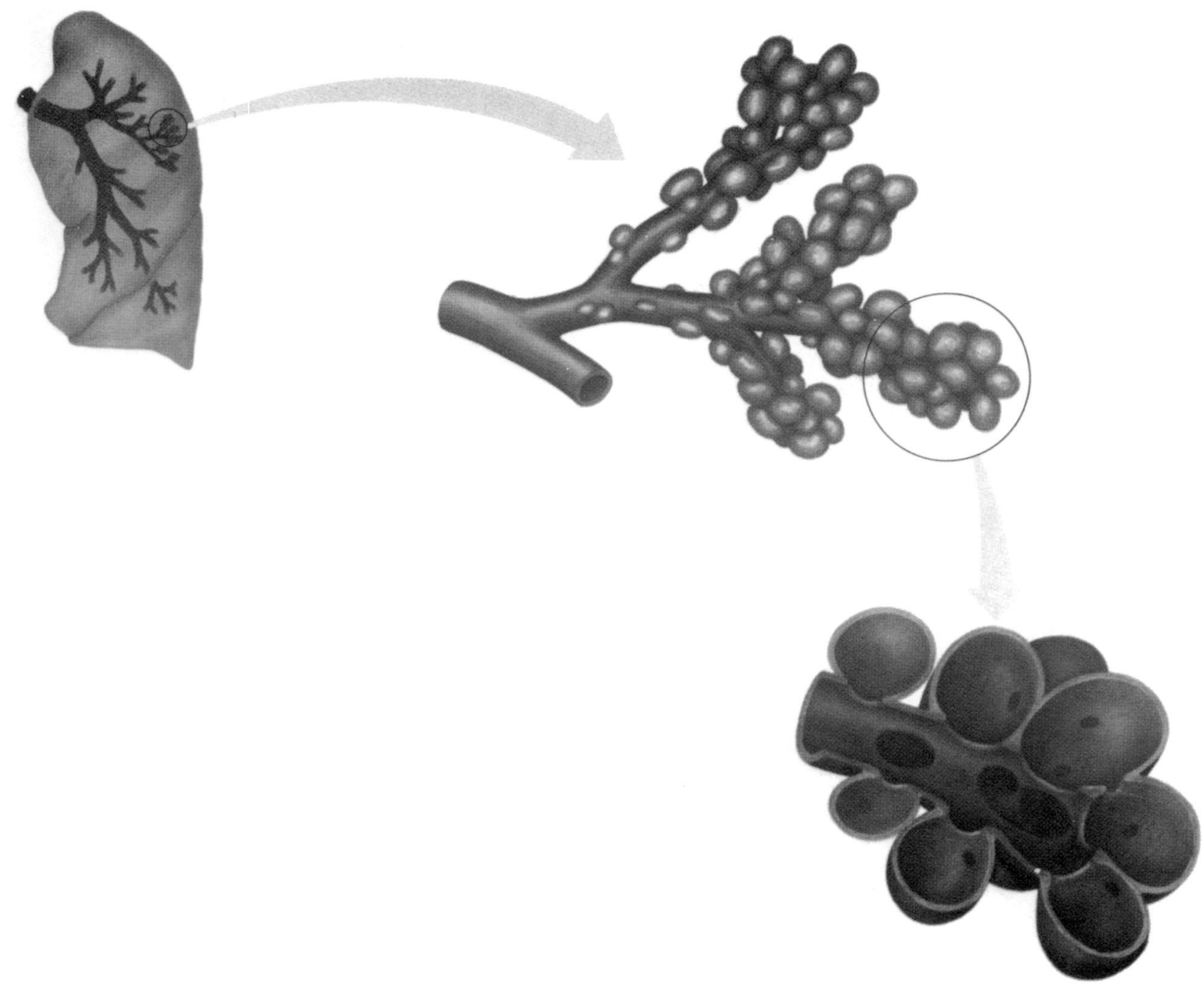

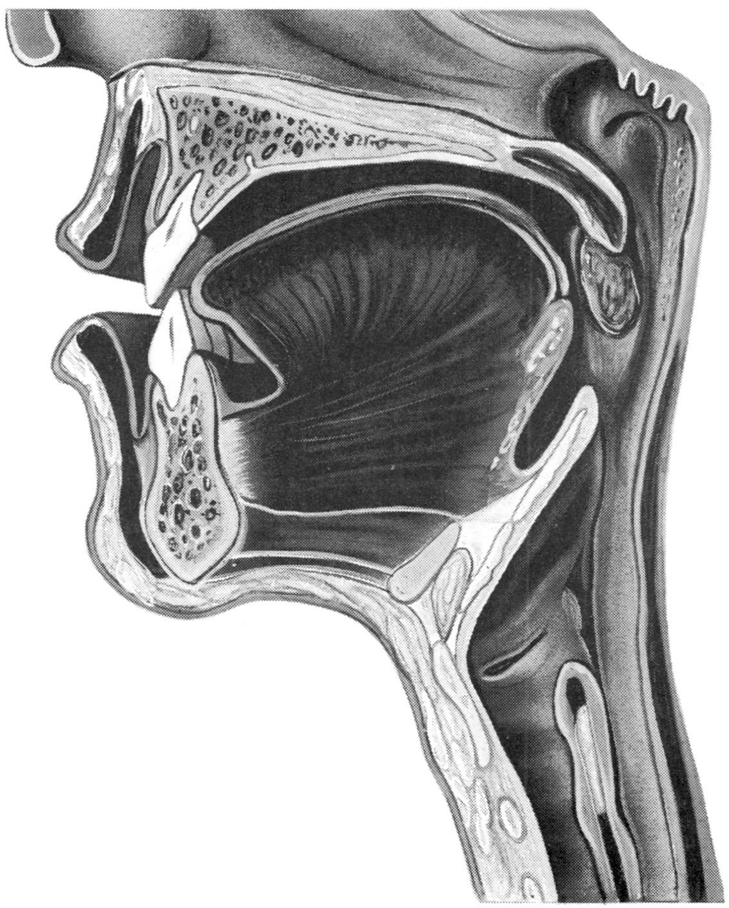

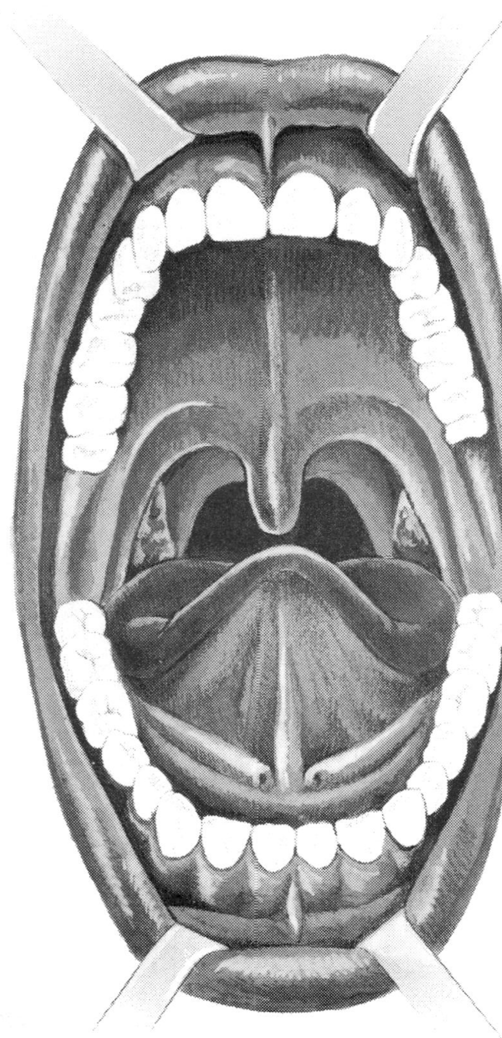

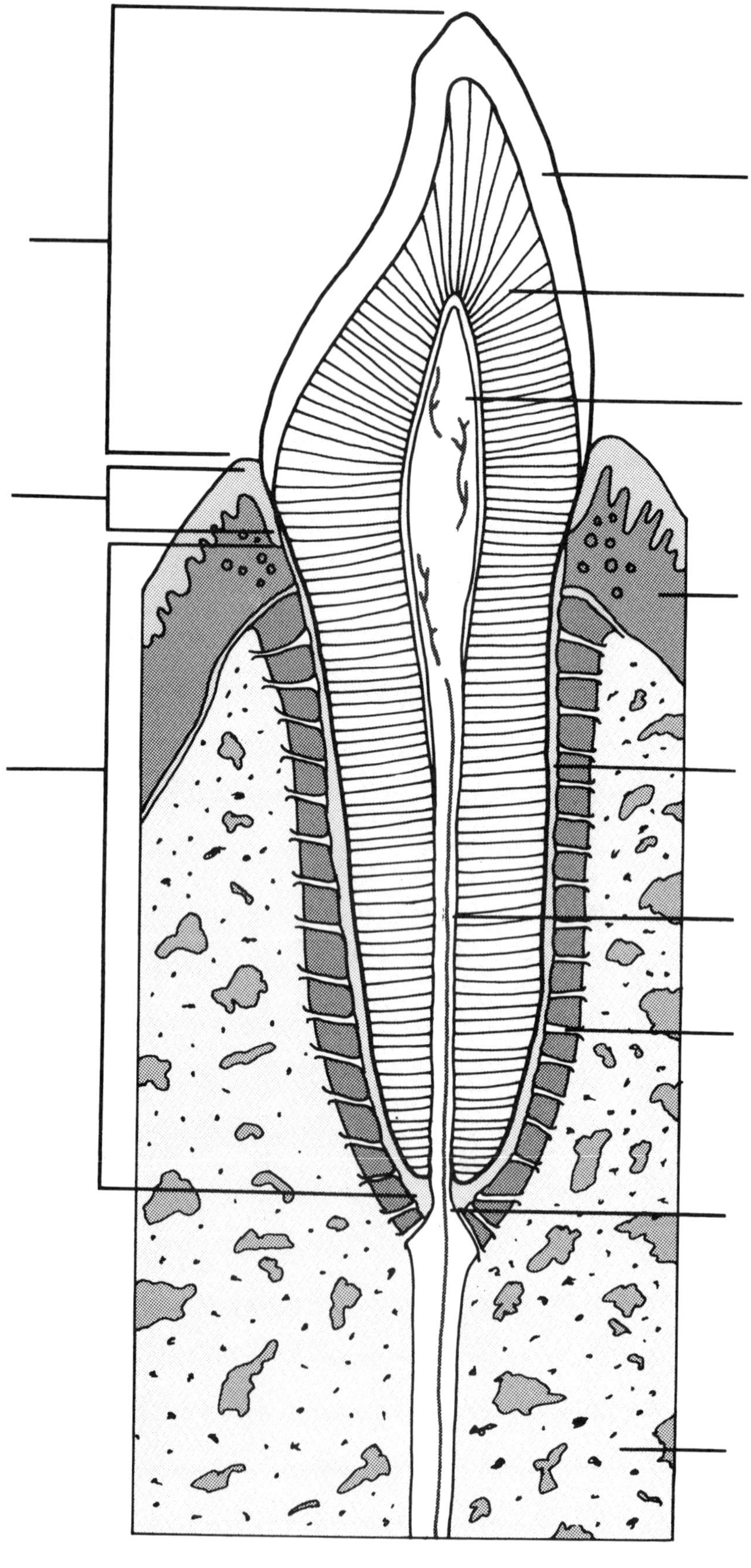

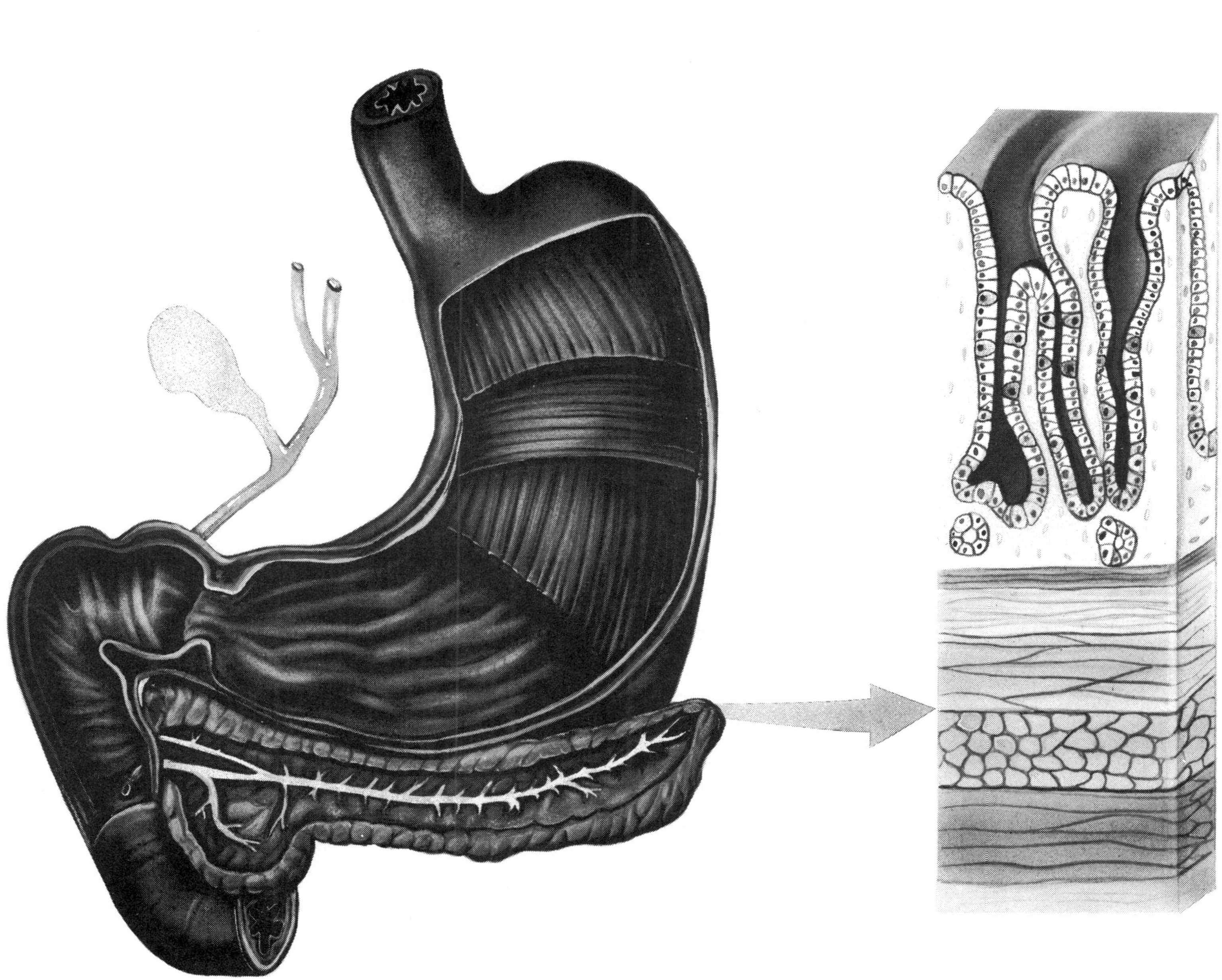

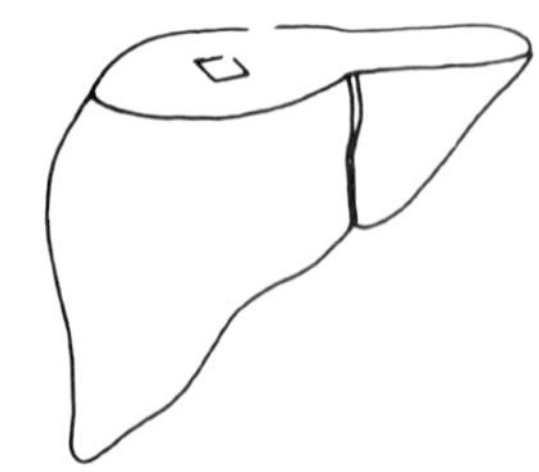

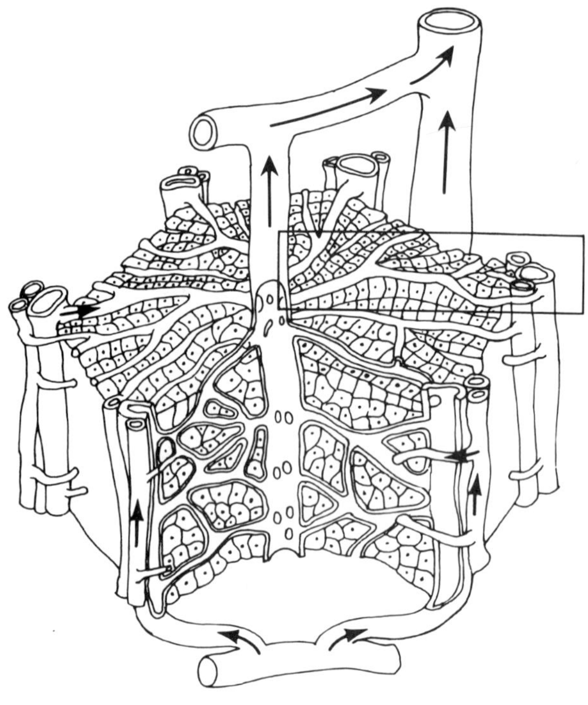

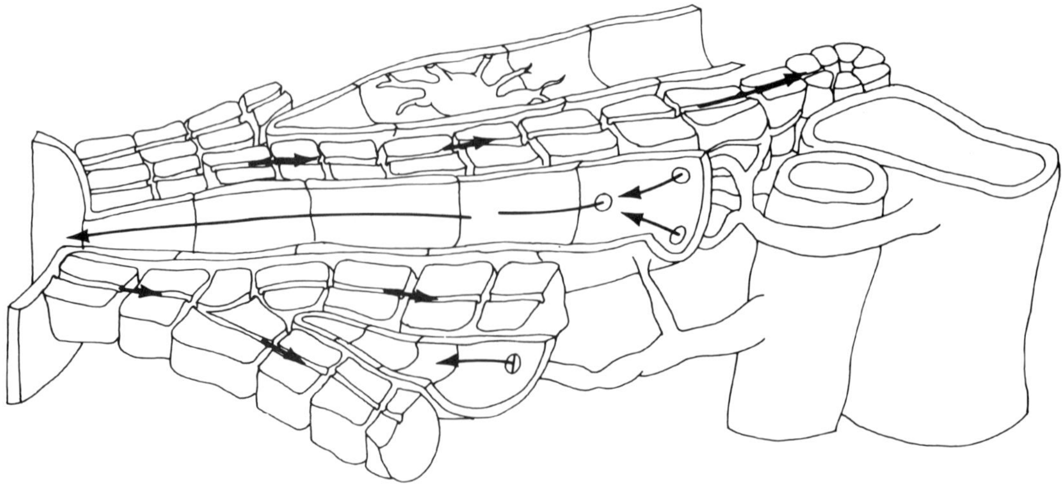

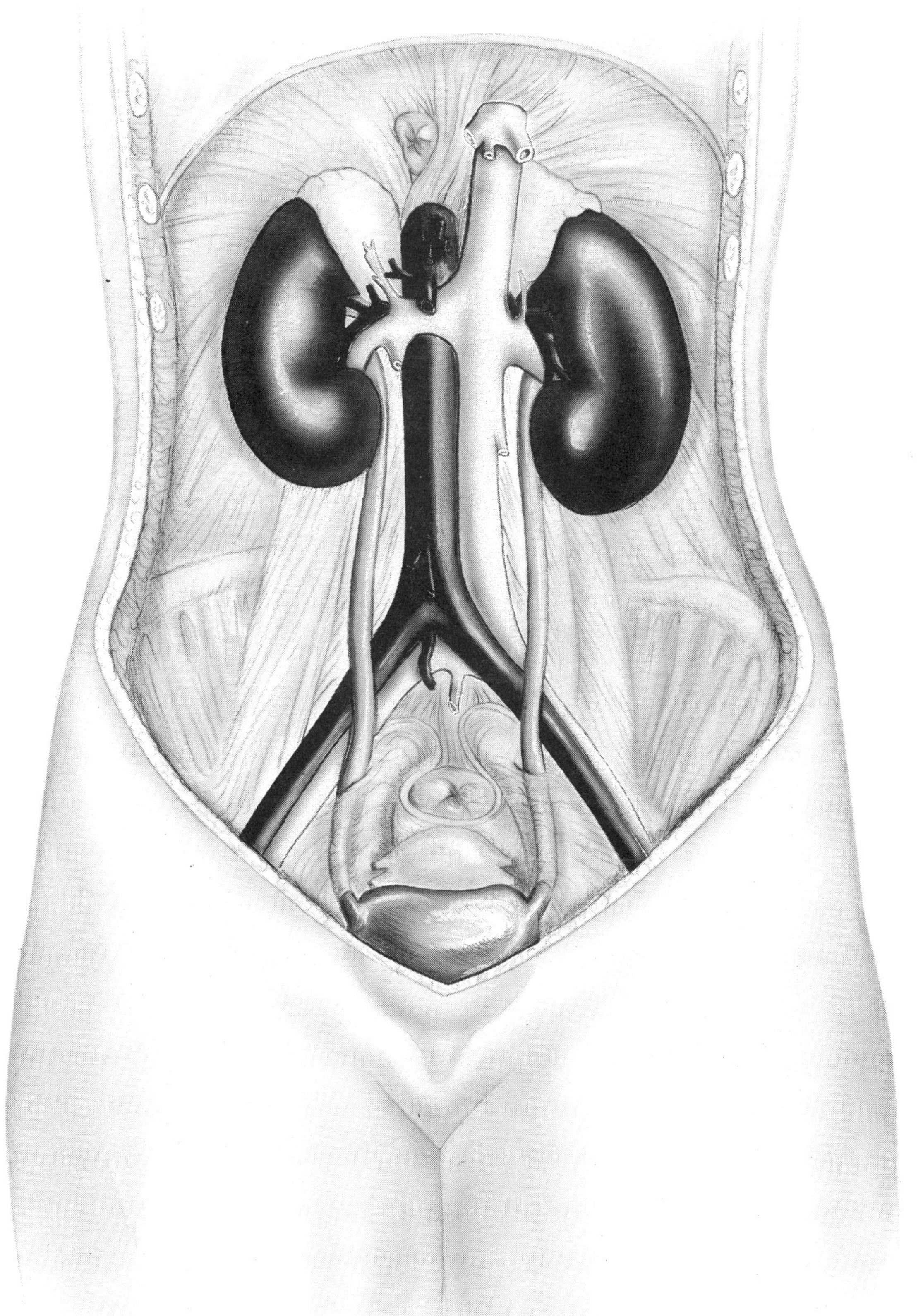

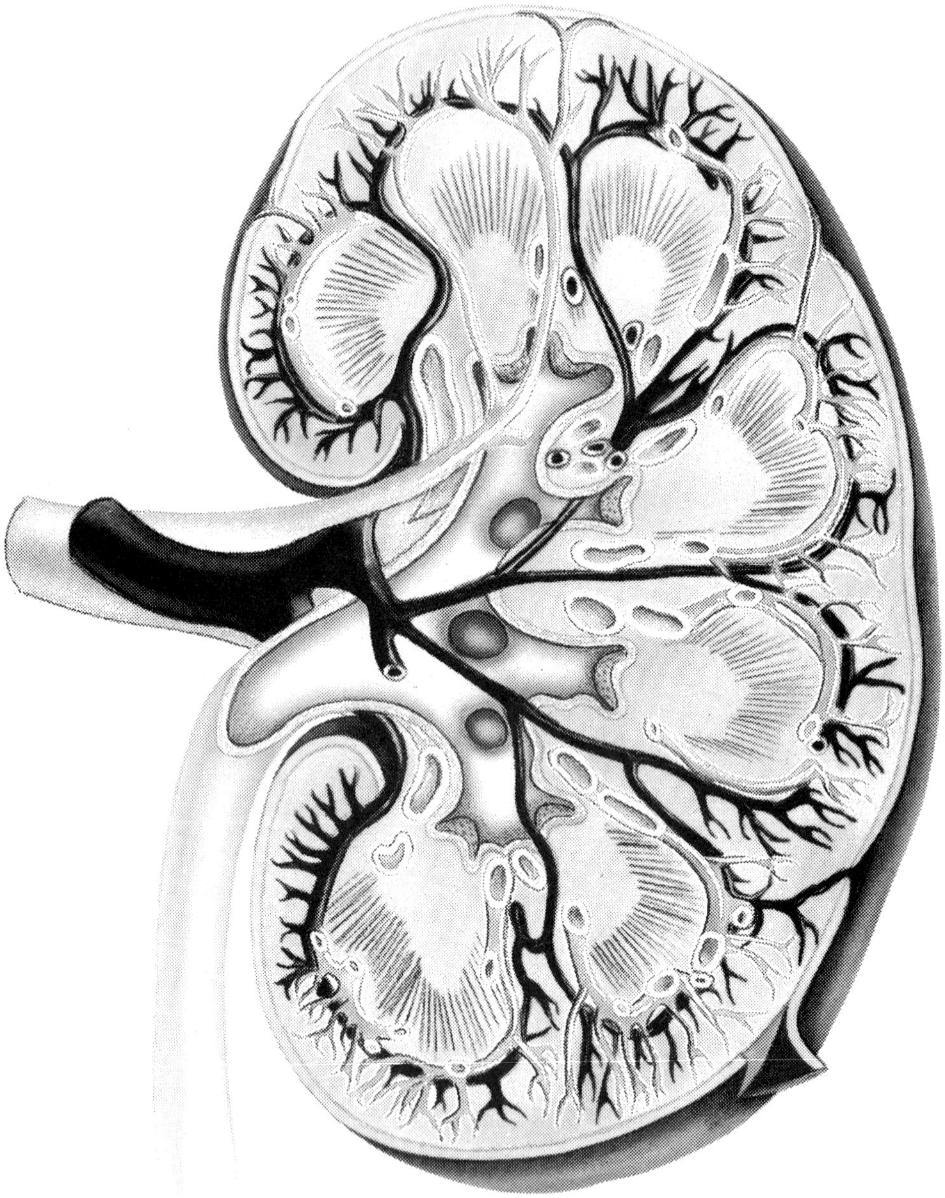

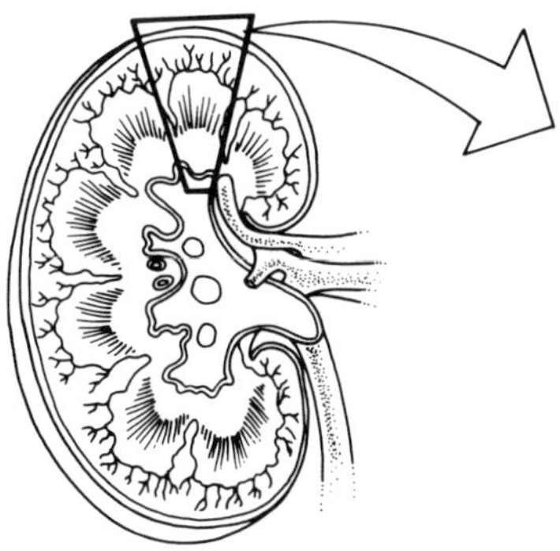

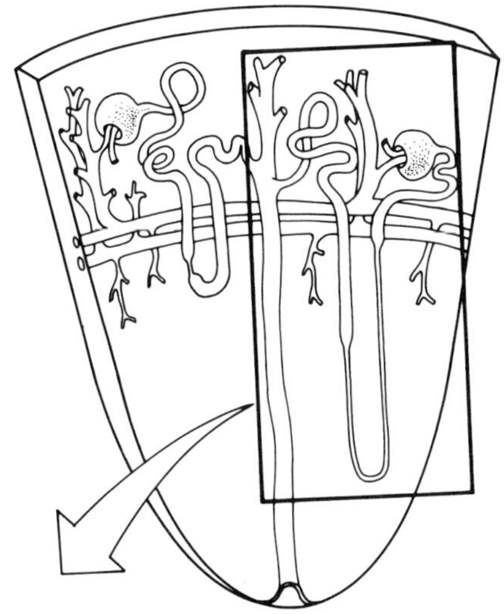

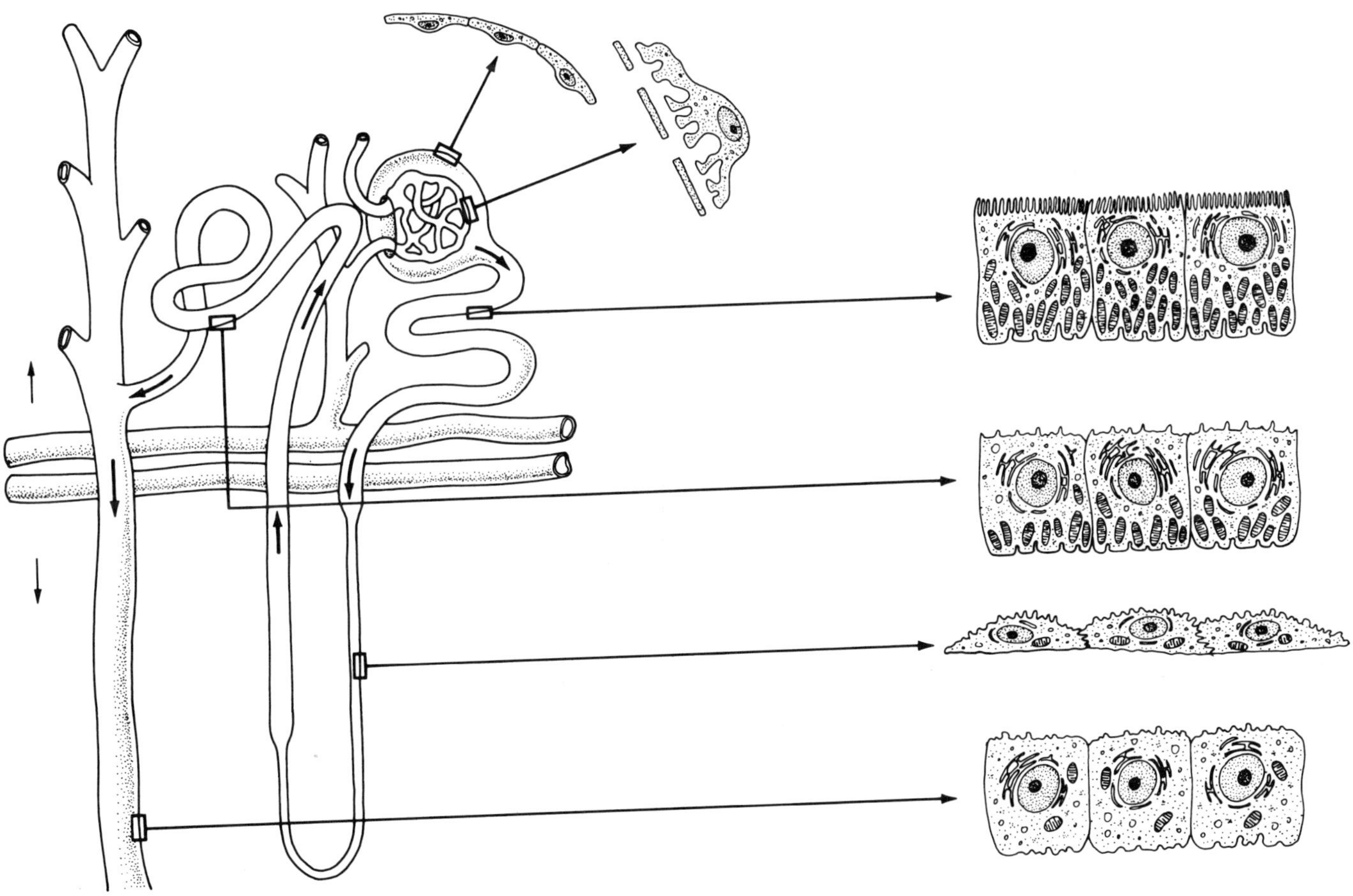

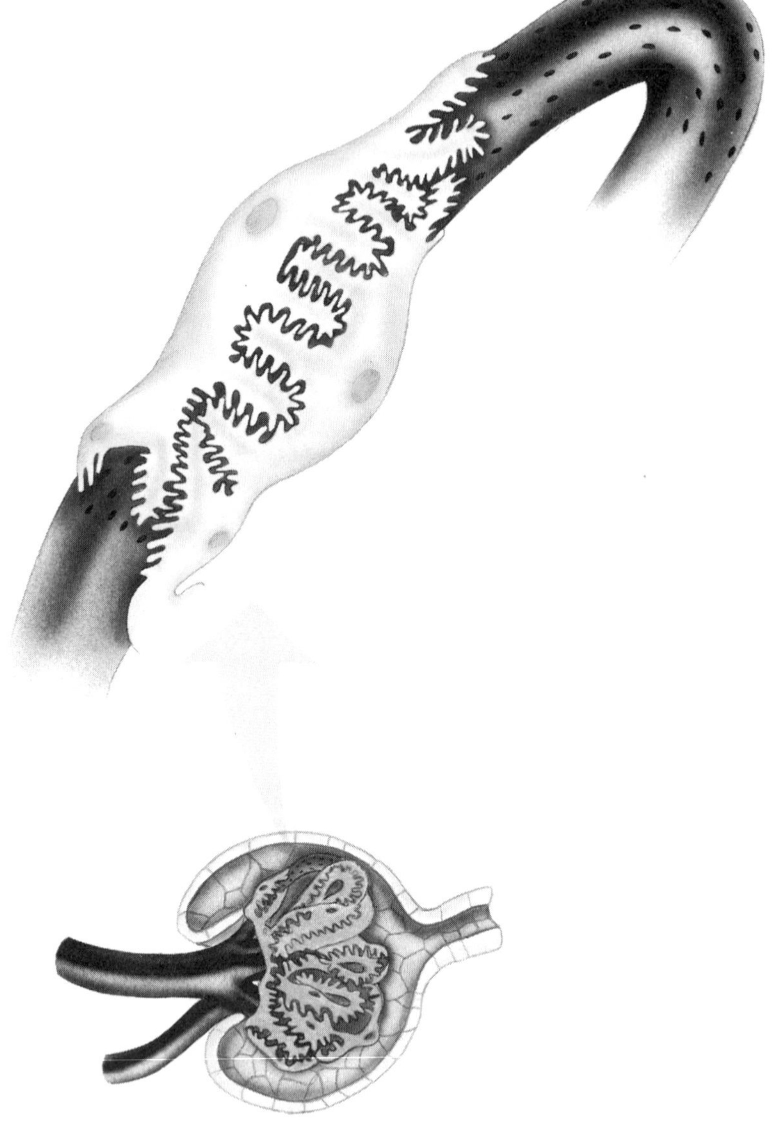

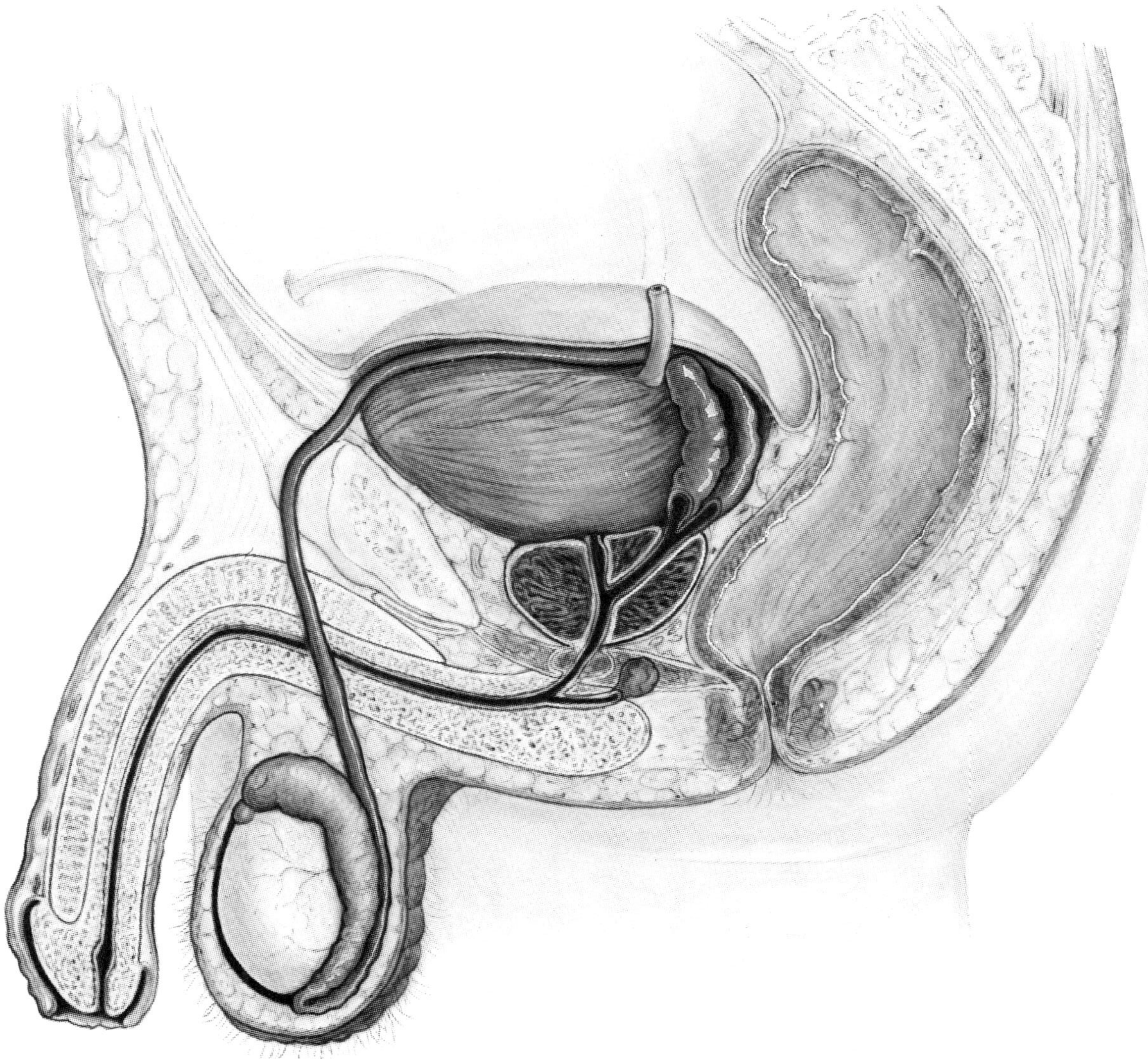

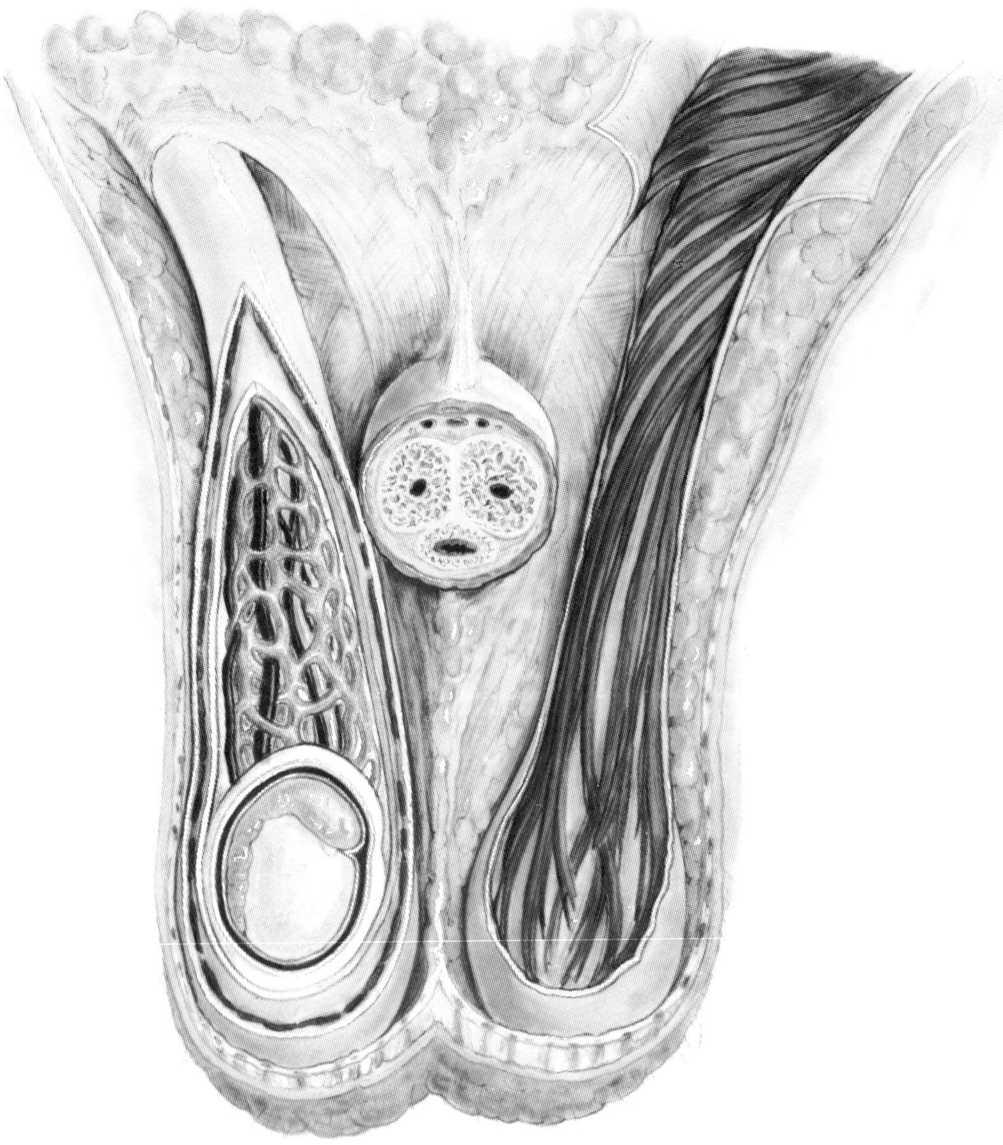

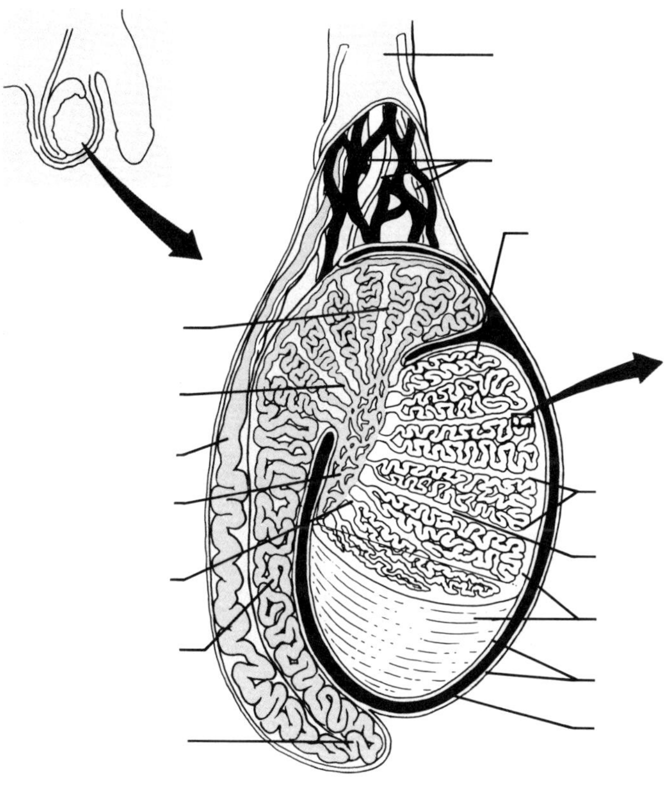

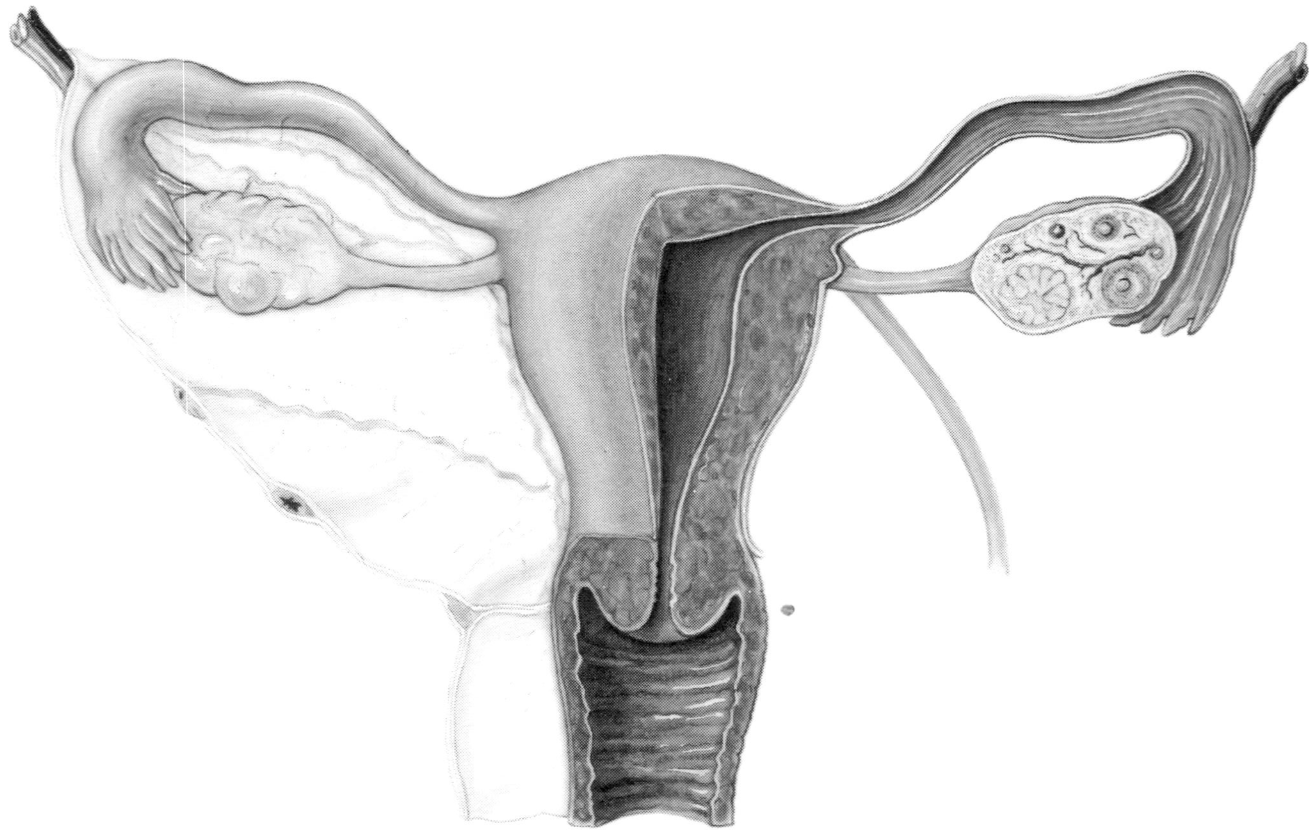

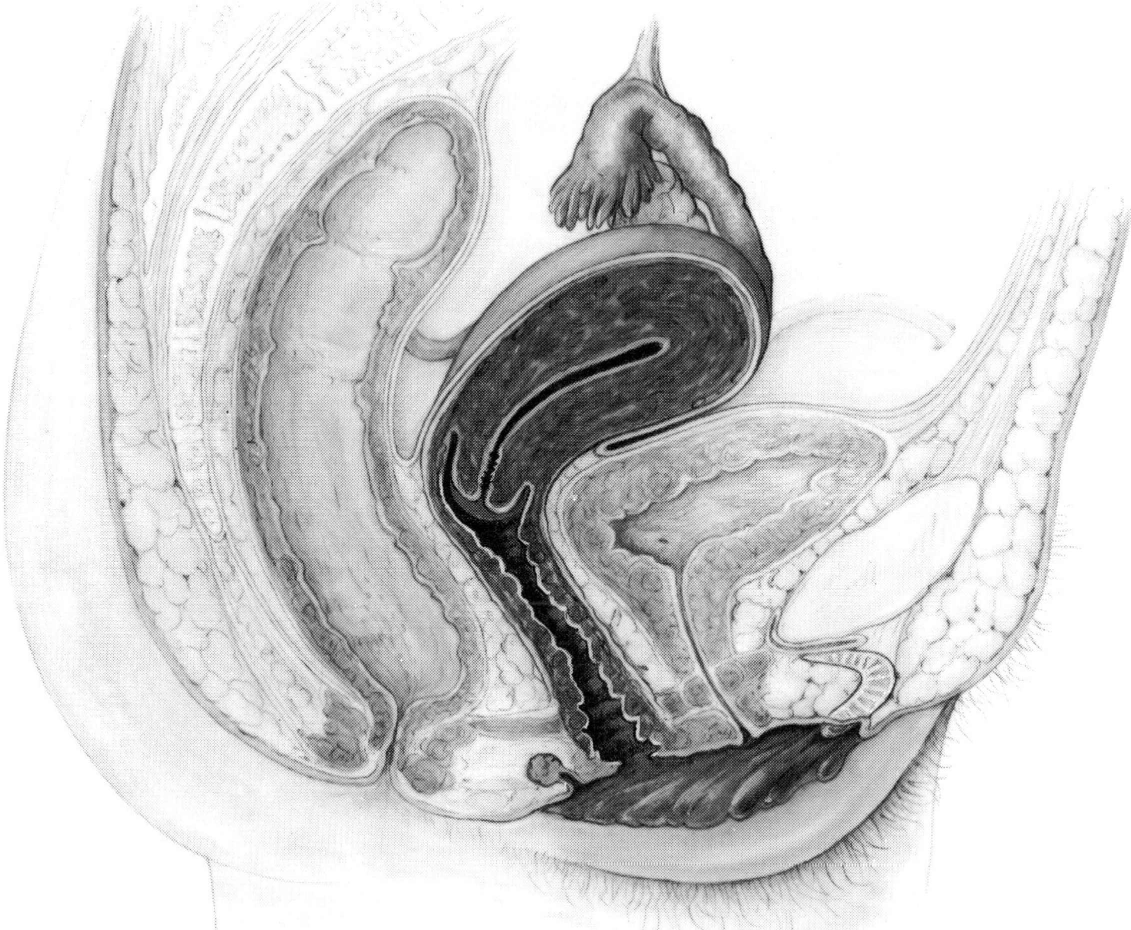

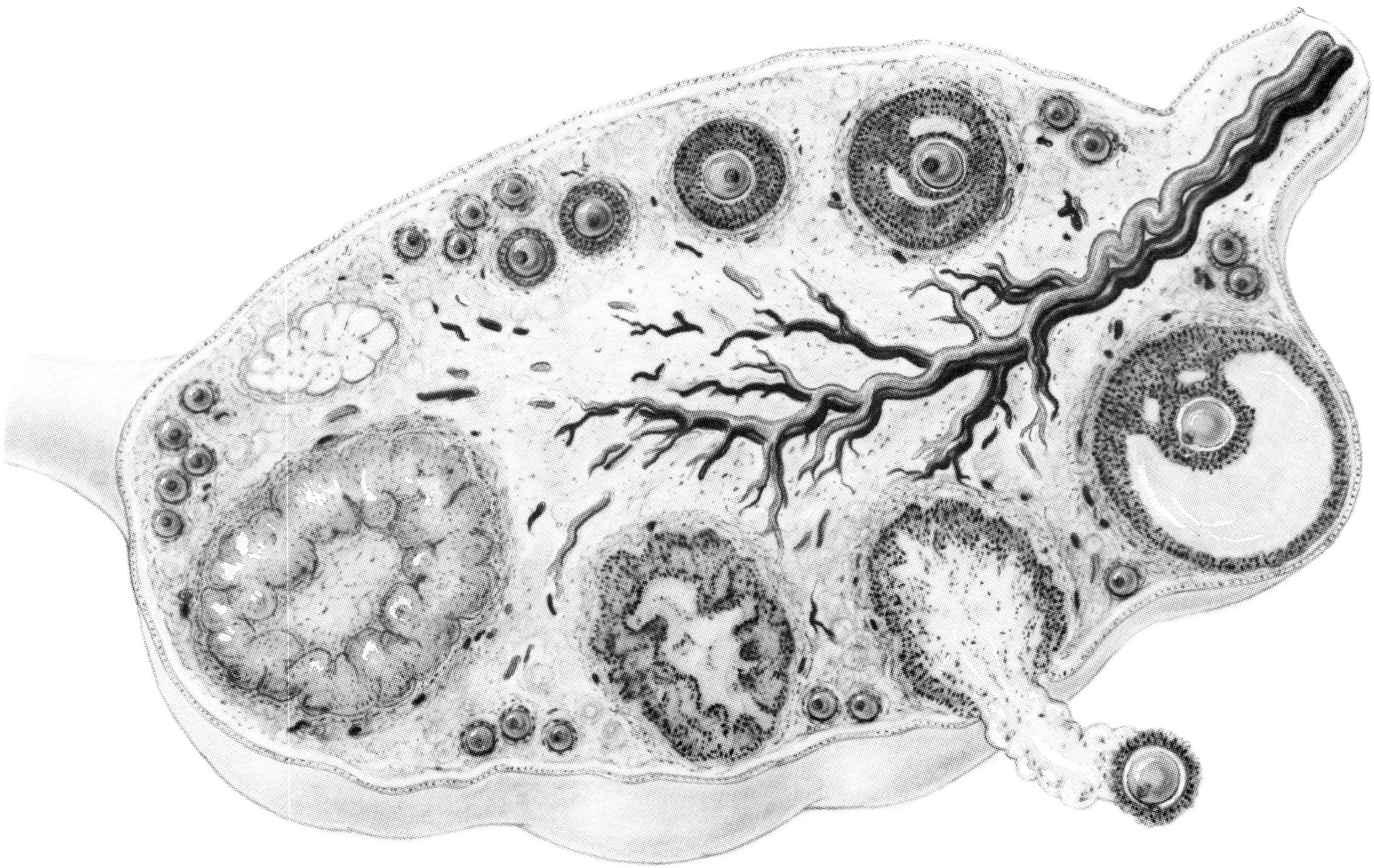

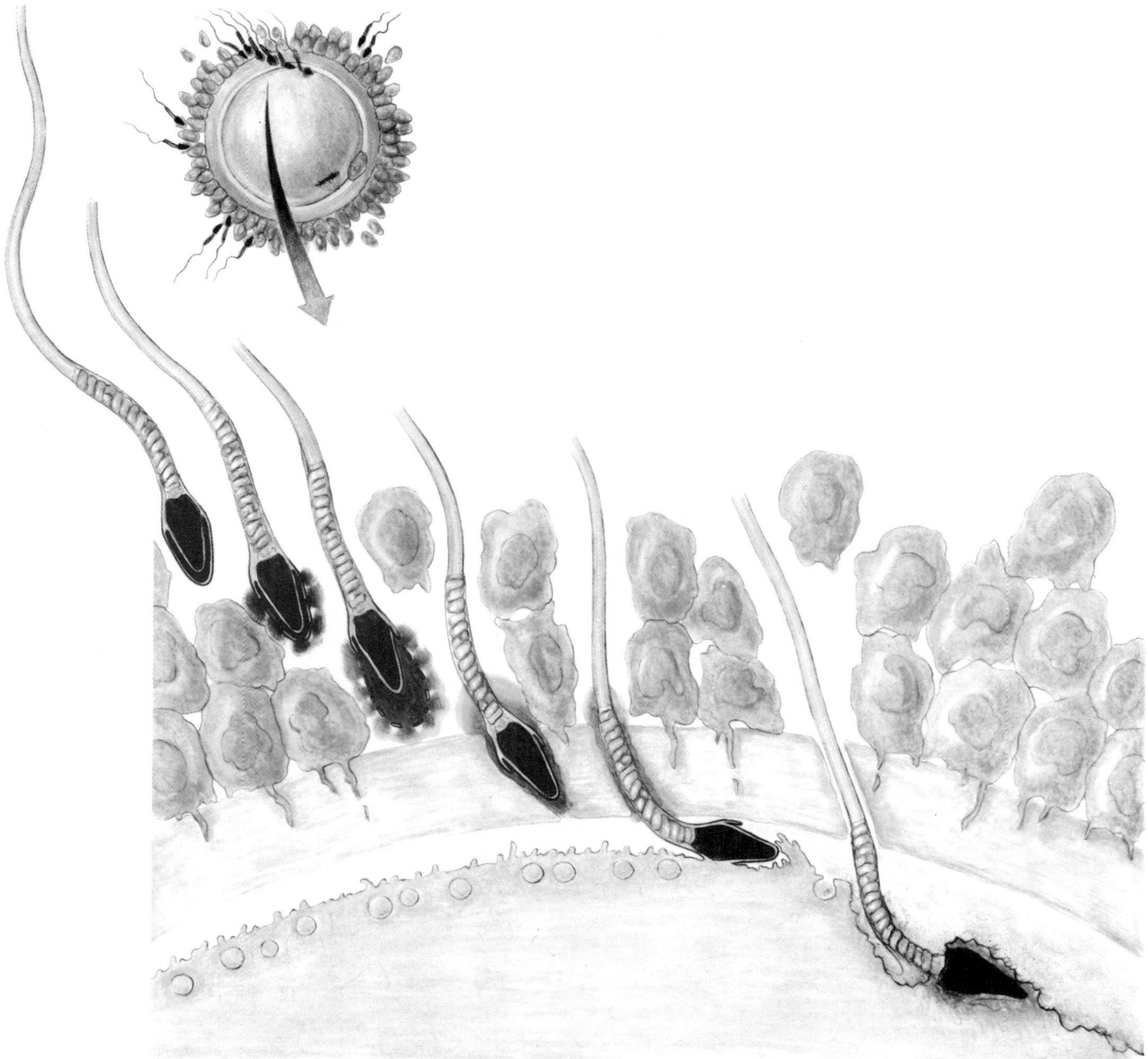

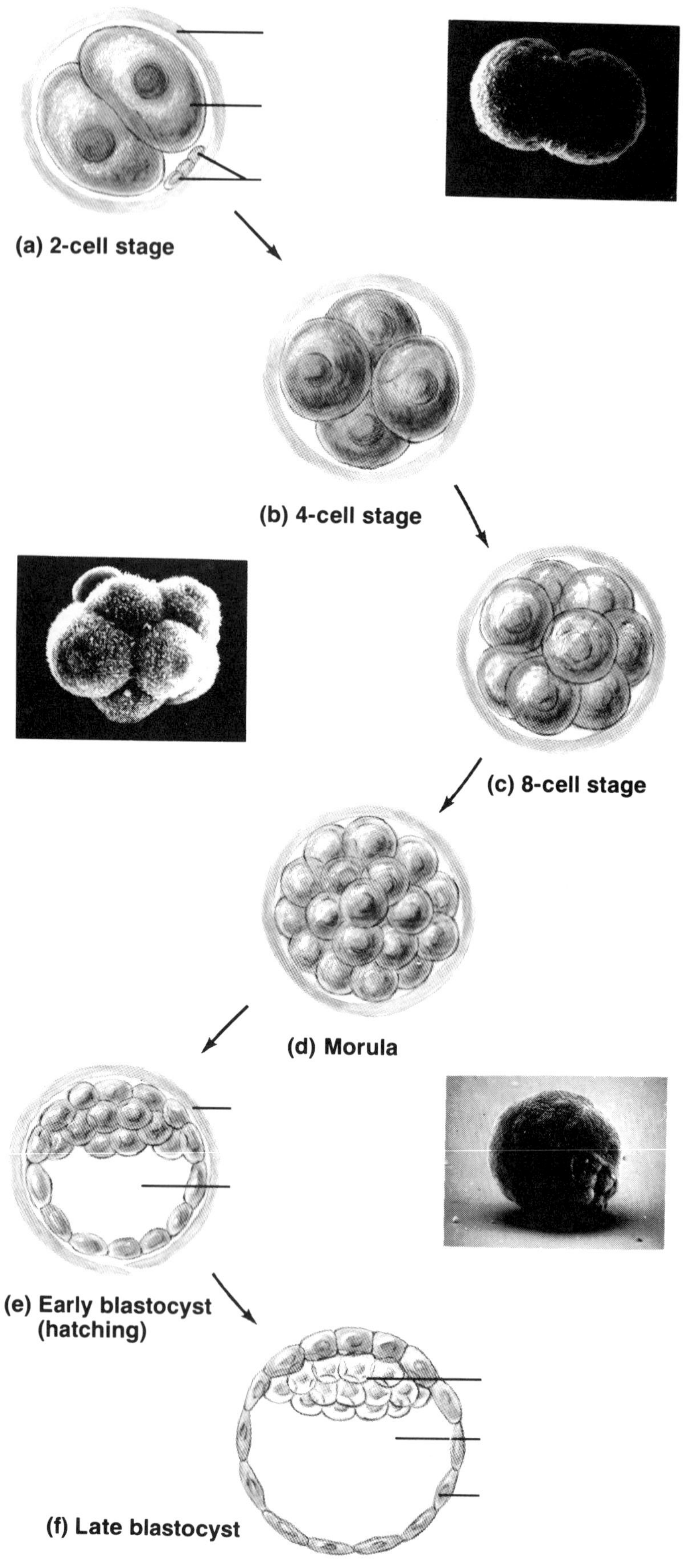
(a) 2-cell stage
(b) 4-cell stage
(c) 8-cell stage
(d) Morula
(e) Early blastocyst (hatching)
(f) Late blastocyst

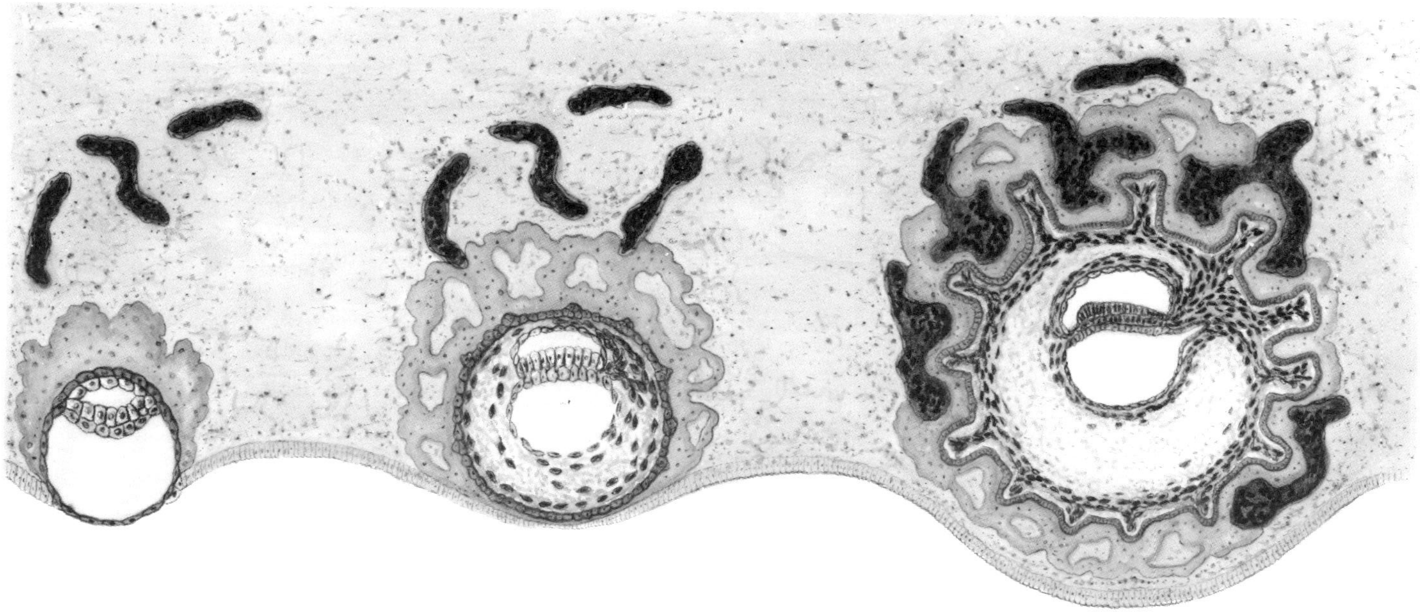

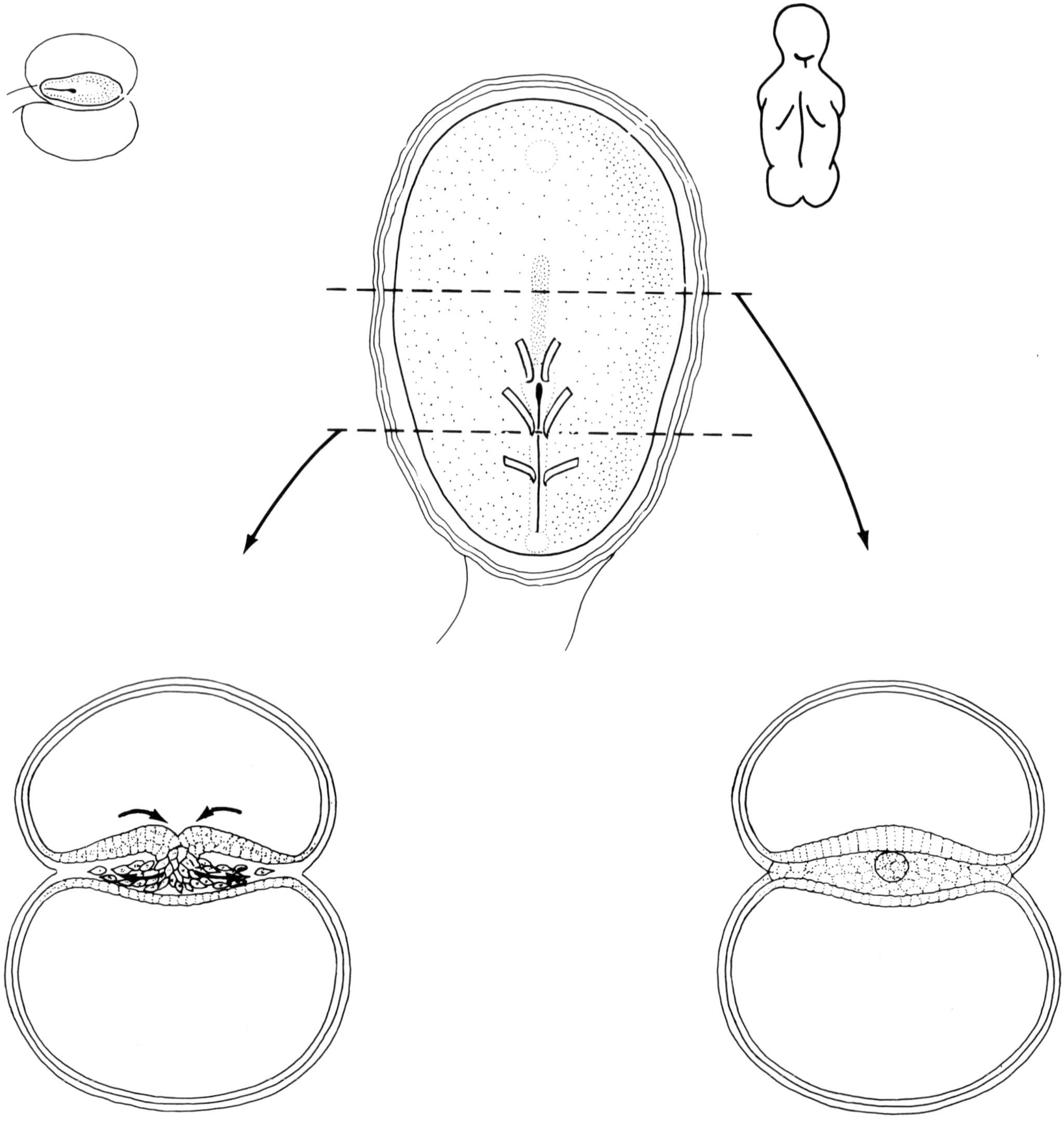

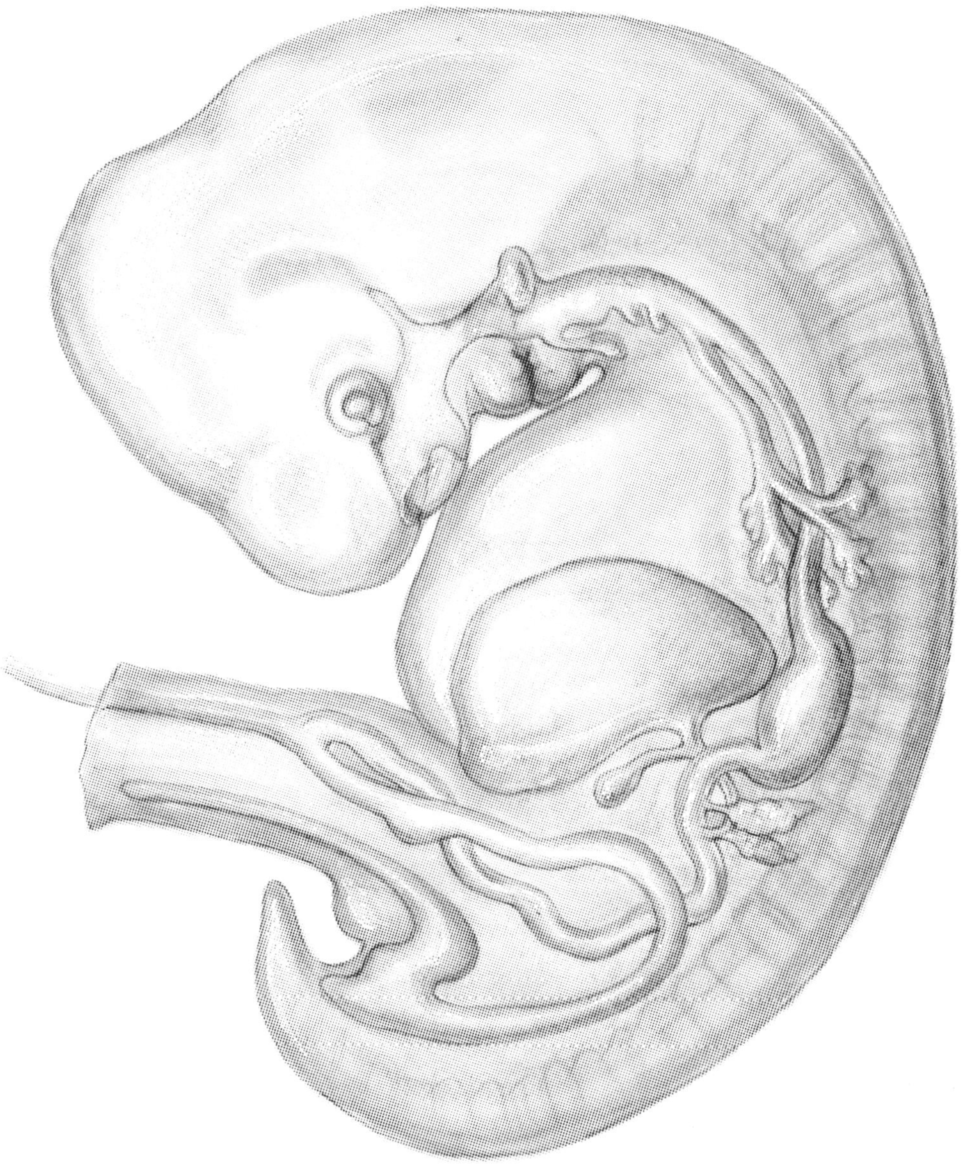

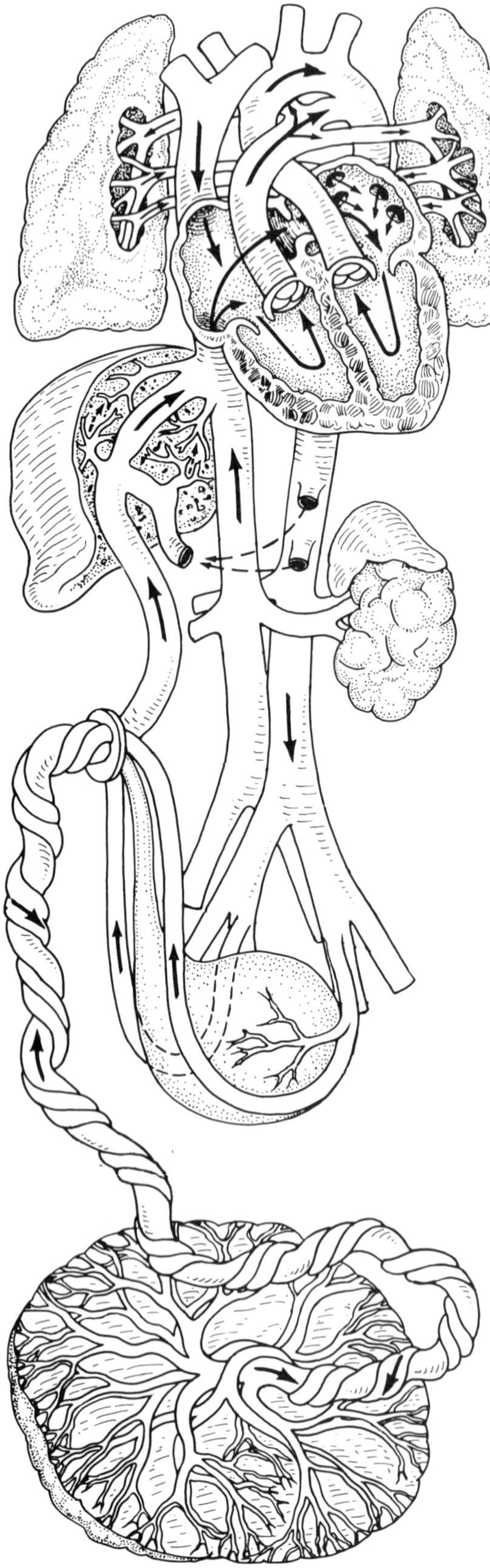

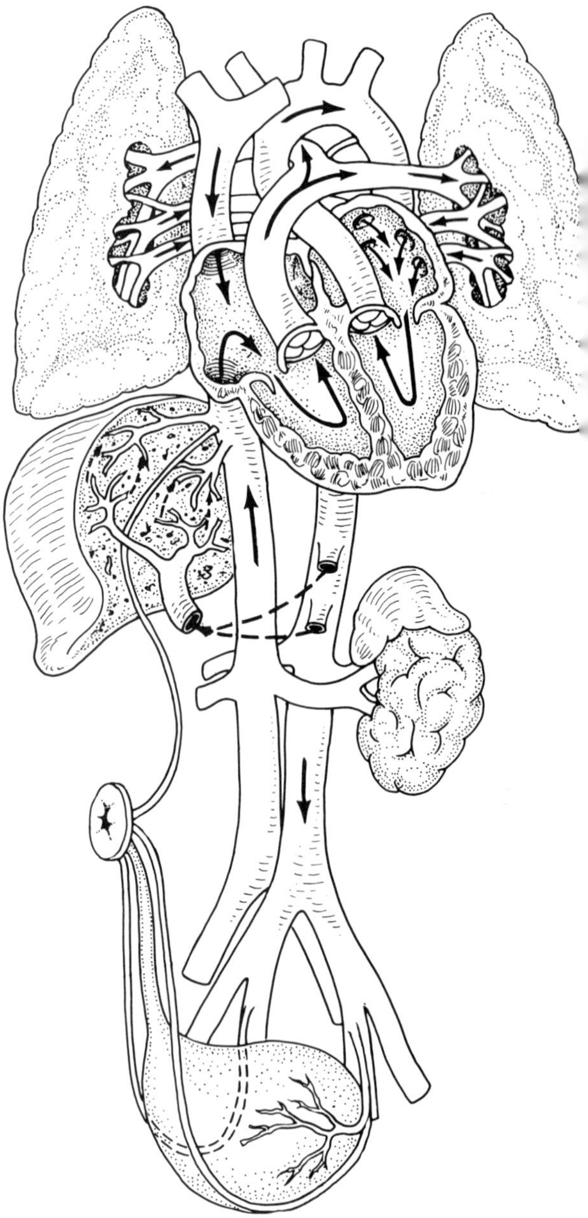

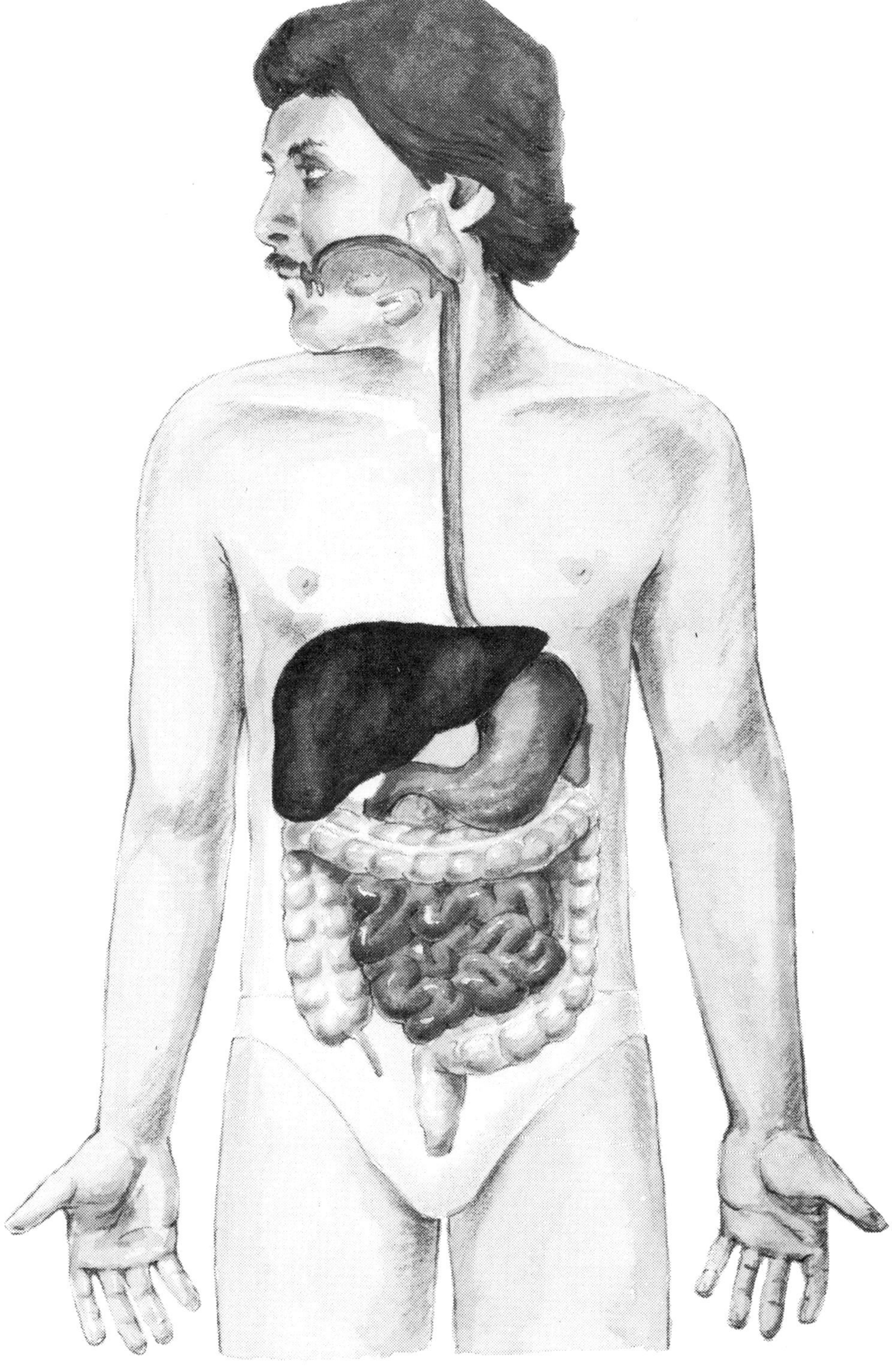

The Benjamin/Cummings Publishing Company, Inc.

ISBN 0-8053-4287-7

90000

9 780805 342871